THE

EVOLUTION

OF

PSYCHOANALYTIC

TECHNIQUE

St. Wolfgang baut ein Kirchelein.
Der Teufel reißt ihm's immer ein,
Und statt den heiligen Mann zu irren,
Muß er den Steuer und Stein zuführen.
Ein Bild für jeden braven Mann,
Der seinen Teufel zwingen kann.

Reprinted courtesy of Margaret Nunberg

THE EVOLUTION

OF

PSYCHOANALYTIC

TECHNIQUE

EDITED BY

Martin S. Bergmann

AND

Frank R. Hartman

WITH A FOREWORD BY EDITH JACOBSON

Basic Books, Inc., Publishers

NEW YORK

Library of Congress Cataloging in Publication Data
Main entry under title:

The Evolution of psychoanalytic technique.

   Includes index.
   1.   Psychoanalysis—Addresses, essays, lectures.
I.   Bergmann, Martin S., 1913–   II.   Hartman,
Frank R., 1936–   [DNLM: 1.   Psychoanalysis—History
—Collected works.   2.   Psychoanalysis—Collected works.
WM460 E93]
RC509.E9   616.8′917′09   75–7258
ISBN 0–465–02126–3

TO

the memory of my father

Shmuel Hugo Bergmann

(1883–1975)

*Professor Emeritus of Philosophy
and First Rector of
the Hebrew University in Jerusalem,
who died shortly before
this dedication could reach him*

M.S.B.

FOR

Daniel X. Freedman, M.D.

mentor and friend

F.R.H.

CONTENTS

Part One
An Introduction to Psychoanalytic Technique

Part Two
Contributors to the Development of Psychoanalytic Technique

Part Three
Early Papers on Technique

Part Four

The Controversy Around Ferenczi's Active Technique

Part Five

Freud's Case Histories in the Light of Later Knowledge

Part Six

Psychoanalytic Technique in the Twenties

Part Seven
The Controversy Around Wilhelm Reich's Character Analysis

Part Eight
Psychoanalytic Technique in the Thirties: Controversy and Synthesis

FOREWORD

EDITH JACOBSON

I ACCEPTED the invitation to write a foreword to this excellent book with great pleasure. From my seminars at the New York Psychoanalytic Institute I am very aware how little my students know about the years when our current psychoanalytic technique was developed and, besides, how little they have read the authors of the past who have made the most significant contributions to this field.

What increased my interest in this book was quite naturally the fact that my own training in psychoanalysis took place in the 1920s at the Berlin Psychoanalytic Institute. This was the time when analysts such as Eitingon, Abraham, Rado, Alexander, Horney, Simmel, Fenichel, and Reich were our training analysts and our co-discussants, and when Freud's papers on the new anxiety theory, the death instinct theory, and the new structural theory had just been published and were under discussion. In those years we naturally "devoured" not only all the old but especially these new papers and discussed them vividly. The younger training analysts discussed clinical, theoretical, and technical problems in a group that met twice a month privately for discussion under the guidance of Fenichel, with Schultz-Henke usually as an opponent. He left the Society later on and developed his special type of theory and technique. Even more important was a small discussion group, consisting of Fenichel, Wilhelm and Annie Reich, Gerö, and myself, which discussed, on the basis of clinical material, Wilhelm Reich's new ideas on technique which resulted in his book on *Character Analysis*, a book that every young analyst should still read today. All of the group, including Fenichel, were at that time greatly influenced by his ideas, which certainly contributed considerably to the development of the analysis of the defenses and to Anna Freud's book on the ego and its defenses. Of course, Freud's new anxiety theory and his new structural theory exercised the greatest influence on all of us. I personally developed the keenest interest in the study of the conflicts between ego and superego, all the more so since I analyzed children at that time and had a difficult adult patient with severe, recurring depressions—cases which lent themselves to the study of superego development and conflicts between super-ego and ego. The papers that I wrote on these subjects have unfortunately never been translated.

Anyhow, these were the times when we all had the experience that our analytic technique underwent considerable changes. We began to discover that direct early id interpretations did not have sufficient emotional impact unless we first worked on the defenses and that the deep id material would only slowly rise to the surface with corresponding emotional impact. Thus, we discovered that an analysis would require a longer time than it had taken in the years prior to these new discoveries. My own training analysis lasted three years, which at that period was regarded as quite a long time.

I hope that my report on those years, during which I was trained in Berlin, illustrates the great significance of this book, which naturally covers a much broader field.

I admire particularly the idea of first presenting the two excellent essays, which reveal the amount of past and current literature the authors have studied, and then giving excerpts of the writings of those authors who made the most important contributions to this field.

I believe that this book deserves to be, and will be, read by a great number of psychoanalysts, especially the younger ones, as well as by psycho-analytically oriented psychiatrists, psychologists, psychotherapists and psy-chiatric social workers, and educated lay persons.

EDITORS' PREFACE

MARTIN S. BERGMANN

FRANK R. HARTMAN

THIS IS NOT A TEXTBOOK on psychoanalytic or psychotherapeutic technique; there are many already. A textbook, by its very nature, must gloss over divergences, uncertainties, and questions which have not yet found an answer, ultimately becoming a document of the time in which it was written. This book is a sourcebook, one which records an era and contains the controversies, debates, and discoveries in the original writings of the pioneers of psychoanalytic technique.

The editors see it as a contribution to the history of ideas, that branch which deals with man's attempt to understand his unconscious and, if possible, to change the ancient dictum that character is destiny.

The idea for the book emerged from a course which the senior editor (M. S. B.) had given to psychologists, psychiatrists, and social workers; and the therapist-reader can learn from it something not available from a textbook. For here he is asked to go back to a past era and relive with the participants the excitement, the uncertainty, and the clash of ideas that emerged as analysts were trying to conquer a new terrain. It is hoped that the therapist-reader will empathize with the pioneers, identify with them, and thus become himself a carrier of the psychoanalytic tradition. He will then be able to see his current clinical problems with a deepening historical perspective.

This book traces the evolution of psychoanalytic technique. It brings together the significant publications of the pioneers of psychoanalysis. The emphasis is on technique rather than theory—that is, the problems encountered, the unexpected difficulties, and their solutions. Where there were controversies, divergent views are represented, including those ultimately rejected by Freudian psychoanalysis; thus the reader is able to make his own evaluations. By reviewing the historical record, he will learn to appreciate the kernel of truth that each of the participants brought to the argument.

Freud laid the foundation of psychoanalytic technique during the interval between 1910 and 1919. After that he left the field of technique to others until 1937, when he contributed two additional papers. We have not reprinted any of his papers because they are easily available. Rather, we have reprinted contributions of other analysts that were published between 1919 and 1939 and that are assembled here for the first time. Many of them appeared in journals no longer easily accessible. Two have been translated from the German for the first time. The book ends with the outbreak of World War II and Freud's death. In bringing these articles together we hope to create the illusion that the reader is participating in an extensive symposium in which, as it were, the pioneering psychoanalysts speak to one another. To this end the senior editor has provided an introduction and footnotes to every article stating the reason for its inclusion and explaining its historical significance. We have supplied cross-references in the editors' footnotes. To facilitate smooth reading, quotations from Latin and foreign languages have been translated. We have also explained archaic terms and allusions to literary works familiar to the German-speaking intellectual world but not necessarily to the English-speaking reader. Notes that provide commentary and which are designed to aid the reader's understanding of the text have been placed at the bottom of the page. Parenthetical citation notes refer to the reference list at the back of the book. To distinguish between each author's original notes and notes added by the editors, editors' notes have been placed in brackets. Finally, the index has been designed to facilitate cross-references among the pioneering analysts on the crucial issues in the field of technique, such as free associations, negative therapeutic reactions, and the currently controversial treatment of narcissistic character disorders.

Of the two introductory chapters, one is an introduction to the conceptual structure of psychoanalysis; the other traces the history of psychoanalytic technique from the beginnings of Freud's work to 1939. It is our hope that, taken together, the two chapters will provide the reader with the necessary background for a deeper understanding of the issues discussed in the papers.

Interest in technique began very early at the first Congress at Salzburg. In 1908, Stein read a paper with the illuminating title, "How Is the Libido Freed by Analysis To Be Guided into Therapeutically Favorable Channels?" Papers on other subjects of technique, particularly dream interpretation, followed. However, in this book we decided to concentrate on the period after Freud's papers on technique were written.

The relationship between a man's thought and his biography is admittedly controversial. We have nevertheless included biographical sketches in the hope that they will enhance the interest of the papers. The reader can judge for himself to what extent the life of an author influenced his ideas.

Most of the papers appeared before World War II. We have made one major exception in Part Three, where we have reprinted later articles which illuminate Freud's case histories in the light of subsequent knowledge. We have also included a later article by Sterba evaluating, in retrospect, the influence of Wilhelm Reich on a certain period of psychoanalysis.

To those who believe that Freud discovered once and for all an ideal technique for the treatment of neurotic disorders, this book will be a disappointment. But Freud himself once said: "When you think of me, think of Rembrandt, a little light and a great deal of darkness." (This quotation is not in any of Freud's writings; it was told to the senior editor over a quarter of a century ago by his teacher, Paul Federn, who was a member of Freud's earliest circle.)

The frontispiece deserves explanation. Hanns Sachs, in a tribute to Freud (1944, p. 168), reports a meeting of the Vienna Psychoanalytic Society:

> The highlight of the evening was, of course, Freud's discussion of the topic. I remember especially well one occasion when Dr. Nunberg gave a profound, but very speculative, paper. Freud opened his remarks by reminding us of a picture well known in Austria, by Moriz von Schwind, representing an episode from the legend of St. Wolfgang. It shows the devil who has made a contract with the Saint to provide the stones with which a church was to be built (he is, of course, in the end cheated of his reward) pushing a great load of rocks on a wheelbarrow uphill, while the Saint is seen in the background in his bishop's vestments, praying in dignified repose. "Mine," said Freud, "was the devil's lot. I had to get the stones out of the quarry as best I could and was glad when I succeeded in arranging them willy-nilly so that they formed something like a building. I had to do the rough work in a rough way. Now it is your turn and you may sit down in peaceful meditation and so design the plan for a harmonious edifice, a thing that I had never a chance to do."*

The paper under discussion was *On the Synthetic Functions of the Ego*, which has been reprinted in this volume. At the next meeting, Freud gave Nunberg a reproduction of the picture. This volume could have been entitled "Of Bishops and Devils."

We are most indebted to Mrs. Herman Nunberg, the translator of the meetings of the Vienna Psychoanalytic Society, for an excellent translation of the papers by Landauer and Kaiser, as well as for permission to reproduce the frontispiece picture. We also wish to thank Mrs. Katherine M. Epler and Miss Paula Gross for editorial assistance and Mrs. Halina Bialokur, Mrs. Sherry M. Warters, and Mr. Michael P. Bergmann for assistance in the preparation of the manuscript.

* The legend of the frontispiece reads thus:

> St. Wolfgang builds a tiny church.
> The devil tears it down.
> The holy man is not confused.
> He makes the devil serve him by carrying stones.
> This is a model for every upright man
> Whom stupid devils do assail.

Part One

An Introduction to Psychoanalytic Technique

CHAPTER 1

MARTIN S. BERGMANN

WHAT IS PSYCHOANALYSIS?

An Examination of the

Assumptions and the Language

of Psychoanalysis

FOR CENTURIES man has been dimly aware of forces within him operating beyond his conscious control. Homer had an explanation for the outbreak of irrational behavior for those moments when a man is no longer himself, or at least no longer what he liked to believe himself to be. King Agamemnon explained conduct he was ashamed of by saying "Zeus put wild ate* in my understanding . . . so what could I do, Deity will always have its way" (Dodds 1957). But it remained for Freud to discover a way to study the irrational:

> The subject matter of psychoanalysis is the irrational, the method of investigation is rational (Fenichel 1939, p. 13).

Self-knowledge has traditionally been seen as the key to mastering the irrational. Throughout history, we can distinguish three ways of achieving it. The first, rational introspection, was practiced by epistemological philosophers from Socrates to Schopenhauer and Nietzsche; the command "Know Thyself" is said to have been inscribed over the door of the temple to Apollo

---

* Dodds translates "ate" as "divine temptation of infatuation; it causes men to commit acts they neither understand nor condone."

at Delphi. The second, in origin even older than systematic introspection, aims at gaining such knowledge by altering states of consciousness, either with the aid of hallucinogenic drugs or by concentrated meditation. The third, and most recent in origin, is psychoanalysis. Unlike the other two, psychoanalysis did not begin with a quest for self-knowledge but more modestly as a technique of cure for neurotic symptoms that resisted other medical endeavors. In the next chapter I have described in some detail how in Freud's hands psychoanalysis grew from a technique of understanding the symptom to a method of understanding the person as a whole.

In an encyclopedia article Freud defined psychoanalysis as follows: "Psychoanalysis is the name (1) of a procedure for the investigation of mental processes . . . ; (2) of a method [based upon that investigation] for the treatment of neurotic disorders; and (3) of a collection of psychological information . . . which is gradually being accumulated into a new scientific discipline" (Freud 1923a, p. 235).

This tripartite character of psychoanalysis, consisting of a therapeutic process, an investigative technique, and a scientific discipline, supplies the basic rationale for many of the procedures that Freud recommended, such as the invisibility and the anonymity of the analyst, and his refraining from giving advice. The specific cure at which psychoanalysis aims is best attained when a quasi-experimental situation prevails (Kris 1956a; Eissler 1953). This unique combination of therapeutic procedure, investigation, and theory reflects the genius of Freud. However, it has also tended to isolate psychoanalysis from other disciplines. Strictly as a therapeutic procedure, psychoanalysis could have been compared in efficacy with other techniques. As a method of investigation, it could have been compared in its rigor and experimental proof with other scientific methods. Finally, as a scientific theory it could have been compared with other disciplines in the way that its "operational definitions" are set up and "correspondence rules" formulated (Nagel 1959). The difficulty in synthesizing psychoanalytic theory with clinical observation is frequently reflected in the fact that the case material quoted in the psychoanalytic papers too often neither proves nor disproves the theoretical assumptions it is called upon to illustrate (G. Klein 1973, p. 154).

It seems that the complexity inherent in the tripartite structure also predestined psychoanalysis to remain the creation of one man. If it had been only an investigative method like archaeology, or a practical application of basic sciences like medicine, psychoanalysis would have remained open-ended, permitting the inclusion of numerous men. The tripartite structure leaves room for only one man of genius—its founder.

Common to the three aspects of psychoanalysis is the genetic view which Freud had defined earlier as the essence of psychoanalysis:

Not every analysis of psychological phenomena deserves the name of psychoanalysis. The latter implies more than the mere analysis of composite phenomena into simpler ones. It consists in tracing back one psychical structure to another which preceded it in time and out of which it developed. Medical psycho-analytic procedure was not able to eliminate a symptom until it had traced that symptom's origin and development. Thus, from the very first psycho-analysis was directed towards tracing developmental processes (Freud 1913c, pp. 182–183).

This passage deserves reflection, for it highlights the fact that the first discovery of psychoanalysis was the connection between the neurotic symptom and infantile sexuality. Many of the discoveries to follow were, in essence, also discoveries of hitherto unsuspected connections. A great deal of the effort in any analysis of an individual was found to be devoted to the uncovering of such surprising connections. This gave rise to an interesting discussion of the model for psychoanalysis: was it to be found in archaeology and was its aim to rediscover what had once been consciously available and had become scattered by the process of repression and isolation, or was it to follow the pattern of science and attempt to discover connections that had never been made before? The extract from Hartmann's book of 1939, reprinted in this volume, deals with this question.

In 1914, Freud redefined the essence of psychoanalysis as follows:

It may thus be said that the theory of psycho-analysis is an attempt to account for two striking and unexpected facts of observation which emerge whenever an attempt is made to trace the symptoms of a neurotic back to their sources in his past life: the facts of transference and of resistance. Any line of investigation which recognizes these two facts and takes them as the starting-point of its work has a right to call itself psycho-analysis, even though it arrives at results other than my own (Freud 1914a, p. 16).

Writing shortly after the break with Adler and Jung, Freud hoped to define psychoanalysis in such a way as to distinguish between treatment approaches within psychoanalysis and those which should be regarded as beyond its borders. The reader can see from this quotation that once more the genetic point of view predominates, the task is to trace the symptom back to its roots in infancy. The definition of psychoanalytic technique, however, is limited by two additional "musts": the analysis of resistance and the analysis of transference.

In the encyclopedia article from which I have quoted earlier, Freud offered another definition of psychoanalysis:

The assumption that there are unconscious mental processes, the recognition of the theory of resistance and repression, the appreciation of the importance of sexuality and of the Oedipus Complex—these constitute the principal subject-matter of psychoanalysis and the foundations of its theory. No one who cannot accept them all should count himself a psycho-analyst (Freud 1923a, p. 247).

The reader will note a particular difficulty here. Unlike other disciplines, such as embriology or genetics, psychoanalysis is delineated by Freud in terms

of three of its major findings: (1) the existence of unconscious mental processes; (2) the workings of resistance and repression; and (3) the pivotal
significance of infantile sexuality and the Oedipus Complex. There is no logical
reason to stop at these three; more will be added, thus leading both to controversy and orthodoxy. Moreover, psychoanalysis is unique and distinct from
any other healing process in that the investigatory or research phase is never
completed before the therapeutic procedure begins, but must be taken up
afresh in each individual case.

Technique of investigation and method of treatment, however, coincided
only at the beginning of psychoanalytic history. As the techniques of psychoanalysis became more subtle, the two diverged. For example, it could be
profitable, from a scientific point of view, to analyze every dream exhaustively
over many sessions, but Freud recognized that, as a therapeutic procedure,
such an undertaking would deprive the analysis of vital contact with the
everyday life of the patient (1911a). He, therefore, advocated that the
analyst be satisfied with a partial understanding of the dream and wait for
fresh dreams to carry on the work of interpretation.

Like other sciences, psychoanalysis has its own data of observation.
Unlike other sciences, its data fall into two distinct groups. The first group
comprises those data given to an outsider, in this case the psychoanalyst, by
the analysand: his life situation, current problems and ways of coping with
them, aspirations, memories, fantasies, dreams, data about people who were
or are important to him, how these people relate to him and he to them, and
so forth. The second category is very different. Here, the data emerge from
the psychoanalytic situation itself: the fluctuations in the feelings of the patient toward the analyst and the variations in cooperation and resistance that
the analysand shows in the analysis. In the first case, the analyst is the
recipient of what is told to him; in the second, the data emerge directly from
the interaction between him and the analysand. One of Freud's greatest discoveries was the extent to which the two types of data interact. What the
analysand tells the analyst influences the analysand's attitude toward the
analyst. When a previously repressed or suppressed feeling or a thought about
the analyst becomes conscious, new memories and other data become free
from their unconscious connection with the analyst; they thereby have the
opportunity to become reattached to the original object to whom they once
applied and from whom they were detached by repression or isolation.

In the papers collected in this volume, the reader will find that some
psychoanalysts tend to favor a model of the analyst as an observer and a
purveyor of interpretation, while others rely more on what they learn from the
interaction itself. Although Freud discovered transference and although, at a
certain period in his life, he was enthusiastic about the possibility of translating the original neurosis into a transference neurosis (Freud 1914c, p. 154),
he nevertheless is himself a representative of the first model. As soon as the

transference storms would subside during an analysis, he would return to the calm atmosphere of more detached observation and interpretation. Ferenczi is the leading analyst to favor the interaction model; others were Rank, Reich, and Strachey. Although these differences often took the form of controversies about technique, probably they went back to personal predilections.

The mass of data supplied by the analysand in the course of the analysis can become chaotic unless the interrelationships between them are recognized and the data transformed into a meaningful Gestalt. Some of these connections could be made by a gifted psychologist, who is unfamiliar with psychoanalytic technique. Indeed, the recognition of certain patterns by the prospective analysand is frequently his reason for seeking analytic help. Many people, for example, recognize spontaneously that their love affairs or job careers follow a repetitive pattern. In the course of an analysis they will relive the same pattern. However, unlike other persons in their daily lives, the analyst renounces involvement in both loving and hating for the sake of understanding. Because of this failure of the analyst to respond in the way expected, the pathology-producing process is interrupted, and the way is paved for a higher form of psychic organization (Loewald 1971). During this process, the psychoanalyst is primarily an interpreter, a maker of connections. The timing, the sequence, and the manner in which these connections should be made are important questions discussed in the papers collected here.

The hierarchical structure of psychoanalytic concepts has caused much confusion to students of psychoanalysis. I will attempt in what follows to clarify some of this confusion. Waelder classifies what he calls the "essentials of psychoanalysis" into six broad groups according to their distance from the clinical data and the level of abstraction to which they belong (1962, pp. 619–622). The first level includes all the facts the analyst has gathered about the analysand, facts such as biographical data—all that was conscious to the analysand—as well as new information that emerged in the course of the analysis (the varieties and intensities of the transference reaction, the type and strength of resistances, the predominant sexual fixations, and so forth). Together they constitute *the level of observation*. The word *fact* is, therefore, interpreted in its broadest sense to include childhood memories, screen memories, dreams, fears, wishes, and so forth.

In the course of an analysis, new connections are made among the various data of observation. These connections are usually made by the analyst and, broadly speaking, are called interpretations. In addition, under the impact of psychoanalytic ego psychology, the analysand is encouraged to make such connections himself. When the raw data become organized into meaningful constellations, we deal with the second level, *the level of clinical interpretation*.

From groups of data organized on the second level, generalizations can be made; for example, about the structure of the anal character (Freud

1908), or about differences in development between boys and girls, such as the paths by which they reach the Oedipus Complex (Freud 1925a). This is the third level, that of *clinical generalizations*. Concepts, such as repression, the return of the repressed, the repetition compulsion, and the transference neurosis, constitute the fourth level, that of *clinical theory*.

Many creative scientists would have been content to stop at this level, but Freud, even before the discovery of psychoanalysis, was searching for a psychology that would represent psychic processes abstractly and in quantitative terms. In a footnote, Freud defined the purpose of this metapsychology as follows: "to clarify and carry deeper the theoretical assumptions on which a psychoanalytic system could be founded" (1917b, p. 222n). To Waelder, *psychoanalytic metapsychology* represents the fifth level, which he describes as an attempt to construct a physicalistic-mechanistic model of the personality that would reflect the motivational theory of psychoanalysis (1960).

Waelder draws a further distinction between metapsychology and Freud's *philosophy* that constitutes the sixth level of abstraction. He believes that Freud's philosophy was a personal matter with little relevance to the rest of psychoanalysis. Further reflection will show that he underestimates the dependence of psychoanalysis on Freud's philosophical point of view. Only Freud's rational, humanistic philosophy could have allowed him to combine a method of investigation with a technique of therapy. As Sherwood has pointed out (1969), truth is not a necessary ingredient of therapeutic efficacy. The making of the unconscious conscious as a technique of treatment rests on far-reaching philosophical assumptions of the ultimate rationality of man.

From the point of view of philosophy of science, Waelder's hierarchies are also open to criticism. The level of clinical abstraction follows the pattern of an *abstractive theory* (Nagel 1961, p. 125; Wilson 1973, p. 321). Such a theory is a shorthand for observational data and can be translated without a remainder back into the language of observation. By contrast, the levels of clinical theory and metapsychology are *hypothetical theories*. They introduce inferred entities and these are not reducible without surplus meaning back to data of observation. Freud employed both abstractive and hypothetical methods without clearly differentiating the two. The discovery of the Oedipus Complex is an example of abstractive reasoning. The concept of the ego is an example of a hypothetical way of reasoning (Wilson 1973). The two modes of reasoning are different, but not subordinated to one another.

Abstractive and hypothetical methods are used in normal science (Kuhn 1962). However, Freud's special genius in formulating concepts was not limited to these two techniques of reasoning. He had a special capacity to see similarities in phenomena no one before had previously compared—as, for example, dreams and symptoms. He had ways of transposing concepts from a limited area to broader ones. Before Freud, narcissism was synonymous with a sexual perversion. Freud extended the term to explain psychosis, hypochondriasis, homosexuality, and even sleep (Freud 1914b). By a similar process of abstraction, he enlarged the term masochism from being a definition of a particular perversion to explain certain character structures (moral masochism) and certain aspects of feminine sexuality (Freud 1924b). This approach, unique to Freud, has given to psychoanalysis some of its most important concepts. They were arrived at intuitively; it was only later that Freud used the language of metapsychology to bring such diverse phenomena under one conceptual roof.

The level of metapsychology has caused many students difficulties. Some further clarification of this level seems advisable. Rappaport and Gill (1959) have suggested a classification similar to that of Waelder. However, theirs is a fourfold classification. They write:

> Systematic studies in metapsychology, however, will have to distinguish between *empirical propositions, specific psychoanalytic propositions, propositions of the general psychoanalytic theory,* and *propositions stating the metapsychological assumptions.* . . . For instance; *empirical proposition:* around the fourth year of life, boys regard their fathers as rivals; *specific psychoanalytic proposition:* the solution of the Oedipal situation is a decisive determinant of character formation and pathology; *general psychoanalytic proposition:* structure formation by means of identifications and anti-cathexis, explains theoretically the consequences of the "decline of the Oedipus Complex"; *metapsychological proposition:* the propositions of the general psychoanalytic theory which explain the Oedipal situation and the decline of the Oedipus Complex involve dynamic, economic, structural, genetic and adaptive assumptions (Rappaport and Gill 1959, p. 796).

The five propositions of psychoanalytic metapsychology are often referred to in the papers here assembled. The dynamic point of view refers to psychic forces operating within a person often in conflict with one another, for example, the conflict between love and hate.

The economic point of view deals with constructs about the nature of the psychic energy involved. Thus, a repressed idea was conceptualized by Freud as having attached to it a special quantum of energy called cathexis, with which the repressed idea pushes forward into consciousness. It is held in repression by a counterforce exercised by the ego and conceptualized as anticathexis.

The structural point of view assumes that within a person there are structures—organized entities—that relate to other organized entities in a dynamic way. The ego, id, and superego are the primary structures. Some analysts have also considered the repetition compulsion an independent agency (Waelder 1936).

The genetic point of view deals with the origin of psychological phenomena. Historically it was the first, and remains the most important, of the analytic discoveries. The statement that an adult neurosis has at its nucleus the infantile neurosis is an example of a genetic statement.

The adaptive point of view stresses the fact that solutions, even neurotic ones, are based on interaction with the environment. This point of view is associated with Hartmann (1939). It gained importance with the growing understanding of the role of the personality of the mother in the development of the child (Mahler 1968). From early infancy on, the child adapts himself as best he can to the environment, which, in the first and most decisive period, is usually his mother.

Within metapsychology itself the economic assumptions have caused the greatest difficulties, because energy can neither be seen nor measured. It is generally assumed in psychoanalysis that the correct interpretation saves energy, because ideas kept in repression or isolation use up anticathectic energy of the ego. When, in the course of a psychoanalysis, the defenses of the ego become conscious and are relinquished by the now freer ego, energy is saved and can be used for new, productive purposes.

In the papers on technique here assembled, metapsychology is omnipresent, especially the economic point of view. The reader will find, for example, that Ferenczi, in his 1919 paper on "active technique," argued that the prohibitions he proposed would raise the pressure of the unconscious energy, enabling it to overcome the resistance of the censorship. To Ferenczi, this was not a metaphor, but a literal fact, even though the pressure of unconscious energy could not be measured. Fenichel's definition of neurosis is also economic in emphasis: "A neurosis is a discharge of dammed-up instinctual energy occurring in defiance of the wishes of the ego" (1937a, p. 133). The metapsychological statement that neurosis is the result of dammed-up instinctual energy has therapeutic implications. "If the analysis succeeds in establishing possibilities for the satisfaction of sexuality, the remaining conflicts are so weakened in their economic impact that practically it amounts to a disappearance of the conflict" (Fenichel 1937b, p. 110). The reader will note that Reich and Fenichel were convinced that once the genital phase had been attained and the capacity for orgasm had become possible, the neurosis will not survive. Freud, although himself the originator of the "damming up" explanation, did not share this optimistic view, particularly in the last paper on technique (1937b). He was more impressed by the limitations imposed by the aggressive drive and the danger of the negative therapeutic reaction (see this volume, pp. 32–34).

The view that metapsychology is of a higher order of abstraction than clinical theory was challenged by George Klein (1973). He argued that metapsychology represents nothing more than Freud's philosophy, a philosophy demanding that all scientific explanation be purged of teleological explanations. On the other hand, he argued, psychoanalytic clinical theory does not deal with the how, but with the why; with purpose, significance and function, of, let us say, symptoms—that is, with concepts considered nonscientific in Freud's generation. Clinical theory and metapsychology are mutually contradictory. The concepts of clinical theory are no less abstract than those of metapsychology, but are differently constructed. Klein goes on to suggest that psychoanalytic language differentiates sharply between intraphenomenological concepts, which describe how the patient feels or thinks, and extraphenom-enological concepts, which deal with the inferences that the analyst makes on the basis of this behavior. Psychoanalytic theory should consist of the phe-nomenological concepts, the logic of the analyst's inferences and the extra-phenomenological concepts used. Historically speaking, psychoanalysis could never have been created on the model suggested by Klein. To Freud, the libido theory formed the core of psychoanalytic nosology. The vicissitudes of the libido accounted for the "choice of the neurosis" for the differences between neurosis and psychosis. Abraham's well known classificatory table (1924a), still considered valid by Fenichel (1945, p. 101), is based entirely on the libido theory and therefore on a metapsychological assumption.

Following Klein, it is possible to hypothesize that psychoanalysis would have avoided certain pitfalls if it had been free from the shackles of metapsy-chology. But it is not possible, at this date, to separate the two without doing violence to the whole edifice of psychoanalysis. For better or for worse, this was the historical and intellectual atmosphere in which Freud made his con-tributions, and metapsychology was the language he spoke.

Let me illustrate by an example from Freud speaking of melancholia:

> An object-choice, an attachment of the libido to a particular person, had at one time existed; then owing to a real slight or disappointment coming from this loved person, the object-relationship was shattered. The result was not the normal one of a withdrawal of the libido from this object and a displace-ment of it on to a new one, but something different, for whose coming about various conditions seem to be necessary. The object-cathexis proved to have little power of resistance and was brought to an end. But the free libido was not displaced on to another object; it was withdrawn into the ego. There, how-ever, it was not employed in any unspecified way, but served to establish an identification of the ego with the abandoned object. Thus, the shadow of the object fell upon the ego . . . (Freud 1917a, pp. 248–249).

This statement, translated into ordinary English, would read: "There are some people who, when disappointed in those they love, do not go on to find others whom they can love, but withdraw from future love relationships. They behave as if they themselves were the persons who abandoned them." My translation certainly lacks the lustre of Freud's original insight, but it is no less explana-tory. However, Freud and the majority of psychoanalysts, even those writing today, employ the language of psychoanalytic metapsychology. And those who wish to read psychoanalytic texts "in the original," that is, without inter-mediary popularizers, have to master the rudiments of this language.

The hierarchical conceptualization of psychoanalysis here delineated con-firms Fenichel's dictum quoted earlier, that the subject matter of psychoanaly-sis is irrational, but not the method. Psychoanalysis has an Aristotelian structure, in the sense that "essences are grasped by intellectual abstraction and involve the ability to conceptualize" (Coltrera 1962, p. 171). It is this tendency toward abstraction that arouses the ire of existentialist psychotherapists, who consider each man unique and understandable to other men only in an empathic communication.

Understanding the hierarchical organization of psychoanalytic concepts, including metapsychology, is an indispensable tool without which the great majority of analytic writings remain unintelligible. It must be admitted, however, that it is not always easy to decide on which level of abstraction a term is being used. Familiar terms like transference are used, at times, on the level of clinical generalization, and at other times on the level of clinical theory. Similarly, the term death instinct will be seen as part of Freud's metapsychology to those who believe in it, while those who dispute it will relegate it to Freud's philosophy. One must also keep in mind that Waelder's, as well as Rappaport and Gill's, categorizing efforts are the result of a later attempt at clarification. Neither Freud nor the psychoanalytic pioneers, whose articles are here reprinted, were aware of such distinctions. They passed easily from one level of abstraction to the next and back again. In the footnotes, I have attempted to draw attention to such jumps.

So far we have dealt with the theory of psychoanalysis, but the situation is similar in the field of technique. The main tool of the psychoanalyst consists in the interpretation of the analysand's free associations. As the theory of technique developed, increasing attention was paid to the kind and the timing of such interpretations, and the sequence in which they should be given to the analysand. It is around these problems that many of the controversies on technique center. The sequence in which interpretations are offered shows a similar hierarchical application. For example, Loewenstein describes how an interpretation proceeds stage by stage:

> The analyst's task is then to show the patient that all these events in his life have some elements in common. The next step is to point out that the patient behaved in a similar way in all these situations. The third step may be to demonstrate that this behavior was manifested in circumstances which all involved competitive elements and where rivalry might have been expected. A further step, in a later stage of the analysis, would consist, for instance, in pointing out that in these situations rivalry does exist unconsciously, but is replaced by another kind of behavior, such as avoiding competition. In a still later stage of the analysis, this behavior of the patient is shown to originate in certain critical events of his life encompassing reactions and tendencies, as for example, those which we group under the heading of the Oedipus Complex (Loewenstein 1951, p. 4).

The reader will note that the interpretations of the analyst take place on the level of "clinical interpretations," as defined earlier. But whether the analyst is conscious of it or not, his interpretations are derived from "clinical theory." The first two interpretations, that the events in the patient's life have something in common and that he behaves in a typical way in all of them, are based on Freud's discovery of the repetition compulsion (1941c). When the analyst points out that expected feelings of rivalry have failed to emerge, his observation is again delivered from clinical theory. This time, he assumes that a defense mechanism is inhibiting the appropriate and expected affect.

As in chess, there are a number of openings available. Another analyst might prefer to demonstrate to the analysand that his response to a given situation, be it anxiety, rage, or depression, is in excess of what the facts in the situation warrant; that the event symbolizes another, as yet repressed event to which the affect really belongs.

Following Loewenstein, we have dealt with one unit in the interpretative process. It takes time and effort to bring to consciousness the various parts that belong to this interpretative unit and to undo the repression and isolation of these parts. To illuminate an entire neurosis is a still greater task, for a neurosis is composed of a large number of such units of behavior.

To illustrate: Anna Freud considers it important, in the psychoanalysis of male homosexuality, to deal with: (1) the fear of the castrating woman; (2) the fear of oral dependency on the woman; (3) the disgust of the woman as an anal object, based on the unconscious equation of vagina and anus; and (4) the fear of the homosexual's own aggression toward women. The analysis of these attitudes Anna Freud considers necessary, if the homosexual is to dare approach a woman sexually.

These preliminary interpretations are not specific to homosexuality. What is specific is the projection of the homosexual's phallic wishes on the partner, or the projection of the feminine wishes in the case of the active homosexual. Under the domination of the primary process, what has been projected is recaptured back into the self, in the sexual act. The change in the direction of the libido, however, will become possible only after the homosexual understands the nature of his projections on the partner (see Wiedeman 1962, p. 404). It is evident that Anna Freud does not deal with specific biographical and interpersonal events in the life of a homosexual. She, therefore, does not mention such items as the lack of interest of the father in the boy or the mother's seductive behavior. She deals only with the intrapsychic consequences of such pathology-producing interactions, and she states her conclusions in the language of clinical theory.

In principle, all neurotic entities could be similarly divided into general and specific mechanisms found in a given neurotic group. Fenichel's book (1945) is the best attempt to organize psychoanalytic knowledge along such lines.

I would like once more to return to a model definition of psychoanalysis, for which I am again indebted to Anna Freud:

> According to an old "classical" definition, a therapeutic procedure has the
> right to be called psychoanalytic if it recognizes and works with two proces-

ses in the patient's mind: *transference and resistance,*—a pronouncement which is neither as loose nor as permissive as it sounds. It means, translated into other words, that our "orthodox" or "classical" technique is based on an idea which is at the same time dynamically, genetically, qualitatively and structurally determined. We assume that the patient suffers from a conflict, not with the environment, but within the structure of his own personality; that this conflict has arisen in his remote past and is not accessible to his consciousness; further, that the instinctual strivings, lodged in the id, though unconscious, are possessed of a permanent urge to rise upwards and manifest themselves in the present, making use for that purpose of any available object in the outer world. Here the analytic technique comes to his help by offering him such an object in the person of the analyst, on whom the past unconscious experience is made conscious and re-lived. On the other hand, the analytic technique deals with the resisting counter-force lodged in the patient's ego which is intent on keeping down the id-strivings and on preventing them from becoming manifest. Here the technique is directed at making the patient aware of his defensive devices, thereby depriving them of much of their efficiency. Therefore, to deal with both transference and resistance brings about inevitably the revival of all those moments in infancy and childhood when id and ego-forces have clashed with each other; it implies re-activating these conflicts and finding new solutions for them (Anna Freud 1952, p. 45).

From this overall definition of analytic technique Anna Freud derives the rationale for free association, dream and transference interpretation, the use of the couch, and other procedures customary in psychoanalytic practice.

I would like to submit Anna Freud's definition of psychoanalysis to further scrutiny. It is evident that the definition is based on a number of assumptions. Some of these draw the line of demarcation between psychoanalysis and other therapies, while others state the assumptions of psychoanalysis in metapsychological terms.

The reader will have noticed that Anna Freud begins with Freud's definition of 1914, given earlier in this chapter. From there, she proceeds to state the four dimensions of psychoanalytic metapsychology, although she employs the term "qualitatively," instead of the usual "topographically." Her statement is beguiling in its apparent simplicity. If we attempt to explicate the assumptions underlying it, we obtain the following:

First Assumption: The conflict must be intrapsychic. This assumption raises the problem of those who seek analysis because of marital conflicts, conflicts with their children, or an inability to decide between love objects. Such conflicts when presented are interpersonal. It follows from Miss Freud's definition that interpersonal conflicts, if they are to be treated psychoanalytically, must be transformed into intrapsychic conflicts. Indeed, among those who seek psychoanalytic help, only a minority can acknowledge from the beginning that their problems are intrapsychic. The great majority see their problems as interpersonal and consider themselves innocent victims of another's neglect or hostility. It takes time, and in certain cases an inordinate amount of time, for such analysands to see their problems as intrapsychic. Those who cannot make this transition must be classified as "unanalyzable."

Second Assumption: A conflict that psychoanalysis attempts to treat goes back to the early years of life. Here, Miss Freud recapitulates Freud's 1913 statement about the special significance of the genetic point of view. Here too, a number of further problems call for clarification. Psychoanalysis does not assume that all conflicts go back to early infancy, but only those that have caused a neurosis. In practice, however, this can turn out to be a difficult distinction particularly in the early stages of a psychoanalysis. Life, as dramatists and novelists know so well, is full of conflicts—conflicts between one's ideals and reality pressures; between loyalty and a new love; between conscience and ambition. We might say that a conflict is in the psychoanalytic domain when (1) it cannot be resolved over some period of time; (2) when the solution to the conflict resulted in a heavy burden of guilt; or (3) when the solution obtained required the formation of neurotic symptoms. In any of these situations, the synthetic functions of the ego are not equal to the task (see Nunberg 1930, this volume, Chapter 22).

Third Assumption: The conflict is not accessible to consciousness. Here, Miss Freud stresses the topographic aspect. One may raise the question: does the entire conflict have to be unconscious, or only certain aspects of it? If the former is the case, then it would seem that analysis would paradoxically be less helpful if the analysand already has some insight into his problem. In actual experience, some "return of the repressed" has usually taken place before the analysis has begun, weakening the defensive system of the ego, causing anxiety, and driving the analysand to ask for help.

Fourth Assumption: The instinctual strivings possess a permanent urge to rise upward. Unlike the first three assumptions that limit the scope of psychoanalysis, this one is a metapsychological assumption and belongs to a different universe of discourse.

Fifth Assumption: These urges make use of whatever objects they can find in the outer world. In transference, the analyst becomes available as such an object. The analyst here has a function similar to that of the day residue in the dream. The assumption is built upon the previous one; however, while the previous one is metapsychological, this one has clinical implications. It is the explanation for transference phenomena, as well as the rationale for the fostering of the transference neurosis.

Sixth Assumption: The ego has counterforces at its disposal, which aim to keep the id strivings down. Like the fourth assumption, this is a metapsychological generalization to explain certain types of resistances called ego resistances (see p. 37).

Seventh Assumption: The analysand's awareness of his own counterforces deprives them of much of their efficiency; to deprive them of their power has a freeing effect. This is once more a metapsychological assumption, designed to explain the relief analysands feel when they give up some of the defenses they have habitually clung to.

Eighth Assumption: When transference and resistance are analyzed, childhood conflicts are inevitably reactivated. Striking here is the word "inevitably." This statement appears as a statement of fact. However, since there is no independent way of testing whether the transference and resistances have been analyzed, the use of the word "inevitably" reduces this statement to a definition. For if childhood conflicts fail to be reactivated, the statement will imply that analysis of the transference and resistances has not taken place sufficiently.

Ninth Assumption: A reactivation of the old conflicts and the finding of new solutions are coupled by Anna Freud in one sentence. But are they really one and the same process? Some psychoanalysts assume that new sublimations follow automatically out of the analysis of the infantile conflict. That such results are, in fact, obtained in a number of successfully analyzed cases is not under dispute, but what is less certain is the assumption that such sublimations *must* follow. This certainty is derived from the theory of psychic energy, not from the observed facts. This assumption, once more, comes close to an article of faith.

The explication of Anna Freud's definition leads, I believe, to the conclusion that psychoanalytic theory is not simply derived from the facts of observation; nor can one say that it represents only a process of abstraction from the facts. Rather, strewn among the various assumptions, there are hidden statements of belief to support the practitioner and give him a sense of security—or even a feeling of inevitability—at times beyond what the data at his disposal can offer. Unlike in Freud's definition of 1913, Anna Freud includes the metapsychology of psychoanalysis in its very definition.

I have referred earlier to the tripartite definition of psychoanalysis given by Freud in 1923. The unique combination of investigative method, therapeutic technique, and theory, has made it difficult to define the borders of psychoanalysis. At times, Freud came close to equating psychoanalysis with the psychology of the unconscious. This was, indeed, his earliest definition of metapsychology (Freud 1901, p. 259).

While Freud was alive, the limits of psychoanalysis coincided more or less with the limits of his genius. I am aware of only one basic concept which was not generally adopted, and one administrative decision in which Freud's views did not prevail. The majority of psychoanalysts did not follow Freud's lead in the acceptance of the death instinct theory (see for example, Fenichel 1935b). On an administrative level, the American Psychoanalytic Association, contrary to Freud's explicit wishes, confined its membership to physicians. Otherwise, Freud's views were interpreted in various ways by competing schools within psychoanalysis, but were not contradicted.

After Freud's death, the limits of psychoanalysis became harder to define.

When an innovator decided to leave the fold, as did Adler, Jung, Rank, Reich, and Horney, the problem was solved administratively. But when the innovator, or his disciples, insisted on working within the framework of Freudian psychoanalysis, dissension became inevitable. Generally speaking, independent, new clinical observations, like those of Spitz (1945) or Mahler (1968), can be absorbed, even though they are far reaching in their implications and may lead to significant modifications in theory or technique. It is different when new subsystems are created, like those of Federn, Hartmann, Melanie Klein, or Erikson. The inclusion of such subsystems cannot be noncontroversial. (For instructive examples of such controversies, see Waelder 1960; Glover 1966.) Ultimately, all theoretical differences become questions of technique. It is in this area that the most interesting discussions and controversies are still with us, though often in a different guise. The issues raised in the realm of technique took place in Europe between 1919 and 1939. This book deals with this period.

CHAPTER 2

MARTIN S. BERGMANN

NOTES ON THE HISTORY OF PSYCHOANALYTIC TECHNIQUE

PSYCHOANALYTIC TECHNIQUE is charged with the task of removing the deleterious effects of a neurosis in the most thorough and expeditious way. Ideally, such a technique should have been based on an earlier and independent discovery of how the neuroses were formed and the psychic forces that kept them in operation. However, as this chapter will demonstrate historical development proceeded differently. The technique came first, and it was through application of the technique that the structure of the neuroses was illuminated. On the other hand, the nature of the curative factor in psychoanalysis remained controversial, and most of the papers presented in this volume deal in one way or another with this problem (see also Salzburg Symposium 1925; Marienbad Symposium 1936; and, more recently, Loewald 1960).

In the clinical papers written between 1910 and 1915, Freud created the model for the psychoanalyst as technician. In studying subsequent developments, it will be useful to keep in mind the changes that took place in the model structure itself, and to differentiate these from variations introduced by psychoanalysts when they treated patients with pathology that required a deviation from the model, such as children, psychotics, or patients of advanced age (see Eissler 1953). The evolution of the model before 1939 will be discussed in detail here. However, the variations introduced in the treatment of patients not analyzable by the classical technique have been omitted, lest they confuse the unfolding picture. Throughout this chapter I have followed developments in chronological order. However, from time to time, when a particular subject became central I have broken the historical sequence and pursued the topic to the present time. Hence the frequent references to works that were written after 1939.

For purposes of exposition, I have divided the history of psychoanalytic technique into five stages;* in fact, however, the line of demarcation between them is less rigid than my divisions would indicate. The technique of the first stage, which Freud described as the forerunner to psychoanalysis (1914a), is known as the cathartic procedure and extends from 1893 to 1898. The second, lasting roughly from 1898 to 1910, can be described as the period of the basic discoveries. The third, extending from 1910 to 1919, can be designated as a period of consolidation. The fourth, from 1919 to 1926, is the period of the reformulation of the instinct theory and the formulation of the structural theory. The fifth extends from 1926 to 1939, and can be described as the period during which psychoanalytic ego psychology came into its own.

The method that Freud employed during the stage preceding psychoanalysis was the so-called "pressure" method, which he described as follows:

> . . . I decided to start from the assumption that my patients knew everything that was of any pathogenic significance and that it was only a question of obliging them to communicate it. Thus when I reached a point at which, after asking a patient some questions such as "How long have you had this symptom?" or "What was its origin?" I was met with the answer: "I really don't know," I proceeded as follows: I placed my hand on the patient's forehead or took her head between my hands and said: "You will think of it under the pressure of my hand. At the moment at which I relax my pressure you will see something in front of you or something will come into your head. Catch hold of it. It will be what we are looking for.—Well, what have you seen or what has occurred to you?"
>
> On the first occasions on which I made use of this procedure . . . I myself was surprised to find that it yielded me the precise results that I needed. And I can safely say that it has scarcely ever left me in the lurch since then (Breuer and Freud 1893–1895, pp. 110–111).

In spite of this optimistic proclamation, Freud encountered certain situations in which the pressure technique failed. This failure led to the discovery of transference. Freud described the conditions that produced failure thus:

> (1) If there is a personal estrangement—if, for instance, the patient feels she has been neglected, has been too little appreciated or has been insulted, or if she has heard unfavourable comments on the physician or the method of treatment. . . .
> (2) If the patient is seized by a dread of becoming too much accustomed to the physician personally, or losing her independence in relation to him, and even of perhaps becoming sexually dependent on him. . . .

* The reader may wish to compare this account of the history of psychoanalytic technique with the accounts of Coltrera and Ross (1967) and Lipton (1967).

(3) If the patient is frightened at finding that she is transferring on to the figure of the physician the distressing ideas which arise from the content of the analysis. This is a frequent, and indeed in some analyses a regular occurrence. Transference on to the physician takes place through a false connection . . . (Breuer and Freud 1893–1895, pp. 301–302).

This is the first appearance of the term "transference" in the psychoanalytic literature. It is historically interesting that Freud regarded only the third item as transference. During the next period points one and two would also come to be regarded as transference manifestations.

Breuer and Freud's use of hypnosis for cathartic purposes seemed to have little in common with psychoanalysis until Lewin demonstrated that the psychoanalyst is not so far removed from the hypnotist as he would like to believe (1954 and 1955). Lewin noted that psychoanalytic technique originally consisted of putting the patient into a sleeplike state and encouraging the dreamlike production of the "talking cure." Later on, under the impact of ego psychology, the technique laid greater emphasis on the patient remaining awake. Nevertheless, the analysand who free associates can be seen, according to Lewin, as a quasi-sleeper, and the psychoanalyst who offers interpretations as a quasi-awakener. Thus, the patient is alternately put to sleep and awakened. It is the situation of quasi-sleep which is responsible for two forms of resistance. Some patients resist by remaining "excessively awake." They look at their watches, are concerned with outside noises and in extreme situations get up from the couch. Other patients show their resistance by refusing to be awakened. Typically, they come to the psychoanalytic hour with a dream so long that it cannot be interpreted. Or else they have so much to tell the analyst that they resent the interruption of their talk.

Psychoanalysis came into being as a result of a unique interaction of several events. First, the pressure method was abandoned for that of free associations. The second was Freud's disappointment in, and subsequent relinquishment of, his theory that neuroses were the result of sexual seduction of children by adults. The third was the discovery of the Oedipus Complex as the nucleus of the neurosis.

Freud told the story of the collapse of the seduction theory in a number of papers (including 1914a), but it was not until the publication of the *Letters to Fliess* (Freud 1892–1899), that we learned the exact date of that event. The decisive letter is Letter 69,* written on September 21, 1897, in which Freud

---

* There is an interesting slip of the pen in this letter. Freud relates that he has no feeling of shame, but adds, "Of course I would not tell it to Dan and talk about it in Askelon." This is a reference to the biblical quotation: "Tell it not in Gath, publish it not in the streets of Askelon, lest the daughters of the Philistines rejoice, lest the daughters of the uncircumsized triumph." Dan, however, unlike Gath, is not one of the five cities of the Philistines, but an Israeli tribe. The slip may therefore indicate that Freud was concerned not only with his enemies, but also with an inner struggle.

informed Fliess that he had lost faith in his *neurotica*. "Disheartening technical failures" and "interruptions of therapy" were mentioned, but above all Freud could not accept the fact that the father was so often described as the seducer. Freud concluded the letter with the important insight that the unconscious cannot distinguish between reality and affectively loaded fiction. Less than a month later, he announced to Fliess that he had discovered the Oedipus Complex. This is the closest we can come to a birthdate for psychoanalysis.*

Thus begins the second, and perhaps Freud's most creative, period (Waelder 1960), a period extending from the date of the above letter to 1910. It is during the thirteen years following the realization of the significance of the Oedipus Complex that Freud made the crucial discoveries that led to the interpretation of dreams, the significance of infantile sexuality, the laws governing unconscious thinking, and the technique of free association. *The Interpretation of Dreams* (1900) and *Three Essays on the Theory of Sexuality* (1905b) are the outstanding contributions of this period.

The interpretation of dreams and the technique of free association were developed by Freud simultaneously. In a significant passage he says:

> I need say little about the interpretation of dreams. It came as the first-fruits of the technical innovation I had adopted when, following a dim presentiment, I decided to replace hypnosis for free association (Freud 1914a, p. 19).

The reader will note that the discovery of the technique came first.

If we follow the development of Freud's theories in chronological order, we can understand better why he assigned such a crucial role to wish fulfillment in the interpretation of dreams. To borrow a biblical expression, the stone which the master builder had neglected during the first stage of development had become the cornerstone of the second stage. Since he had failed to see the role of wish fulfillment in the symptom formation of hysteria, he now gave it the central role in the formation of the dream. (Only much later did Freud revise the theory of dream formation [1933, Chapter 29], but by that time he was working within the framework of the structural theory.)

From the point of view of technique, dream interpretation can be divided into three distinct phases. The first is the dream itself, as told by the analysand to the analyst. This is referred to as the manifest content of the dream. The second phase consists of the associations of the patient to the dream as a whole, or to the various details in the dream. This stage usually includes the day residue of the dream. The third stage is reached when the associations yield a meaningful, even though unexpected, interpretation of the dream. This stage Freud called the dream thought, or the latent content, of the dream.

Censorship can intervene in any one of the three phases. In the first, the dream, or significant parts of it, are temporarily or permanently forgotten. In

---

* For a significant reevaluation of this episode in Freud's development, see Schimek (1975).

the second stage, no associations, or only very sparse ones, occur to the dreamer. In the third stage, the associations are plentiful, but no discernible pattern emerges. While Freud included all three under the heading of censorship, we are dealing with different kinds of inhibition. In the second stage, the access to the unconscious is temporarily or permanently barred. On the third level, the integrative functions of the ego are not strong enough to overcome the diffused associations. Although the locus of the resistance may shift from dream to dream, some difficulties are typical of certain analysands. We can differentiate those who habitually do not report dreams from those who do report dreams but cannot associate to them, and finally those who associate but passively leave it to the analyst to make sense out of the dream. Such resistances confront the analyst with a dilemma. He may either guess at the meaning of the dream to the best of his knowledge, or according to his own preconceived ideas, or else he may leave the dream uninterpreted.

It was in *The Interpretation of Dreams* that Freud formulated for the first time what was to become the basic rule of psychoanalysis:

> My patients were pledged to communicate to me every idea or thought that occurred to them in connection with some particular subject; [The German is even more emphatic: "Meine Patienten, die ich verpflichtet hatte"] amongst other things they told me their dreams and so taught me that a dream can be inserted into a psychical chain that has to be traced backwards in the memory from a pathological idea. It was then only a short step to treating the dream itself as a symptom and to applying to dreams the method of interpretation that had been worked out for symptoms (Freud 1900, pp. 100–101).

When, at the turn of the century, Freud first formulated the basic principles of psychoanalysis, the relationship between theory and technique can be said to have been harmonious. The adult neurosis sprang from one of two sources: either from a traumatic experience occurring early in life, or from a powerful wish that was considered dangerous and therefore had to be repressed. In either case, the neurotic symptom had to be traced back—by free association, the analysis of transference, and the resistance—to early infancy. At first, Freud believed that repression itself was the cause of the neurosis. Only later did he differentiate between successful and unsuccessful repression, with neurosis breaking out when the repression failed. (For a fuller discussion, see Freud 1915c and also Brenner 1957). When repression is not entirely successful, the original memories (in the case of a traumatic neurosis), or the original wishes, possess a dynamic power of their own and force their way back to consciousness. In Chapter 7 of *The Interpretation of Dreams*, Freud formulated the metapsychological principle that accounts for the upward mobility of these unconscious wishes:

> . . . I consider that these unconscious wishes are always on the alert, ready at any time to find their way to expression when an opportunity arises for allying themselves with an impulse from the conscious and for transferring their own great intensity on to the latter's lesser one (Freud 1900, p. 553).

The fact that thoughts or wishes can transfer their intensity to other thoughts or wishes was conceptualized by Freud as endowed with hyper-cathexes (1900, pp. 594–597; see also 1915b). In a footnote, Freud added:

> If I may use a simile, they are only capable of annihilation in the same sense as the ghosts in the underworld of the Odyssey—ghosts which awoke to new life as soon as they tasted blood (Freud 1900, p. 553).

Making the unconscious conscious is what the "ghosts" need in order to awaken to new life. Sachs used the same metaphor:

> . . . they are recognized as the new flowering of wilted memories, of long forgotten epochs of the mind and of wishes which died unborn. Like a pro-cession of resurrected dead they pass through the analysis; when these shadows, after absorbing a drop of life's blood by way of the transference, have verbalized for the first time their grievances and frustrations, they can be laid and sent to their rest (Sachs 1948, p. 36).

Wishes have to be hypercathected if they are to pass through the barriers that separate the unconscious from the preconscious, and the latter from con-sciousness. The hypercathexis is necessary in order to overcome the psychic energy which bars their re-entry and which is conceptualized as anticathexis. The unconscious wishes find it easier to pass the barrier in a state of sleep, when anticathexis is weakened. But even in a state of sleep, forces are present that attempt to block the entrance of the prohibited wishes and thus protect sleep. In the case of the dream, these protective forces were conceptualized collectively as censorship, which, if it does not succeed in barring the entrance of prohibited wishes entirely, at least disguises them, thus altering the latent into the manifest content of the dream.

In a neurosis, an analogous process takes place. Symptoms were con-ceptualized as psychic equivalents of dreams and the elucidation of dreams through the technique of free association frequently throws light on the symptoms.

Since symptoms were conceptualized as compromises between the dy-namic unconscious wishes or memories seeking re-entry into consciousness and the forces opposing them, treatment should aim to make conscious the nature of this compromise. These psychic compromise formations do not follow a rational logic, but have a logic of their own, called by Freud "the logic of the primary processes," expressed in a special language which analysts and analy-sands together have to decipher anew in every treatment situation. If this is successfully deciphered, the symptoms should no longer be useful as a com-promise and should wither away. In his writings on technique up to 1919, Freud assumed that the adult patient could tolerate the re-entry of infantile wishes and memories that could not have been tolerated by a child. He thought that once made conscious, these wishes could be dealt with without the need to resort to new repression, or the formation of new symptoms.

Following *The Interpretation of Dreams*, the next important step in the field of technique was made by the publication of Freud's case histories between 1905 and 1919; during this period Freud's mastery of technique grew and the papers on technique were published. These case histories are unique and different from one another. Written in Freud's inimitable style, they read at times like masterpieces of literature. The cases breathe the life of characters in novels and yet the narrative is frequently interrupted by theoretical considerations and vigorous arguments in defense of psychoanalysis. A significant literature has grown around them. The most important of these essays have been reprinted in Part Five of this volume. The cases, as well as the comments on them, are an excellent source for the understanding of the development of psychoanalytic technique. What a motley crew these case histories have proven to be!

The patient, Dora, the heroine of the first case (Freud 1905a), was an eighteen-year-old girl "in the first blossom of her youth," when her father, in Freud's words, "handed her over for psychotherapy." The treatment lasted only eleven weeks and was broken off by Dora. Her symptoms were not impressive by the standards of hysterical symptoms prevalent at that time. She suffered from migraine, dyspnea, nervous cough, loss of voice, and depression. Freud classified her as "a case of petite hysteria." Her hysterical symptoms did not go back to infancy, as the theory of the time demanded, but only to her fourteenth year. She developed them as a way of coping with contradictory wishes aroused in her by the sexual advances of a family friend whose wife was Dora's father's mistress. The case deals essentially with the interpretation of two dreams. Since Dora was analyzed a year after the publication of *The Interpretation of Dreams*, the case illustrates how dreams were used in Freud's clinical work. By current standards, the interpretations were given too early and this error alone would have accounted for Dora's flight. There were other factors to complicate the situation. Dora was an unwilling patient, her neurosis had not yet crystallized (Blos 1972), and Freud was socially too closely identified with Dora's father. However, Erikson in a perceptive essay has shown that Freud failed to understand Dora's "Actuality" (1962, this volume, Chapter 12).

On the theoretical side, Freud advanced an idea destined to play a major role in later developments. Hysterics not only "suffer from reminiscences," as he had pointed out in the *Studies on Hysteria*—they are not in possession of their true biography. The sequence of events has been falsified, lacunae in the life history have appeared, and important connections have been broken off. Upon completion of the treatment, the patient comes into possession of his true biography (Marcus 1974). At this point, Freud goes beyond the cure of the symptoms; a humanistic theme of self-knowledge enters the medical endeavors. At the time, Freud still believed that normal persons are in possession of their unfalsified biography. Kris (1956c) showed the "myth of autobiography" extends beyond hysteria.

In the postscript to the case, Freud discussed the role of transference in psychoanalytic therapy. The difference between the views he held in 1895 and in 1905 is striking:

> What are transferences? They are new editions or facsimiles of the impulses and phantasies which are aroused and made conscious during the progress of the analysis; but they have this peculiarity, which is characteristic for their species, that they replace some earlier person by the person of the physician (Freud 1905a, p. 116).

Freud concluded: "I did not succeed in mastering the transference in good time" (p. 118). Transference, which had been regarded ten years earlier as a false connection and an obstacle to cure, was now to become a major technical tool.

The second case history was that of Little Hans, a phobic boy of four years (1909a). It is of special interest not only because it was the first case of child analysis, but also because it was conducted largely by proxy—through the patient's father. In reading this case one can appreciate how certain ideas, later to become commonplace, struck Freud for the first time:

> Certain details which I now learnt—to the effect that he was particularly bothered by what horses wear in front of their eyes and by the black round their mouths—were certainly not to be explained from what we knew. But as I saw the two of them sitting in front of me and at the same time heard Hans's description of his anxiety-horses, a further piece of the solution shot through my mind, and a piece which I could well understand might escape his father. I asked Hans jokingly whether his horses wore eyeglasses, to which he replied that they did not. I then asked him whether his father wore eyeglasses, to which against all the evidence, he once more said no. Finally I asked him whether by "the black around the mouth" he meant a moustache; and I then disclosed to him that he was afraid of his father, precisely because he was so fond of his mother (Freud 1909a, pp. 41–42).

The case history of Little Hans also contains ideas basic to Freud's philosophy of treatment that will be elaborated later:

> . . . we endeavour rather to enable the patient to obtain a conscious grasp of his unconscious wishes. And this we can achieve by working upon the basis of the hints he throws out, and so, with the help of our interpretative technique, presenting the unconscious complex to his consciousness *in our own words* [Freud's italics]. There will be a certain degree of similarity between that which he hears from us and that which he is looking for, and which, in spite of all resistances, is trying to force its way through to consciousness; and it is this similarity that will enable him to discover the unconscious material (Freud 1909a, pp. 120–121).

This technique of working Freud called "giving the patient anticipatory ideas" (1910a), and still later (1937a) he called these anticipatory ideas "constructions." Freud was convinced that the wrong construction did no harm beyond the waste of time (1937a, p. 261) (see also Bergmann 1968). An important essay by Glover (1931) challenging this approach to treatment is reprinted in this volume as Chapter 23.

The third case was that of a young officer known as "The Rat Man," who presented symptoms of a severe obsessional neurosis (1909b). The article by Kanzer (1952, this volume, Chapter 10) shows that the analysis remained incomplete because Freud did not yet know how to mobilize the transference for therapeutic purposes. The essay by Zetzel (1966, this volume, Chapter 11) highlights the fact that Freud did not then know how to bring the mother into the analysis, even though she appeared often in Freud's notes to the case. Zetzel also shows that the "Rat Man" underwent what we would call today psychoanalytically-oriented psychotherapy, but not an analysis. One might also add that he received a massive dose of indoctrination (Kris 1951). Since the patient was killed during World War I, the permanence of the results obtained by Freud was impossible to verify.

The fourth case, known as "The Wolf Man" (Freud, 1918), was that of a Russian millionaire reduced to utter helplessness by a psychic disturbance. His case was studied over many years, first by Freud and later by two generations of psychoanalysts. His case has been under psychoanalytic observation longer than any other (see Gardiner 1971). He was helped by Freud, but not sufficiently, and, after World War I, in treatment with Brunswick, he developed symptoms as ominous as those he evinced when he first came to Freud (Brunswick 1928).

The case is of further interest because Freud felt it necessary to employ a number of parameters (Eissler 1953)—that is, departures from the generally accepted classical technique. At one time, he promised the "Wolf Man" a total cure of a symptom, and at another he used the threat of termination of the treatment in order to overcome the "Wolf Man's" resistances. This departure from traditional technique, in turn, became an important stimulus for Ferenczi's active technique (see Part Four).

Toward the end of his life, Freud re-evaluated the use of the threat of termination in an interesting passage:

> I have subsequently employed this fixing of a time-limit in other cases as well, and I have also taken the experiences of other analysts into account. There can be only one verdict about the value of this blackmailing device: it is effective provided that one hits the right time for it. But it cannot guarantee to accomplish the task completely. On the contrary, we may be sure that, while part of the material will become accessible under the pressure of the threat, another part will be kept back and thus become buried, as it were, and lost to our therapeutic efforts (Freud 1937b, p. 218).

The title of this work, "From the History of an Infantile Neurosis," is significant, for in choosing it Freud emphasized one of the basic tenets of psychoanalysis, the importance of the infantile neurosis for the adult neurosis.

Infantile experiences themselves are sufficient, he stated, to produce a neurosis in an adult (p. 54). The infantile neurosis in this case was brought into analysis fifteen years after its termination (p. 8). Crucial to this infantile neurosis, and to many others, was the witnessing of the primal scene at the age of eighteen months. As a rule, an infantile neurosis is repressed during the latency period and re-emerges in a changed form in the adult neurosis. It was only after World War II that it became evident that the adult neurosis is seldom a replica of the infantile neurosis (Hartmann 1954). The transference neurosis that develops in the course of an analysis forms the bridge between the infantile and the adult neurosis. Doubts about the infantile neurosis as a real event in childhood have recently been raised (Tolpin 1970; Blos 1972).

The case histories, taken together, fail to illustrate the efficacy of psychoanalysis as a technique of therapy, but we are still surprised and delighted, if not by the cures obtained, then by the utterly unexpected connections that Freud was able to make.

I have already called the period between 1910 and 1919 a period of consolidation. It is during this decade that Freud wrote the papers on technique that every writer on the subject has used as his starting point. It is worth noting that during the same decade, Freud also wrote the papers on metapsychology, and indeed, as I have shown in the previous chapter, there is a close connection between technique and metapsychology. The paper given at the Weimar Congress (1910a) reflected Freud's high optimism at the time. A year earlier, he had been honored in America, thus receiving his first public recognition. Every year brought new converts to psychoanalysis, and the inner schisms had not yet taken place. Freud believed that in the not too distant future the neuroses would disappear and, like many other once menacing diseases, would be relegated to the past:

> . . . the psychoneuroses are substitutive satisfactions of some instinct the presence of which one is obliged to deny to oneself and others. Their capacity to exist depends on this distortion and lack of recognition. When the riddle they present is solved and the solution is accepted by the patients these diseases cease to be able to exist. . . . Sick people will not be able to let their various neuroses become known—their anxious over-tenderness which is meant to conceal their hatred, their agoraphobia which tells of disappointed ambition, their obsessive actions which represent self-reproaches for evil intentions and precautions against them—if all their relatives and every stranger from whom they wish to conceal their mental processes know the general meaning of such symptoms, and if they themselves know that in the manifestations of their illness they are producing nothing that other people cannot instantly interpret (Freud 1910a, p. 148).

It is interesting to note that Freud counted heavily on a sense of social shame to aid in the elimination of neuroses.

In the meantime, the need of guidelines for other analysts became urgent. Freud never wrote a textbook, but the papers on technique written between

1910 and 1915 are generally acknowledged as the foundation of psychoanalytic technique (1911a, 1912a, 1912b, 1913a, 1914c, and 1915a). Since they are easily available, we have not reprinted them here, but they are a prerequisite for the understanding of the papers here collected.

The dream had been Freud's royal road to the unconscious. It was therefore logical that the first paper on technique should bear the title, "The Handling of Dream Interpretation in Psychoanalysis" (1911a). The advice Freud gave must have been hard for the author of *The Interpretation of Dreams* to follow. He demanded that the analyst refrain from pursuing dream interpretation as an art for its own sake. In practice, this meant that only a few dreams could be fully interpreted in the course of an analysis. Instead, the analyst should always begin anew wherever the free associations of the patient lead. One should also be careful not to overvalue the contribution of dreams to the analysis, as they might then become a special target of the resistances; patients who know that their dreams are wanted by the analyst might dream profusely in a state of positive transference, and very little in a state of negative transference.

The next paper, "The Dynamics of Transference," illustrates the extent to which technique is dependent on metapsychological assumptions (1912a). Freud divided the libidinal needs of the analysand into two groups, those that are satisfied in reality and those that remain unsatisfied. It is from the unsatisfied libidinal needs that the transference emerges. Neurotics were seen as people with many libidinal needs left unsatisfied and therefore particularly prone to the development of transference. Freud concluded:

> These circumstances tend toward a situation in which finally every conflict has to be fought out in the transference . . . [p. 104]. It cannot be disputed that controlling the phenomena of transference presents the psychoanalyst with the greatest difficulties. But it should not be forgotten that it is precisely they that do us the inestimable service of making the patient's hidden and forgotten erotic impulses immediate and manifest. For when all is said and done, it is impossible to destroy anyone *in absentia* or *in effigie* (Freud 1912a, p. 108).

This conclusion was the starting point for those analysts who emphasized that transference interpretations are the only mutative ones. Strachey, whose paper (1934) is reprinted in this volume as Chapter 24, is the foremost advocate of this point of view.

In the next paper, "Recommendations to Physicians Practicing Psychoanalysis" (1912b), Freud advocated that every analyst undergo psychoanalysis. He added the important demand that the analyst remain opaque to his patients and, like a mirror, show his patients only what is shown to him. The psychoanalyst should refrain from indoctrination, should keep his social, political, and literary views out of the analysis, refrain from sharing with the analysand his own problems and even his own biography. The therapeutic relationship should be kept free of personal, business, and other entanglements. It has often been held against Freud that he himself did not live up to this requirement, but was in fact a vivid and even argumentative therapist.

The mirror model was a favorite target of critics of psychoanalysis. Even among psychoanalysts, it has often been criticized (Loewald 1960). What has frequently been overlooked, however, is the fact that the mirror model was only an ideal, like free association, to be approximated, but never fully realized. As an ideal, it is important, for it acts as a restraint on the psychoanalyst, who, like other men, has his own needs for contact. This restraint is essential if the transference is to remain free to develop according to the needs of the analysand; that is, free from any concern about the analyst's reaction to his verbalizations. Only under such conditions can the analysis proceed to a successful termination.

The real personality of the psychoanalyst is transmitted to the analysand through a variety of clues: his manner of dress, his manner of speech, his handling of everyday situations, the decor of his office, and many more. In response to these clues the analysand has either positive or negative feelings to the real personality of the analyst. However, as the analysis progresses, the analysand tends to displace onto the psychoanalyst the attitude he had toward important persons in his past. Such persons are usually the mother, father, siblings, or spouse, but may also be other important personages, such as grandparents, teachers, physicians, previous therapists, or others. In addition to these displacements, the analysand also projects onto the psychoanalyst, from time to time, his own unconscious wishes, needs, or guilt feelings. Transference to the therapist takes place as a result either of displacement or of projection (see Nunberg 1951, for an interesting discussion of these mechanisms).

Freud's paper is of further importance in the history of psychoanalytic technique, because it was here that Freud formulated the concept of "the evenly suspended attention," as the psychoanalyst's counterpart to the patient's free associations (see the introductory note to Reik's essay, this volume, p. 370). In the course of subsequent developments the idea of the evenly suspended attention became the focus of an important controversy between Reik, as the advocate of surprise (1933, this volume, Chapter 26) and Fenichel, who emphasized a more orderly sequence of interpretations (1935a, this volume, Chapter 30; see also Fenichel 1939, pp. 3–6).

Among the papers on technique, the one written in 1914 (1914c) is of special significance. Freud had gained a new understanding of the transference, for he saw it now as a specific case of a more general compulsion to repeat. The recognition of the repetition compulsion made it possible to differentiate between two stages in the development of transference. The first consisted of transient displacements of feelings from the original parents or other persons onto the analyst. This stage can be called the stage of transference manifestations. At a certain point in the analysis, when the transference manifestations become more intense, the repetition compulsion takes over and transference manifestations are converted into a transference neurosis, which represents the working of the repetition compulsion in the analytic situation:

> . . . As long as the patient is in the treatment he cannot escape from this compulsion to repeat; and in the end we understand that this is his way of remembering. . . . [p. 150] We render the compulsion harmless, and indeed useful, by giving it the right to assert itself in a definite field. We admit it into the transference as a playground in which it is allowed to expand in almost complete freedom and in which it is expected to display to us everything in the way of pathogenic instincts that is hidden in the patient's mind. Provided only that the patient shows compliance enough to respect the necessary conditions of the analysis, we regularly succeed in giving all the symptoms of the illness a new transference meaning and in replacing his ordinary neurosis by a "transference-neurosis" of which he can be cured by the therapeutic work (Freud 1914c, p. 154).

In 1917, Freud repeated the idea in a more forceful manner:

> . . . when, however, the treatment has obtained mastery over the patient, what happens is that the whole of his illness's new production is concentrated upon a single point—his relation to the doctor. Thus the transference may be compared to the cambium layer in a tree between the wood and the bark, from which the new formation of tissue and the increase in the girth of the trunk derive. When the transference has risen to this significance, work upon the patient's memories retreats far into the background. Thereafter it is not incorrect to say that we are no longer concerned with the patient's earlier illness but with a newly created and transformed neurosis which has taken the former's place (Freud 1916–1917, p. 444).

Once more, and to my knowledge for the last time, Freud had embraced a new technical discovery with premature enthusiasm. Events did not justify either Freud's optimism or his arboreal metaphor. In 1955, Glover observed, still within the arboreal metaphor, that the transference neurosis does not replace the old neurosis, but seems rather to be grafted onto it: The two then co-exist. Glover also observed that many categories of patients, notably the character neuroses, failed to develop a transference neurosis. In addition, clinical evidence indicated that in those cases where the transference neurosis became all-consuming, the patient's ego was prone to regression, and the autonomous functions of the ego weakened (Blum 1971). It also appeared that the transference neurosis could be accentuated by an excessive reliance on transference interpretation to the exclusion of extra-transference interpretations. (Strachey, whose 1934 paper is reprinted in this volume as Chapter 24, is the outstanding advocate of this approach.) Nevertheless, and paradoxically in the light of these findings, the fostering and resolution of the transference neurosis were generally considered the distinguishing marks of psychoanalytic technique. Gill's definition of psychoanalysis is a typical expression of this point of view:

Psychoanalysis is that technique which, employed by a neutral analyst, results in the development of a regressive transference neurosis and the ultimate resolution of this neurosis by the technique of interpretation alone (Gill 1954, p. 775).

Freud did not refer to the transference neurosis after 1920, a fact which was taken to mean that he had lost his faith in it (Blum 1971). However, this seems unlikely. When Freud changed his mind he usually said so openly; typically, at such a point he would conquer new psychic territory. No central concept appeared, however, to replace the transference neurosis. Without openly contradicting Freud, Loewald suggested a more modest approach to the transference neurosis, one which does not demand that the original neurosis be replaced by the transference neurosis:

The transference neurosis in this sense is the patient's love life—the source and crux of his psychic development—as relived in relation to a potentially new love object, the analyst, who renounces libidinal-aggressive involvement for the sake of understanding and achieving higher psychic organization. From the point of view of the analyst, this means neither indifference nor absence of love-hate, but persistent renunciation of involvement, a constant activity of uninvolving which tends to impel the patient to understand himself in his involvement instead of concentrating exclusively, albeit unconsciously, on the object (Loewald 1971, p. 63).

How much Freud's psychoanalytic optimism, so prevalent in the period between 1910 and 1914, was a reflection of the optimism of the era is a matter for conjecture. There can be little doubt, however, of a different attitude in the paper on technique of 1919. Freud goes on to explain that the patient's suffering must not come to an end too soon, or he will leave treatment with a moderate amelioration of his symptoms, but before any real psychic change has taken place. One must be on guard lest the patient substitute an organic illness or an unhappy marriage for the neurotic symptoms which the analyst has eliminated. Almost every type of neurosis poses special problems. Thus, some phobic patients will not give up their phobias through interpretation alone. The therapist must use the power of the positive transference and, if need be, the threat of termination to force such reluctant patients to face the phobic situation. Compulsive patients present other problems:

Their analysis is always in danger of bringing a great deal to light and changing nothing (Freud 1919, p. 166).

In 1915 Freud formulated another principle that was destined to play a crucial role in the history of psychoanalytic technique, and to become the focus of another controversy—the abstinence rule:

I shall state it as a fundamental principle that the patient's need and longing should be allowed to persist in her in order that they may serve as forces impelling her to do work and to make changes, and that we must beware of appeasing these forces by means of surrogates (Freud 1915a, p. 165).

In the paper delivered at the Budapest conference the implications of the abstinence rule were stated:

> Any analyst who, out of the fullness of his heart, perhaps, or of his readiness to help, extends to the patient all that one human being may hope to receive from another, commits the same economic error as that of which our non-analytic institutions for nervous patients are guilty. Their one aim is to make everything as pleasant as possible for the patient so that he will feel well there and be glad to take refuge there again from the trials of life. In so doing they make no attempt to give him more strength for facing life and more capacity for carrying out his actual tasks in it. In analytic treatment all such spoiling must be avoided. As far as his relations with the physician are concerned, the patient must be left with unfulfilled wishes in abundance. It is expedient to deny him those satisfactions which he desires most intensely and expresses most importunately (Freud 1919, p. 164).

The theoretical status of the concept of abstinence was ambiguous, since if, as Freud assumed, the patient fell ill as a result of frustration, why should further frustration be curative? It was not self-evident that frustration supplies the energy for cure. Once more the energy model became a source of difficulty. Disregarding metapsychology, experience usually confirms the belief that gratification lulls the analysand while frustration increases the negative transference and hastens the formation of the transference neurosis.

Ten years later, Ferenczi, in one of his last contributions to technique, advanced the opposite principle—that of indulgence (1930, this volume, Chapter 21). A wide variety of phenomena are included under the rubric of Frustration vs. Indulgence. Ferenczi stressed "indulgences," such as extension of the analytic hour; one may add use of telephone contacts, answering questions rather than demanding that the analysand associate to the question, speaking when the analysand is silent out of defiance, seeing relatives of the analysand, reading his books, seeing the analysand's exhibitions or performances, waiving payment for missed hours when the analysand is not at fault, and a host of others. Ferenczi also included physical intimacies, usually, but not always, restricted to touch. Freudian analysis, generally speaking, rejected the expediency of Ferenczi's indulgences, but allowed greater latitude in psychotherapy than in psychoanalysis.

The question was an agitating one until Eissler introduced the concepts of analyzable and nonanalyzable parameters (1953). Any indulgence interferes with the capacity of the analysand to express negative feelings to the analyst and acts as a block. The unanalyzed indulgences are believed to block the analysis more than those that are later subjected to analysis. Eissler showed that the ability of an analysand to receive no other gratification than the correct interpretation depends on the structure of his ego. One may be forced to introduce a "parameter," such as Freud advocated for phobia cases, but the impact of the parameter must be analyzed as quickly as possible, if the analysis is not to be replaced by suggestion.

After 1919, Freud's interest veered away from technique. It turned to broader, more philosophical or metapsychological problems, thus initiating the fourth phase, that of the dual instinct and structural theories (Freud, 1920, 1921, 1923b). He did not return to problems of technique until 1937 (Freud, 1937a, 1937b). The field of technique was thus left open to other analysts for two decades.

The first important papers by other analysts made their appearance in 1919 (see Abraham's paper, this volume, Chapter 4, and Ferenczi, this volume, Chapter 5). This was also the year when Ferenczi began his experiments with "active technique." The fact that Freud himself found it necessary to introduce modifications in the classical technique probably stimulated Ferenczi's more radical innovations. These innovations created the first controversy that was restricted to the realm of technique. Part Four of this volume is devoted to it. It initiated a creative, vigorous, and highly controversial period in the field of technique, which continued until the outbreak of World War II.

Ferenczi's innovations were soon followed by those of Rank, who devised not only a new theory of neurosis but also a new technique (1924). He believed that the basic trauma underlying all neuroses is the trauma of birth; the task of treatment, therefore, is to help the analysand to relive and work out that trauma. Like pregnancy, the analysis itself must come to an abrupt end after nine months. Freud, at first, was captivated by Rank's bold speculation, but in his 1926 book (1926a), he turned against it. The paper (1925) by Alexander in this volume (Chapter 6) gives a good picture of the influence of Rank's thinking on psychoanalysts of the 1920s.

Among the writings on technique in this decade, the slender volume by Ferenczi and Rank (1924) occupies a special place. Like Reich's *Character Analysis*, which appeared a few years later, it was a blueprint for conducting an analysis. Going back to Freud's paper (1914c), Ferenczi and Rank took as their point of departure the new insight that acting and experiencing are a form of remembering; but unlike Freud, who would have sought to convert the experience back into memory, Ferenczi and Rank believed that in transference the analysand relives not his repressed memories, but his primal neurosis (German: *Urneurose*), which was repressed immediately without having been conscious. It is, therefore, not only uneconomical, but impossible to expect the analysand to convert his reliving into remembering.

Every analysis falls into two distinct phases. During the first stage the libido unfolds in the analytic situation in which the analyst is essentially a passive recipient, interrupting only where a correction is needed. The task of the analysand's ego is to gain, with the aid of the analyst, the necessary courage to let all the derivatives of the libido into consciousness. As soon as this phase is completed, the second stage, that of the weaning of the libido, begins. This stage is achieved by the setting of a termination date by the analyst who lets the analysand vent his disappointment and anger on the "weaner"—the analyst conceptualized as if he were a weaning mother. The task of the ego during this phase is to assist the analyst by recognizing the realistic necessity for weaning. Like Strachey a decade later, Ferenczi and Rank felt that what is not experienced in transference is therapeutically inert.

From the recently published correspondence between Freud and Jung (McGuire 1974) we learned that Freud was interested in the problem of sadism and the negative therapeutic reaction long before there was any evidence for it in his published writings. Thus, in 1909 he writes to Jung:

> In my practice, I am chiefly concerned with the problem of repressed sadism in my patients; I regard it as the most frequent cause of the failure of therapy. Revenge against the doctor combined with self punishment. In general, sadism is becoming more and more important to me (Letter 163).

From the published writings it is possible to generalize and say that during the teens of the century, Freud made clinical observations that in the late twenties he transformed into general and even philosophical concepts. This is what happened to the repetition compulsion, which was stated in a clinical context in the paper of 1914 (1914c), and enlarged in the book of 1920. For six years the concept of the repetition compulsion had apparently been dormant in Freud's writings, but now it became the impetus behind the reformulation of his instinct theory. It led him to postulate the existence of the aggressive drive as an antagonist to the libidinal drive. This formulation came to be known as the dual instinct theory.

Similarly, the negative therapeutic reaction was extended from a clinical observation into a theoretical concept. It was observed clinically in the case of the "Wolf Man" (1918), and later made the basis for the discovery of the superego (1923b). Even the workings of the superego were described in a clinical paper in 1916, when Freud noted the tendency of a group of neurotics to break down at the moment when their "cherished wishes" came to fulfillment. He called members of this group "Those Who Are Wrecked By Success." The workings of the superego were also noted in another group—men whose sense of guilt was so oppressive, although unconscious, that they committed crimes in order to attach their guilt to something tangible (Freud 1916, p. 332). Freud called them "Criminals Out of a Sense of Guilt." The sense of guilt, observed clinically in these two groups, was described in his paper, but the term "superego" had not yet been coined.

The harmony between theory and technique received a jolt with the discovery of the negative therapeutic reaction, that forced Freud to conclude that making the unconscious conscious does not always result in the amelioration of the symptoms, but may lead to their exacerbation. In 1923 he wrote:

> There are certain people who behave in quite a peculiar fashion during the work of analysis. When one speaks hopefully to them, or expresses satisfaction with the progress of the treatment, they show signs of discontent and

their condition invariably becomes worse. One begins by regarding this as defiance and as an attempt to prove their superiority to the physician, but later one comes to take a deeper and juster view. One becomes convinced, not only that such people cannot endure any praise or appreciation, but that they react inversely to the progress of the treatment. Every partial solution that ought to result, and in other people does result, in an improvement or a temporary suspension of symptoms produces in them for the time being an exacerbation of their illness; they get worse during the treatment instead of getting better. They exhibit what is known as a "negative therapeutic reaction" (Freud, 1923b, p. 49).

The negative therapeutic reaction was the impetus for the delineation of the superego as a psychic structure and thus led to the structural theory. It led Freud to add to the hitherto topographic view, which divided the psyche into three layers—unconscious, preconscious, and conscious—and a structural view which divided the psyche into three psychic structures—ego, id, and superego.

The impact of the structural theory on technique was great. Psychoanalysts began to look upon a dream, a fantasy, a symptom, or even an analytic hour, as a compromise among the three structures (ego, id, and superego), and to analyze dreams or fantasies into three components. The transition from topographic to structural thinking took a long time, particularly in the case of the dream, where it was postulated only in 1964 (see Arlow and Brenner 1964, pp. 114–143). Implicit in the structural view, although not recognized until made explicit by Eissler (1958), was also a new understanding of the nature of free associations. They are productive only when a triologue between ego, id, and superego can take place. When one of these structures is in absolute control, free associations become monotonous and unproductive. For example, in melancholia, when the superego dominates the psychic field, free associations yield nothing but barren self-accusations. On the other hand, as Eissler points out, if delinquents were given the opportunity to free associate, the associations would yield nothing but unabashed id wishes.

The immediate impact of the structural theory on technique was to suggest that the psychoanalyst could attempt to undo the pathological introjection of the superego by encouraging the analysand to reproject the superego onto the analyst; he would then hopefully introject the analyst as a rational superego figure that would not be at odds with the realistic demands of the ego and the realizable wishes of the id. This approach originated with Franz Alexander in his 1925 paper (this volume, Chapter 6), and it dominated discussions of the metapsychology of the process of cure throughout the 1920s. It was still prevalent in the 1930s, when it was challenged increasingly by the advocates of psychoanalytic ego psychology (see Ferenczi 1928a, this volume, Chapter 17; Nunberg 1928, this volume, Chapter 15; Strachey 1934, this volume, Chapter 24; and Riviere 1936, this volume, Chapter 28). See also the introductory note to Alexander's paper, this volume, p. 99). Since this work of projection and reintrojection of the superego was assumed to be a relatively easy task, the analyses based on an approach via the superego were expected to be short.

After the hegemony of the dual instinct theory and the structural theory was established, Glover could say in 1931, "The trend of modern psychoanalytic therapy is in the direction of interpreting sadistic symptoms and guilt reactions." (See this volume, Chapter 23.) The former takes cognizance of the aggressive drive; the latter emphasizes the role of the superego. At this stage new questions emerged. What would be the impact of interpretations that emphasized primarily repressed sexual and libidinal fantasies or, by contrast, interpretations that centered their attention exclusively or predominantly on derivatives of the aggressive drives? The controversy between Reich and Nunberg can be viewed along similar lines. Reich regarded all manifestations of the positive transference in the initial phase of analysis as hypocritical concealment of latent aggression, while Nunberg believed in fostering the positive transference in the initial phases of treatment.

Both the dual instinct theory and the structural theory increased the options available to the psychoanalytic therapist, thus fostering controversies in the realm of technique. Many who longed for firmer and more authoritarian guidance to psychoanalytic technique found these controversies difficult to accept; they felt they lowered the value of psychoanalysis. To those who see this development in a historical perspective and have an interest in the evolution of ideas, however, these controversies seem not only inevitable, but, what is more important in the long run, they lead to a deepening understanding of the psychoanalytic process.

Historically speaking, the theory and treatment method of Melanie Klein and her followers can also be seen as a response to the dual instinct theory and the growing significance of the superego. Going beyond Freud in the scope of her assumptions, she postulated that at birth the aggressive drive is much stronger than the libidinal drive; she drew far reaching conclusions from her assumption:

> In the very first months of the baby's existence it has sadistic impulses directed, not only against its mother's breast, but also against the inside of her body: scooping it out, devouring the contents, destroying it by every means which sadism can suggest. The development of the infant is governed by the mechanisms of introjection and projection. From the beginning the ego introjects objects "good" and "bad," for both of which the mother's breast is the prototype—for good objects when the child obtains it, for bad ones when it fails him. But it is because the baby projects its own aggression on to these objects that it feels them to be "bad" and not only in that they frustrate its desires: the child conceives of them as actually dangerous—persecutors who it fears will devour it, scoop out the inside of its body, cut it to pieces, poison it—in short, compassing its destruction by all the means

which sadism can devise. These imagos, which are a phantasically distorted picture of the real objects upon which they are based, become installed not only in the outside world but, by the process of incorporation, also within the ego (Klein, 1934, p. 282).

In accordance with these assumptions, Klein and her followers tended to emphasize the importance of analyzing the superego (see Strachey's contribution to the Marienbad Symposium, 1936). Strachey's article (1934), reprinted in this volume (Chapter 24), will give the reader an overview of Klein's ideas, particularly in the section entitled "Introjection and Projection." The reader is also referred to the article by Riviere (1936, this volume, Chapter 28). Klein and her collaborators developed a theory of neurosis (as well as a technique) in adult analysis that differed significantly from the approach taken by the psychoanalytic ego psychologists. While psychoanalytic ego psychologists seek the most appropriate interpretation and tend to make one interpretation at a time, Kleinian analysts can, if necessary, offer as many as eight or nine basic interpretations in a single session (see Rosenfeld 1958).

Only three years separate the publication of *The Ego and the Id* (Freud 1923b), when the structural view was formulated, from that of *Inhibitions, Symptoms, and Anxiety* (Freud 1926a), the book that ushered in psychoanalytic ego psychology. There is a marked difference in the role assigned to the ego in the two books. The former stresses the weakness of the ego vis-à-vis the superego; in the latter, the ego, because it can give the anxiety signal, is conceptualized as more powerful. While we know the clinical reason that led Freud to find the topographic view insufficient and to supplement it with the structural one—the discovery of the negative therapeutic reaction—we are less clear about the reasons that led him to reformulate the problem of anxiety.

In the new formulation of 1926, all anxiety was considered to be a response to real danger or to danger judged real by an immature ego. So it is that separation anxiety or castration anxiety, although not real by objective standards, is judged to be real. In order to avoid the unpleasure of anxiety, the ego builds a complex defense mechanism. Anxiety, therefore, is at the core of every neurosis (see Nunberg 1928, this volume, Chapter 15; also Waelder 1967). If our focus is then on anxiety, we can differentiate between two groups. The first comprises those patients who come to analysis in an acute state of anxiety. The very fact that they suffer from anxiety shows that an intrapsychic conflict is raging and none of the three psychic structures is strong enough to settle the conflict in its favor. In the other group are those patients who have succeeded in building a more or less functioning structure of defense mechanisms against the outbreak of anxiety, but the defensive structure itself has become a heavy burden. Once they enter treatment, these patients become afraid that the treatment itself will arouse the anxiety that they had such difficulty in repressing.

Within the framework of the structural theory it becomes possible to differentiate three types of resistances to treatment. The resistances of the id manifest themselves in an unwillingness to give up pleasure, even if the costs now or later are great, or, stated differently, even if the ego or the superego disapproves of the pleasure. These are different from the resistances of the superego, which manifest themselves in a persistent feeling of guilt and in the negative therapeutic reaction. Finally, different again are the resistances of the ego, which manifest themselves in the employment of unconscious defense mechanisms. In the 1930s analysts questioned how much of the psychoanalyst's energy should be devoted to undoing the ego's defense mechanisms and how much to making the unconscious conscious. The controversy over this issue is reflected in many of the articles reprinted in this volume.

The resistances of the ego were, in turn, subdivided into three: repression resistances, transference resistances, and resistances due to secondary gains from illness, thus making a total of five groups of resistances (Freud 1926a, p. 160). Strachey made the further point that in negative transference the resistance comes from the id, while in positive transference it comes from the ego (1934, this volume, Chapter 24). When the analysand is afraid to confess love or is afraid that his love will be rejected, it is the ego that is resisting. On the other hand, when he refuses his cooperation because his offer of love has been rebuffed, it is the id that is resisting. Kaiser added the valuable observation that it is not the transference per se, but the rationalization with which it is denied that causes the transference to become a resistance (1934, this volume, Chapter 27).

The impact of this classification of the resistances on psychoanalytic technique was considerable. Whereas the early papers treated resistance as one global phenomenon, the analyst was now charged with the new task of deciphering the particular directions from which the resistances originated at any particular moment during treatment. He had to pay attention to the everchanging nuances of the resistances with which he was working. While thus engaged he was less likely to blame the analysand for resisting and therefore was less likely inadvertently to reinforce the analysand's already powerful superego (see Nunberg 1928, this volume, Chapter 15).

With this emphasis on the resistances it was to be expected that a school of therapy would emerge that would emphasize resistance analysis as the most important or the only relevant part of psychoanalysis. In the chapter reprinted from Wilhelm Reich's book (1933, this volume, Chapter 18), the reader can see that resistance analysis takes precedence over dream analysis, analysis of the superego, and that of the id. The same idea is advanced in an even more radical form by Kaiser in his essay (1934, this volume, Chapter 27). Fenichel, on the other hand, insisted on a balance between resistance analysis and what came to be called content analysis (1939, this volume, Chapter 19). The issues involved in this controversy are of great significance and a special section of this book is devoted to it (Part Eight).

The 1930s show a sharpening division of opinion between the ego psychologists on the one side, and, on the other side, those who wished to interpret defenses only (see Kaiser 1934, this volume, Chapter 27), those who wished to rely exclusively on interpretation of transference (see Strachey 1934, this volume, Chapter 24), those who, like Reik, continued to regard psychoanalysis as primarily an intuitive art (1933, this volume, Chapter 26), and, finally, the followers of Melanie Klein, who stressed the predominance of aggressive wishes in early infancy and the need to analyze these aggressive wishes first. In time, psychoanalytic ego psychology also reevaluated the significance of the lifting of infantile amnesia. In the early years it was common to measure progress in analysis by the intensification of the transference and the recall of infantile memories. With a more comprehensive understanding of the structure of the ego, the significance of recall lost its central position. Mere remembering can be deceptive, and when the main defense is isolation rather than repression, it is more important to recapture the affect and to understand the significance of a memory than to accumulate new memories (Kris 1956b).

In the late 1920s and early 1930s, psychoanalysis reached the textbook stage with Glover as the pioneer (1927). He was followed in 1930 by Ella Sharpe, whose lectures were influenced by Melanie Klein. On the continent, Wilhelm Reich was the dominant figure; his textbook appeared in 1933. While the foundations of ego psychology were laid by Freud in 1926, it was the controversy with Reich that provided the impetus for two important books on technique, Anna Freud's *The Ego and the Mechanism of Defense* (1936), and Fenichel's book on technique, from which an excerpt is reprinted in this volume (1941, Chapter 19).

Reich's *Character Analysis* is of great intrinsic interest, and a further dimension is added to this work by the fact that without it psychoanalytic ego psychology might have taken a different turn. A whole section of this book has been devoted to Reich and his critics (Part Seven). Reich's position is summarized in the opening statement to the reprinted chapter. Here, therefore, it is only necessary to emphasize that his basic concept of character resistance is not synonymous with Freud's resistances of the ego. The structural division of the personality into ego, id, and superego meant little to Reich (see Kris 1951; and Hartmann 1951); to him, character resistance was a global phenomenon, spreading through the entire personality.

Like Freud's papers on technique, Anna Freud's book is easily available. We have accordingly not excerpted from it. The contemporary reader may, however, have difficulty with the metaphorical language, where psychic forces are described as if they were army commanders full of cunning:

> The instinctual impulses continue to pursue their aims with their own peculiar tenacity and energy, and they make hostile incursions into the ego, in the hope of overthrowing it by a surprise-attack. The ego on its side becomes suspicious; it proceeds to counter-attack and to invade the territory of the id. Its purpose is to put the instincts permanently out of action by means of appropriate defensive measures, designed to secure its own boundaries (Anna Freud 1936, pp. 7–8).

The book, however, is required reading for anyone who wishes to understand the history of psychoanalytic technique in the 1930s.

Anna Freud's outstanding contributions can be summarized as follows: she was the first to realize that an important implication of psychoanalytic ego psychology was the relaxation of the demand that the analysand associate freely from the beginning of his analysis. She understood that the way the analysand resists free association may be as informative as the associations themselves. This formulation alone enabled analysts to avoid much disappointment and removed much of the angry countertransference that characterized the relationship between the early psychoanalysts and their resisting analysands. Thus the frantic search of Ferenczi and Reich to discover devices that would supplement free associations became superfluous:

> Even to-day many beginners in analysis have an idea that it is essential to succeed in inducing their patients really and invariably to give all their associations without modification or inhibition, i.e., to obey implicitly the fundamental rule of analysis. But, even if this ideal were realized, it would not represent an advance, for after all it would simply mean the conjuring-up again of the now obsolete situation of hypnosis, with its one-sided concentration on the part of the physician upon the id. Fortunately for analysis such docility in the patient is in practice impossible. The fundamental rule can never be followed beyond a certain point (Anna Freud 1936, p. 13).

Also basic was Anna Freud's differentiation between the transference of libidinal impulses and the transference of defense:

> It may happen in extreme cases that the instinctual impulse itself never enters into the transference at all but only the specific defence adopted by the ego against some positive or negative attitude of the libido, as, for instance, the reaction of flight from a positive love-fixation in latent female homosexuality or the submissive, feminine-masochistic attitude, to which Wilhelm Reich has called attention in male patients whose relations to their fathers were once characterized by aggression (Anna Freud 1936, p. 20).

Her book also contains the important warning that an analyst who concentrates excessively on the translation of symbols (Stekel is the outstanding example), or an analyst who concentrates unduly on the analysis of transference, is in danger of shifting the balance in favor of the id, at the expense of the analysis of the defense mechanisms of the ego. She thus took a position opposite to that of Strachey (1934, this volume, Chapter 24), who thought that the only effective interpretations are transference interpretations. With the publication of Anna Freud's book, psychoanalytic ego psychology came into its own. An interesting retrospective account of this phase in psychoanalytic development is given by Sterba (1953, this volume, Chapter 20).

As one would expect, there was much debate among his followers as to who were Freud's true heirs. However, if the line of reasoning followed in this chapter is correct, all the participants could lay claim to this heritage. They all used free association and analyzed transference and resistance. They differed in the kind, the dosage, and the timing of interpretations. What to one analyst was a phase in Freud's development became to another the cornerstone of his whole approach. To Reik, the evenly suspended attention was crucial; to Reich, the analysis of resistance; to Strachey, the analysis of transference; to Alexander and Klein, the analysis of the superego. To Anna Freud, equidistance toward ego, id, and superego—and the neutrality of the analyst toward those three structures—was crucial. This book does not aim to convert the reader to any particular point of view; the object of its interest is the history of the development of psychoanalytic technique.

In 1939 World War II broke out; also in that year Freud died, psychoanalysis was expelled from central Europe, and its center moved to the English-speaking world. The post–World War II period brought new problems, new formulations, and a new generation of participants. The new developments deserve a volume of their own.

Part Two

Contributors to the

Development of

Psychoanalytic

Technique

CHAPTER 3

FRANK R. HARTMAN

BIOGRAPHICAL

SKETCHES

Karl Abraham (1877–1925)

KARL ABRAHAM was born on May 3, 1877 in Bremen, Germany, the second of two sons, to Nathan and Ida (née Oppenheimer) Abraham. Theirs was a rather poor, Orthodox Jewish family. Nathan Abraham (1842–1915) started as a teacher of Jewish religion and law, but gave up this profession in 1873 and began a wholesale drapery business in order to earn enough to marry. He remained "orthodox without ever being obsessional and intolerant."[1]

Ida Abraham "was devoted, kind [and] fussy . . . she had a forceful character and was universally loved."[2] She was also ". . . in some respects the prototype of the over-anxious Jewish mother" (H. Abraham 1974).

According to Hilda Abraham, Karl's daughter, Nathan and Ida were first cousins, and there were probably other intermarriages in the family. She thus concludes that Abraham's papers, "The Significance of Intermarriage Between Close Relatives" (1909) and "On Neurotic Exogamy" (1913a) are autobiographical.

Both Karl and his elder brother, Max, who was born in 1874, had asthma as children. Max was sickly and delicate from birth and in later childhood was restricted from any play or sports. In adulthood he became "very neurotic." By contrast, Karl was a healthy, happy and contented infant, whose nursing was a joy to his mother (H. Abraham 1974). When he was still small his mother had a late miscarriage of a girl (H. Abraham 1974). Although healthier than Max, Karl had the same restrictions placed upon his activities out of "fairness" to his brother. His childhood was thus isolated and lonely.

Hilda Abraham considers the following passage from "Oral Erotism and Character" (1924) to be "purely autobiographical" (1974).

In certain other cases, the person's entire character is under oral influence, but this can only be shown after a thorough analysis has been made. According to my experience we are here concerned with persons in whom the sucking was undisturbed and highly pleasurable. They have brought with them from this happy period a deeply-rooted conviction that everything will always be well with them. They face life with an imperturbable optimism which often does in fact help them to achieve their aims.

Certainly throughout his life Abraham was lively, cheerful, and optimistic. He rarely became angry, but when he did his outbursts were like the explosion of a "powder-keg." He liked to eat and to play a game with his children in which they took turns stealing a lump of sugar from him while he pretended to be asleep. When Freud inquired of Abraham why he had not given more attention to "hysterical anorexia" in his paper "The First Pregenital Stage of the Libido" (1916), Abraham acknowledged a blind spot and responded, "I know from experience that my reaction to unpleasant events regularly makes itself felt by loss of appetite" (H. Abraham and E. Freud 1965).

The Abraham household included two unmarried paternal aunts, one of whom was psychotic, and the maternal grandparents. On the Sabbath and religious holidays the family was joined by the paternal uncle Adolf, "peculiar," "neurotic," and also unmarried (H. Abraham 1974).

Abraham began elementary school at age six and by the next year was a top student. He was also given religious instruction. He entered the *Gymnasium* at age nine and early showed talent for, and interest in, languages. This interest intensified during his adolescence and remained throughout his life. The study of the subject was considered impractical and financially out of the question. When the time came to consider study beyond the *Gymnasium*, a family conference was held and the career of a dentist (not requiring university training) was urged upon him. Abraham refused this suggestion but did agree to study dental medicine (which did require a university education). Uncle Adolf agreed to pay for his education.

In 1895 Abraham entered the University in Würzburg. After one semester he switched to the study of medicine, which he pursued further in Berlin and Freiburg-im-Breisgau. He received his M.D. in June, 1901. Excused from military service because of emphysema, he applied for a position under Liepmann at the Berlin municipal mental hospital at Dalldorf. More than a choice of career was involved; before accepting this position he wrote his father that he would be unable to observe the Sabbath and the dietary laws there. His father answered that he would have to make such an important decision on his own.

While in Berlin Abraham met Hedwig Burgner (b. 1878), the daughter of an assimilated Jewish family. Her mother had died when she was seventeen; as the eldest child she took charge of her younger siblings, one of whom underwent a five-year psychiatric hospitalization, beginning with symptoms of anorexia. She was discouraged from university study by her father, but she audited courses and shared Abraham's interest in languages.

Bored with the neuropathological approach at Dalldorf, Abraham resigned in the spring of 1904 and sought a position at Burghölzli in Zurich under Bleuler and Jung, where he began work December 8, 1904. Delighted with the new and stimulating atmosphere, Abraham hoped to remain there permanently and, despite financial difficulties, he and Hedwig were married January 23, 1906.

While at Burghölzli, Abraham began studying the works of Freud early in 1907. They started to correspond in the late spring of 1907. Fortunately, most of the letters have been published; they provide fascinating insight into both men. By October of that year, it had become clear to Abraham that his hope of promotion in Zurich would not be fulfilled. He therefore decided to move to Berlin where he set up practice as a specialist in nervous and mental diseases. He also hoped to begin a psychoanalytic practice. Having moved his family to Berlin, he visited Freud for four days in Vienna in December 1907. Thus began an association and collaboration of great importance for psychoanalysis and psychiatry.

For many years Abraham was to remain the only analyst in Berlin, although he followed Freud's example and held weekly meetings in his home for those interested in the subject. This group became the Berlin Society, the first to join the International Psycho-Analytic Association. Following Jung's resignation, he became the acting president (1914–1918) of the International Association.

Abraham was drafted after the outbreak of World War I and assigned to light duty because of his previous rejection. He was made a surgeon, and his unit was first stationed on Berlin's outskirts, but in 1915 was moved to Allenstein in East Prussia. In 1916 Abraham was permitted to form a psychiatric unit, enabling him to make his contributions to the study of war neurosis.

While in Allenstein, he was able to continue the part-time practice of analysis. He undertook the simultaneous treatment of a woman in her fifties and her pre-adolescent nephew with a learning block. The aunt's treatment yielded the case material for "A Particular Form of Neurotic Resistance Against the Psychoanalytic Method" (1919b), this volume, Chapter 4, as well as for his paper on the applicability of analysis in advanced age (1919a). In an earlier paper on technique, "Should Patients Write Down Their Dreams?" (1913b), Abraham reported that when patients recorded their dreams, they generally could not make out what they had written. Consequently, he advocated that in general one should prohibit the practice.

While in Allenstein, Abraham suffered a recurrence of his childhood asthma after a bout of severe bronchitis. He also contracted dysentery, which subsequently recurred. Nevertheless, when he returned to Berlin following the war he was soon actively organizing the first psychoanalytic training facility with the help of Eitingon, as financial backer and administrator, and Hanns Sachs, who became the first "training analyst." Simmel and Boehm were also on the original faculty. The Berlin Polyclinic was opened on February 14, 1920, and was renamed the Berlin Institute in 1924. It was finally to provide Abraham with a circle of analysts, but he remained rather reserved with the students. He was elected president of the International Psycho-Analytic Association in 1924 and 1925.

In the years following the war, Abraham continued to have bronchitis and occasional attacks of dysentery. In May, 1925, he choked on a fish bone which lodged in his lung and developed a pulmonary abscess. Attempts to control or cure the abscess proved fruitless; he finally developed a subphrenic abscess from which he died with continual hiccoughs[3] on December 25, 1925.

Abraham's works are to be found in his *Selected Papers* (1960) and *Clinical Papers and Essays on Psychoanalysis* (1955). His major contributions are not on the subject of technique. His writings on pregenital phases of development (especially orality), his linking of character formation to developmental phases, and his early contributions to the subject of manic-depressive and other psychoses are but a few of his valuable contributions. He is considered by many as second only to Freud.

His mother, wife, and children survived him and fled Nazi Germany. His daughter Hilda (1906–1971) became an analyst practicing in England and editor and translator of her father's works. His son, born in 1910, lives in England. His brother Max and his wife died in an extermination camp in Poland.[4]

NOTES

1. K. Abraham's son; personal communication, 1973.
2. Ibid.
3. Ibid.
4. Ibid.

Sandor Ferenczi (1873–1933)

SANDOR FERENCZI was born to a Jewish middle-class family in Miskolc, Hungary, on July 16, 1873. Both parents were of Polish origin. His father emigrated from Cracow to Hungary at the age of eighteen and shortly thereafter took an active, if minor, part in the 1848 revolt of the Hungarians against Hapsburg rule. Following the failure of the revolution (1849), Ferenczi was allowed to settle in Miskolc. He Magyarized his name from Fraenkel to Ferenczi (Jones 1955, p. 157) and became a proprietor of a bookshop, lending library, and theatrical booking agency. Eventually, he became a publisher as well, counting among his authors Michael Tompa, the Hungarian revolutionary poet (Balint 1970, p. x).

The Ferenczi household was a very busy place; in addition to housing eleven children (seven boys and four girls) the home served as a gathering place for much of the intellectual and artistic circle of Miskolc. Sandor, who was the third (Balint 1970) or fifth (Lorand 1966) son (there is a disagreement among biographers on this point), was to prefer the company of such companions most of his life. The bookshop is acknowledged to have served as a "second nursery" for the children, but it is not reported under which parent's supervision. He idolized his father, who died when Ferenczi was fifteen, and had an intensely ambivalent relationship with his mother. All sources agree that he had an inordinate, almost insatiable, need to be loved (Jones 1955, p. 158; Balint 1970, p. xi; Lorand 1966, p. 32).

In 1890, at the age of seventeen, Ferenczi began the study of medicine in Vienna; he completed his courses in 1895 and passed the qualifying examinations for his M.D. in 1896. After a year of military service, he pursued further training in psychiatry and neurology in the city hospitals of Budapest. He became a prolific writer in these fields, entered private practice in 1900, and was appointed a psychiatric expert to the Royal Court of Justice in 1905.

When asked to review *The Interpretation of Dreams*, Ferenczi paid it only casual attention and found it "unscientific." However, when he heard of the word-association experiments at Burghölzli, he bought a stop watch, which he carried at all times, and began testing all those with whom he came in contact (Balint 1970, p. xi). These "experiments" led him to an appreciation of psychoanalysis as a whole, and he read the existing literature in 1907. He wrote Freud and visited him in Vienna on February 2, 1908.

An immediate bond was established between the two men, which was to result in Ferenczi's being asked to vacation with Freud later that year and to accompany him along with Jung on his visit to Clark University in 1909. Freud addressed Ferenczi as "My dear son" and expressed the wish that he might become a son-in-law.

Upon his return to Budapest, Ferenczi began the practice of psychoanalysis and soon became an outstanding clinician and prolific writer. He was brilliant, intuitive, and empathetic as well as being a keen observer. He reported that he soon developed a reputation for working with hopeless cases. He had a remarkable ability to speak in or reproduce the language of children. Socially he was warm, charming, gregarious, witty, and perhaps impulsive. He was quick tempered, but also quick to forgive. He loved jokes and was prone to making slips of the tongue, earning the title "King of Parapraxes" (Jones

1955, p. 158; Jones 1959, p. 199). All sources agree he was a hypochondriac (Lorand 1966, p. 32).

Outwardly, Ferenczi continued to lead the life of a confirmed bachelor, active in the intellectual coffeehouse society of Budapest. Only his closest friends were aware that for some ten to fifteen years he maintained a relationship with Gizella Palós, née Altshol (1863–1949), an attractive and charming married woman ten years his senior and the mother of two daughters. Although they had been separated for many years, Gizella's husband refused to allow her a divorce. With the passing of each year the chances of Ferenczi and Gizella being able to marry and have children became more remote. This produced a good deal of conflict in Ferenczi because he longed to become a father (Balint 1970, p. xii).*

In 1910 Ferenczi, at Freud's instigation, proposed the founding of the International Psycho-Analytic Association at the Nuremberg Congress. The presidency went to Jung because Freud believed that the surest future for psychoanalysis lay in the important psychiatry of Zurich, which was predominately Christian. In 1913, Ferenczi founded the Budapest Society. In that same year he undertook the analysis of Ernest Jones while the latter's mistress of many years (twelve years his senior) was in analysis with Freud (Jones 1959, p. 197).

With the outbreak of World War I, both Freud and Ferenczi found themselves virtually without patients, and Ferenczi underwent an analysis with Freud while he was waiting for his orders to report to his military unit. After he was assigned as physician to a regiment of hussars in Papa, his leaves in Vienna and visits by Freud to his post were used to continue the analysis.

In 1916, Ferenczi was assigned to part-time work as a psychiatrist in a Budapest military hospital. From this experience came his contributions on war neuroses. It also enabled him to resume his private practice of analysis. Soon he was to refer a friend, Anton von Freund, heir to a large fortune in Budapest, to Freud for analysis. In 1918, von Freund donated a large sum of money to found an independent psychoanalytic publishing house and a clinic and training facility in Budapest. Freud now believed Budapest to be the future center of analysis.

The Budapest Congress of September 1918 further heightened the air of optimism. The governments of Germany, Austria, and Hungary sent official observers and entertained the thought of establishing clinics for the treatment of war neuroses. Ferenczi was elected president of the International Psycho-Analytic Association. Forty days later the Hapsburg empire collapsed. Economic chaos and political turmoil soon followed; the border between Austria and Hungary was closed, and Ferenczi was isolated in Budapest.

A thousand students at the University of Budapest petitioned that

* Although Balint says Mrs. Palós was seven years older than Ferenczi, recent evidence of her birth year indicates that she was ten years his senior (McGuire 1974).

Ferenczi be made the world's first professor of psychoanalysis. The new liberal regime granted the chair. And under the new liberalized divorce laws, marriage for Ferenczi and Gizella was made possible. However, the sudden death of Gizella's husband allowed them to wed in 1919 without further delay.

The Bolshevik revolution of Béla Kun, its subsequent suppression, and the installation of the reactionary and anti-Semitic Horthy Regime by the Rumanians, complicated matters. The chair was abolished, and for a time Ferenczi was afraid to be seen on the streets. Within months he was asked to surrender the presidency of the International Psycho-Analytic Association to Ernest Jones because of his isolation. He thought seriously of moving to Vienna, but von Freund was now dying and under the terms of his bequest, Ferenczi was the only person who could oversee its use. Ultimately, all but a small sum was tied up by bureaucratic restrictions until its value was destroyed by inflation.

Before the unsettled year 1919 was out, Ferenczi had encouraged his analysand, Melanie Klein, to work with children; demonstrated his mastery of the subject of classical technique with "On the Technique of Psychoanalysis" (1919a, this volume, Chapter 5), which Freud praised as "pure analytic gold"; and published his first experiment with activity, "Technical Difficulties in the Analysis of a Case of Hysteria" (1919c, this volume, Chapter 7). Other important papers on the subject of activity followed, among them "On Forced Fantasies" (1924a, this volume, Chapter 8). These contributions established Ferenczi as a technical innovator and demonstrated a continuance of his interest in experimentation.

Meanwhile, Ferenczi and Rank began work on *Developments in Psychoanalysis*. The publication of this work in 1924, with its emphasis on transference and abreaction in the transference as opposed to the recollection of childhood memories (see editor's comments, this volume, pp. 32–33), produced further controversy on technique.

"The Problem of the Termination of the Analysis" (1927, this volume, Chapter 16) and "The Elasticity of Psychoanalytic Technique" (1928a, this volume, Chapter 17) return to more classical technique. Concerning this last paper, Freud wrote: ". . . the 'Recommendations on Technique' I wrote long ago were essentially of a negative nature. I considered the most important thing was to emphasize what one should *not* do. . . . Almost everything positive that one *should* do I have left to 'tact,' the discussion of which you are introducing. The result was that docile analysts did not perceive the elasticity of the rules I had laid down, and submitted to them as if they were taboos. . . . All those who have no tact will see in what you write a justification for arbitrariness, *i.e.* subjectivity, *i.e.* the influence of their own unmastered complexes" (Jones 1955, p. 241). Balint (1967), however, dates the beginning of the last phase of Ferenczi's technical experiments, which he calls "pointers to the future," to this same year (1928). "The Principles of Relaxation and Neocatharis" (1930, this volume, Chapter 21) is from this period.

This time his interest in helping "dried up" cases and/or those with early traumata led him to attempt to correct the early trauma by pregenital play such as holding or kissing, with the idea of thus facilitating further personality growth to be followed by analysis in the more classical form. (The results of many experiments that Ferenczi undertook remain unpublished.) These experiments did not meet with Freud's approval, and a cooling in their friendship resulted.

This time when Freud wrote to Ferenczi, he stated:

> I see that the differences between us have come to a head in a technical detail which is well worth discussing. You have not made a secret of the fact that you kiss your patients and let them kiss you. . . . Now when you decide to give a full account of your technique and its results you will have to choose between two ways: either you relate or you conceal it. . . . What one does in one's technique one has to defend openly. . . . We have hitherto in our technique held to the conclusion that patients are to be refused erotic gratifications. You know too that where more extensive gratifications are not to be had milder caresses very easily take over their role. . . .
>
> Now picture what will be the result of publishing your technique. . . . A number of independent thinkers in matters of technique will say to themselves: why stop at a kiss? Certainly one gets further when one adopts "pawing" as well, which afterall doesn't make a baby. And then bolder ones will come along who will go further to peeping and showing—and soon we shall have accepted in the technique of analysis the whole repertoire of demiviergerie and petting parties, resulting in an enormous increase of interest in psychoanalysis among both analysts and patients. The new adherent, however, will easily claim too much of this interest for himself, the younger of our colleagues will find it hard to stop at the point they originally intended and God the Father Ferenczi gazing at the lively scene he has created will perhaps say to himself: maybe after all I should have halted in my technique of motherly affection *before* the kiss (Jones 1957, p. 163).

Balint comments: "How far this was a legitimate experiment and how far it was only a symptom of the immense desire in Ferenczi for love and affection is impossible to decide since he died before his experiments could be concluded. Knowing his character, the most likely answer will be that it was both . . ." (1967, p. 164).

During 1932 Ferenczi became progressively weaker from pernicious anemia. He gave up his practice in December 1932 and died of that disease on May 22, 1933. He was subsequently to become a subject of controversy on account of Jones's assertion that he was psychotic in the last years of his life. This is disputed by Lorand (1966, p. 31) and Balint (1970, p. xiv).

Ferenczi's importance to psychoanalysis cannot be disputed. However, the voluminous Ferenczi–Freud correspondence remains unpublished and no definitive biography has as yet been written. Thus, much remains either unknown or contradictory.

Ferenczi's many valuable and original contributions to psychoanalysis can be found in the following volumes: *Sex in Psychoanalysis* (Volume I of the *Selected Papers of Sandor Ferenczi, M.D.*, 1950), *Further Contributions to the Theory and Technique of Psychoanalysis* (Volume II, 1952), *Final Contributions to the Problems and Methods of Psychoanalysis* (Volume III, 1955), *Thalassa: A Theory of Genitality* (1924b), *Psychoanalysis of the War Neuroses* (with K. Abraham, E. Simmel, and E. Jones, 1921a), and *The Development of Psychoanalysis* (with O. Rank, 1925).

Franz Alexander (1891–1964)

FRANZ GABRIEL ALEXANDER was born January 22, 1891 to Professor Bernard and Regina (née Broessler) Alexander in Budapest. He had four sisters (two older) and one brother. The family was nominally Roman Catholic; his parents, although born Jewish, had been baptized, as was common practice, in order to obtain the father's professorship.[1] According to his autobiography (Alexander 1960), he idealized both his parents. His mother was always there for support and comfort; she was the person who kept the household running smoothly at all times, somewhat behind the scenes. He felt himself to be her favorite. He states that he had practiced psychoanalysis many years before he could comprehend an ambivalent attitude toward a mother.

Alexander wrote much about his father, a professor of philosophy, and his own wish to unite science and the humanities in order to please the father who had wanted him to pursue an academic career in archaeology. In the *Gymnasium*, Alexander was a good student, but he was nearly expelled when caught reading Wedekind's *Spring's Awakening* (Alexander 1960, p. 90) in Latin class shortly after having attended a student socialist meeting.

In choosing medicine, however, he was strongly influenced by an identification with a rich and powerful uncle. In medicine, which he studied in Göttingen and Budapest, he still aimed at an academic career and, before obtaining his M.D. in 1913 from the University of Budapest, he published a paper on the metabolism of the brain. The experiments were done under Geza Revesz who was to become his brother-in-law.

While Alexander was serving his compulsory year of military service, World War I was declared. He served on the Italian front and in 1918 was put in charge of a field laboratory dealing with malaria control.

Following the armistice, Alexander joined the staff of the psychiatric hospital of the University of Budapest with the idea of pursuing his earlier physiological experiments rather than clinical psychiatry. Reluctant contact with patients reminded him of an earlier reading of *The Interpretation of Dreams*. "The book was crazy; but the patient was too. . . ." Maybe he could learn something (Alexander 1960, p. 52). His interest in psychoanalysis began to grow rapidly with his understanding of patients, but so, too, did his conflict about entering the field. It would mean no professorship, and his father considered it "a descent into a spiritual gutter" (Alexander 1960, p. 55).

In 1919, Alexander moved to Berlin to begin the study of psychoanalysis and postgraduate psychiatry. After a period of indecision, he was the first student at the Berlin Polyclinic. His training analysis with Hanns Sachs lasted approximately six months.[2] In 1921, he received Freud's annual prize for his paper "Castration Complex and Character" and became an assistant at the polyclinic. Soon thereafter he became an instructor, and in 1925 a training analyst at the renamed Berlin Institute. In 1927, Alexander's first book, *Psychoanalysis of the Total Personality* (English ed., 1930) was published. During this period, he had won Freud's high regard and the referral of many patients personally important to Freud.

In 1921, at the age of thirty, Alexander had married Anita Venier of Venice. She had a difficult and irascible personality as well as an impressive lineage. Her father, Count Venier, who was the direct descendant of the last Doge of that city-state and heir to his honors,[3,4] had died before her birth. Her mother, from the ancient family of Cherubini, remarried soon afterward and placed her, at the age of three, in a convent school for girls of the nobility. She remained there until shortly before her marriage, developing talent and competence as a painter. In 1927, the same year that *Psychoanalysis of the Total Personality* was published, Max Liebermann awarded her a first prize in a national show of paintings. Despite their individual successes, the Alexander marriage was far from happy or peaceful, though it endured. They had two children, Silvia, born in 1921, and Francesca Alexander Levine, born in 1926 and now a professor of sociology working in the field of juvenile delinquency.[5]

In 1930 Alexander moved to the United States where he spent the first year as a visiting professor of psychoanalysis at the University of Chicago and the following year in Boston pursuing further psychoanalytic studies of crime and juvenile delinquency, a subject which had interested him in Berlin (see *The Criminal, the Judge, and the Public*, with Staub, 1931).

In 1932 he founded the Chicago Institute for Psychoanalysis, which was to become a leading American center for the teaching of psychoanalysis during his twenty-five years as director. During these years he developed more fully his concept of the "corrective emotional experience," a concept that emphasizes the opportunity for relearning in the analytic process through the transference and the exposure to new superego attitudes. The historical origin of this concept is found in "A Metapsychological Description of the Process of

Cure" (1925, this volume, Chapter 6). His fear of the development of a full-blown transference neurosis in analysis led him to experiment with interrupted analyses. In so doing, he attempted to respond countertransferencially to the unconscious of the patient in such a way as to manipulate the transference and shorten the analysis.

Alexander was an important contributor to the study of psychosomatic medicine and introduced the multidisciplinary approach to its problems in his organization of the research teams of the Chicago Institute. With others he founded the journal, *Psychosomatic Medicine*. He also introduced the concept of "Vector Analysis" and wrote on psychotherapy.

In 1956 he left Chicago to become head of the department of psychiatry and director of research at Mt. Sinai Hospital in Los Angeles, and was awarded a grant from the Ford Foundation for further research. According to Grinstein, he wrote 238 papers and wrote or edited 18 books. The papers reprinted here, as well as many others reflecting his development over the years, are to be found in *The Scope of Psychoanalysis* (1961). Other works include *Psychosomatic Medicine: Its Principles and Application* (1965), *Fundamentals of Psychoanalysis* (1948), *Psychoanalysis and Psychotherapy* (1956), and *The Western Mind in Transition* (1960).

He died on March 8, 1964.

NOTES

1. R. Bak; personal communication, 1973.
2. R. Loewenstein; personal communication, 1973.
3. *Chicago Daily News*, February 3, 1936.
4. F. A. Levine; personal communication, 1973.
5. Ibid.

Edward Glover (1888–1972)

EDWARD GLOVER, the youngest of three sons, was born on Friday, January 13, 1888, to Matthew and Elizabeth (née Shanks) Glover in Lesmahagow, Scotland. His mother had been raised and educated at home, in accordance with the harshly puritanical beliefs and practices of the Reformed Presbyterian Church, by a great-aunt and uncle. The latter was a minister of this sect, and Edward was born and raised in his manse. Edward's mother broadened her horizons, however, and was a "competent wife and house-keeper and a restrained but doting mother"[1] with "occasional outbursts of obsessional anxiety" (Kubie 1973).

His father was a country schoolmaster who emphasized "straight thinking" and its disciplined expression. He was a Darwinian agnostic, brilliant, intellectual, and able to teach a broad curriculum including many languages.

Glover began school at the age of three-and-one-half. Shortly thereafter his older brother John, his mother's favorite, died at about the age of six (Kubie 1973, p. 86).[2] Edward was, in his own words, "reluctant, rebellious, contumacious, and obstinate"[3] as a school boy, particularly disliking the subjects of religion and arithmetic. He further described himself as a "prig," which he defined as one who is "holier-than-thou"[4]—a good little boy when under observation who more than makes up for the goodness in secret. He was compared unfavorably with his brilliant eldest brother James (b. 1882), whose first short story was published when he was fourteen, and his parents became concerned that he would disgrace the family.

At the age of eleven he entered senior school where his father became his teacher. An abrupt change occurred; he became an excellent student. He thought of becoming a teacher himself, but as he put it, "in the long run an identification with my eldest brother . . . prevailed, or shall we say the need of the young brother to copy and surpass his older sibling rival."[5] He therefore chose medicine, passing the university medical matriculation examination at age fourteen. For the next twenty-four years he and James were close friends and loving brothers.

At sixteen, Glover took and passed the arts matriculation examination, merely for intellectual pleasure, and began the study of medicine at the University of Glasgow. These were the happiest years of his life. He excelled academically, pursuing his studies with ease and pleasure, and achieved an "Edwardian emancipation"[6] from the Presbyterian remnants of his childhood. He graduated M.B., Ch.B. at the age of twenty-one, as had James. He was still brash and obstinate "with an obsessional character derived in part from my parents but in part from my own developed techniques for dealing with frustration and anxiety" (Kubie 1973, p. 87). Between 1909 and 1918 he pursued further training in internal medicine, with particular emphasis on tuberculosis and diseases of the chest, in Glasgow, London, and Birmingham. He also learned the fundamentals of medical research and earned his M.D. in 1915, at the age of twenty-seven.

Meanwhile, James, who had been in poor health for some time, developed diabetes. During World War I, he worked part-time as an ophthalmologist near Edward, introducing his brother to Freud's writings.

Edward Glover and Christine Margaret Spiers, the daughter of a clergyman, were married in 1918. She died eighteen months later from complications following the stillbirth of a son. This loss doubtlessly had some effect upon Edward's decision to join James in going to Berlin in 1920 to seek psychiatric and psychoanalytic training at the newly founded polyclinic. The Glover brothers were both analyzed by Abraham and accompanied him on vacation. Edward Glover called this experience "more an apprenticeship than an analysis" (Kubie 1973). In Berlin he found himself much attracted to metapsychological speculation; yet he also felt the need to restrain himself and seek confirmation.

He became a member of the British Society in late 1922 and began the practice of analysis and a collaboration with James in which he expressed his "drive to balance the 'more exuberant exercises' of his brother's imagination" (Kubie 1973, p. 90). He regarded his brother as a genius, as did others, but considered him to be entirely too theoretically speculative. Later he was also to criticize the theoretical formulations of Klein, Jacobson, and Hartman as too speculative. However, Glover's own theory of early childhood development has proven to be susceptible to the same criticism.

Glover's first contribution to technique was "Active Therapy and Psychoanalysis" (1924, this volume, Chapter 9). He attributes his interest in the subject of technique not only to his reaction to Ferenczi but also to the "general confusion" he found on the subject and his own "uncertainties" as to how to proceed with his own early cases.[7]

In 1924 Edward Glover married his second wife, Gladys Blair. In 1926 their only child, a high-grade mongoloid daughter, was born; she was raised at home with great care and devotion by the Glovers. In that same year James died in diabetic coma on vacation in Spain. Edward volunteered to take over the course of lectures on technique planned by James, and this provided further stimulation for Glover to undertake the most thorough treatment of the subject after Freud. The book which resulted, *The Technique of Psychoanalysis* (1955), is still considered a classic. Further thought on psychotherapy and psychoanalysis produced "The Therapeutic Effect of the Inexact Interpretation" (1931, this volume, Chapter 23). Glover listed this as his second best paper[8] of the more than two hundred he produced. His work also resulted in six additional major books, including *The Early Development of the Mind* (1956), a selection of his papers. He was also interested in juvenile delinquency and criminology.

For many years Glover held posts in the British Institute and Society, but resigned from them in 1944 in disagreement with the compromise between the Kleinian and Freudian factions. He died on August 16, 1972.

NOTES

1. Oral history: The Reminiscences of Edward Glover. The Psychoanalytic Movement (1965). In the Oral History Collection of Columbia University, New York.
2. M. Crosbie; personal communication, 1974.
3. Oral history. op. cit.
4. Ibid.
5. Ibid.
6. Ibid.
7. Ibid.
8. Ibid.

Mark Kanzer   (b. 1908)

MARK KANZER was born on December 6, 1908 in Brooklyn, New York. He received his bachelor's degree from Yale in 1928, a master's degree in psychology from Harvard in 1929, and his M.D. from the University of Berlin in 1934. He interned at Montefiore Hospital in the Bronx. Moving south to the borough of Manhattan, he received psychiatric training at Bellevue Hospital and neurological training at Mt. Sinai Hospital (1935–1939).

Kanzer was trained at The New York Psychoanalytic Institute from 1938 to 1943. He then entered the army and rose to the rank of major while an instructor at the School of Military Neuropsychiatry at Fort Sam Houston. A member of the American Psychoanalytic Association since 1945, he was a founding member of the Downstate (or Brooklyn) Psychoanalytic Institute (1949) and became its director from 1962 to 1965. Since then, he has been director of its postgraduate training program.

Kanzer has served on the editorial boards of the *Annual Survey of Psychoanalysis, Journal of the American Psychoanalytic Association,* the *Glossary of the American Psychoanalytic Association,* and the *International Journal of Psychoanalytic Psychotherapy.* Presently he is co-editor of *American Imago.* He has been both Freud Lecturer (1967) and Nunberg Lecturer (1970).

Kanzer has made a large number of original contributions to the field of psychoanalysis in his many articles. (Grinstein lists fifty-six, not including reviews.) "The Transference Neurosis of the Rat Man" (1952, this volume, Chapter 10) fits into two of his dominant interests—transference and the history of psychoanalysis. He is currently at work on a sequel to this paper which will take into account Freud's notes (published 1954) on the case and the more recent introduction of the term "therapeutic alliance" (Zetzel 1956). The new paper will appear in the *Downstate Twenty-Fifth Anniversary Volumes,* which he is editing. His final comment on the paper is that: "Otherwise, after twenty-two years, the 'Transference Neurosis of the Rat Man' is one of the few of my older papers that I would change little, if at all, if I were writing it today."[1]

Mark Kanzer is actively engaged in private practice in New York and Westchester.

NOTE

1. Personal communication; 1974.

Elizabeth R. Zetzel (1907–1970)

ELIZABETH ZETZEL (née Rosenberg) was born on March 17, 1907, in New York City. Both her parents were from wealthy, prominent Jewish families known for their philanthropic work. Her mother, Bessie Herman Rosenberg, who is still living, is the daughter of a Boston shoe manufacturer. James Naumberg Rosenberg, her father, was the grandson of a rabbi who emigrated from Germany to Pittsburgh in 1848. He was an international lawyer and also a writer of prose, poetry, and drama. He served many philanthropic causes and in 1947 played an important role in drafting the genocide convention that was adopted by the United Nations. He devoted the last thirty years of his life to painting and died at the age of ninety-five in 1970.

In addition to Elizabeth, the family consisted of a sister, Ann, two years younger, and a brother, Robert, fifteen years younger. She attended the Friends Seminary and the Ethical Culture School, with the exception of the year 1921–1922, which she spent in Vienna while her father served as European chairman of an organization for relief and rehabilitation of Eastern European Jews.

Zetzel attended Smith College and graduated Phi Beta Kappa with highest honors in economics in 1928. That fall she went to England to study at the London School of Economics, and soon began an analysis with Ernest Jones, who was to remain the dominant influence upon her subsequent career.[1]

With Jones's encouragement she decided to study medicine at the University of London, from which she received her M.B. in 1937 and her M.D. in 1941. Meanwhile, she had become a candidate at the British Psycho-Analytical Institute in 1935. Margorie Brierly was her instructor in theory, and Ella Sharpe taught her dream interpretation. Melanie Klein's lectures (together with those of Joan Riviere and Susan Isaacs) made an important and lasting impression upon her. Another important influence was D. W. Winnicott (Zetzel 1969). Like Winnicott, she tried to bridge the gap between Klein and Anna Freud.

Zetzel graduated from the institute in 1938 and instead of starting a practice sought psychiatric training at the Maudsley Hospital. Subsequently, she became a full-time psychiatrist at the Mill Hill Emergency Hospital. She attended seminars given by Anna Freud and Grete L. Bibring and the important meetings of the British Society, which sought to clarify the Viennese and Kleinian positions. She presented her paper on war neuroses to the British Society, as a prerequisite for full membership, during an air raid in 1943.

Following her service as a major in the Royal Army Medical Corps (1943–1946) she became assistant director of the clinic of the British Institute, an instructor at the institute, and subsequently secretary of the society. In the meantime she and Dr. Eric Guttman, a neurologist and psychiatrist and a

German refugee, whom she had met at Maudsley, were married in 1944. Their son James Eric was born in 1947. Dr. Guttman died of rheumatic heart disease in 1948.

While visiting her ill grandmother in Boston in 1948, she met Dr. Louis Zetzel, himself a widower with two daughters. They were married in 1949, and she moved to Boston and began her active career of teaching psychiatry at Harvard and psychoanalysis at the Boston Institute. Her interests in clinical psychiatry, in selection of cases for training purposes, and in contributing to the theory and clinical practice of analysis are reflected in her more than forty papers. An important selection of these and a complete bibliography can be found in *The Capacity for Emotional Growth* (1970). Among her contributions was the introduction of the term, "therapeutic alliance" (1956). In addition, she made a number of contributions to psychoanalytic technique and to psychotherapy. She died in her sleep on November 22, 1970 (Rangell 1971).

NOTE

1. L. Zetzel; personal communication, 1974.

Erik H. Erikson (b. 1902)

ERIK H. ERIKSON was born on June 15, 1902, in Frankfurt, Germany, to Danish parents. His mother was predominantly Jewish, his father Protestant (Coles 1970).* They separated before his birth, and when he was three his mother married his pediatrician, Dr. Theodor Homburger, who was Jewish. Erikson grew up in Karlsruhe believing Dr. Homburger (whose last name he used) to be his biological as well as his psychological father.

Erikson attended the *vorschule* (public elementary school) from the ages of six to ten and the *Gymnasium* until age eighteen. He was not a good student except for the subjects of ancient history and art. He sought to overcome his "Danishness" by espousing German nationalism. Nonetheless, he was taunted at school as a Jew. At synagogue he was called "the goy" because of his pronounced Nordic features. The "loving deceit," as he later termed his parents' handling of his paternity, and the aforementioned experience are the stuff of which particularly intense "identity crises" or "diffusion" are made. However, when Erikson was living it, such terms were not available—he had not yet formulated them.

Having graduated from the *Gymnasium*, Erikson chose not to pursue a

* All additional biographical information is based on this source.

university education, but rather spent some seven years in restless wandering through Europe, largely on foot.

These were years of "morbid sensitivity," years spent reading, sketching, and keeping a journal. He paused in his travels to study art in Karlsruhe and Munich, and he achieved a showing of his work in Munich's Glaspalast. These years spent in search of a career and final identity certainly seem to constitute a prolonged "adolescent moratorium," but again this concept had to await Erikson's formulation.

In 1927 Erikson returned to Karlsruhe and seemed relatively settled into a career of studying and teaching art. However, he was invited by Peter Blos, Sr., then the private tutor for Dorothy Tiffany Burlingham's four children, to come to Vienna. A small, informal, and "progressive" school for some twenty children in a building owned by Eva Rosenfeld had evolved under Mrs. Burlingham and Anna Freud's general direction. Another teacher was needed and the job was offered to Erikson. He accepted.

Soon he was engaged in becoming a certified Montessori teacher (one of two men among the many women) and in obtaining psychoanalytic training. Anna Freud was his analyst, and his teachers were August Aichhorn, Edward Bibring, Helene Deutsch, Heinz Hartmann, and Ernst Kris.

In 1930 he and Joan Mowat Serson, of American and Canadian ancestry, were married. She had gone to Europe to do research for a doctoral dissertation on the origins of the various schools of modern dance or body movement that had proliferated in Germany after World War I. After their marriage she joined her husband in teaching. Their sons Kai, a sociologist, and Jon, a photographer, were born in Vienna. A third child, Sue, a social anthropologist, was born in New Haven in 1938.

Erikson graduated from the Vienna Training Institute in 1933 and was made a full member of the Vienna Psychoanalytic Society. This enabled him to join the International Psycho-Analytic Association and thus to emigrate immediately to the United States. (As he has *no* academic degree, practice would have been difficult without this credential.) Initially settling in Boston, with Hanns Sachs's help, he was subsequently to hold academic appointments at Yale, the University of California at Berkeley, and Harvard, where he became a University Professor. Association with numerous other institutions provided him with a breadth of experience and a source of observation from which he has drawn in making his contributions. Erikson has retired from his academic appointment and lives in California.[1]

Erikson's work has achieved a special place in American intellectual thought as well as in psychoanalytic thought. He has made major contributions to post–World War II ego psychology. In the area of analytic technique he has emphasized the importance of actuality (see this volume, Chapter 12) and has contributed to the meaning and use of the dream (see "The Dream Specimen of Psychoanalysis," 1954).

His formulation of the concept of "epigenesis of the ego" added to theory. His extension of the phases of psychosocial development to include the whole life cycle within the greater context of society stimulated new insights. His outline of psychosocial development includes new and useful concepts; among them are basic trust, identity vs. identity diffusion, generativity, and integrity. For Erikson, each stage of growth is connected with a "normal crisis." Thus, it is ironic that "identity crisis" has so often been heard as a chief complaint.

He has contributed widely to psychoanalytic literature. *Childhood and Society* (1950) has reached a large general audience. His other books, *Insight and Responsibility* (1964), *Identity: Youth and Crisis* (1968), *Young Man Luther* (1958), and *Gandhi's Truth* (1969) are also well known. The serious student should also take note of a selection of papers, *Identity and the Life Cycle: Selected Papers* (1959).

NOTE

1. E. Erikson; personal communication, 1974.

Karl Landauer (1887–1945)

KARL LANDAUER was born on October 12, 1887, in Munich, Germany. He was the youngest child and only son among three children of a long-established, prosperous German Jewish family. His father, like his father before him, was a banker. He died after a long illness when Landauer was fourteen.

His mother was a well-educated and distinguished woman who subsequently died in a Nazi extermination camp. The family spent summers in the Bavarian Alps; as a result Landauer had a lifelong love of mountains and nature study. An avid reader, he was also interested in art history, painting, sculpture, architecture, and anthropology.

Landauer completed his early education in Munich; he studied medicine in Freiburg, Berlin, and Munich; at the latter university, he studied under Kraepelin. He became interested in psychoanalysis when one of his teachers of psychiatry called his attention to Freud because he found Landauer's thinking similar.

He probably received his M.D. in 1910. After a year of compulsory military service, he went to Vienna in 1912 for analysis and analytic training with Freud and further psychiatric and neurological training under Wagner-Jauregg. He attended Freud's Saturday night lectures and began attending the meetings of the Vienna Psychoanalytic Society on October 8, 1913. In the ensuing months he read two papers on the psychoses and gave an analytic case report. His contributions to the discussions also reveal interest in direct observation of and therapeutic work with children. He is listed as a member of the society for the year 1913–1914 (Nunberg and Federn 1962–1975, vol. 4, pp. 309–310).*

With the outbreak of World War I, Landauer was called to duty with the German army and so left Vienna. He saw action as a medical officer on both the Western and the Russian Fronts. His work with patients and his personality so impressed Heinrich Meng, a fellow medical officer, that Meng decided on a career in psychoanalysis (Freidemann 1966, p. 337).

In 1917, while stationed in Heilbronn-am-Neckar he met and married Karoline Kahn. In 1919, they settled in Frankfurt-am-Main, where he worked in the psychiatric hospital and continued psychiatric and neurological work in his private practice, gradually limiting himself to the use of psychoanalytic technique. By 1924, when "Passive Technique" (this volume, Chapter 13) appeared, his practice was almost entirely psychoanalytic in character; but his early interest in the treatment of the psychoses persisted, as is evident in this paper.

The Landauers had three children: two girls and a boy, who are now, respectively, a child analyst, a director of a day-care center, and a management consultant. Landauer maintained a correspondence with Freud and visited Vienna regularly in the 1920s and 1930s for further consultation and study. During the same period, he also visited Berlin and attended meetings of the Berlin Society.

In 1928, Landauer began planning for the founding of the Frankfurt Psychoanalytic Institute and invited Meng to join him as cofounder and co-director (Meng 1971, p. 77). It was officially opened in February, 1929, as a "Guest Institute" of the University of Frankfurt in close association with its Institute for Social Research. Frieda Fromm-Reichmann was the third original faculty member;† Erich Fromm and S. H. (Fuchs) Foulkes came somewhat later (Meng 1971, p. 78).

Erich Fromm, whose early analytic work was supervised by Landauer, remembers him as a "very kind, intelligent, sensitive person."[1] Foulkes, who was the first director of the institute clinic, adds that Landauer's emphasis on minute details of his own as well as the analysand's "body language" in his discussions of psychoanalytic technique "became very important [to him in] group analysis."[2]

Throughout his career, Landauer was interested in analytic technique and

* All information other than that specifically cited was kindly supplied by Dr. Landauer's family in 1974.

† See "Bulletin of the International Psycho-Analytic Association," *Int. J. Psycho-Anal.* 11 (1930):246.

in a wide range of theoretical subjects. Grinstein lists forty-one original contributions; among them are papers on drives and affects, the psychoses and their treatment, stupidity, pseudostupidity, and motility.

In 1933, the Nazi ascent to power led to the destruction of the institute and the emigration of its faculty. The family moved to Amsterdam where Landauer continued his analytic practice and teaching. In 1936, he was invited to read a paper, in Vienna, "On Affects and Their Development," at Freud's eightieth birthday celebration.

Karl Landauer was taken from his home during a systematic house-to-house search by the Germans in June of 1943 and sent to the Dutch transit camp. He, his wife, and a daughter were shipped to Bergen-Belsen in April, 1944, where he died of starvation in January, 1945.

NOTES

1. E. Fromm; personal communication, 1974. His recollection is that the official name of the institute was the "Sudwestdeutsches Psychoanalytisches Institut."
2. S. H. Foulkes; personal communication, 1974.

Herman Nunberg (1884–1970)

HERMAN NUNBERG was born on January 23, 1884 in Bendzin, in the Russian section of Galicia, to Ludwik and Auguste (née Kalman) Nunberg. He was the second child and elder son among five children. According to Nunberg (1970, p. 3) his father, Ludwik, was an affectionate, creative and talented man who had wished to pursue a career in opera. When he had approached *his* father, "a giant of a man and a tyrant," with this idea, however, he was physically dissuaded from this choice[1] and obediently, if unhappily, went into the family lumber business with his brothers. He was not a financial success.

Auguste Nunberg was delicate, sickly, and devoted to her children. The marriage was not a happy one and Herman sided with his mother (Nunberg 1970, p. 3). While he was still young, the family moved to Czenstochova where he was tutored at home. When he was twelve, his mother died and shortly thereafter he was sent to Cracow for his secondary schooling. His father sought to direct him toward a technical career at that time, but he subsequently decided to enter the University of Cracow.

Once enrolled in the university, Nunberg was drawn to both history and medicine, but chose the latter. He was active at this time in the revolutionary Social Democratic Party of Poland and took part in underground activities, such as transporting ammunition.[2] In the meantime, his father became remarried to Helen Rosenkranz Nunberg, and three half-siblings (two girls and a boy) were born. This marriage was no happier than the first.[3] In 1906, the family moved to Zurich, Switzerland.

About this time, Nunberg entered the University of Zurich; throughout medical school he was a student of Bleuler's, both at the University and then later at Burghölzli. His doctoral dissertation grew out of work with Jung and Bleuler. He graduated from medical school in 1910 and received further psychiatric training, first at the mental hospital in Schaffhausen and then at Waldau, the psychiatric hospital of the University of Berne. His introduction to psychoanalysis was Bleuler's suggestion that he read *The Interpretation of Dreams*.

Nunberg became a member of the Swiss Psychoanalytic Society before returning to Cracow in 1912 to take a post at the University Psychiatric Clinic. He also began a professional association and friendship with Ludwig Jekels which was to endure over the years.

With the onset of World War I, he became an "enemy alien" in Cracow and chose to move to Vienna to become a student of Freud's. He obtained a psychiatric position with Wagner-Jauregg and Pötzl, transferred from the Swiss to the Vienna Society, and began making regular use of Freud's consultation hour to discuss cases. When he suggested to Freud that every analyst ought to be required to undergo analysis, Freud called upon him to advance this idea at the Budapest Congress of 1918. The resolution was stoutly opposed by Rank and Tausk and was not adopted until 1926, but Nunberg is credited as the originator of "training analysis." He taught one of the first series of courses in Vienna and was a member of the original training committee of the institute.

He reports that at the time of the beginning of Freud's cancer in 1923, there was speculation in the psychoanalytic circle on a psychogenic cause and possible cure of cancer, one that would depend upon the patient's "wish" to recover. This stimulated Nunberg to look at his case material and write "The Will to Recovery" (1926, this volume, Chapter 14; Nunberg 1970, p. 41).

When he first presented "The Synthetic Function of the Ego" to the Vienna Psychoanalytic Society (1930, this volume, Chapter 22), a paper which he states was written partly in reaction to Jung (Nunberg 1970, p. 54), Freud responded with an allusion to the fable of St. Wolfgang and the Devil (see the frontispiece to this volume and p. xiv).

In 1929, Margaret Rie left the stage, a career she had pursued with some success, and she and Nunberg were married. She was the daughter of Dr. Oskar Rie, Freud's old friend and collaborator and the "Otto" of the dream of Irma's injection. The Nunbergs honeymooned at the Oxford Congress, where he again read "The Synthetic Function of the Ego."

The years 1930–1933 were mixed and difficult for the Nunbergs. Their first child was stillborn; Mrs. Rie died, and then Dr. Rie; Henry, Nunberg's only full brother, a psychoanalyst in Switzerland, also died suddenly of a streptococcal infection. In 1931, their daughter, Melanie, was born. In the meantime Nunberg's textbook, *Principles of Psychoanalysis* (1932, English ed., 1955), which had grown out of his lectures in Vienna, was published and he accepted an offer made on the recommendation of Ferenczi to teach for a year in Philadelphia. A son Henry, now a practicing analyst, was born in 1933 while the family was in the midst of deciding to settle permanently in the United States.

After a further brief stay in Philadelphia, where he was instrumental in the founding of the Philadelphia Psychoanalytic Association, Nunberg moved to New York in 1934 and began private practice. He was not admitted to the New York Society until 1940 because of his refusal to sign a statement opposing the training of lay analysts. Soon after his admission, however, he was made a training analyst and member of the Educational Committee of the New York Institute. He subsequently became president of the New York Society for two terms.

Paul Federn, who had brought the minutes of the Vienna Society to the United States, appointed his son Ernst and Nunberg as the editors of the manuscript in his will. The two undertook the task with the assistance of Margaret Nunberg and the complete four-volume *Minutes of the Vienna Society* have now been published in English. The papers reprinted here, together with many other valuable contributions, are to be found in Volumes I and II of *Practice and Theory of Psychoanalysis* (1948 and 1965). Nunberg's last work was his brief autobiography, *Memoirs, Reflections, Ideas* (1970). He continued a limited private practice until his death on May 20, 1970.

NOTES

1. L. Nunberg; personal communication, 1973.
2. Henry Nunberg; personal communication, 1974.
3. L. Nunberg. op. cit.

Wilhelm Reich (1897–1957)

WILHELM REICH was born on March 24, 1897 in Dobrznyica, in Austrian Galicia, the first child of Leon and Cecilia (née Roniger) Reich. The family was of the assimilated Jewish upper-middle class. Shortly after his birth, the Reichs moved to a large, financially successful agricultural estate in Jujinetz, in Bukovina, the Austrian Ukraine.

Leon Reich was apparently a violent, passionate man, given to outbursts of rage. Harsh with his peasants and tyrannical with his family, he was deeply jealous of his wife. She in turn is described as warm, affectionate, cowed, and submissive to her husband. Throughout his life, Wilhelm frequently referred to his mother in the warmest terms but was reluctant to discuss his father (I. O. Reich 1969, pp. 3–4). The cook Zosia was also an important figure in the household.

A brother, Robert, was born when Reich was three. The two boys were educated at home and were not allowed to play with the children of the peasants or with the few Yiddish-speaking children in the nearby village.

When he was eight, Wilhelm developed a rash on his elbow. This was the beginning of the psoriasis from which he was to suffer for the rest of his life. He was hospitalized for six weeks in Vienna, which must have been at best difficult, if not traumatic, for the boy.

While he was intensively preparing for his entrance to the German *Gymnasium* in Czernowitz, Reich, then probably thirteen, concluded that his mother was having an affair with a tutor. He informed his father and his mother committed suicide. The father was deeply troubled by the loss of his wife, insured his own life heavily, and began exposing himself to the Ukrainian winter. He died of tuberculosis in 1914 (I. O. Reich 1969, p. 4). Reich was thus orphaned at seventeen. These events must have been severely traumatic.

He passed his *Matura* in 1915 and joined the Austrian army, becoming an officer in 1916. Robert, not yet of military age, lived with relatives in Vienna. Reich saw action on the Italian front during World War I.

Following the armistice, the Reich brothers were reunited in Vienna. Deprived by the war of the wealth of their father's estate and unable to collect his life insurance, they were penniless. It was decided that Robert would work while Willy studied. He chose medicine and soon was able to earn a little money by tutoring students in courses he had just passed.

In 1919, Reich became interested in psychoanalysis and in 1920 (aged twenty-three) he became a member of the Vienna Psychoanalytic Society and began the practice of psychoanalysis. He received his M.D. in 1922 and pursued further psychiatric training under Wagner-Jauregg and Paul Schilder (1922–1924). At some point he had brief analyses with Sadger (I. O. Reich 1969, p. 11) and Federn (Raknes 1970, p. 15).

According to Grinstein, Reich published some thirty papers between 1920 and 1930. In 1922, he became one of the assistants to Hitschmann at the Vienna Polyclinic and in 1923 joined Federn, Hitschmann, Nunberg, Reik, and others in teaching lecture courses offered by the Vienna Psychoanalytic Society. His subject was "Clinical Psychoanalysis" (Hitschmann 1932, p. 248). With the founding of the Vienna Training Institute in 1925, he became one of the four members of the Training Committee.*

From the beginning of his career, Reich was interested in technique and

---

* The others were Federn, Nunberg, and Hitschmann. The directors were Helene Deutsch, Anna Freud, and Siegfried Bernfeld (Hitschmann 1932, p. 248).

character analysis. Character neuroses were considered particularly difficult clinically. Freud confided to Abraham in a letter of March 30, 1922: "I find character analysis with pupils much more difficult in many respects than analysis with professional neurotics, but admittedly I have not worked out the new technique" (H. Abraham and E. Freud 1965, p. 330). Helene Deutsch suggested that a seminar be set up with her presenting a case and Reich discussing it; the case was that of "Hysterical Fate Neurosis," later published in *Neuroses and Character Types* (Deutsch 1965).[1] *Character Analysis* (1933, this volume, Chapter 18) was a direct outgrowth of the seminar. This was the beginning of what is now known as a "continuous case seminar" in technique; Reich was to conduct it from 1924 until 1930. He was an excellent teacher. Sterba recalls him as a brilliant clinician with a remarkable ability to summarize cases.[2]

Reich married Annie Pink, one of his first patients, in 1921. They and Otto and Claire Fenichel were close friends and frequently vacationed together in various mountain regions. In addition to their mutual interests in biology and sexology, both were Marxist politically and members of the Social Democratic party. Reich was particularly active politically.

In 1924 Eva Reich, now a physician, was born. In 1926, Reich's brother Robert died of tuberculosis. Reich became depressed and sought an analysis with Freud in 1927, which the latter refused. Reich was deeply hurt by this rejection; in addition, he developed tuberculosis for which he was hospitalized in Davos for some months in 1927. While there, he wrote *The Function of the Orgasm*. In 1928, the Reich's second daughter, Lore, now a psychoanalyst, was born. In that same year, Reich joined the Communist party and became active in organizing clinics for the dissemination of information on birth control, sex education, and child rearing for workers.

In 1930, Reich moved to Berlin for an analysis with Rado, which—like the previous analyses—proved unsatisfactory (I. O. Reich 1969, p. 20). Reich began presenting the material from which he was writing *Character Analysis* to a seminar consisting of Jacobson, Gero, Fenichel, Wilhelm Kemper, Hellmuth Kaiser, Annie Reich, and others.[3] Concerned with the growing Nazi threat, he intensified his activities on behalf of the Communist party and sexual freedom and enlightenment. His writings of this period reflect these interests.

In 1933, his first marriage ended in divorce, and he moved to Denmark where he was joined by the dancer, Elsa Lindenberg, his second, common-law wife. Also in that year, the Verlag refused to publish *Character Analysis*; it was subsequently published by Reich's own organization, Sexpol Verlag.

His advocacy of sexual freedom led to his expulsion from the Communist party in 1933. His previous affiliation with the Communists, however, began to cause problems for the Berlin Society, which had become a target of the Nazis. Reich's name was removed from the Berlin membership list over his protest. His linking of Freud's "death instinct" to capitalism caused further controversy for the International Psycho-Analytic Association. According to Jones (1957, p. 191), he resigned; according to Reich he was expelled from the International Psycho-Analytic Association in 1934 (I. O. Reich 1969, pp. 30–31).[4]

From 1934 to 1939, Reich lived in Oslo where he began a series of experiments to verify and measure sexual energy in the sexual organs. These and other experiments led ultimately to Orgonomy, which he considered the basic biophysical life force. Paralleling this, he developed a new technique of therapy, at first called "vegetotherapy" and later orgone therapy. This technique involves direct attack on the muscular armoring of the body, thereby causing release of spasm and liberating affect. According to Gladys M. Wolfe,[5] widow of the translator of *Character Analysis*, Reich followed the somatic work with discussion and interpretation of the affects and memories thus evoked.

An important stimulus to the development of this technique was probably Claire Fenichel's work as a body movement therapist. She was a student of Elsa Gindler, and Wilhelm and Annie Reich, as well as Elsa Lindenberg, worked with her.[6] While they were in Oslo, a break painful to both occurred between Reich and Fenichel (see excerpt from Fenichel's "Problems of Psychoanalytic Technique," 1939, this volume, Chapter 19).

Reich emigrated to the United States in 1939. This move coincided with a separation from Elsa Lindenberg. On December 25 of that year he and Ilse Ollendorff were married. A son, Peter, now a writer, was born in 1944. In the United States, Reich continued his experiments, began research into orgonomic treatment of cancer, and trained disciples. It is interesting to note that as Freud had been reproached for not analyzing negative transference (Freud 1937b) in his disciples, so Reich was reproached for not analyzing the positive transference in his (I. O. Reich 1969, p. 46).

In 1951, Reich suffered a severe heart attack. Following his illness, he began to show signs of increasing irritability, fear, and progressive distrust of his wife, Ilse Ollendorff Reich, who finally left him in 1954. His work on cancer, particularly that with the orgone accumulator (orgone box) became controversial and a court injunction was obtained by the government against not only the sale and interstate distribution of the orgone box* but also against all literature published by the Orgone Institute Press, including *Character Analysis*.

Reich refused either to obey the injunction or to retain further legal counsel, preferring to seek justice before some higher tribunal, perhaps poster-

---

* A simple device which Reich felt concentrated orgone energy, thus accounting for his observations of a higher temperature in the accumulator than in the surrounding environment. Reich was deeply hurt when Albert Einstein preferred a more mundane explanation of this phenomenon (I. O. Reich 1969, pp. 58–59).

ity. At his trial of May 3, 4, 5, and 7, 1956, he chose to act as his own counsel. He was found guilty of contempt of court and his orgone boxes and his publications including *Character Analysis* were burned, his scientific foundation was fined $10,000, and he was sentenced to two years imprisonment.

On March 12, 1957, Reich began serving his prison sentence at the Federal Penitentiary in Danbury, Connecticut. After 10 days, at Danbury, on the basis of a psychiatric recommendation, he was transferred to Lewisburg Federal Penitentiary because of its psychiatric treatment facilities. There, however, the psychiatrists refused to declare him psychotic. He died on November 3, 1957.

Today wherever and whenever one speaks of character, character armor, or begins psychotherapy with direct physical work with the body, a debt is owed to Wilhelm Reich. A selected bibliography, including an introduction to the basic works on Orgonomy, can be found in Ilse Ollendorff Reich's *Wilhelm Reich—A Personal Biography*. A second biographical source, Peter Reich's beautiful *A Book of Dreams* (1973), should be read by the serious student. The movement founded by Reich is still flourishing.

NOTES

1. H. Deutsch; personal communication, 1973.
2. F. H. Parcells. Video interview of Richard F. Sterba. Courtesy of Dr. Parcells.
3. E. Jacobson; personal communication, 1973.
4. E. Jacobson supports Reich's view (personal communication, 1973).
5. G. M. Wolfe; personal communication, 1973.
6. C. Fenichel; personal communication, 1973.

Otto Fenichel (1897–1946)

OTTO FENICHEL was born December 2, 1897 to Leo and Emma (née Braun) Fenichel in Vienna. Leo Fenichel had emigrated from Tarnow, Poland; in Vienna he became a successful and wealthy lawyer. Otto was the youngest of three children. His elder brother followed the father into law and joined him in practice. His sister, a poetess, married a Czech industrialist and died at an early age in an automobile accident.

Leo Fenichel was an authoritarian and demanding father and placed great importance on education.[1] Mrs. Fenichel was matriarchal and Victorian in her attitude and was devoted and loving to her children.[2] The Fenichels were devotees of the opera and on Saturdays Otto and his brother would set up their own play stage to give family performances.[3]

Fenichel was a brilliant student in the *Gymnasium*. He was active in the

left-wing youth movement and wished to become a biologist. Influenced by his
father, he decided to enter medical school, where he developed an interest in
the biology and psychology of sexology.[4] At the age of seventeen, while still a
medical student, he decided to become a psychoanalyst; in 1918 he presented
his first paper, "The Derivatives of the Incest Conflict," to the Vienna Psychoanalytic Society. He and Wilhelm Reich became friends about this time. He
received his M.D. in 1921.

In 1922, Fenichel moved to Berlin for psychoanalytic training at the
polyclinic and further psychiatric experience under Bonhoeffer and Cassierer.
His training analyst was Sandor Rado. In 1923, he became an assistant at the
polyclinic (second to Hans Lampl and senior to Rudolph Loewenstein), and
in 1925 a training analyst. His analysands include Edith Jacobson and Ralph
Greenson. In 1926, he and Claire Nathansohn were married. Their daughter,
Dr. Hanna Pitkin, is now an assistant professor of political science at the
University of California at Berkeley.

Politically, Fenichel remained a Marxist and maintained his interest in
sociology and social change. He did not join the Communist party but joined
Reich in party activities in 1930. He maintained that "psychoanalysis was
dialectical materialism in psychology."[5] He led a seminar of younger analysts
that devoted a good share of time to these subjects. The reproach of a senior
colleague concerning the content of the seminar led to his comment: "What of
it? If you don't like the way we do it—let us be naughty children."[6] As a
result, the group was thereafter known as "the children's seminar."[7] Like
many of his generation, he soon became disillusioned with communism and
the Soviet experiment.

Always nearsighted, Fenichel was tall and slender in his youth. His
memory was remarkable and friends would call him for train schedules instead
of telephoning the railroad. He was also fascinated by maps. He loved travel,
particularly hiking trips, and managed before his emigration to visit the greater
part of Europe's mountain ranges. On such trips he loved to check his map,
rush ahead, and then explain the location, view, and so forth, to his companions. Once while walking he was so preoccupied with finding a landmark
on his map that he actually fell over it.[8]

Fenichel was devoted to learning and teaching. He drew up and circu-
lated bibliographies for courses in psychoanalysis which can still be found in
institute libraries. In discussion, his comments could be devastating to a paper
being examined or bring to it a new clarity. Said to have been without personal
rancor on such occasions, he nonetheless shared his friend Reich's intolerance
of small talk and did not "suffer fools lightly." His reviews of others' papers
often became important essays in their own right; his paper (1935a) reprinted
in this volume, Chapter 30, is an example of this. In 1932 he became co-
editor of *Imago*.

In 1933, Fenichel left Germany and went to Oslo to help establish psychoanalytic training there. In 1934, his *Outline of Clinical Psychoanalysis* was published and immediately became the standard reference book for psychoanalysis and related fields. He was invited to Prague in 1935 to set up the teaching of psychoanalysis; in 1938 he moved to Los Angeles and became a training analyst and assistant editor of the "Psychoanalytic Quarterly." His first marriage ended in divorce, and in 1940 he married Dr. Hanna Heilborn, a lay analyst. His short book *Problems of Psychoanalytic Technique* appeared in 1941; an excerpt is reprinted in this volume, Chapter 19.

The Psychoanalytic Theory of Neurosis, an encyclopedia of extraordinary completeness, was published in 1945. It replaced his earlier book and remains the standard reference work. Dr. Joseph Sandler of London has assembled a large number of analysts to bring this work of one man up to date. Fenichel's *Collected Papers* (1953, 1954) were published posthumously. He died on January 22, 1946, midway through his internship at Cedars of Lebanon Hospital.

NOTES

1. H. Fenichel; personal communication, 1973.
2. C. Fenichel; personal communication, 1973.
3. C. Fenichel. op. cit.
4. Reider; personal communication, 1974.
5. Ibid.
6. Ibid.
7. Ibid.
8. E. Jacobson; personal communication, 1973. C. Fenichel (op. cit.) and I. O. Reich (personal communication, 1973) concur.

Richard F. Sterba   (b. 1898)

RICHARD F. STERBA was born on May 6, 1898, in Vienna. He was the second of two sons born to Joseph and Mathilde (née Fischer) Sterba. The family belonged to the Catholic middle class. Joseph Sterba, the son of a Czech tailor, was a secondary-school teacher of mathematics and physics. According to Sterba,[1] he was a father in the Germanic tradition—strict, authoritarian, distant, and unaffectionate with his two boys Oscar (fourteen years older) and Richard. His interest in them was limited to their academic achievement. He was also a Germanophile in favor of the union of Germany and Austria. Mrs. Sterba, the daughter of a Viennese schoolteacher, was warm, affectionate, and devoted to her children, particularly to Richard, her favorite.[2] Sterba has described his brother as the "torturer" of his childhood (Sterba 1970, p. 120).

Sterba was educated in the Viennese public schools. In the second grade he asked for violin lessons; he became an accomplished musician and has remained so. During prepuberty he became rebellious, as manifested by learning inhibitions in Latin and mathematics. These were so severe in the *Gymnasium*, under a teacher who very much resembled his father, that he feared he would never pass the *Matura*. World War I intervened, and he was declared to have passed the *Matura* without taking it (*Kriegs-Matura*) so he could enter the army at the age of eighteen. Nonetheless he had a repetitive "examination dream." His paper on this subject (Sterba, 1928) is autobiographical.[3] While in the *Gymnasium* he developed an interest in psychology.

In officer candidate school, while still vacillating between the careers of music and medicine, he found himself with older, predominantly Jewish intellectuals, among them Arthur Rössler, the writer, and Karl Alder, the second son of the leader of the Austrian Social-Democratic Party. Here for the first time he heard of Freud and his writings. He later saw action on the Italian front as a lieutenant.

While studying medicine at the University of Vienna after the war, he was impressed with Paul Schilder's lectures in psychiatry utilizing psychoanalytic explanations of case histories. After getting his M.D. in 1923 he decided to become an analyst and began a two-year residency in psychiatry under Wagner-Jauregg and Schilder. In 1924, he and Grete L. Bibring became the first students at the Vienna Training Institute.[4] His analyst was Eduard Hitschmann, and his supervisors were Helene Deutsch and Wilhelm Reich, whom he considered a "brilliant clinician." They, Nunberg, and Anna Freud were his teachers. He considers the best source for his learning of technique to have been the small, informal postgraduate seminars of the Vienna society which he attended for some twelve years.

In 1926, Sterba and Editha Rodanowicz-Harttmann were married. She had a doctorate in musicology but later became an analyst. In 1927, Sterba graduated from the institute and was made a training analyst in 1929; his first teaching assignment led him to write and publish *Introduction to the Psychoanalytic Theory of the Libido* (1942). Their daughters, Monique, now a writer, and Verena, now a psychiatric social worker, were born in Vienna.

Following Hitler's rise to power, Sterba's brother, a general practitioner, became a Nazi, while Sterba took a strong anti-Nazi stand. This led to a break in relations between the brothers. Sterba refused to participate in the Aryanization of the Vienna Training Institute and fled to Switzerland following the Nazi partition of Austria in 1938. A former analysand, Fritz Redl, and others ultimately helped the Sterbas to settle in Detroit.

Sterba reports that the reaction of his colleagues to "The Fate of the Ego in Psychoanalytic Therapy" (1934, this volume, Chapter 25) was very negative, with only Anna Freud defending its concept of a therapeutic ego split.[5]

He has made over seventy contributions to the literature of psychoanalysis, either singly or in coauthorship with Editha Sterba. Their psychoanalytic study of Beethoven (1954) deserves special mention. In addition, they are at present preparing a book-length study of Michelangelo.

N O T E S

1. R. F. Sterba; personal communication, 1974.
2. Ibid.
3. Ibid.
4. F. H. Parcells. Video interview of Richard F. Sterba. Courtesy of Dr. Parcells.
5. R. F. Sterba. op. cit.

James Strachey (1887–1967)

JAMES BEAUMONT STRACHEY was born on September 26, 1887 in London. He was the youngest of thirteen children of whom ten survived infancy.

His father, Lieutenant General Sir Richard Strachey (1817–1908), was a descendant of *the* Stracheys, a large and illustrious family dating from the sixteenth century. Sir Richard was not only a career soldier and administrator but also an explorer, geologist, geographer, and mathematician (Holroyd 1967, pp. 14–16). He was seventy when James was born.

Jane Maria Lady Strachey (1840–1928), his mother, was the daughter of Sir John Peter Grant, a distinguished Indian administrator and in the Stuart line of succession to the English throne. She was an early champion of women's rights and wrote verses and nursery rhymes for children. She was the dominant force within the home, made the important decisions concerning the children, and handled the finances.

Lady Strachey was "at home" on Sundays to a varied and interesting group of people. Relatives, who were eccentric or notable, or both, and family members with swarms of children, together with fashionable guests, filled the house. Bertrand Russell, visiting on such an occasion, observed so many similar-appearing children that he began to doubt his sanity (Holroyd 1967, p. 31).

An elder sister became head of Newnham College at Cambridge and another was Secretary of the Fawcett Society, a women's rights organization. Lytton Strachey, the eminent biographer, was his brother. Lytton was the favorite of Lady Strachey's younger children, and James grew up rather in his shadow.

James's education began at the Hillboro school, where he and Rupert

Brooke, the poet, became friends; this intimate friendship was to last for many years. He prepared for Cambridge as a day student on scholarship at St. Paul's School in London (with academic distinction) and followed Lytton to Trinity College, Cambridge. His early years there were apparently unhappy. He withdrew into daydreaming and inactivity. His indecisiveness about a career and his poor academic performance led his family to doubt that he would amount to much. He joined the Apostles, a famous club, became a Fabian Socialist, and a member of the Bloomsbury circle.

In 1909, he began a literary career as a secretary to his cousin, St. Loe Strachey, editor of *The Spectator*. He contributed weekly articles to this magazine, kept up his Bloomsbury social life, and became interested in Freud through the Society for Psychical Research. He graduated from Cambridge in 1912.

James's pacifism and his activities on behalf of the pacifist movement during World War I caused his cousin to fire him in 1915. Family influence, however, obtained him a wartime position with the Quakers, and he avoided jail. During the years of World War I his interest in analysis grew. He contacted Ernest Jones about training, was advised to enter medical school, and did so. After a few uninspired weeks in basic science, however, he dropped out to become drama critic of the *Athenaeum*. Nonetheless, he wrote to Freud and was invited to Vienna. In 1920, he and Alix Sargant-Florence, an intellectual ex-Newnham student, were married. They both then went to Vienna for analysis with Freud.

The Stracheys began translating Freud's work while in Vienna, and on vacation in London in 1921 joined Jones and Riviere on the "Glossary Committee," which developed the standard English nomenclature of psychoanalysis.

In 1922, James and Alix Strachey returned to London, where they became practicing analysts and members of the British Society. In 1924, James prevailed upon Leonard Woolf of the Hogarth Press to publish the International Psycho-Analytical Library.

Strachey was impressed at an early date with the important contributions of Melanie Klein. The paper (1934, reprinted in this volume, Chapter 24), usually called the "Mutative Interpretation," is his best known and was clearly influenced by Mrs. Klein.

Winnicott (1969, p. 130) used the language of this paper to point out that Strachey's anger "was often not obvious at the point of urgency." Basically shy and quiet, Strachey was also atheistic and cultured. In addition to psychoanalysis, he was active in writing on music, particularly that of Haydn, Mozart, and Wagner.

From his first translation of Freud, Strachey's interest in the subject grew as he tracked down errors and slips in the original as well as in the English renderings. In 1946, he began making plans for *The Standard Edition of the*

Complete Psychological Works of Sigmund Freud; in 1953 he and Alix gave up their practices and retired to the country to devote full time to translating and editing. Although completion of the monumental twenty-four volumes was threatened by repeated retinal detachments that endangered his sight, he was able to finish all but Volume 24, the index.

In his later years, he also closely supervised Holroyd's interesting biography of Lytton Strachey, which deals candidly with his brother's homosexuality. This work (Holroyd 1967, 1968, and 1972) provides much further information on James Strachey and his family.

James Strachey died of a heart attack on April 25, 1967.

Theodor Reik (1888–1969)

THEODOR REIK was born in Vienna on May 12, 1888, the sixth of seven children, to Max and Caroline Reik. At the time of his birth, his older brothers Hugo and Otto were fifteen and fourteen years old and there was a sister of about six. Another sister had died at the age of one, and there had been a stillborn sibling as well.[1] Max Reik was an Austrian civil servant assigned to the railroad. He had a cheerful and outgoing disposition in spite of the burden of a large family and a small income. He was often concerned and disapproving of Theodor's academic performance, which was particularly weak in mathematics, and often erratic in general.

The mother was moody, rather withdrawn, and often depressed. According to Reik (1949, p. 477), she was severely depressed and withdrew to a darkened bedroom for some months following the death from diphtheria of the older sister when he was less than two years old. He reports in an anecdote that his mother emerged from her bedroom only when she observed a servant mistreating him. We may assume that he derived comfort from this reparative story. Shortly thereafter another sister, Margaret, was born. Reik described himself as his mother's favorite child, but felt that his "struggle" to overcome his identification with "a melancholy and self-defeating mother" (Reik 1949, p. 478) had been a severe burden for him.

A further source of family unhappiness was the presence in the home of Reik's paternal grandfather, who is described as "strange." His preoccupation with prayer and Talmudic study produced a good deal of conflict between him and Reik's relatively "assimilated" and religiously nonobservant father. Young Theodor identified with his father in the conflict. The grandfather was eventually committed to a mental institution.

When Reik was eighteen and frantically studying to pass his secondary school examinations (*Matura*) with high marks in order to please and impress his father, whose health had been deteriorating, the father died. Reik responded with obsessional symptoms, including reading every word Goethe ever wrote. His mother withdrew into depression, and her health slowly failed; she died when he was twenty-two. In the meantime Reik had passed the *Matura* and entered the University of Vienna. There he worked at odd jobs while studying psychology and French and German literature.

While a student, he began to court Ella Oratsch, a childhood playmate. Her mother was Jewish and lived in terror of her husband, who was a Gentile and anti-Semitic. He had threatened to kill any man approaching either of his daughters. The courtship was thus carried on in secret for some years; in its course it became clear that Ella suffered from rheumatic heart disease. Reik has written at length about his sex life with other women during his engagement and the resulting conflict. He has likewise detailed the unhappiness of the subsequent marriage and his extramarital affairs (Reik 1949, pp. 478–479).

In 1910, the year of his mother's death, while still a student, Reik met Freud, who became his patron, protector, critic, and teacher for many years. In 1912, he received the first Doctor of Philosophy degree for a dissertation written from a psychoanalytic point of view (on Flaubert) from the University of Vienna. In 1914 Reik became a member of the Vienna Psychoanalytic Society, married Ella over the outraged objections of her father and, provided with a monthly allowance by Freud, began analysis with Karl Abraham in Berlin, for which no fee was charged. A son, Arthur, was born in 1915. Shortly thereafter Reik was called to service in the Austrian army. He saw action in Montenegro and Italy and was decorated for bravery.

In 1918, Reik began the practice of psychoanalysis in Vienna, where he was awarded Freud's annual prize for the best nonmedical paper presented up to the time of the Budapest Congress—"Puberty Rites among Savages" (1915a). When the Vienna Psychoanalytic Society began its first courses for students, in 1923, he lectured on religion and obsessional neurosis. He succeeded Rank as secretary of the society.

However, his marriage with Ella was deteriorating, as was her health. Following Arthur's birth, her rheumatic heart disease began to worsen, and in 1920 she developed subacute bacterial endocarditis. She became a querulous and demanding invalid with whom sexual relations were forbidden. His constant pampering of his wife and his sacrifices for her and her family led him to discover that in behaving masochistically he had a feeling of "moral superiority."

Reik became involved in a malpractice suit based on the Austrian equivalent of practicing medicine without a license. The incident stimulated Freud to write "The Question of Lay Analysis" (1926b). In 1928, Reik moved to Berlin, where he was appointed to the faculty of the Berlin Institute. He soon began to suffer new neurotic symptoms, consisting mainly of attacks of dizziness, chest pain, and a conviction of impending death. These led to a brief second analysis with Freud. According to Reik, Freud remained silent most of the last hour and then asked quietly, "Do you remember the novel *The Murderer* by Schnitzler?" This question "surprised" Reik on many levels, brought home to him his wish to kill his wife, and resulted in the disappearance of his symptoms (Reik 1949, p. 426). This experience doubtless influenced his subsequent emphasis upon surprise as basic in analytic technique.

Beginning with "New Ways in Psychoanalytic Technique" (1933, this volume, Chapter 26), he began to emphasize his insistence that psychoanalysis rested almost totally upon the unconscious interplay between analyst and analysand, in contrast to the orderly and systematical approach of others, such as Reich and Fenichel.

The Reiks fled from Berlin to the Hague in 1934. Shortly thereafter, Arthur married and emigrated to Palestine. Ella followed her son and died soon thereafter. Theodor Reik and Maria Cubelic married in the Hague, and a daughter, Theodora Irene, was born there in 1936. A second daughter, Miriam, was born in 1938 shortly after the Reiks arrived in New York in flight from Nazism. This marriage proved happier than the first. During these years Reik wrote *Masochism in Modern Man* (1941).

In New York Reik was deeply disappointed at the refusal of the New York Psychoanalytic Society to accept him. However, in 1948, together with a group of disciples, he founded his own training organization, the National Psychological Association for Psychoanalysis.

Over the years, Reik wrote some 40 books; Grinstein lists 261 original contributions by him. In *Listening with the Third Ear* (1948), a book for the general reader, Reik extended his emphasis that in every analysis the analyst continues his self-analysis and arrives at insights about the analysand through his own resolution of corresponding difficulties within himself.

Theodor Reik died on October 31, 1969, at the age of eighty-one.

NOTE

1. A. Reik and M. Reik; personal communication, 1973.

Hellmuth Kaiser   (1893–1961)

HELLMUTH KAISER was born on November 3, 1893 in Heidelberg, Germany (Enelow and L. Adler 1965).* His father was a professor of physiology at the University of Heidelberg and his mother was gifted both intellectually and artistically; the family was Jewish.

Kaiser was something of a child prodigy or "Wunderkind," learning

* All biographical information, unless otherwise noted, is based on this source.

number theory and solving advanced mathematical problems by the age of five. He wrote poems, dramas and novels, and had memorized all of Goethe's *Faust* by the age of twelve.

In 1901 his father, in conflict with the head of his department at the university, resigned and moved to Berlin, where his previous inventions secured for him a position as director of research in a large electrical firm.

Kaiser passed the *Abiturium* and entered the University of Göttingen in 1912, initially studying law but shortly abandoning this to study mathematics and philosophy. Soon he came under the influence of the philosopher Leonard Nelson whom he idealized. (Nelson as a "neo-Kantian" believed he had found "*the* true answer to the question of how people should act.") He was drafted into the German army in 1915, and saw action with the field artillery in Flanders and Verdun. Afterward he served as a noncombatant because of illness.

In 1919 he returned to Göttingen and his allegiance to Nelson, who planned a "leader school" (*Führer-Schule*) to educate teachers and children according to his philosophy that would help attain his ultimate goal of a "dictatorship of reason." Kaiser shared his teacher's "convictions and attitudes in every way," and thus it was a terrible blow when Nelson arbitrarily dropped him for reasons Kaiser never understood.

Kaiser continued his study of mathematics and philosophy, and received his Ph.D. from the University of Munich in 1922 with a thesis on "The Theory of Probability Judgement." He had married in the meantime and had two daughters. The inflation of postwar Germany soon devalued the capital that would have enabled Kaiser to pursue an academic career, and like his father, he took a position as director of the bureau of statistics of a large electrical manufacturing concern. His job was dissatisfying. Bored and soon depressed, Kaiser entered analysis.

Kaiser decided to seek training at the Berlin Institute but was rejected because he was not a physician. He nonetheless wrote a psychoanalytic study of Kleist's "The Prince of Homburg," which was published in *Imago*. Freud wrote Kaiser a lengthy letter in praise of the work, and he was admitted to the Berlin Institute as a result. His analyst was Gustav Bally, and his supervisors were Sandor Rado, Karen Horney, Hanns Sachs, and Wilhelm Reich. He graduated in 1929, entered private practice, joined the staff of Simmel's psychiatric hospital, and became a member of Fenichel's "children's seminar" (Kemper 1973), which from 1930 to 1933 included Wilhelm Reich and was preoccupied with the latter's ideas and the material from which he was writing *Character Analysis*.

Frustrated by the lack of progress in one of his early patients, Kaiser presented the case to four different training analysts, only to receive four differing opinions. This loss of authority, as it were, became a major stimulus for his "Problems of Technique" (1934, this volume, Chapter 27). A further impetus for this paper was the idea learned from Reich that "how the patient speaks and behaves is more important than what he says." The paper was severely criticized by Fenichel and more gently criticized by Reich (see editor's introductory note, p. 383–384). However, its continued influence on later thought is demonstrated by the frequency with which it is cited in the ensuing literature.

Kaiser was hurt by the reaction to his paper, but by the time it was published he had already fled Nazi Germany to Mallorca. There he began occupying himself with woodworking. As a result of the Spanish Civil War he was expelled from Mallorca, and after a dangerous stay in France without a visa he finally managed to emigrate to Palestine in 1938. Shortly thereafter, his first marriage ended in divorce and he married his second wife, Ruth, with whom he had a son. Kaiser earned his living making recorders and his wife did weaving. She commented about her husband, "Hellmuth is a rebel and a professional outsider."[1]

In 1949, Kaiser was brought to the United States by Karl Menninger and became a training analyst at the Topeka Institute. Kaiser, however, continued to move further from classical psychoanalysis, developing a unitary theory of neurosis and many interesting and stimulating ideas on psychotherapy. He moved to Hartford, Connecticut in 1954, and in 1955 "The Problem of Responsibility in Psychotherapy," reflecting many of his new ideas, was published. By now he objected to the very term "technique" as suggesting that the therapist was doing something to the patient. Instead he emphasized the importance of the dyadic encounter and of increasing the patient's understanding of his responsibility for his own situation. The therapist, however, should require nothing more of the patient than his presence (even if he remains silent) and that he pay.

His last paper, "Emergency," appeared in 1962. Written in the form of a play, it represents an interesting situation in which the nominal therapist (Dr. Porfirio) is actually the patient of the nominal patient (Dr. Terwin). Kaiser here underlines his view that the only necessity for psychotherapy is that patient and therapist be together in the same room. These and his other papers are collected in the posthumous volume, *Effective Psychotherapy: The Contribution of Hellmuth Kaiser* (Fierman 1965).

Kaiser died in California on October 12, 1961.

NOTE

1. Letter of R. Kaiser to T. A. Munson, M.D. April 5, 1959.

Joan Riviere (1883–1962)

JOAN RIVIERE (née Joan Hodgson Verrall) was born on June 28, 1883, the eldest of three children, to Hugh John and Anna Hodgson Verrall, in Brighton, Sussex, the famous English resort. The Verralls were an old Sussex family centered in the town of Lewes and belonged to the professional middle class.[1] A paternal uncle was the famous Cambridge classicist, Arthur Woolgar Verrall. Hugh Verrall was a third-generation solicitor. He became a magistrate, as had his father, was active in local politics, and was known for his intellectual bent.

Anna Verrall was the daughter of the Reverend John Willoughby Hodgson, Vicar of Chalgrave, Bedfordshire. She was orderly, efficient, inventive, and practical. Her second child was Mary Willoughby Verrall, who became one of the first "lady almoners" (hospital social workers) and married a physician. The youngest in the family was Hugh Cuthbert Verrall, who, following his service in World War I, emigrated to Canada.

Little is known of Joan's early childhood except for an interest in painting. Her major school experience was some years at Wycombe Abbey, one of the first girls' "public" boarding schools. This experience "she *very* much disliked."[2] At age seventeen she went abroad for a year to Gotha, Germany, where she began to acquire her mastery of German. Returning to England, she spent the next years sketching, designing dresses, and working for a time as a dressmaker. Sewing and fashion were to remain important interests throughout her life. She was a tall, impressive, distinguished, and very beautiful woman. She was also severe, openly critical, uncompromising, and, when angry, "very alarming" (Strachey 1963; Heimann 1963).[3]

In 1906, she married Evelyn Riviere, the son of a well-known painter and member of the Royal Academy. Evelyn Riviere was a Chancery barrister and successful in his career, becoming a "Bencher" at Lincoln's Inn. He died in 1945. The Rivieres' only child, Diana, was born in 1908.

Joan Riviere began an analysis with Ernest Jones in 1914 or 1915. She had no intention of becoming an analyst herself, but with the increased demand for analysis at the end of the war she reluctantly agreed, at Jones's urging, to accept two patients on a trial basis.[4] The experiment worked. She became a founding member of the British Society, a training analyst, and translations editor of the *International Journal of Psycho-Analysis*. She produced the first English translations of Freud that were of high quality and edited and translated his *Collected Papers*. In translating *Das Unbehagen in der Kultur* she rejected Freud's suggested transliteration, "Man's Discomfort in Civilization," and supplied the English title, *Civilization and Its Discontents*.

Jones, Riviere, and the Stracheys, as the "Glossary Committee," developed the standard English vocabulary of psychoanalysis, inventing such words as "cathexis" and "superego."

In 1922, she went to Vienna for a year's analysis with Freud. Thereafter, she corresponded with him frequently and probably returned for brief periods.[5]

Beginning about 1930, Joan Riviere became an important and respected member of the "Kleinian" or "English" school of psychoanalysis. In the partisanship of the succeeding years, Strachey mentions his fear of her and attributes the same feeling to Jones (Strachey 1963, p. 230). Glover, the chief antagonist of the Kleinians, makes clear his respect for her as an independent thinker and important theorist of the Kleinian School.[6]

Grinstein attributes to her fifteen contributions to the literature. In addition, she edited or contributed significantly to three important books presenting the Kleinian position: *Developments in Psycho-Analysis* (1952), *Love, Hate and Reparation* (1937), and *New Directions in Psycho-Analysis* (1955). She maintained a limited practice until a few weeks before her death on May 20, 1962, at the age of seventy-eight.

NOTES

1. D. Riviere; personal communication, 1974. This brief biography is based upon information kindly supplied by Miss Riviere.
2. Ibid.
3. Ibid.
4. Ibid.
5. Ibid.
6. Oral history: The Reminiscences of Edward Glover. The Psychoanalytic Movement (1965). In the Oral History Collection of Columbia University, New York.

Part Three

Early Papers

on

Technique

CHAPTER 4

KARL ABRAHAM

A PARTICULAR FORM OF NEUROTIC RESISTANCE AGAINST THE PSYCHO-ANALYTIC METHOD

INTRODUCTORY NOTE

This early article by Abraham, frequently quoted in the subsequent literature, is among the most influential in the early history of psychoanalytic technique. Abraham depicts most vividly the difficulties of the early analysts as they struggled with the application of the basic rule. He found that the difficulty in free associations often symbolized an anal fixation and resistance to defecation on the part of the child. But the article goes deeper. Abraham delineates the narcissistic character so clearly that this paper is second only to Freud's (1914b) as a study in narcissism. It was, therefore, a pioneering paper in the understanding of narcissism. It deals more directly with problems of technique than Freud's classic paper.

Originally entitled "Uber eine besondere Form des neurotischen Widerstandes gegen die psychoanalytischen Methodik," *Internationale Zeitschrift für ärztliche Psychoanalyse* 5 (October 1919):173–180. Reprinted in *Selected Papers on Psychoanalysis*, trans. Douglas Bryan and Alix Strachey (New York: Basic Books, 1953), chap. 15. Reprinted by permission of Basic Books, Inc., Publishers.

WHEN WE BEGIN to give a patient psycho-analytic treatment we make him acquainted with its fundamental rule to which he has to adhere unconditionally. The behaviour of each patient in regard to that rule varies. In some cases he will easily grasp it and carry it out without particular difficulty; in others he will frequently have to be reminded of the fact that he has to make free associations; and in all cases we meet at times with a failure to associate in this way. Either he will produce the result of his reflected thoughts or say that nothing occurs to him. In such a situation the hour of treatment can sometimes pass without his producing any material whatever in the way of free association. This behaviour indicates a 'resistance', and our first task is to make its nature clear to the patient. We regularly learn that the resistance is directed against allowing certain things in the mind from becoming conscious. If at the commencement of the treatment we have explained to the patient that his free associations give us an insight into his unconscious, then his refusal to give free associations of this kind is an almost obvious form for his resistance to take.

Whereas in most of our cases we meet with a resistance of this kind which appears and disappears by turns, there is a smaller group of neurotics who keep it up without interruption during the whole of their treatment. This chronic resistance to the fundamental rule of psycho-analysis may obstruct its progress very much and even preclude a successful result. The question has hitherto received little consideration in our literature, like so many other questions of technique. I have met with this difficulty in a number of cases, and other psycho-analysts tell me that they have had the same experience. There is therefore a practical as well as a theoretical interest in investigating more strictly this kind of neurotic reaction to psycho-analysis.

The patients of whom we are speaking hardly ever say of their own accord that 'nothing occurs' to them. They rather tend to speak in a continuous and unbroken manner, and some of them refuse to be interrupted by a single remark on the part of the physician. But they do not give themselves up to free associations. They speak as though according to programme, and do not bring forward their material freely. Contrary to the fundamental rule of analysis they arrange what they say according to certain lines of thought and subject it to extensive criticism and modification on the part of the ego. The physician's admonition to keep strictly to the method has in itself no influence on their conduct.

It is by no means easy to see through this form of behaviour. To the physician who is not experienced in recognizing this form of resistance the patients seem to show an extraordinarily eager, never-wearying readiness to be psycho-analysed. Their resistance is hidden behind a show of willingness. I must admit that I myself needed long experience before I was able to avoid the danger of being deceived. But once I had correctly recognized this systematic resistance its source also became clear to me. For although neurotics of this type, of whom I have treated a certain number, exhibited great variety as regards illness and symptoms, in their attitude towards psycho-analysis and the physician they all produced a certain number of characteristics with astonishing regularity. And I should like to make those characteristics the subject of discussion in the following pages.

Under the apparent tractability of these patients lies concealed an unusual degree of defiance, which has its prototype in the child's conduct towards its father. Whereas other neurotics will occasionally refuse to produce free ideas, these patients do so continually. Their communications are superabundant in quantity, and, as we have said, it is this fact that blinds the inexperienced physician to their imperfection as regards quality. They only say things which are 'ego-syntonic'. These patients are particularly sensitive to anything which injures their self-love. They are inclined to feel 'humiliated' by every fact that is established in their psycho-analysis, and they are continually on their guard against suffering such humiliations. They furnish any number of dreams, but they adhere to the manifest content and understand how to glean from the dream-analysis only what they already know. And they not only persistently avoid every painful impression but at the same time endeavour to get the greatest possible amount of positive pleasure out of their analysis. This tendency to bring the analysis under the control of the pleasure principle is particularly evident in these patients and is, in common with a number of peculiarities, a clear expression of their narcissism.* And it was in fact those among my patients who had the most pronounced narcissism who resisted the fundamental psycho-analytic rule in the way described.

The tendency to regard a curative measure merely as an opportunity for obtaining pleasure and to neglect its real purpose must be regarded as a thoroughly childish characteristic. An example will illustrate this. A boy of eight was ordered to wear spectacles. He was delighted with this, not because he was to be relieved of an unpleasant visual disturbance by their use, but because he was to be allowed to wear spectacles. It soon turned out that he paid no attention to whether the trouble was removed by means of the spectacles or not; the fact of possessing them and being able to show them at school pleased him so much that he forgot all about their therapeutic value. The attitude of the class of patients we are discussing to psycho-analysis is exactly the same. One expects from it interesting contributions to the autobiography which he is writing in the form of a novel; another hopes that psycho-analysis will advance him to a higher intellectual and ethical level, so that he will be

---

* [We owe to K. R. Eissler (1958) the observation: "Even if the delinquent wanted to confide in the analyst, he would verbalize almost nothing, but that the content of his unfulfilled desires and his plans for gratification, since in accordance with the structural defects of the superego, his capacity for self observation has scarcely been developed" (p. 230). When Abraham was struggling with this problem, the relationship between psychic structure and the ability to free associate was not yet understood (see editor's introduction, p. 34).]

superior to his brothers and sisters towards whom he has hitherto had uncomfortable feelings of inferiority. The aim of curing their nervous disabilities
retreats into the background in proportion as such narcissistic interests
predominate.

The narcissistic attitude such patients adopt towards the method of treatment also characterizes their relations to the analyst himself. Their transference on to him is an imperfect one. They grudge him the rôle of father. If signs
of transference do appear, the wishes directed on to the physician will be of a
particularly exacting nature; thus they will be very easily disappointed precisely in those wishes, and they will then quickly react with a complete withdrawal of their libido. They are constantly on the look-out for signs of
personal interest on the part of the physician, and want to feel that he is
treating them with affection. Since the physician cannot satisfy the claims of
their narcissistic need for love, a true positive transference does not take
place.*

In place of making a transference the patients tend to identify themselves
with the physician. Instead of coming into closer relation to him they put
themselves in his place. They adopt his interests and like to occupy themselves
with psycho-analysis as a science, instead of allowing it to act upon them as a
method of treatment. They tend to exchange parts, just as a child does when it
plays at being father. They instruct the physician by giving him their opinion
of their own neurosis, which they consider a particularly interesting one, and
they imagine that science will be especially enriched by their analysis. In this
way they abandon the position of patient and lose sight of the purpose of their
analysis. In particular, they desire to surpass their physician, and to depreciate
his psycho-analytical talents and achievements. They claim to be able to 'do it
better'. It is exceedingly difficult to get them away from preconceived ideas
which subserve their narcissism.† They are given to contradicting everything,
and they know how to turn the psycho-analysis into a discussion with the
physician as to who is 'in the right'.

The following are a few examples: A neurotic patient I had not only
refused to associate freely but to adopt the requisite position of rest during the
treatment. He would often jump up, go to the opposite corner of the room and

* [Abraham grasped the brittle nature of the transference in a narcissistic character
neurosis. (For later developments on this subject, see Kohut 1971; Kernberg 1975.)]

† [Abraham, in keeping with the times, is critical in his attitude toward the narcissistic character and sees it mainly in a negative light. With a better understanding of early
development, psychoanalysts became less critical of those patients who need to identify
themselves with the analyst, criticizing, or even reversing roles with him. Such patients,
Kohut (1971) pointed out, develop special forms of transference that must be understood.
Abraham's patient could be seen in Kohut's terms as having a "mirror transference." In
addition to Kohut's work, Jacobson (1964) threw light on such patients when she observed that in them self and object representations are not entirely separated. Important
for understanding such narcissistic patients is also Mahler's work (1968) on symbiosis and
separation-individuation.]

expound, in a superior and didactic manner, his self-formed opinions about his neurosis. Another of my patients displayed a similar didactic attitude. He actually said straight out that he understood psycho-analysis better than I did because it was he and not I who had the neurosis. After long-continued treatment he once said, 'I am now beginning to see that you know something about obsessional neurosis'. One day a very characteristic fear of his came out. It was that his free associations might bring to light things that were strange to him but familiar to the physician; and the physician would then be the 'cleverer' and superior person of the two. The same patient, who was much interested in philosophical matters, expected nothing less from his psycho-analysis than that science should gain from it the 'ultimate truth'.

The presence of an element of *envy* is unmistakable in all this. Neurotics of the type under consideration grudge the physician any remark that refers to the external progress of their psycho-analysis or to its data. In their opinion he ought not to have supplied any contribution to the treatment; they want to do everything all by themselves. This brings us to a particularly striking characteristic which all these patients show, which is that they make up at home for their failure to associate freely during the hour of treatment. This procedure, which they very often call 'auto-analysis', contains an obvious depreciation of the physician's powers. The patients actually see in him a hindrance to progress during the hours of treatment, and are exceedingly proud of what they imagine they have achieved without his assistance. They mix the free associations obtained in this way with the results of considered thought, classify them according to some definite idea, and produce them in this state to the physician next day. In consequence of severe resistances, one of my patients thought that he was making very little progress during a succession of hours, and finally that he was making none at all. The next day he came to me and said he had had to 'work' for many hours alone at home. Naturally I was meant to infer from this the weakness of my own abilities. One element in such an 'auto-analysis' is a narcissistic enjoyment of oneself; another is a revolt against the father. The unrestrained occupation with his own ego and the feeling of superiority already described offers the person's narcissism a rich store of pleasure. The necessity of being alone during the process brings it extraordinarily near to onanism and its equivalent, neurotic day-dreaming, both of which were earlier present to a marked degree in all the patients under consideration. 'Auto-analysis' is for them a form of day-dreaming, a substitute for masturbation, free from reproach, since it is justified and even prescribed on therapeutic grounds.

I may say that the cases to which I refer belong chiefly to the obsessional neuroses. One case was an anxiety-hysteria mixed with obsessional symptoms, and in another there was a paranoid disturbance. In view of the more recent results of psycho-analysis we shall not be surprised to find pronounced sadistic-anal traits in all the cases. Their hostile and negative attitude towards the physician has already been mentioned; and anal-erotic motives explain the rest of their behaviour. I will give a few examples of this. In these as in other neurotics with strong anal erotism, talking in the analysis, by means of which psychic material is discharged, is compared to emptying the bowels. (I may say that some identify free associations with flatus.) They are persons who have only with difficulty been taught in childhood to control their sphincters and to have a regular action of the bowels. They used to refuse to empty their bowels at a specific time, so that they could do this when it suited their convenience; and they now behave towards psycho-analysis and the physician in the same way from unconscious motives. Tausk [1919] has recently pointed out the fact that small children like to deceive adults with regard to emptying their bowels. They appear to try very hard to satisfy the demands of their mother or nurse, but they have no motion. Tausk adds that this is perhaps the earliest occasion on which the child becomes aware that it can take grown-up people in. The neurotics under discussion continue this tradition of infantile behaviour. They pride themselves, as it were, upon being able to decide whether, when, or how much they will give out from their unconscious psychic material. Their tendency to bring perfectly arranged material to the analytic hour shows not only an anal-erotic pleasure in systematizing and cataloguing everything but exhibits yet another interesting feature. Freud [1918] has recently drawn attention to the unconscious identification of excrement and gifts. Narcissistic neurotics with a strong anal disposition such as we are dealing with here have a tendency to give presents instead of love [see Chapter 14 (Abraham 1917)]. Their transference on to the physician is incomplete. They are not able to expend themselves unconstrainedly in free associations. As a substitute they offer their physician gifts; and these gifts consist of their contributions to psycho-analysis which they have prepared at home and which are subject to the same narcissistic over-estimation as the products of the body. The narcissistic advantage for them is that they keep the power of deciding what they are going to give.

One of my obsessional patients who suffered from brooding and doubting mania contrived to make psycho-analysis itself, its methods and its results, the subject of his brooding and doubts. He was almost entirely dependent on his family, and he used to plague himself among other things with doubts as to whether his mother or Freud was 'in the right'. His mother, he said, had often advised him, in order to improve his constipation, not to dream in the closet but only to think of the process of defæcation itself; whereas Freud, on the contrary, gave exactly the opposite rule, namely, to associate freely and then 'everything comes out of itself'. It was a long time before the patient carried out his psycho-analysis no longer according to his mother's method, but according to that of Freud.

The well-known parsimony of anal-erotics seems to be in contradiction with the fact that these patients readily make material sacrifices for the treatment, which, for the reasons above mentioned, is a protracted one. This behaviour, however, is explicable from what has already been said. The patients are making a sacrifice to their narcissism. They are all too apt to lose sight of the fact that the object of the treatment is to cure their neurosis. It is another consideration which enables them not to pay attention to expense. To paraphrase an old anecdote, it might be said that nothing is too dear for their narcissism.

The character-trait of parsimony is, besides, found elsewhere in them. They save up their unconscious material. They are prone to build on the belief that one day 'everything will come out all at once'. They practise constipation in their psycho-analysis, just as they do in the sphere of bowel activity. Evacuation is to take place on some occasion after a long delay and to give them particular pleasure. This finale is again and again postponed, however.

The analysis of patients of this description presents considerable difficulties. These difficulties reside in part in the pretended compliance with which the patients cloak their resistance. For analysis is an attack on the patient's narcissism, that is, on that instinctual force upon which our therapeutic endeavours are most easily wrecked. Everyone who is acquainted with the situation will therefore understand that none of my cases gave quick results. I must add that in no case did I obtain a complete cure, though I did succeed in effecting an improvement of some practical value, which in a few cases was of a far-reaching character. My experiences perhaps give a too unfavourable picture of the therapeutic prospects. When I treated my first cases I lacked a deeper insight into the peculiar nature of the resistances. It must be remembered that it was not until 1914 that we got our first knowledge of narcissism, thanks to Freud's classical study. I certainly have the impression that it is easier to overcome such narcissistic resistances now that I make known to these patients the nature of their resistances at the very beginning of the treatment. I lay the greatest stress on making an exhaustive analysis of the narcissism of such patients in all the forms it takes, and especially in its relation to the father complex. If it is possible to overcome their narcissistic reserve, and, what amounts to the same thing, to bring about a positive transference, they will one day unexpectedly produce free associations, even in the presence of the physician. At first these associations come singly, but with the advance of the process described they become more abundant. Therefore, though I have, to begin with, called special attention to the difficulties of the treatment, I should like in conclusion to issue a warning against making an entirely unfavourable prognosis for all such cases.

CHAPTER 5

SANDOR FERENCZI

ON THE TECHNIQUE
OF PSYCHO-ANALYSIS

INTRODUCTORY NOTE

Freud's papers on technique laid down the basic principles or the grand strategy of this procedure. Ferenczi's papers deal with more specific subjects. They resemble reports to headquarters from the front line. He reports candidly what he found, the difficulties encountered, and the remedies he was forced to devise. The first section, on the abuse of free associations, deals with the most vexing problem the pioneering analyst discovered—the inability or unwillingness of many patients to follow the "basic rule"—to associate without suppression.

I / *Abuse of Free Association*

THE WHOLE METHOD rests on Freud's 'fundamental rule of psycho-analysis', on the patient's duty to relate everything that occurs to him in the course of the analytical hour. Under no circumstances may an exception be made to this rule, and everything that the patient—from whatever motive—endeavours

Originally entitled "Die psychoanalytischen Technik," *Internationale Zeitschrift für ärtzliche Psychoanalyse* 5 (1919):181–192. Reprinted in *Selected Papers of Sandor Ferenczi, M.D.*, vol. 2, *Further Contributions to the Theory and Technique of Psychoanalysis*, comp. John Rickman, trans. Jane Isabel Suttie et al. (New York: Basic Books, 1952), pp. 177–189. Reprinted by permission of Basic Books, Inc., Publishers. This paper was initially read before the Hungarian Psychoanalytical Society (Freud Society) in Budapest.

to withhold, must be unrelentingly brought to light. It may happen, however, that when the patient has with no little pains been educated up to literal acquiescence in this rule that his resistances take possession of it and endeavour to defeat the doctor with his own weapon.

Obsessional neurotics sometimes have recourse to the evasion of relating *only* senseless associations, as though deliberately misunderstanding the doctor's request that they should recount everything, senseless things as well. If they are let alone and not interrupted, in the hope that in time they will weary of the proceeding, this expectation is often disappointed, until one is convinced that they are unconsciously displaying a tendency to reduce their doctor's request to an absurdity. With this kind of superficial association they usually deliver an unbroken series of word associations, the selection of which, of course, betrays the unconscious material which the patient wishes to avoid. It is quite impossible, however, to achieve a thorough analysis of any particular ideas, for if by chance certain striking concealed traits are pointed out, instead of simply accepting or rejecting the interpretation—one is merely presented with more 'senseless' material.

There is nothing for it but to make the patient aware of his tenacious behaviour, whereupon he will not fail to remark triumphantly, 'I am only doing as you asked, I am telling you all the nonsense that occurs to me'. At the same time he may perhaps make the suggestion that the strict adherence to the 'fundamental rule' might be relaxed, the conversations be systematically arranged, definite questions be put to him, and the forgotten material searched for methodically or even by means of hypnosis. The reply to these objections is not difficult: the patient, it is true, is asked to say everything, even the senseless things, that occurs to him, but certainly not to repeat only meaningless or disconnected words. This behaviour contradicts—so we explain to him—just that rule of psycho-analysis that forbids any critical choice of ideas. The quick-witted patient will thereupon retort that he cannot help it if only nonsense occurs to him, and propounds perhaps the illogical question whether from now on he should withhold all the nonsense. One must not get annoyed, otherwise the patient would have attained his object, but must keep him to the continuation of the work. Experience shows that the admonition not to misuse free association usually has the result that from then on it is no longer only nonsense that occurs to the patient.

It is only in the rarest cases that a single explanation of this matter suffices; if the patient again becomes resistant to the doctor or the treatment, he starts the meaningless associations again, he even puts the difficult question of what he should do when not even entire words but only inarticulate sounds, animal noises, or instead of words, melodies, occur to him? He is requested to repeat such sounds and tunes confidently, like everything else, but he is told of the bad intention that is concealed behind his apprehension.

Another well-known manifestation of the 'association resistance' is that 'absolutely nothing occurs' to the patient. The possibility of this can be granted without further discussion. But if the patient is silent for a more prolonged period, it usually signifies that he is withholding something. A patient's sudden silence must therefore always be interpreted as a *passagère symptom.**

A long silence is often due to the fact that the request to relate *everything* has not yet been taken literally. If the patient is asked after a considerable pause what was in his mind during the silence, he will perhaps reply that he had *only* contemplated an object in the room, had a sensation or a paræsthesia† in this or that part of the body, etc. There is often nothing for it but to explain all over again to the patient that he must relate everything that goes on inside him, sensations, therefore, as well as thoughts, feelings, and impulses. But as this enumeration can never be complete, the patient will always—when he relapses into a resistance—discover fresh possibilities for rationalizing his silence and secretiveness. Many, for instance, say they withhold something because they had had no clear thoughts but only indistinct confused sensations. Of course this is a proof that, in spite of the request to the contrary, they are still subjecting their ideas to criticism.

If explanations have no result, it must be assumed that the patient wants to beguile one into detailed instructions and explanations and in this manner to obstruct the work. In such cases it is best to encounter the patient's silence with silence. It may happen that the greater part of the hour passes without the doctor or the patient having said a single word. The patient finds it very difficult to endure the doctor's silence; he gets the impression that the doctor is annoyed with him, that is, he projects his own bad conscience on to the doctor, and this finally decides him to give in and renounce his negativism.

One must not let oneself be misled even when one or other patient should threaten to go to sleep from sheer boredom; in a few instances the patient actually has fallen asleep for a short time, but I had to conclude from their speedy awakening that the preconscious had kept a grip of the situation even during sleep. The danger that the patient may sleep away the whole hour does not therefore arise.‡

---

* [Passagère Symptom = Transitory Symptom. Earlier, Ferenczi (1912) devoted an article to the question of transitory symptoms occurring during the psychoanalytic session. There he said: "Free association and the analytic scrutinizing of the incoming thoughts is not infrequently interrupted in hysterics by the abrupt appearance of somatic phenomena of a sensory or motor nature. . . . They really are representations, in symptom form, of the unconscious feeling and thought-excitations which the analysis has stirred up from their inactivity (state of rest, equilibrium) and brought near to the threshold of consciousness, but which before becoming quite conscious—in the last moment, so to speak—have been forced back again on account of their painful character (to consciousness), whereby their sum of excitation, which can no longer be quite suppressed, becomes transformed into the production of somatic symptoms."]

† [Paraesthesia = abnormal spontaneous sensation, for example, numbness.]

‡ The fact that the doctor at many interviews pays little heed to the patient's associa-

Many a patient makes the objection to free association that too much occurs to him at once and he does not know what to relate first of all. Should one allow him to determine the sequence himself he will perhaps reply that he cannot decide to give one or other idea the preference. In such a case I had to have recourse to letting the patient relate everything in the sequence in which it had occurred to him. He replied that in that case he was afraid lest while he followed out the first thought of the series he would forget the others. I soothed him with the hint that everything of importance—even if for the moment it seems to be forgotten—will come to the surface later of itself.*

Little peculiarities, too, in the manner of associating have their significance. So long as the patient introduces every idea with the phrase, 'I think that', he shows that he is inserting a critical examination between the perception and the communication of the idea. Many prefer to cloak unpleasant ideas in the form of a projection upon the doctor, by saying perhaps, 'You are thinking that I mean that . . .', or 'Of course you will interpret that to mean . . .'. To the request to omit any criticism, many reply, 'criticism is ultimately also an idea' which one must acknowledge without further debate, but not without drawing attention to the fact that if the fundamental rule were strictly kept to it could not happen that the common communication of the criticism should precede or indeed replace that of the idea itself.

In one case I was forced, in direct contradiction to the psycho-analytic rule, to insist that the patient should always complete any sentence he had begun. For I noticed that whenever a sentence took an unpleasant turn he never completed it, but switched off in the middle with a 'by the way' on to something unimportant and beside the mark. It had to be explained to him that the fundamental rule did not, it is true, demand the *thinking* out of an idea, but certainly the complete *utterance* of what had been thought. He required many admonitions, however, before he learnt this.

Quite intelligent and otherwise sensible patients sometimes try to reduce the methods of free association to absurdity by putting the question, 'What,

---

tions and only pricks up his ears at certain statements also belongs to the chapter 'on counter-transference'; dozing may happen in these circumstances. Subsequent scrutiny mostly shows that we were reacting unconsciously to the emptiness and worthlessness of the associations just presented by the withdrawal of conscious excitation; at the first idea of the patient's that in any way concerns the treatment we brighten up again. The danger, therefore, of the doctor falling asleep and leaving the patient unobserved is not great. (I owe the full confirmation of this observation to a verbal discussion of the subject with Prof. Freud.)

* It is probably hardly necessary to state expressly that the psycho-analyst must avoid any untruth in relation to his patient; this holds, of course, equally in respect of matters concerning either the doctor's methods or person. The psycho-analyst should be like Epaminondas, of whom Cornelius Nepos tells that he 'nec ioco quidem mentiretur' [not to deceive, even in jest]. Of course the doctor may and must withhold a part of the truth at first, for instance, what the patient is not yet prepared for; that is, he must himself determine the speed of initiation.

though, if it occurred to them to get up suddenly and run away, or to maltreat the doctor physically, to kill him, to smash a piece of furniture, etc.?' If one then explains that they were not told to *do* everything that occurred to them but only to *say* it, they usually reply that they are afraid they could not distinguish so sharply between thinking and doing. We can reassure these over-anxious folk that this fear is only a reminiscence of childhood when they actually were not yet capable of such a differentiation.*

In rare cases patients are overwhelmed by an impulse, so that instead of continuing to associate they begin to *act* their psychic content. Not only do they produce 'transitory symptoms' instead of ideas, but while fully conscious they carry out complicated activities, entire scenes, of whose transference or reconstruction nature they have not the least conception. Thus one patient at certain exciting moments in the analysis jumped up suddenly off the sofa, walked up and down the room and ejaculated abusive words. The historical basis for the movements as well as for the abusive words was then revealed by the analysis.

An hysterical patient of the infantile type surprised me after I had succeeded in weaning her temporarily of her childish seduction artifices (constant imploring contemplation of the doctor, striking or exhibitionist apparel) by an unexpected direct attack; she jumped up, demanded to be kissed, and finally came actually to grips. It goes without saying, of course, that the doctor must not lose his attitude of benevolent patience even in the face of such occurrences. He must point out over and over again the transference nature of such actions, towards which he must conduct himself quite passively. An indignant moral rebuff is as out of place in such cases as would be the agreement to any of the demands. Such a reception, it will be found, rapidly exhausts the patient's inclination for assault, and the disturbance—that is to be interpreted psycho-analytically in any case—soon settles iself.

In a paper 'Über obszöne Wörte' (Ferenczi 1928b) I have already insisted that one must not spare patients the effort of overcoming the resistance to saying certain words. Easing of the difficulty, as by permitting certain communications to be set down in writing, contradict the purposes of the treatment, which consist essentially in the patient's mastering his inner resistances by continuous and progressive practice. When, too, the patient is endeavouring to remember something of which the doctor is quite aware, he must not just be helped out at once; otherwise the possibly valuable substitute ideas will be lost.

Of course this withholding of help on the doctor's part cannot be absolute. If for the moment one is less concerned about exercising the patient's

---

* [Here Ferenczi betrays a lack of sufficient clinical experience. There are indeed adult patients who cannot draw the line between thinking and acting. It would be a mistake to reassure such people who have had ample reason to mistrust their capacity to maintain this demarcation line.]

psychic powers than with hastening certain understandings, then one simply puts into words the ideas one supposes him to have, but which he lacks the courage to utter, and thus obtains a confession from him. The doctor's position in psycho-analytic treatment recalls in many ways that of the obstetrician, who also has to conduct himself as passively as possible, to content himself with the post of onlooker at a natural proceeding, but who must be at hand at the critical moment with the forceps in order to complete the act of parturition that is not progressing spontaneously.

II / *Patient's Questions—Decisions during Treatment*

I MADE IT A RULE, whenever a patient asks me a question or requests some information, to reply with a counter interrogation of how he came to hit on that question.* If I simply answered him, then the impulse from which the question sprang would be satisfied by the reply; by the method indicated, however, the patient's interest is directed to the sources of his curiosity, and when his questions are treated analytically he almost always forgets to repeat the original enquiries, thus showing that as a matter of fact they were unimportant and only significant as a means of expression for the unconscious.

The situation becomes particularly difficult, however, when the patient appeals to one, not with some question or other, but with the request that some matter of personal significance, such as the choice between two alternatives, be decided for him. The doctor's endeavour must always be to postpone decisions till the patient is enabled, by a growing self-reliance due to the treatment, to deal with matters himself. It is well, therefore, not to accept too easily the patient's stressing of the urgency for an immediate decision, but to consider also the possibility that such apparently very real questions have perhaps been pushed into the foreground unconsciously by the patient, whereby he is either clothing the analytic material in the garb of a problem, or his resistance has taken this means of interrupting the progress of the analysis. In one patient's case this last manœuvre was so typical that I had to explain to her, in the military phraseology in vogue at the time, that when she could find no other way out, she flung problems at me like gas-shells, in order to confuse me. Of course a patient may really on occasion have to decide an important matter without delay during treatment; but it is as well if on these occasions, too, the doctor plays as little as possible the part of spiritual guide, after the fashion of a *directeur de conscience*, contenting himself with that of analytic

---

* [For further discussions of the role of questions in psychoanalysis, see Olinick 1954 and 1957.]

confesseur who illuminates every motive, those, too, of which the patient is unconscious, as clearly as may be from every side, but who gives no direction about any decisions and actions. As far as this is concerned, psycho-analysis is diametrically opposed to every psycho-therapy as yet practised, to suggestion as to treatment by persuasion.

In two circumstances the psycho-analyst is in the position of interfering uncompromisingly in the patient's career. First, when he is convinced that the patient's vital interests really demand an immediate decision of which he is as yet incapable by himself; in this case, however, the doctor must be aware that he is no longer dealing as a psycho-analyst, indeed that certain difficulties in the prosecution of the treatment may arise from his interference; for instance, an undesirable strengthening of the transference relation. Secondly, the analyst can, and must from time to time, practise 'active therapy' in so far as he forces the patient to overcome the phobia-like incapability of coming to a decision. By the change in the affective excitations that this overcoming will occasion he hopes to obtain access to as yet inaccessible unconscious material.

III / 'For Example' in Analysis

IF A PATIENT presents one with some generalization, whether it be a manner of speaking or an abstract statement, he should always be asked what occurs to him in connection with that generalization. This question has become so fluent with me that it occurs almost automatically as soon as the patient begins to speak in too general terms. The tendency to pass from the general to the more and more particular dominates the whole of psycho-analysis; it is this alone that leads to the fullest possible reconstruction of the patient's life history, to the filling of his neurotic anamnesias. It is therefore wrong, following the patient's inclination for generalization, to co-ordinate one's observations about him too soon under any general thesis. In real psycho-analysis there is little room for moral or philosophical generalizations; it is an uninterrupted sequence of concrete facts.

A young patient gave me the confirmation in a dream that the phrase 'for example' is really the proper technical method for guiding the analysis from the remote and unessential directly to the imminent and essential.

She dreamt: *'I have toothache and a swollen cheek; I know that it can only get better if Mr. X. (formerly engaged to me) rubs it; for this, however, I must get a lady's permission. She really gives me permission, and Mr. X. rubs my cheek with his hand; at this a tooth jumps out as though it had just grown and as though it had been the cause of the pain'.*

Second dream fragment: '*My mother asks me what happens at psycho-analysis. I tell her: one lies down and must say everything that passes through one's head. But what passes through one's head? she asks again. All sorts of thoughts, even the most incredible ones. What, for example? For example, that one has dreamt that the doctor has kissed one and . . .* this sentence remained unfinished and I woke'.

I shall not enter into the details of the interpretation, and need only remark that we are dealing here with a dream whose second part *interprets* the first. The interpretation is set about quite methodically. The mother, who evidently takes the place of the analyst, is not satisfied with the generalizations by means of which the dreamer attempts to get herself out of the affair, and will not be content till, in reply to the question, 'what, *for example?*' occurs to her, the latter concedes the only correct sexual interpretation of the dream.

What I maintained, therefore, in a paper on 'The Analysis of Compari-sons' (see Ferenczi 1919c; Freud 1919), namely, that just the most significant material is concealed behind comparisons apparently thrown out in passing, holds good also for ideas that the patients evolve in reply to the question, 'what, for example?'.

IV / *The Control of the Counter-Transference*

PSYCHO-ANALYSIS—to which, generally speaking, the task of exposing mysticism seems to have fallen—succeeded in laying bare the simple, one might say naïve, rule of thumb that lies at the bottom of even the most complicated medical diplomacy. It discovered the transference to the doctor to be the effective agent in all medical suggestion, and showed that such a trans-ference ultimately only repeats the infantile-erotic relationship to the parents, to the indulgent mother or to the stern father, and that it depends upon the patient's experience of life or his constitutional tendency whether or how far he is susceptible to the one or the other kind of suggestion (Ferenczi 1909).

Psycho-analysis thus discovered that nervous patients are like children and wish to be treated as such. Doctors with a gift of intuition knew this even before us, at least they behaved as though they knew it. The vogue of many a 'downright' or 'kindly' sanatorium doctor is to be explained in this way.

The psycho-analyst, however, may no longer be gentle and sympathetic or downright and hard according to inclination and wait till the patient's soul moulds itself to the doctor's character; he must understand how to *graduate* his sympathy. Indeed he may not even yield inwardly to his own affects; to be influenced by affects, not to mention passions, creates an atmosphere unfavorable for the taking up and proper handling of analytic data. As the doctor, however, is always a human being and as such liable to moods, sympathies and antipathies, as well as impulses—without such susceptibilities he would of course have no understanding for the patient's psychic conflicts—he has constantly to perform a double task during the analysis: on the one hand, he must observe the patient, scrutinize what he relates, and construct his unconscious from his information and his behaviour; on the other hand, he must at the same time consistently control his own attitude towards the patient, and when necessary correct it; this is the mastery of the *counter-transference* (Freud).*

The pre-condition for this is of course the analysis of the doctor himself; but even the analysed individual is not so independent of peculiarities of character and actual variations of mood as to render the supervision of the counter-transference superfluous.

It is difficult to generalize about the way in which the control of the counter-transference should interfere; there are too many possibilities. To give some conception of it, it would probably be best to adduce some examples from actual experience.

At the beginning of psycho-analytic medical activities one naturally suspects least of all the danger that threatens from *this* side. One is in the blissful mood into which a first acquaintance with the unconscious transports one, the doctor's enthusiasm transfers itself to the patient, and the psycho-analyst owes surprising cures to this happy self-assurance. There is no doubt that these results are only in a small degree due to analysis, but are for the most part purely suggestive, that is, are the results of the transference. In the elevated mood of the honeymoon months of analysis, one is miles from considering, let alone mastering, the counter-transference. One yields to every affect that the doctor-patient relationship may evoke, is moved by the patient's sad experiences, probably, too, by his phantasies, and is indignant with all those who wish him ill. In a word, one makes all their interests one's own, and is surprised when one or other patient in whom our behaviour may have raised irrational hopes suddenly breaks out in passionate demands. Women demand that the doctor shall marry them, men that he shall support them, and they construct arguments for the justification of their claims out of his utterances. Naturally one gets out of these difficulties easily enough during the analysis; one falls back upon their transference nature and employs them as material for further elaboration. In this way, however, one gets an insight into the cases of non-analytic or 'wild' analytic therapy that eventuates in accusations or legal proceedings against the doctor. The patients are simply unmasking the doctor's unconscious. The enthusiastic doctor who wants to 'sweep away' his patient in his zeal to cure and elucidate the case does not observe the little and big indications of fixation to the patient, male or female, but they are only too

* [These ideas are an extension of Freud's "evenly suspended attention," 1912b, p. 111.]

well aware of it, and interpret the underlying tendency quite correctly without guessing that the doctor himself was ignorant of it. In such arraignments, therefore, both the opposing parties, remarkably enough, are right. The doctor can swear that he—consciously—intended nothing but the patient's cure; but the patient is also right, for the doctor has unconsciously made himself his patient's patron or knight and allowed this to be remarked by various indications.

Psycho-analytic discussion protects one, of course, from such inadvertencies; nevertheless it does happen that the insufficient consideration of the counter-transference puts the patient into a condition that cannot be altered and which he uses as a motive for breaking off the treatment. One must just accept the fact that every new psycho-analytic technical rule costs the doctor a patient.

If the psycho-analyst has learnt painfully to appreciate the counter-transference symptoms and achieved the control of everything in his actions and speech, and also in his feelings, that might give occasion for any complications, he is threatened with the danger of falling into the other extreme and of becoming too abrupt and repellent towards the patient; this would retard the appearance of the transference, the pre-condition of every successful psychoanalysis, or make it altogether impossible. This second phase could be characterized as the phase of resistance against the counter-transference. Too great an anxiety in this respect is not the right attitude for the doctor, and it is only after overcoming this stage that one perhaps reaches the third, namely, that of the control of the counter-transference.

Only when this has been achieved, when one is therefore certain that the guard set for the purpose signals immediately whenever one's feelings towards the patient tend to overstep the right limits in either a positive or a negative sense, only then can the doctor 'let himself go' during the treatment as psychoanalysis requires of him.

Analytic therapy, therefore, makes claims on the doctor that seem directly self-contradictory. On the one hand, it requires of him the free play of association and phantasy, the full indulgence of *his own unconscious*; we know from Freud that only in this way is it possible to grasp intuitively the expressions of the *patient's unconscious* that are concealed in the manifest material of the manner of speech and behaviour. On the other hand, the doctor must subject the material submitted by himself and the patient to a logical scrutiny, and in his dealings and communications may only let himself be guided exclusively by the result of this mental effort. In time one learns to interrupt the letting oneself go on certain signals from the preconscious, and to put the critical attitude in its place. This constant oscillation between the free play of phantasy and critical scrutiny presupposes a freedom and uninhibited motility of psychic excitation on the doctor's part, however, that can hardly be demanded in any other sphere.

CHAPTER 6

FRANZ ALEXANDER

A METAPSYCHOLOGICAL

DESCRIPTION OF THE

PROCESS OF CURE

INTRODUCTORY NOTE

*From a historical as well as clinical perspective this article by Franz Alexan-
der is of special significance. In it, for the first time, the idea was advanced
that psychoanalytic technique should center its energy on the amelioration of
the superego. For nearly ten years this theory dominated psychoanalytic technique. Written barely two years after the formulation of this concept (Freud
1923b), it represented one of the earliest responses of psychoanalytic technique to the structural theory. The therapeutic plan proved naïve. Alexander
overestimated the ease with which an analysand would be willing to exchange
his own punitive superego for the more benign one offered to him by the
psychoanalyst. Although modified and made more sophisticated in later developments, Alexander's influence can be traced in many papers, including
Ferenczi 1928a (this volume, Chapter 17), Nunberg 1928 (this volume, Chapter 15), Strachey 1934 (this volume, Chapter 24), and Riviere 1936 (this*

Originally entitled "Metapsychologische Darstellung des Heilungsverganges," *Internationale Zeitschrift für Psychoanalyse* 11 (1925):157–178. English translation initially
published in *International Journal of Psycho-Analysis* 6 (1925):13–34. Reprinted in *The
Scope of Psycho-Analysis: 1921–1961, Selected Papers of Franz Alexander* (New York:
Basic Books, 1961), pp. 205–224. Copyright © 1961 by Basic Books, Inc., Publishers. Reprinted by permission. The first part of the article is devoted to model building in the
tradition of psychoanalytic metapsychology. Since this part has no relevance to technique,
it has been omitted.

volume, Chapter 28). When writing this paper, Alexander was under the influence of Rank's ideas of the birth trauma. The reader can see in this article how Rank's views were woven into interpretations given to the analysands at that time.

THE NEUROTIC ACTIVITY of the super-ego is two-fold: it disturbs and inhibits ego-syntonic behaviour, which is *a priori* in conformity with the requirements of reality, by equating this, as the result of faulty reality-testing, with actions which it has learned to criticize in the past and by dealing with it in the way it dealt with them. At the same time, by means of self-punishment, it permits autoplastic, symbolic gratification of precisely those condemned wishes. In the form of impotence, for example, all exogamous wishes are equated with incest-wishes and as such are interfered with. The super-ego behaves, in short, like a dull-witted frontier guard who arrests everyone wearing spectacles, because he has been told that one particular person is wearing spectacles. It behaves like a reflex which can only produce one innervation. The corneal reflex is almost always an expedient reaction which protects the eyes from foreign bodies, yet on occasion it can prove a hindrance, as during medical examination by an eye-specialist. It would be simpler if this reflex action could be avoided by conscious effort, instead of having to be overcome by the use of eyelid retractors. In this case some communication between consciousness, which tests reality, and the 'spinal mind' is desirable: in other instances the reflex defence is more prompt and more certain. In a similar way the strict categorical imperative of a super-ego which is functioning well is frequently adapted to the requirements of social life: nevertheless there are occasions when, owing to new situations and alterations in reality, a more direct relation between the reality-testing faculty and the instinctual world is necessary, between the ego and the *id*, excluding the super-ego which is out of touch with reality.

It may at first seem paradoxical that this rigid inhibiting institution, the super-ego, should actually enable instinctual gratifications which have been condemned by the ego to be realized. We know, however, that the super-ego can easily be hoodwinked; once its punishing tendencies are gratified, its eyes remain shut. It is one of the oldest findings of psycho-analysis that a symptom represents a compromise between the need for punishment and the crime itself.* It is in principle a matter of indifference whether these two tendencies are gratified in one phase as in hysteria, or in conjunction as in the obsessional

---

* [In a book published two years later, *The Psychoanalysis of the Total Personality* (1927), Alexander described the corruptibility of the superego, i.e., the superego punishes first and then permits once more some prohibited gratification in an unending cycle. This is what he means by "autoplastic symbolic gratification." Freud himself has held similar

neurosis, or in two stages as in the manic-depressive neuroses. It is striking to observe how meticulously the conscience of the obsessional neurotic records, like a careful shopkeeper, all debts and claims, all punishments and aggressions; with what sensitiveness it demands new punishments when the limits of the wrong-doing that is covered by punishment are overstepped. Similarly, in the melancholic phase of manic-depressive neurosis, conscience gives expression to acts of glaring tyranny and injustice only to be thrown over without any guilt-feeling during the maniacal phase. We might compare it with a struggle between two utterly antagonistic political parties, where one provokes the other to excesses in order to compromise the latter and encompass its destruction with some show of justification.

Herein lies the two-fold rôle of the super-ego: knowing nothing of reality, it frequently inhibits activities that are actually ego-syntonic and, by over-severity towards the inner world, it permits condemned instinctual gratification along the autoplastic route of symptom-formation. The results of its activity constitute the expression of the Breuer-Freud principle. As an automatic organ, as the tonic deposit of by-gone adaptations to reality, it obviates fresh testing of reality, and when it becomes neurotically diseased these by-gone attempts prove inefficient protection against the regressive tendencies of the *id*.

The super-ego, therefore, is an anachronism in the mind. It has lagged behind the rapid development of civilized conditions, in the sense that its automatic, inflexible mode of function causes the mental system continually to come into conflict with the outer world. This is the teleological basis for the development of a new science, that of psycho-analysis, which, be it said, does not attempt to modify the environment but, instead, the mental system itself, in order to render it more capable of fresh adaptations to its own instincts. This task is carried out by limiting the sphere of activity of the automatically-functioning super-ego, and transferring its rôle to the conscious ego. This is no light task; it implies the conscious creation of a new function. The ego of those living under conditions of Western civilization has been instituted solely for the purpose of testing reality. It is an appreciable increase of the burdens of consciousness to take over the investigation and regulation of instinctual activities, to learn the laws and speech of the *id* in addition to the laws of reality. Quantities of energy which are tonically 'bound' in the automatic function of the super-ego must once more be converted into mobile energy, a part that is now body must again become mind. The resistance against this reversal is well

---

ideas in the 1916 paper, "Criminals Out of a Sense of Guilt" (1916, pp. 332–336). See editors' introduction, p. 33. It was also the central idea in Freud's study of Dostoevsky (1928). However, to Freud, these were special cases. For Alexander, they became the norm. To Freud, the ego was the repressing agency, although the ego often represses on behalf of the superego. Alexander conceptualized the superego as the repressing agency.]

known to us from the analytic resistances during treatment and the general resistance against the science of psycho-analysis.*

Here we have the solution of the problem set in this paper. The curative process consists in overcoming resistances to the ego's taking over of the function of the super-ego. Neurotic conflict, that state of tension arising from repudiation of the symptom, can be solved in two ways only: either the ego's rejection of the symptom must cease, in which case it must abandon reality-testing, together with those forms of instinct-mastery which are already adapted to reality, and take part in homogenizing all the mechanisms of instinct-mastery in the direction of disease; or it must put into force the point of view adapted to reality. This homogenization of the mental system in the direction of disease is familiar to us in the psychoses, where the ego abandons reality-testing and re-models reality in an archaic sense, in accordance with the stage of instinctual gratification preferred.† Psycho-analytic treatment drives in the opposite direction: it seeks to effect a homogeneous system by bringing the whole system nearer to the conscious level, by opening communication between the ego and the *id*, which had been previously barred by the super-ego. The ego is now called upon to settle the claims made by instinct, to *accept* or *reject* them in accordance with the results of reality-testing. As Freud expressed it, the aim of treatment is to substitute judgement for repression. The repressive activity of the super-ego only bars the road to motor discharge of any instinctual demand: it does not imply the abandonment of that demand. On the contrary, it allows a secret gratification. For the ego there are two possibilities only: *accept and carry out* or *reject and abandon*. The task during treatment is to eliminate gradually the repressing institution, the super-ego: from the two component-systems, the ego and the super-ego, a homogeneous system must be constructed—and this must have a two-fold perceptual apparatus, one at the outer surface directed towards reality, and one at the inner boundary directed towards the *id*. Only in this way can a mastery of instinct be achieved which is free from conflict and directed towards a single end.

The transfer to the ego of the rôle of super-ego takes place in two phases during treatment. Making use of the transference, the analyst first of all takes over the part of super-ego, but only in order to shift it back on to the patient again when the process of interpretation and working through has been carried out; this time, however, the patient's conscious ego takes it over. The achievement of analysis is a topographical one involving dynamic expenditure; it displaces the function of testing and regulating instinct to a topographically

---

* [In keeping with the psychoanalytic writings at that time, Alexander moves on from the level of clinical theory when he describes the superego, to the level of meta-psychology when he talks about bound energy and back to clinical theory when he talks about resistances. See editors' introduction, pp. 7–8.]

† [This interpretation of psychosis was formulated by Freud in the essay of 1924a.]

different part of the mental apparatus, viz., the conscious ego. To do this, it must overcome the inertia-principle, i.e. the objection to exchange an automatic function for a conscious activity.* The rôle of the analyst therefore consists in at first taking over the supervision of instinctual life, in order to hand back this control gradually to the conscious ego of the patient. By means of the transference he gains the patient's confidence and produces the original childhood situation during which the super-ego was formed. So long as the whole mental system of the patient is freed from the supervision of instinctual life, so long as the analyst is responsible for the entire instinctual life, the process goes on without interruption. Once the rôle of super-ego has, with the help of this projection mechanism, been taken over in entirety, so that the previous intra-psychical relation between *id* and super-ego has been converted into a relationship between the analyst and the *id*, the more difficult dynamic task begins, namely, to shift back on to the patient once more this rôle of supervision. This returning of the rôle of super-ego takes place for the most part during the period of becoming detached from the analyst. In terms of this schema the psychological processes involved in treatment are very easily described, but it falls to me yet to go more fully into the universal applicability of this description.

The nature of transference from this point of view is that the intrapsychical relations between *id* and super-ego are transferred from super-ego to analyst.† To understand this, we must think of the origin of the super-ego and compare it with the phenomena of transference. Put briefly, the super-ego is an organ of adaptation which has arisen through a process of introjection of such persons (or, more correctly, of the relationships to such persons) as in the first instance enforced adaptation. By this process a formation is set up in the mental system which represents the first requisitions of reality; this consists of introjected educative parental regulations. The super-ego is made up to an important degree of parental commands and prohibitions; hence it is mainly an acoustic formation, as the auditory hallucinations of melancholics show. The commands and prohibitions were conveyed through the auditory apparatus. As Freud has shown us, the relations between *id* and super-ego are nothing more or less than a permanent crystallization of the by-gone relations between the child and its parents. This can be best studied in the case of personalities which are neurotically split, the super-ego functioning as an ut-

---

* [The inertia principle, also called the nirvana principle, was first formulated by Freud in 1920. In the essay of 1924b, Freud said: "It will be remembered that we have taken the view that the principle which governs all mental processes is a special case of Fechner's 'tendency towards stability,' and have accordingly attributed to the mental apparatus the purpose of reducing to nothing, or at least of keeping as low as possible the sums of excitation which flow in upon it. Barbara Low has suggested the name of 'Nirvana Principle' for this supposed tendency, and we have accepted the term."]

† [On the basis of theory alone, Alexander assumed that since the superego was once introjected, it can be reprojected, in toto, on the psychoanalyst. This is not usually possible in neuroses.]

terly foreign body. The entire complicated symptomatic structure of the obsessional neurosis is a play enacted by an obstinate, untrained child and its parents; and just as all French comedies deal with monotonous regularity with the theme of adultery, so in every neurosis we come across the identical theme in varying guise. Even the methods of the '*id*-child' remain unchanging— always to provoke the parents, the super-ego, to unjust and over-severe punishment, in order to do what is forbidden without any feeling of guilt, precisely as in the triangle play the conduct of one partner is represented in a way which seems to justify the adultery of the other.

In the course of transference this intrapsychical drama is converted into a real one between the *id* and the analyst. It is not necessary to enter into further details: the patient seizes with extraordinary alacrity the opportunity of realizing in relation to the analyst his former relations with his parents, which he has been forced to introject only because he was unable to realize them in reality. In this way he is able to cancel that piece of adaptation to reality which has been forced upon him and is represented by the super-ego. He soon observes, however, from the attitude of the analyst—who works counter to the pleasure-principle—that whilst these tendencies can be understood they are not gratified, and this in no case from personal motives. The new educative process then begins. The demands of reality are, however, not communicated by means of orders and prohibitions, as previously happened under the sway of the super-ego, but by a 'super-personal' method, by logical insight, by accurate testing of reality. In this way the re-living of his past becomes abandoned by the patient himself, and the original instinctual demands, which in consequence of a personal judgement can no longer be experienced in the transference-situation, appear in the mind as memories. Tonic discharge is blocked, automatic repetition is prevented; the former demands of instinct become active once more; they become problems of the immediate present, and as such form part of the content of consciousness. Instead of automatic repetition, memory appears, indeed, more, the demands of instinct are more active and necessitate new means of discharge. From now on this discharge must be not only ego-syntonic but in accordance with the demands of reality, since it can only be effected in agreement with the organ of reality-testing.

So events run in theory, but not in practice. Every analyst has, time after time, observed that when a transference-situation has been resolved and brought into a genetic relation with the original childhood situation, in no instance does an immediate orientation in the direction of normal libido-control occur, but instead a regression to still earlier stages of instinctual life. The libido eludes analytic endeavours by a backward movement, and retires to positions it had previously abandoned.* Each fresh interpretation brings about

* [In keeping with the language of psychoanalytic metapsychology, regression is conceptualized as a backward movement of the libido. The metaphor is based on the hydraulic model.]

a still deeper regression, so much so that the beginner often imagines he has driven a hysteric into a state of schizophrenia. I must confess that the desire to be clear in my own mind as to the nature of these processes was stimulated to a large extent by certain uncanny moments during analytic work, when to my dismay symptoms of conversion-hysteria which had already been carried over into the transference gave place to paranoidal and hallucinatory symptoms.* Further progress in the analysis, however, showed that each new symptom is carried over into a new transference-situation, so that every deep analysis runs through a whole gamut of artificial neuroses, ending regularly, as Rank has shown us, in a reproduction of the prenatal state (1924). I have been able to trace the same gradation of regression in the contemplative states of Buddhism.

Analysts cannot logically dispense with recognizing and appraising this ultimate mode of regression also, in order to be able to drive the libido from this most inaccessible hiding-place forwards in the direction of genital exogamy. We owe much to Rank for having called attention to the general significance of this deepest form of regression; above all that he has shown this regression during treatment to be an affective repetition of actual experience and not perhaps a pre-conscious phantasy. I cannot emphasize too strongly that those who oppose themselves to this view of it are making the same mistake that Jung made many years ago. One would have just as much right to regard all oral or anal-erotic regressions as the products of regressive phantasy-creation.

On the other hand, it is clear from the foregoing considerations that this regressive movement ensuing upon analysis of the transference-situations—which arise spontaneously and are characteristic in each individual case—is to be regarded as resistance.† Observation during treatment of this continually backward-flowing regression provides us with an extraordinary picture, one which lays bare the entire complicated process of the construction of the super-ego. The picture is made up of a consecutive series of transference-situations, in which the analyst plays ever-changing rôles taken over from the super-ego. The consecutive series of regressive transference-rôles is a picture of the layers of the super-ego seen upside-down. It is a gathering together of imprints from the various stages of development. The deepest layer represents the biological relation between mother and child, and merges gradually more and more into social relations with the father. The mother represents the first demands in instinct-development: through the act of birth she first demands abandonment of the state of passive nutrition by the

* [In the light of current knowledge, the patients described by Alexander would now be diagnosed as borderline or psychotics, with a neurotic facade hiding this psychotic core.]

† I wish to lay the greatest stress on this point in contradistinction to Rank's point of view; in his presentation, the resistance-character of intra-uterine regressions is by no means clear.

blood-stream and requires the substitution of nutrition through the alimentary canal and active employment of mouth and lungs. Later she calls for the abandonment of breast-feeding and is usually the first to disturb the child's autocratic command over its excretions. Gradually the father and the whole father-series take over the larger part in the education of instinct and represent the demands of the community. The father, however, takes on the earlier mother-rôle not only in regard to frustration experiences but in a positive way: just as the mother was the source of bodily nourishment, so the father provides mental pabulum. The passive-homosexual relation to the father found in every analysis is the repetition and substitute for the passive suckling situation; the paternal penis is the substitute for the breast, as Freud showed already in his analysis of Leonardo da Vinci. We find the most strongly repressed ideas of oral incorporation of the penis and of the father as a whole, in a form with which we have been familiarized by Abraham's (1924a) accurate descriptions. Roheim (1923) has shown us in his admirable study of primeval history how the sons tried to transfer the mother-rôle to the father, by devouring him and defæcating on his grave, on the parallel of suckling at the maternal breast(1923). The same history in reverse order is faithfully reproduced during treatment. The father-rôle, which at the beginning is invariably transferred to the analyst, is more and more displaced by the mother-transference. On this point I can fully confirm Rank's observation. Attempts in a progressive direction disturb the picture often enough, nevertheless the regressive tendency predominates. Although not really free from conflict about the father the patient regresses to times when the latter was not a source of disturbance and when the only battle he had to fight was a biological one with the mother. The cause of the regression is now clear: it is the expression of the Breuer-Freud principle, the automatizing tendency to solve new problems according to the old plan. The mind attempts to solve the father-conflict on the model of the suckling-situation: the father is to be destroyed by way of oral incorporation, in this way providing new strength for the struggle for existence, just as the mother's milk provided strength for physical development.

The patient is under the influence of the same tendency to automatize when he attempts to meet the task of detaching himself from the analyst by a phantasy-reproduction of the birth-trauma. For the most part he has already solved the problem of birth with its transposition to extra-uterine life: the most conclusive evidence for this is that he is alive. Before the end of treatment, however, he is faced with the entirely unsolved problem of doing without analytic aid. It is small wonder that he feels this to be similar to the severance from the mother's body. On that occasion also he had to learn the use of organs entirely *ab initio*, when taking over the nutritive rôle of the mother. Now at the termination of treatment his consciousness, which hitherto has been adapted only to testing reality, has to face new tasks. Having learned during treatment the language of instinct, it must take over responsibility for the regulation of instinctual activities, a regulation which has previously been exercised by the super-ego operating automatically. During treatment the analyst has thought and interpreted instead of the patient: indeed, by reconstructing the past he has done some remembering in his stead. From now on all this must be the patient's own concern. In bidding good-bye to his super-ego he must finally take leave of his parents, whom by introjection he had captured and preserved in his super-ego. He has indeed been ignominiously hood-winked in analysis. The analyst seduced him into giving up the introjected parents, by himself taking over the rôle of the super-ego, and now he wants to saddle the patient with the burden. The latter protests and attempts in return to score off the analyst by sending him in the long-since-closed account for his birth, and this often by way of somatic symptoms. He feels, as did one of my patients, a circular constriction round his forehead, the pressure of the pelvic canal by which his head was so shamefully disfigured at birth: he is breathless and feels a heavy pressure round the chest. Only when all this has been proved mere resistance against detaching himself from the analyst, against independence, does he consciously attempt to do without further analytic help. The patient is not overcoming his birth-trauma by means of these birth-scenes; on the contrary, he is countering detachment from the analyst with them; he is substituting action and affective reproduction of the birth which is done with, for the separation from the physician with which he is faced. He reproduces the past instead of performing the task in front of him. Even after treatment he will not have overcome the birth-trauma.* Rank himself has shown us in the most convincing way that man never gives up the lost happiness of pre-natal life and that he seeks to re-establish this former state, not only in all his cultural strivings, but also in the act of procreation. These forms of representation are, however, ego-syntonic; in analysis the patient must give up only such attempts at repetition as are autoplastic and dissociated from reality; he must give up symptoms, relations to the super-ego in which he has perpetuated his whole past and which finally he aimed at rescuing for good in the analytical transference-situations. In the same way as he repeats in analysis the severance from the mother's body, he repeats all other difficult adaptations of his instinctual life which have been forced upon him during development all with one end in view, to avoid a new adaptation to actual reality.

We are at one with Ferenczi and Rank (1924) in thinking that every subsequent stage of libido-organization is only a substitute for the abandoned intra-uterine state: we have already accepted this idea in the analysis of the

---

* [The concepts of symbiosis and separation-individuation developed by Mahler (1968) were not available to Alexander. Today, we know that at the end of an analysis, the patient recapitulates the difficulties he had during the separation-individuation phase, but not the trauma of birth.]

castration complex (1921). Nevertheless each successfully established stage of organization represents a fixation-point: the intra-uterine state is the first, but, dynamically speaking, by no means always the most significant of the long series of fixation-points. The period at which an individual utters the negation which sets up a neurosis varies widely; yet it is precisely this point which determines the form of his subsequent neurosis. When in the course of treatment his special fixation is analysed out, subsequent regression represents resistance against the consequences of this analytic solution, against the demands of the ego, against the activity directed outwards.

We have here corroborated in principle Rank's significant conception but have had to amplify it by a needful quantitative (economic) valuation of intra-uterine fixation. For analytic treatment the task remains to convert the tonic energy 'bound' in automatic repetitions into the labile energy of conscious mental activity, in order that the struggle with reality may be taken up. The energies 'bound' in the acquired automatisms of the super-ego are freed through recollection. To compare memory-material with the testing of reality is the highest achievement of the mental apparatus. Only the ego can remember: the super-ego can only repeat. The dissolution of the super-ego is and will continue to be the task of all future psycho-analytic therapy.*

* I am aware that in the foregoing presentation the concept of the 'super-ego' has been somewhat schematic and therefore more narrowly defined than in Freud's descriptions. I limit the 'super-ego' to the unconscious alone, hence it becomes identical with the unconscious sense of guilt, with the dream-censorship. The transition to conscious demands, to a conscious ego-ideal, is nevertheless in reality a fluid one. We might regard these parts of the 'super-ego' which project into consciousness as the most recent and final imprints in the structure of it, as constituents of the 'super-ego' *in statu nascendi*. They are not so fixed as the categorical, unconscious constituents of the conscience, and are more accessible to conscious judgement. This schematic presentation has been adopted in order to throw into sharper relief the dynamic principles concerned. I have compared extremes, the completely mobile apparatus of perception with the extremely rigid unconscious part of the 'super-ego'. Freud's conception and description, which takes into account the complete 'super-ego' system, is nevertheless the more correct psychologically.

[A philosophical section of this paper has been omitted.]

Part Four

---

*The Controversy
Around Ferenczi's
Active Technique*

CHAPTER 7

SANDOR FERENCZI

TECHNICAL DIFFICULTIES IN THE ANALYSIS OF A CASE OF HYSTERIA

(Including Observations on Larval Forms of Onanism and "Onanistic Equivalents")

INTRODUCTORY NOTE

As indicated in the editors' introduction, by 1919 Freud's optimism about the gradual disappearance of the neuroses was over. The publication of the case of the Wolf Man (Freud 1918) a year earlier than this paper demonstrated that at times the analyst may be forced to depart from his role as the interpreter and resort to other measures, including the threat of termination. Ferenczi had no quarrel with Freud's understanding of the nature of the

Originally entitled "Technische Schwierigkeiten einer Hysterieanalyse," *Internationale Zeitschrift für ärztliche Psychoanalyse* 5 (1919):34–40. Reprinted in *Selected Papers of Sandor Ferenczi, M.D.*, vol. 2, *Further Contributions to the Theory and Technique of Psychoanalysis*, comp. John Rickman, trans. Jane Isabel Suttie et al. (New York: Basic Books, 1952), pp. 189–198. Reprinted by permission of Basic Books, Inc., Publishers.

neurosis, but, as indicated in the other publication of that year (this volume, Chapter 5), he was disillusioned in the capacity of many analysands to free associate productively. Ferenczi may, therefore, have had reason to believe that his own innovations with recalcitrant analysands would meet with approval, or at least that they were not fundamentally different from those advocated by Freud.

We have reprinted in this section two of Ferenczi's essays and a rebuttal by Glover. The reader who wishes to go into greater detail can also read Ferenczi (1919b, 1920, and 1925).

A PATIENT who was endeavouring with great intelligence and much zeal to carry out the directions for psychoanalytic treatment, and who left nothing to be desired in the way of theoretical insight, nevertheless, after a certain degree of improvement, probably due to the first transference, made no progress for a long time.

As the proceedings made absolutely no headway, I decided on extreme measures and fixed a date up to which I would continue to treat her, in the expectation that by this means I should provide her with an adequate incentive to effort. Even this, however, proved only of temporary assistance;* she soon relapsed into her former inactivity, which she concealed behind her transference love. The hours went by in passionate declarations of love and entreaties on her side, and in fruitless endeavours on mine to get her to understand the transference nature of her feelings, and to trace her affects to their real but unconscious object. On the completion of the period set I discharged her uncured. She herself was quite content with her improvement.

Many months later she returned in a quite desolate condition; her earlier troubles were returning with all the old violence. I yielded to her request and again undertook the treatment. After a short time, as soon as the degree of improvement previously attained was once more established, she began the old game again. This time extraneous circumstances interrupted the treatment, which again remained incomplete.

A fresh exacerbation and the disappearance of the extraneous difficulties brought her to me for the third time. This time, too, we made no progress for a long time.

In the course of her inexhaustibly repeated love phantasies, which were always concerned with the doctor, she often made the remark, as though by the way, that this gave her feelings 'down there'. That is, she had erotic genital

* [The technical device of fixing an artificial termination date was tried by Freud in the case of the Wolf Man. At that time, Freud believed that the threat of a termination date would force unconscious material to emerge and remove the danger of a stalemate in the analysis due to the secondary gains of pleasure from the gratification of passive needs. The subsequent history of the Wolf Man did not justify this optimism. (See Eissler 1953 for further discussion.)]

sensations. But only after all this time did an accidental glance at the manner in which she lay on the sofa convince me that she kept her legs crossed during the whole hour. This led us—not for the first time—to the subject of onanism, an act performed by girls and women for preference by pressing the thighs together. As on former occasions, she denied most emphatically ever having carried out such practices.

I must confess—and this is characteristic of the slowness with which an incipient new point of view irrupts into consciousness—that even then it was a long time before I hit on the idea of forbidding the patient to adopt this position. I explained to her that in so doing she was carrying out a larval form of onanism* that discharged unnoticed the unconscious impulses and allowed only useless fragments to reach the material of her ideas.

I can describe the effect of this measure as nothing less than staggering. The patient, to whom the customary genital discharge was forbidden, was tormented during the interviews by an almost insupportable bodily and psychic restlessness; she could no longer lie at peace, but had constantly to change her position. Her phantasies resembled the deliria of fever, in which there cropped up long forgotten memory fragments that gradually grouped themselves round certain events in her childhood and permitted the discovery of most important traumatic causes for her illness.

The consequent impetus towards improvement certainly brought about distinct progress, but the patient—although she conscientiously carried out the above rule—seemed to reconcile herself to this form of abstinence and settled down to this stage of knowledge. In other words, she again ceased to exert herself, and took refuge in the sanctuary of the transference love.

Having had my wits sharpened by these previous experiences, however, I could now rout out the hiding-places in which she concealed her auto-erotic satisfaction. It appeared that she did, indeed, carry out the prescribed behaviour *during the hours of analysis*, but constantly transgressed it during the rest of the day. We learned that she knew how to eroticise most of her housewifely and maternal activities by pressing her legs together inconspicuously and unconsciously to herself; naturally she lost herself at the same time in unconscious phantasies which thus protected her from becoming aware of her activities. When the restraint was extended to include the whole day there was another but not even yet definitive improvement.

The Latin phrase, *naturam expellas furca, tamen ista recurrat,*† seemed to be justified in this case.

I noticed frequently in the course of the analysis certain 'symptomatic acts', such as playful squeezing and handling of the most varied parts of the body. After the complete interdiction without any exception of the larval onanism, these symptomatic acts became *masturbation equivalents*. By this I

* [The term "larval onanism" was coined by Ferenczi.]
† ["(Though) you drive nature away with a spear, nevertheless she returns."]

understand apparently harmless stimulations of indifferent parts of the body which are, however, qualitatively and quantitatively substitutes for genital erogenicity. In this case the shutting off of the libido from any other path of discharge was so complete that from time to time it was increased to an actual orgasm at other indifferent parts of the body that are not by nature prominent erotogenic zones.

It was only the impression caused by this experience that enabled the patient to credit my assertion that she was wasting her whole sexuality in those little 'naughtinesses', and to consent for the sake of the treatment to forgo also these gratifications that she had practised since childhood. The annoyance she thus caused herself was great, but well worth while. All abnormal channels of discharge being closed to it, her sexuality found of itself, without any assistance, the way back to its normally indicated genital zone, from which it had been repressed at a certain time in development, as though exiled from its home to foreign countries.

This repatriation was hindered a little by the opposition of a temporary return of an obsessional neurosis experienced in childhood; this, however, was easily translated and easily understood by the patient.

The last stage was the appearance at unseasonable times of a *need to urinate*; the gratification of this was equally interdicted. One day she surprised me with the information that she had experienced so violent a stimulation of the genitals that she could not forbear rubbing the vaginal mucous membrane forcibly in order to get some relief. Though she could not immediately accept my explanation that this confirmed my assertion that she had passed through an infantile period of active masturbation, nevertheless she soon adduced ideas and dreams that did convince her. This masturbatory relapse, however, did not last long. Parallel with the reconstruction of her infantile defence reaction, she achieved, after all these worries, the capacity of obtaining satisfaction in normal sexual intercourse, which—although her husband was unusually potent and had begotten many children by her—had hitherto been denied her. At the same time many of the as yet unsolved hysterical symptoms found their explanation in the now manifest genital phantasies and memories.

From this extremely complicated analysis I have endeavoured to select only what was of technical interest, and to describe the manner in which I came upon the definition of a new rule in analysis. This runs as follows: during treatment one must also think of the possibility of larval onanism and onanistic equivalents, and, where indications of these are observed, abolish them. These apparently harmless activities can easily become hiding-places for the libido which has been driven away from its unconscious excitations by the analysis, and in extreme cases may replace an individual's whole sexual activity. Should the patient notice that these possibilities of satisfaction escape the analyst, he attaches all his pathogenic phantasies to them, short-circuits them constantly by motor discharge, and thus saves himself the irksome and unpleasant task of bringing them to consciousness.

This technical rule has since stood me in good stead in several cases. Long-standing resistances against the continuation of the treatment have been brought to an end by its means.

Observant readers of psycho-analytic literature will perhaps detect a contradiction here between this technical measure and the opinion of many psycho-analysts about onanism.*

The patients, too, with whom I had to employ this technique did not omit to object—'you stated', said they, 'that onanism is not dangerous, and now you forbid it me'. The solution of this contradiction is not difficult. We do not need to alter anything in our opinion of the relative harmlessness, for instance, of the onanism of necessity, and yet maintain the demand for this kind of abstinence. We are here not concerned with a generalization about self-gratification, but with a provisory measure for the purposes and the duration of the psycho-analytic treatment. Besides, a treatment that has been successfully concluded enables many patients to give up this infantile or juvenile form of satisfaction.

Not all, however. Cases do even occur in which the patients during treatment, yield—for the first time in their lives as they say—to the need for masturbatory gratification and date the beginning of the favourable change in their libidinal attitude from this 'courageous deed.'†

This last, however, can only hold for manifest onanism with a conscious erotic phantasy, not for the numerous forms of 'larval' onanism and its equivalents. These are to be regarded as pathological from the first, and in any case require to be cleared up by analysis. This is only to be obtained at the price of at least a temporary resignation of the activity itself, by which means its excitement is directed along purely psychic paths and finally into consciousness. Once the patient has learnt to tolerate the consciousness of his onanist phantasies he may be allowed to deal with himself again. In most instances he will only make use of it in case of necessity.

I take this opportunity of saying something further about larval and vicarious onanist activities. There are many otherwise not neurotic people, especially many neurasthenics,‡ who are, so to say, almost life-long onanists unconsciously. If they are men they keep their hands in their trouser pockets

---

* "Über Onanie." A discussion in the Vienna Psychoanalytic Society, 1912.

[A translation of the paper appears in *Contributions to Psycho-Analysis*, chap. 6; see also Annie Reich 1951].

† [At the time this paper was written, it was not yet understood that during a certain phase in the development of the child, the ability to masturbate indicates a liberation from total libidinal dependence for gratification on the parent, a kind of "Declaration of Independence."]

‡ [See editors' footnote to Wilhelm Reich, this volume p. 230.]

all day, and it is noticeable, from the movements of their hands and fingers, that in doing this they pull, squeeze, or rub the penis. They have 'nothing bad' in their thoughts at the time; on the contrary, they are perhaps sunk in profound mathematical, philosophical, or business speculations.

In my opinion, however, this 'profundity' must not be taken too seriously. These problems certainly arrest their whole attention, but the real depths of the soul life—the unconscious ones—are meanwhile occupied with pure erotic phantasies and procure satisfaction by a short (as it were somnambulistic) path.

Others substitute for the burrowing in the trouser pockets a clonic quiver of the calf muscles, often very annoying to their companions, while women, whose manner of dress as well as sense of decorum forbids such noticeable movements, press their legs together or cross them. They like to create such unconscious 'secondary gain of pleasure', particularly while occupied with absorbing needlework.

Apart from the psychic consequences, this unconscious onanism cannot be held to be quite innocuous. Although or, indeed, because it never issues in full orgasm, but always only in frustrated excitement, it can play a part in the development of an anxiety neurosis. I know cases, however, where this continual excitement is accompanied by very frequent even if minimal orgasms (in men by prostatorrhœa also), and ultimately makes these people neurasthenic and diminishes their potency. Normal potency is only possessed by those who preserve and store up the libidinal impulses for a considerable latent period, and who can discharge them powerfully along genital channels on the occurrence of suitable sexual aims and objects. Constant squandering of small quantities of libido destroys this capacity. (This does not hold in the same degree for conscious, intentional, periodic masturbation.)

A second consideration, which seems to contradict the views previously stated, is the conception of *symptomatic acts*. We learnt from Freud that these manifestations of the psychopathology of everyday life are of use in the treatment as indications of phantasies which are repressed and therefore significant, but otherwise entirely innocuous. Now we see that they too can be intensively charged with libido displaced from other situations, and become onanism equivalents and no longer harmless. Transition stages are here discoverable between symptomatic acts and certain forms of *tic convulsif*, of which as yet we possess no psycho-analytic explanation [see Ferenczi 1921b; see also Mahler 1944]. My expectation is that on analysis many of these tics will blossom forth as stereotyped onanistic equivalents. The remarkable association of *tics* with *coprolalia** (for instance on the suppression of motor manifestations) would then be nothing else but the irruption of the erotic

* [Coprolalia = involuntary use of obscene words.]

phantasies (mostly anal-sadistic) symbolized by the tics into the preconscious, with a spasmodic excitation of the corresponding word memory traces. Coprolalia would then owe its origin to a mechanism similar to the technique employed above which allows certain impulses, hitherto led off in onanistic equivalents, to break through into consciousness.

But after this digression into hygiene and nosology, let us return to the much more interesting technical and psychological considerations raised by the case recounted in the introduction.

I was compelled in this case to give up the passive part that the psycho-analyst is accustomed to play in the treatment, which is confined to the hearing and interpretation of the patient's ideas, and had by active interference in the patient's psychic activities to help over dead points in the work of the analysis.

We owe the prototype of this 'active technique' to Freud himself. In the analysis of anxiety hysterias on the occurrence of a similar stagnation—he had recourse to the method of directing the patients to seek just those critical situations which usually caused them an attack of anxiety; not with the idea of 'accustoming' them to these situations, but in order to free the wrongly anchored affects from their connections. We expect from this measure that the unsatisfied valencies of these free floating affects will above all attract to themselves their qualitatively adequate and historically correlated ideas. Here too we find, as in our case, the ligature of customary, unconscious paths of discharge of excitation and the enforcement of the preconscious cathexis and the conscious hyper-cathexis of the repressed material.

Since our knowledge of transference and of 'active technique' we are able to say that besides observation and logical deduction (interpretation) psycho-analysis has also at command the method of experiment. Just as in experiments on animals the blood pressure in distant parts can be raised by the ligature of large arterial vessels, so in suitable cases we can and must shut off psychic excitement from unconscious paths of discharge, in order by this 'rise of pressure' of energy to overcome the resistance of the censorship and of the 'resting excitation' by higher psychic systems.*

In psycho-analysis as distinguished from suggestion no influence is exercised over the new direction of the current, and we gladly let ourselves be surprised by the unexpected turns taken by the analysis.

This kind of 'experimental psychology' is adapted as is nothing else to convince us of the correctness of Freud's psycho-analytic doctrine of the neuroses and of the validity of the psychology constructed upon it and upon the interpretation of dreams. At the same time we learn the peculiar value of

---

* [Ferenczi's belief in pressure, so basic to his active technique, resulted from a literal application of the hydraulic model. It was one of the unfortunate results of converting a metaphor into a model. See Max Block 1962.]

Freud's assumption of the *psychic 'instances'** and become accustomed to deal with *psychic quantities*† just as with other energy masses.

An example like the one described here shows anew that in hysteria it is not common 'psychic energies' but libidinal or, more exactly, genital impulses that are at work, and that the formation of symptoms ceases when one succeeds in re-directing the abnormally employed libido to the genitals.

* [The term "instance," in German "Instanz," was originally a legal term, cf. "court of first instance"; it is used in psycho-analysis, as in law, in the sense of one of a hierarchy of functions or authorities—J. R., trans.]

† [Ferenczi here ignores the fact that psychic energy cannot be measured.]

CHAPTER 8

SANDOR FERENCZI

ON FORCED FANTASIES

Activity in the

Association-Technique

INTRODUCTORY NOTE

*The suggestions that psychoanalysts force patients to fantasize—and even
more boldly suggest to them the fantasies they should have—represents one
of Ferenczi's most daring innovations and one that has aroused opposition. See
Glover, this volume, Chapter 9. The danger of offering in a disguised form to
the patient one's own fantasies either did not occur to him or did not deter him.
Particularly interesting are Ferenczi's observations on the paucity of fantasy
in a certain class of excessively sheltered children.*

IN MY PAPER (1920) delivered at the Hague Congress [Sixth International
Psycho-Analytical Congress] on the 'active' psycho-analytic technique, I put
forward the view that one is sometimes in a position when one must issue
orders and prohibitions to the patient regarding certain actions for the purpose of disturbing the habitual (pathological) pathways of discharge of ex-

Originally entitled "Über forcierte Phantasien: Aktivität in der Assoziationstechnik,"
Internationale Zeitschrift für Psychoanalyse 10 (1924):6–16. Reprinted in *Selected
Papers of Sandor Ferenczi, M.D.*, vol. 2, *Further Contributions to the Theory and Tech-
nique of Psychoanalysis*, comp. John Rickman, trans. Jane Isabel Suttie et al. (New York:
Basic Books, 1952), pp. 68–78. Reprinted by permission of Basic Books, Inc., Publishers.
This paper was read at the May 1923 meeting of the Hungarian Psycho-Analytical Society.

citations out of the psychical, and that the new distribution of psychical
tension resulting from this interference makes possible the activation of material till then lying hidden in the unconscious and allows it to become manifest in the associations. Every now and then I observe in this connection that
this activity can be extended to influencing the material of associations. If, for
example, one observes signs that a patient is 'misusing the freedom of associations' and one calls his attention to this, or if one suddenly interrupts the
flow of words of the analysand and harks back to something brought forward
earlier, from which the patient with his logorrhœa seeks to fly by means of
'talking past the point', we sin apparently against the 'fundamental rule', but
we remain true to another and even more important regulation of psychoanalysis, which is that the chief duty of the analyst is to unmask the resistances of the patient. We may not make any exceptions to this rule in those
cases in which the resistance employs the fundamental rule of associations to
frustrate the objects of the treatment.

   Still more rarely, as I said at the Hague, have I found it necessary to
extend the prohibition even to the patient's phantasy-activities. Occasionally I
tell patients whose symptoms consist in habitual day-dreaming forcibly to
interrupt these phantasies and to exert all their force in seeking out those
psychical impressions which have been avoided through fear (phobically),
and which have switched the patients over on to the tracks of pathological
phantasy. Such influence I thought and still regard as invulnerable to the
reproach that one is mixing the method of free-association with the procedures of suggestion; our intervention here consists only of an *inhibition*, a
shutting off of certain paths of association, while the products of the analysand himself, which take its place, provide ideas without our having awakened
them in him.

   Since then I perceived that it would be an exaggeration and a pedantry to
introduce this limitation into all cases; indeed, I had to acknowledge that we
had never actually followed this injunction literally. When we interpret the
patient's free associations, and that we do countless times in every analytical
hour, we continually deflect his associations and rouse in him expected ideas,
we smooth the way so that the connections between his thoughts so far as
their content is concerned are, therefore, to a high degree active; meanwhile
we impart to him at the same time association-prohibitions. The difference
between this and the ordinary suggestion simply consists in this, that we do
not deem the interpretations we offer to be irrefutable utterances, but regard
their validity to be dependent on whether they can be verified by material
brought forward from memory or by means of repetition of earlier situations.
Under such conditions, as Freud has established long since, 'suggestibility',
that is, the uncritical acceptance of the propositions of the analyst by the
analysand, is in no way strong; on the contrary, as a rule the first reaction to
an interpretation is resistance in the form of a more or less brusque repudia-

tion, and only much later can we put the confirmatory material to use. Another difference between us and the omnipotent suggestionist is that we ourselves retain a grain of scepticism about our own interpretations and must be ever ready to modify them or withdraw them completely, even when the patient has begun to accept our mistaken or our incomplete interpretations.

Once this is grasped, the principal objection to the somewhat energetic application of the prohibition of associations in analysis falls to the ground, but naturally it is only applied in cases in which without it the work either does not proceed at all or only crawls along slowly.

I turn next to a type of person who both in analysis and life is particularly poor in phantasy, if not actually without them, on whom the most impressive experiences leave no apparent trace. Such persons are able to reproduce in memory situations which according to our reckoning must necessarily rouse the intense affects of anxiety (Angst), revenge, erotic excitement, and so forth, which call for discharge through the pathways of deeds, volition, phantasies, or at least external and internal means of expression, but these people show no trace either of feeling such reactions or of expressing them. Supported by the preconceived opinion that in these cases such conduct is due to repression of psychical material and suppression of affect, I now have no hesitation in forcing the patients to recover the adequate reactions, and if they still persist in saying that they have no ideas, I commission them to discover such reactions in phantasy. To the objection that I am then generally met with that such phantasies are quite 'artificial', 'unnatural', quite foreign to their nature, illogical, and the rest (by which the patient puts aside every responsibility), I am accustomed to retort that he does not have to tell the truth (actuality), but all that comes to his mind without regard for objective reality, and that certainly he is not required to acknowledge these phantasies as completely spontaneous performances. With his intellectual resistance disarmed in such a fashion the patient then tries, usually only very warily, to depict the situation in question, halting and breaking off at every feature, which requires a continuous pressure on the part of the analyst. In the course of time, however, the patient becomes more courageous, his 'fabricated' experiences in phantasy more varied, animated, and full of impressions, so that finally he can no longer regard them in a cool objective way—the phantasy 'transfixes' him. I have found several times that such a 'discovered' phantasy returned in an experience of almost hallucinatory distinctness with the most unmistakable signs of anxiety, rage, or erotic excitement according to the context. The analytical value of such 'forced phantasies', as I would like to call them, is unquestionable. Especially do they furnish a proof that the patient is generally speaking capable of such psychical productions of which he thought himself free, so that they give us a grasp of deeper research into unconscious repression.

In special cases, if the patient in spite of the utmost pressure will produce nothing, I do not stop at laying before him wellnigh directly what he probably ought to have felt in the given situation, or thought or phantasied; and when he finally agrees with my suggestion I naturally lay less weight on the main plot furnished by myself than on the added details supplied by the analysand.

This kind of surprise attack, in spite of the intense strength of the experience of the 'forced phantasy' produced in the hour, tends to mobilize everything (till the next hour) to the undoing as far as possible of its *power of conviction,* and the patient must live through the same or a similar phantasy several times till a modicum of insight remains. In other cases, however, scenes that were unexpected by physician and patient alike are produced or reproduced, which leave an indelible impression on the mind of the patient and the analytic work advances with a bound. If, however, we started our conjectures on the wrong track and furnished the patient with ideas and emotions in continuation of the ideas which we had roused, and if these ideas and emotions contradict those forced up by us, we must freely confess our mistake, although it is not to be excluded that the later analytic material will prove our conjecture to have been right.

The phantasies which I have thus been constrained to force up fall for the most part into three groups: these are (1) positive and negative phantasies of the transference, (2) phantasies recollecting infancy, (3) onanistic phantasies.

I wish to bring to your notice examples from the analytic material of the past few weeks.

A man who was by no means poor in phantasy life, but was strongly inhibited in his own expression of feeling on account of preconceived notions (ideals), was somewhat harshly reminded by his analyst, to whom he had transferred much friendliness and tenderness, towards the end of the analysis of the aimlessness of his attitude towards it, and at the same time a time limit was set at which he must break off whether he was cured or not.* Instead of the expected reaction of anger and revenge, which I wanted to provoke as a repetition of deeply repressed infantile mental processes, there followed several hours that were tedious and without tone or activity, but were also completely free from both affect and from phantasies coloured by it. I put it to him that he must hate me for what had happened, and that it would be unnatural that he should not notice anything. But he repeated unblenchingly that he was only grateful to me, that he only felt friendliness towards me, and so on. I pressed him nevertheless to concoct something aggressive against me. Finally, after the customary attempts at defence and repudiation came timorous, then more vigorous phantasies of aggression, the latter accompanied by signs of obvious anxiety (cold sweats). At last he got to beating phantasies of

---

* [Ferenczi made extensive use of the termination threat. See editors' footnote, p. 25.]

hallucinatory vividness, then the phantasy that he gouged out my eyes which promptly switched over into a sexual scene in which I played the rôle of a woman. During the phantasying the patient had a manifest erection. The further course of his analysis maintained the language of these enforced phantasies which had enabled him to experience with the person of the analyst practically all of the situations of the 'complete Oedipus complex', and enabled the analyst to reconstruct the early infantile developmental history of the libido of the patient.

A patient asserts that she does not know the commonest obscene designation for the genitals and genital processes. I have no ground for doubting her sincerity, but must point out to her that she certainly knew these words in her childhood, had then repressed them, and later on account of repression had let them slip by unnoticed at the moment of hearing them. At the same time I bade her mention to me the words or sounds which came to her mind when thinking of the female genital.* About ten words came first, all having the correct initial letter, then a word containing the first, followed by a word containing the second syllable of the desired expression. In similar fragmentary fashion she told me the obscene word for the male member and for sexual intercourse. In this enforced neologising repressed word memories made their appearance in the same way that knowledge kept secret from consciousness emerges in the surprise attack method in the association experiment.

This case reminds me of another in which the patient brought me an experience of being seduced (which in all probability really happened), but with innumerable variations in order at the same time to confuse both me and herself and to obscure reality. I had again and again to constrain her to 'fabricate' such a scene, and thus new details were established with certainty. I had then to correlate the points that were thus established, first with her general behaviour immediately after the occurrence already mentioned (in her ninth year), during which for months together she suffered from the obsessing idea that she would have to marry someone with a different religious belief from her own, and secondly, with her conduct immediately before her marriage, where she paraded a shocking naïveté, and with the events of the bridal night, during which the absence of difficulty in initiation surprised the bridegroom. The very first of the phantasies indicated above led gradually to establishing the fact of the event, which the patient under the stress of circumstantial evidence had to admit. She employed the general uncertainty of the experience as a last weapon of defence (*i.e.* a kind of scepticism) and finally

---

* [Strictly speaking, to urge the patient not to circumvent the slang name for the genitals is not an example of forced fantasies, provided the analysand understands that his refusal to use the slang name shields him from experiencing the concomitant affect.]

resorted to the philosophical question on the evidential value of sensory impressions (metaphysical mania). 'Indeed one cannot say definitely', she imagined, 'whether the stool standing there is really a stool.' I replied that by giving expression to that idea she had raised the certainty of that memory to the level of a direct sensory experience, and with that degree of certainty we could both rest content.

Another patient suffered from an unbearable 'feeling of tension' in the genitals which often lasted for hours, during which time she was incapable of work and thought; she had to lie down and remain motionless till either the condition passed off or else passed over into sleep, which happened not infrequently—the states never ended with orgastic sensations. When the analysis brought sufficient material concerning the objects of her infantile fixation, and these had come out clearly in repetition in the transference to the analyst, I had to communicate to her the well-founded conjecture that she phantasied in these states unconsciously a (presumably aggressive) sexual act, and indeed with her father, or his surrogate the physician. She remained undiscerning, whereupon I did not hesitate to charge her at the next 'state of tension' to turn her attention consciously to the phantasy I had pointed out. After overcoming the greatest resistance she confessed to me later that she had experienced a phantasy of sexual intercourse, though not an aggressive one, and at the end of it had felt an irresistible impulse to make a few onanistic movements with the pelvis, whereupon the tension suddenly ceased with the feeling of orgastic ease. This then was repeated several times. The analysis showed clearly that the patient entertained the hope unconsciously that the analyst would on hearing the account of these phantasies give effect to them. However, the physician contented himself with making this wish clear to her and searching for the roots in her previous history. From then onwards the phantasy changed: she would then be a man with a conspicuous male member; she made me, however, into a woman. The analyst had then to make clear to her that by so doing she was only repeating the way in which as a child she had reacted to her father's disdain by identification with him (masculine attitude), in order to make herself independent of the favour of her father; this attitude of obstinacy characterized her entire emotional life towards men. There were other variations: phantasies of being teased by a man (with manifest urethral-erotic content), then phantasies of sexual occurrences with her older brother (whom on account of his strength she pretended to love less than the younger brother). Finally she had the quite normal feminine onanistic phantasies, full of resignation, which were surely in continuation of the original loving attitude to the father.

She only brought the smallest part of the phantasies spontaneously; for the most part I had, on the basis of her dreams and the associations she gave in the hour, to provide the direction in which she ought to force her unconscious experiences. A 'period of prohibition' must, however, in every complete analysis follow this 'period of injunctions', that is to say, one must bring the patient to the point of tolerating the phantasies without onanistic discharge, and thereby make conscious the feelings of distress and painful affects related to them (longing, rage, revenge, etc.) without converting them into hysterical 'feelings of tension'.

With the foregoing examples I hope I have illustrated sufficiently the way in which I am able to make use of 'forced phantasies'. My task is now to say something on the indications and contra-indications for this technical stroke. As with 'active' interventions generally, these tasks in phantasying are justified practically always only in the period of detaching, *i.e.* at the end of the treatment. One must add, to be sure, that such detaching never occurs without painful 'deprivations' [Versagungen—frustrations],* that is, without the activity of the physician. So much for the question of the time when it is suitable to apply this technique. As to what phantasies to put to the patient, one cannot in general say; that must be decided by the analytic material as a whole. Freud's aphorism that progress in the analytic technique is to be expected from an extension of our analytical knowledge applies here also. It is necessary to have much experience of 'not-active' analyses and not-forced phantasies before one may allow oneself such interferences, which are always risky, with the spontaneity of the associations of the patient. If the phantasy-suggestion is tried in the wrong direction (which occurs in those most practised in it occasionally) it may unnecessarily protract, though it is intended to shorten, the treatment.

I owe to these researches into the unconscious phantasy life of patients an insight not only into the mode of origin of particular phantasies (in respect to their content) but also—as an incidental gain—an insight into the cause of the animation and torpidity of phantasy life generally. Among other things I made the discovery that the animation of the phantasy stands in direct relation to those childhood's experiences which we call infantile sexual traumata. The greater number of the patients whose phantasy activity I had to rouse and push forward artificially in the manner alluded to belonged to those classes of society or families with whom the comings and goings of the children from earliest infancy onwards were strongly controlled, the so-called childish naughtiness was hindered from the very first, and naughtiness was broken off before it had come to full bloom, and where the children lacked every opportunity to observe, much less to experience, in their surroundings anything sexual. They are such *well-brought-up* children that their infantile-sexual instinctual impulses simply *have not the opportunity to get anchored in the world of reality*. Such an anchoring, in other words a piece of real experience, appears to be a pre-condition for later freedom in phantasy and the psychical

* [Ferenczi here applies Freud's abstinence rule (1919); see editors' introduction, pp. 30–31.

potency connected with it, while the infantile phantasies of the too-well-brought-up child fall into repression—the 'primal repression'*—before they ever reach consciousness. We may say in other words that a certain amount of infantile sexual experience (that is to say, a little 'sexual traumatism') not only does not damage but actually promotes the later normality, particularly the normal activity of phantasy. This fact—which corresponds in detail to the comparison Freud (1916–1917) made of the consequences of upbringing 'on the ground floor and on the first storey'—leads us to place a lessened value on infantile traumata. Originally these were thought to be the origin of hysteria; later Freud himself stripped them of a great part of their significance in that he discovered that the pathological element lay not in real infantile experiences but in unconscious phantasies. Now we find that a great part of the real experiences of childhood actually offers a certain defence against abnormality in the directions of development. Nevertheless the 'actual experience' ought not to exceed a certain optimum; too much, too early, or too strong may bring about repression and with it a poverty of phantasy.

Viewed from the standpoint of the development of the ego, we are able to trace the poverty of phantasy of the too-well-brought-up child (and his tendency to psychical impotence) to this, that children who really have experienced nothing wilt hopelessly in the ever anti-sexual atmosphere of educational ideals, while the others never let themselves become so completely overpowered by education that on its cessation at puberty they do not trace the regressive path to the abandoned objects and aims of infantile sexuality, but can accomplish the precondition of psycho-sexual normality.

* [The term "primal repression" was crucial for Ferenczi and Rank (1924); it refers to the fantasies that were repressed immediately. They are not subject to recall and appear only as acting or experiencing in the analytic situation.]

CHAPTER 9

EDWARD GLOVER

ACTIVE THERAPY
AND PSYCHO-ANALYSIS
A Critical Review

INTRODUCTORY NOTE

Glover's critique of Ferenczi's "active technique" is noteworthy for the clarity of its exposition and cogency of argument. Published in 1924, at the height of the controversy, it has remained the most effective rejoinder to Ferenczi's innovations. In the second section, Glover deals with the problem that Abraham struggled with—the inability of narcissistic individuals to form a reliable transference. To Glover, the safeguarding and fostering of the transference neurosis are more central than to Ferenczi. It is an important article, but not an easy one to read, for it moves back and forth between technique and metapsychology.

PART 1 / Introductory

TO LIMIT A REVIEW of work on active technique to a consideration of the technical suggestions made by Ferenczi would be, as Ferenczi himself sug-

Originally published in *The International Journal of Psycho-Analysis* 5 (July 1924): 269–311. This paper was read before the British Psycho-Analytic Society on February 21, 1923. Reprinted by permission of the Institute of Psycho-Analysis.

gests, to misunderstand the use of the word 'active' and in reality to leave out of account important stages in the history of psycho-analytic therapy.

As he points out, the Breuer-Freud cathartic method was essentially one of great activity. A vigorous attempt was made, under hypnosis if necessary, to awaken memories, i.e. not only was the attitude of the physician an active one, but the patient was called upon to make definite strenuous efforts. Further, the present method is passive only by contrast. It is true that the patient remains passive, but the physician cannot permit the patient's phantasies to continue indefinitely and, when the material is ready to crystallize, the former must abandon his passivity and interpret in order to make easier the associative paths otherwise barred by resistance. During this 'obstetrical thought-assistance', as Ferenczi calls it, the patient remains, as before, passive.

If one follows the development of technique from the time of the cathartic method onwards, it is clear that, not only in stating the aims of psycho-analysis, but in the working out of the dynamics of transference, resistance, etc., most contributions to psycho-analytic literature (and especially those of Freud himself) are contributions to the problem of activity in technique. One might refer, for instance, to Freud's working out of the stages in psycho-analytic therapy where he distinguishes a first phase, during which libido is detached from the symptoms and crowded on to the transference, from the second when the battle rages round this new object, libido is freed, and to prevent withdrawal of this libido to the unconscious, the ego is educated by the interpretative suggestions of the analyst to the point of reconciliation of the two (Freud 1916–1917).

In his work on the dynamics of the transference (1912a), too, Freud lays down conceptions of regression and re-activation with corresponding resistance which are fundamental for the theoretical consideration of active technique and his description of the plasticity of libido and its capacity for collateral circulation is one which Ferenczi uses freely and with effect. Indeed, Freud's early paper on dream-interpretation in analysis is a contribution to the subject of activity in so far as he deprecates the use of interpretation as an art *per se* (i.e., what might be called an arbitrary or active use of interpretation), and lays down that it must be subject to the same rules as treatment in general with the rider that active interpretation can be occasionally followed as a concession to scientific interest (Freud 1911a).

More directly concerned with the transference situation are Freud's remarks on the dangers of 'repetition' and the function of 'working through', in which he points out that the aim of the physician must be the remembering and reproduction in the psychic plane. The physician, he says, must enter into a long-drawn-out fight to prevent the patient discharging impulses in action which should be limited to mental expression. Successful prevention of this nature can be regarded as a triumph and the physician should see to it that the patient does not carry significant repetitions into action (Freud 1914c).

In 1910 Freud laid down that in anxiety-hysteria the patient cannot produce the necessary material as long as he is protected by the condition of the phobia and that, although it is not possible for him to give up these precautionary measures from the outset, one must assist by translation of the unconscious until such time as he can bring himself (*sich entschliessen*) to give up the protection of the phobia and lay himself open to a now much reduced anxiety (Freud 1910a).

After an interval of eight years, and shortly after the publication of Ferenczi's paper on active treatment in hysteria, he returns to the same point with a significant change in the verb. 'One will hardly ever overcome a phobia', he says, 'by waiting until the patient is induced to give it up as the result of analysis. Treated in this way he will never bring up the material so necessary for a convincing solution of the problem. One must adopt other measures. Take, e.g. the case of agoraphobia of which two types are recognized, one slight, the other more severe. The former suffer from anxiety when they walk in the street unaccompanied but they have not altogether given up going by themselves: the latter protect themselves by giving up the attempt. In these latter cases success can only be attained by inducing the patient under the influence of analysis to behave like cases of the slighter type, i.e. to go about alone and to fight down the resultant anxiety. In this way the phobia is slightly weakened and only then will the patient produce associations which will lead to its solution' (Freud 1919).

In the same paper he says that the principle of activity lies in the carrying out of treatment in a state of abstinence; substitute-satisfactions must be denied, especially the most cherished of satisfactions. Not every one, of course, and not necessarily sexual intercourse. The sufferings of the patient should not come to an end too quickly, and when we have alleviated them by breaking up and reduction of symptoms, we must induce sensitiveness at some other point by means of privation. At the same time we must be on the look-out for substitute-formations. Unhappy marriages and bodily ill-health are the most common forms of relief from neurosis. Abstinence originally led to symptom-formation, and it must be the mainspring of the will to health. Again in reference to the obsessional neurosis, 'I have no doubt that in these cases the proper technique lies in waiting until the treatment has itself become a compulsion, and in forcibly restraining the compulsion to disease with this counter-compulsion'. The use of the term 'induce' (*bewegen*) in the case of anxiety-hysteria and of 'forcibly restrain' (*gewaltsam unterdrücken*) in the case of the obsessional neurosis is of significance.

Other writers have worked on the same theme from much the same point of view, as, for example, where Reik (1915b) likens psycho-analysis to the work of a machine for the running of which some degree of friction is indispensable; on the whole, the previous quotations may be taken as representing the general point of view. Now, whilst these observations seem to have been dictated by a combination of clinical expediency and widening of theoretical insight, in Ferenczi's case there seems in addition to run throughout a consistent train of thought, given increasing consideration in an attempt to make the technique more effective in exceptional cases and generally to shorten, if possible, a lengthy procedure.

Referring in a reminiscential vein to his pre-analytic days, Ferenczi tells how a peasant suffering from attacks of loss of consciousness came to consult him. While his history was being noted, which elicited a story of conflict with the father, the patient broke off in a faint in the middle of a sentence, namely, 'I must work like a scavenger whilst——' At this point Ferenczi seized the patient, shook him vigorously, and shouted to him to complete the sentence, which then ran—'whilst my younger brother stays at the home farm.' The loss of consciousness proved to be a flight from reality, and the patient was amazed to find himself completely and immediately cured (Ferenczi 1920).

Passing over intervening stages, we find Ferenczi, in his paper on transitory symptom-formations (1912), regarding such miniature neuroses as points of attack for dealing with the patient's strongest resistances. Such symptoms being affectively experienced in the patient's own person lead, after suitable analysis, to that conviction of the correctness of interpretation which cannot be attained by logical insight alone. They are representations of unconscious feeling stirred up by analysis and forced back, which, no longer capable of complete suppression, are converted into somatic symptoms, an explanation the quantitative factor of which has recently been emphasized by Alexander (1921).

In order not to disturb the case-illustration of Ferenczi's development in technique, his general paper on technique (1919a) may be considered here, although really it follows that on the analysis of hysteria [see this volume, Chapter 5]. It contains many excellent suggestions of a general kind, from which the following, more active, may be selected. The patient can defeat the analyst with the latter's own weapons. Asked to produce associations without regard to content, the former will produce only nonsensical associations and try to reduce both analysis and analyst to absurdity. This must be stopped by interpretation of the underlying intent, the patient's triumphant counter, namely, 'I'm only doing what you ask', being met with the explanation that to produce solely nonsensical associations is in itself a form of thought selection. Sudden silence is a transitory symptom which, if persisting after interpretation, must be met with silence. In some cases a patient breaking off with an 'à propos' can be asked to finish his sentence, since this involves not connected thinking, but connected saying of what is already thought. Obscene words must be spoken, and the compromise of writing them down should be avoided. Do not be content with generalities: concrete representations rather than philosophical speculations constitute the real association form, an interjected 'for example' often getting nearer to the unconscious content. On the question of influencing the patient's decisions his views may be summed up briefly; first find whether the decision is really urgent or whether it is being thrown at the analyst as a gas-bomb to cause confusion. If real and the patient has any capacity for decision, let him decide; if real but the patient is incapable of decision from reality-testing reasons, he may be helped; if real but the incapacity for decision is of the form of a phobia, make the patient come to *some* decision.

Although there is nothing new in the way of theoretical consideration in this paper, or in a short note on influencing the patient during treatment which appeared in the previous *Zeitschrift*, still the general tendency to active interference once ordinary interpretation seems to fail is quite outstanding.

The logical development of these tendencies is to be found in Ferenczi's method of dealing with the analysis of some cases of hysteria [Ferenczi 1919c, this volume, Chapter 7]. On one occasion, observing that a patient's analysis approached a condition of stalemate, he prescribed a certain period within which treatment must be finished. The patient, however, hid her resistance behind a positive transference which was characterized by passionate love declarations; treatment was ended at the stated time, leaving the former quite satisfied with the result. Renewed after an exacerbation of symptoms, analysis again brought about improvement, but just up to the previous stage; beyond that the love-defence was again brought into play and again treatment was ended (this time owing to extrinsic causes). A third attempt was made with an identical result, but now Ferenczi observed that in the perpetual love-phantasies connected with the physician the patient remarked on certain genital sensations. In addition she lay always with the legs crossed. This led to a discussion on masturbation, the performance of which she denied. Finally Ferenczi forbade her to cross the legs, explaining that she thus discharged unconscious excitations in a larval form of masturbation, and the result of this prohibition was immediate increase of bodily and mental restlessness, accompanied by phantasies similar to those of delirium. Infantile experiences and circumstances conducing to illness were remembered in fragments. But again the analysis lingered, and the transference-love masked resistance. Then Ferenczi made the discovery that she eroticized her household activities, as in unconsciously working with the legs pressed together. Prohibition of these extra-mural gratifications led merely to a slight improvement, but also to the performance of various plucking movements during the hour. These were carried out on, so to speak, 'indifferent' parts of the body, but became masturbation equivalents capable of producing orgasm. They had been carried out in childhood, and now, after due suppression, sexuality found its way back to the genital zone, the immediate result of which was the return of an infantile obsessional neurosis. After solution of the latter, an irritation of the bladder made its appearance, usually at times unsuitable for relief. This relief was in turn forbidden, and the patient finally reported an act of genital masturbation, a regressive stage which did not last long and led gradually to pleasure in normal intercourse.

Ferenczi then formulated his new rule, namely, watchfulness for larval forms of masturbation giving cover to libido and possibly displacing the whole sexual activity, i.e. a short way for the discharge in motility of pathogenic phantasies, a short-circuiting of consciousness. These forms must be forbidden when noticed, and in reply to criticism Ferenczi points out that this is a provisional measure. Sometimes the completed treatment renders this form of gratification superfluous, but not always. Masturbation for the first time in a patient's life during treatment is a favourable turn in events, but only if manifest masturbation with conscious erotic phantasies. Larval forms must be analysed, but must first be forbidden, to prevent short-circuiting, and only when the patient can endure these conscious phantasies may he be given freedom to masturbate.

Many larval forms are not neurotic, many are neurasthenic,* and many are unconsciously gratified throughout life, as in the case of persons who, preoccupied in business or metaphysical speculation, with hands deep in the pockets, touch, press, or rub the penis. Similarly, clonic contraction of calf-muscles, and, in women engaged in housework, pressing together of the limbs. The danger is that lack of orgasm leads to anxiety states or that the small discharges obtained disturb potency in a way not occurring in ordinary conscious masturbation. There may be, too, a transference from symptomatic actions to *tics convulsif*, many of which are stereotyped masturbation equivalents.

Ferenczi then sets about a detailed consideration of the rationale of active technique which is available in his paper given at the Hague Congress [Ferenczi 1920], but in the meanwhile he has added to, systematised and differentiated stages in the process.

We have seen that he regards the cathartic method as above all active and the passive technique as containing an active component in the form of interpretation, which is permissible by actual authority of the transference, the patient remaining meanwhile passive. But this activity or passivity is practically limited to mental functions, and apart from the rules about punctual attendance, and the making of decisions without guidance or alternately the shelving of decisions, the *actions* of the patient are not directly interfered with. The experience with anxiety-hysteria, where phobias are brought into actual play with resultant accessibility of new material, is the one exception which demands a category by itself. Here the active interference is not so much on the part of the physician as on the part of the patient; a task is laid upon him which leads to the *doing* of unpleasant things.

Fortified by Freud's declaration of the necessity for carrying out treat-

* [In 1924 the distinction between neurosis and neurasthenia was still maintained. It goes back to Freud's paper of 1895.]

ment in a state of abstinence [Freud 1919], Ferenczi finds occasion for a new variety of task, in cases with masturbatory touching of the genitals, stereotypies, tic-like movements, namely, the giving up of pleasurable activities. Here is his first illustration.

The patient, a musician with phobias and obsessive fears, amongst other inhibitions suffers from stage-fright and attacks of deep blushing. Although able to practise complicated finger exercises when alone, she cannot do so in public, and more, although really gifted, has the obsessive thought that she *must* blame herself for incapacity. Her breasts are large, and thinking herself to be observed much in the street, she is at a loss to know how to conceal her bust, sometimes crossing her arms to press in the breasts. Yet doubt follows all attempts. She is sometimes shy in manner, sometimes bold, unhappy if not noticed, alarmed if any real attention is paid to her. Her mouth smells, she thinks, yet a visit to the dentist can show no abnormality whatever.

After some analysis with Ferenczi she understands her main constructions, yet her condition does not satisfy him. One day she remembers a vulgar street 'catch,' which her elder sister, who, by the way, was rather tyrannical towards her, used to sing. She repeats the *double entendre* and remains silent, whereupon Ferenczi asks her to *sing* the air, which after prolonged delay (two hours in all) she does, hesitatingly at first, but later with a full soprano. The resistance continues, but on hearing that her sister was in the habit of accompanying the song with suggestive gestures, he asks her to reproduce these gestures exactly. Having done so once, she begins to show a taste for repetition, which leads to a countermand. Then for the first time come memories of her brother's birth, singing and dancing before parents who dote on her. An order to conduct part of a symphony leads to the discovery of penis-envy (the baton) and the compulsory playing of a difficult piano part sheds light on her dread of examinations. Her self-blame is on account of the masturbation represented by the finger exercises. Similarly a request to go to the public swimming-bath uncovers the exhibitionistic motive behind her breast-ceremonial, and the discovery that she was passing flatus during the analytic hour in a kind of play, retaining and letting go, led, on the countermanding of this activity, to the tracing of the anal-erotic motive in the mouth-smelling fancy. Finally, treatment was greatly helped by the interpretation of certain movements and gestures whilst on the piano stool: these were carried out and stopped to order, and an unconscious masturbatory practice was revealed.

The technique applies not only in the activation and control of erotic tendencies, but also in the case of highly subliminated activities. A patient whose interest in versification was only partly gratified in puberty is asked to write poetry and displays distinct poetic gift, behind which is the desire for masculine productivity, clitoris-fixation and anæsthesia. When forbidden the new activity, it transpires that really a misuse of talent is in question, the masculine attitude is secondary, a genital trauma having led to displacement to auto-erotism and homosexuality. She only takes to the pen when she fears non-fulfilment of her female functions. The result is a re-established capacity for normal female activity.

Here we have the two stages—'painful' tasks, then 'painful' abstinences, commands and prohibitions. The former render repressed instinct-components into conscious wish-formations and the latter force the awakened excitations back to infantile situations and repetitions. Since these have been subjectively experienced by the patient and objectively observed *in flagrante delicto*,* they cannot be denied. In both stages the mechanism is that of producing a situation of privation.

When of course the patient is already active, masturbates, produces symptomatic acts and transitory neuroses, there is no need for the first stage, forbidding alone is necessary, although sometimes it is advisable to encourage first the full acting out of such situations. Urinary habits, flatus activities, sphincter play in general, various gestures, handling of the face, movements of the legs, shaking of the body, are suitable *points d'appui*.† Even apparent contradictions in theoretical technique are sometimes permissible, as when a patient threatens to cheat in analysis and is encouraged to do so, or when he seems to be associating beside the point and is arbitrarily brought back to connect and complete the broken thread of thought.

Then as to indications: the technique must be used as little as possible, since the passive attitude is best, not only for the patient, but—and this deserves italicizing—also for the physician. It is a therapeutic adjuvant, to be used sparingly like the forceps in midwifery.

1. Beginners are advised against using it. They may easily go wrong or be led into error, and in any case will tend to lose insight into the dynamics of the neurosis.

2. At all events never employ the technique in the early stages of treatment. The transference must be permitted to develop to a sufficient degree of durability—in other words, to a compulsion—otherwise premature action against the pleasure mechanism will lead to the breaking off of treatment.

3. At the end it is often necessary and frequently induces the characteristic last 'present' of unconscious material from certain cases.

4. It can be used in all forms of neurosis, but it is more often indispensable in obsessions and in anxiety-hysteria; in pure conversion-hysteria it is seldom needed.

In this grouping two dangers are present: *a.* the cure may be too rapid, as where an inhibited woman suddenly becomes bold, is surrounded by ad-

* [*In flagrante delicto* (Latin) as in "flagrant delight," is usually used to describe a person caught in the act of sexual intercourse.]

† [*Points d'appui*, literally a fortified position, here used as a suitable point for attack.]

mirers, and breaks off treatment at the end of the first stage; *b.* the resistances encountered may lead to the premature termination of treatment.

Masturbation has already been considered, but Ferenczi adds to this a note on forbidding unsuccessful attempts at satisfaction on the part of impotent persons, although this, he says, is by no means an axiom.

5. Active therapy finds a suitable field in character analysis. Here, as with the psychotic, insight is absent, and we have a private psychosis narcissistically tolerated by the ego.* If these characteristics cannot be melted down in the boiling heat of the transference love-situation (to use Freud's phrase), active technique can be tried; but the resistance is great, the narcissism defending the infantile memories can interfere with the aim of psycho-analysis, and there is always a risk that the patient may break off treatment.

Ferenczi then asks: Can the attitude of the physician be made use of in a more active sense; can the interpretative suggestion which influences the ego in analysis be carried over in some cases in a kind of pedagogic guidance in which some form of praise or blame can be made use of? Leaving this question unanswered, he makes the suggestion that, as the neurotic has something of the child in him, child methods are to a certain extent applicable, more especially in the maintaining of an optimum temperature in the transference situation by a shade of coolness in the heated stages, and of friendliness in the reserved phases.

In the earlier part of his paper Ferenczi differentiated psycho-analytic suggestion from the popular variety, in so far as psycho-analytic suggestion does not say to the patient, 'There is nothing the matter with you,' and also in that the psycho-analytic interpretations are based on memories or repetitions, and not explanatory conversion, as by Dubois.†

He now anticipates possible retaliatory criticisms from Bjerre, Jung and Adler. But Bjerre‡ neglects pathogenic causes, and contents himself with taking the patient's mental and ethical guidance in hand. Jung detaches the patient from the past and links his attention to the tasks of life, whilst Adler concerns himself not with the analysis of libido, but with the nervous character.

Ferenczi, on the other hand, deals with individual or isolated activities, and even then not as an *a priori* moral influence, but merely to counter the pleasure principle, to dam up eroticism (*die 'Unmoral'*), and to remove obstacles to the progress of an analysis of causes. He may, however, in some stages not only tolerate the erotic tendency, but encourage it.

---

* [Glover links Ferenczi's "active technique" to Reich's "character analysis," which at this time is only beginning. Indeed, the connections are interesting. The term "private psychosis" does not strike me as felicitous. Since all psychoses are private, circumscribed psychosis would have been a happier term.]

† [Dubois, Raymond Emil (1818–96). A friend of Freud's teacher Brücke.]

‡ [Bjerre, Poul Kar, introduced psychoanalysis to Sweden. He advocated supplementing psychoanalysis with spiritual guidance. See Ferenczi 1920, p. 211.]

Returning to suggestion, he insists that in active technique certain measures only are presented, apart altogether from the idea of successful outcome, and, indeed, without any certainty of knowing what the outcome will be. No improvement is promised: rather the contrary. The stimulation of a new distribution of psychic energy promises discomfort, and often disturbs the placid torpor of the stagnant analysis.

Catharsis again hoped to awaken memories and thus release affect; active therapy stimulates activities and inhibitions in the hope of attaining secondary unconscious material. Analysis begins where catharsis ends. Catharsis is an aim and end in itself; active therapy is a means to an end. It increases resistances by stimulating the sensitiveness of the ego, and increases the symptoms by increasing conflict; the new condition of tension or increase in tension disturbs hitherto untouched areas. Like the counter-irritant treatment, it not only discovers hidden foci, but increases immunity; the great vessels are tied and circulation flows through the smaller arteries lying deep in the tissues.

PART 2 / *Consideration of the Ferenczi Technique*

In so far as the phrase 'active technique' is associated with the name of Ferenczi, it is necessary to be guided strictly by the indications laid down by Ferenczi himself. From these it will be seen that this technique is by no means to be regarded as a therapy in itself, but rather as a special procedure devised to meet a special analytical situation, namely, where the substitute-gratification of libido-impulses forms a barrier to examination of the underlying unconscious formations. This gratification may be present with comparatively little qualification in numerous larval forms of masturbation, or directly in the form of neurotic character-traits; or, again, it may be qualified by the compromise-formation of the symptom. Hence the application of the procedure may be merely occasional during some analyses or much more constant in others, as, for example, some cases of anxiety-hysteria and in obsessional neurosis. In either instance a prerequisite of its application is the establishment of a durable transference situation where the analyst's active interference is supported by the authority of the imago he represents.

Considering the question merely from this point of view, two criticisms occur, one of general principle, and the other of detail, both of which have been made by Van Ophuijsen (1921). First as to the principle.* Van

* [Van Ophuijsen, J. H. W. A pioneer Dutch analyst, active in his later years in New York.]

Ophuijsen considers that active technique is really an important alteration in so far as the analyst makes use of the transference-situation instead of immediately analysing it. Secondly, that as these resistance states, which necessitate active therapy, may be regarded as 'repetition' phenomena, Ferenczi should have limited his rule in the case of larval masturbation by prohibiting this only when it is the source of resistance *at the time*. In reality, Van Ophuijsen's criticism of detail involves yet another principle, that of the therapeutic part played by the compulsion to repeat and the working through of traumata in the transference-situation, and on these points I should like to offer the following observations:—

Transference.   As far as transference is concerned, the situation might be put as follows: In psycho-analysis a 'transference neurosis' gradually replaces the original neurosis, and this former must be dealt with in turn by repeated analytical interpretation of the repetition-compulsion, as manifested in the transference-resistances. One must ask, therefore: Do not active interferences on the part of the analyst disturb the transference picture as a spontaneous repetition, since the recognition by the patient of transference material *as such* is greatly facilitated by the passive role of the analyst and his impersonality? In other words, when the father-imago is revived by a figure that does not advise, persuade, convert, or command, it is more easily recognized *as such* than when it is anchored to the present by a *real* situation in which a physician actually does advise, persuade, convert or command a patient. From this point of view, too, the possibility of blunders present even in an orthodox analysis is heightened by the hazard of piling up even stronger resistances. Again, since the patient is in a 'transference' neurosis, i.e. an affective relation to the analyst repeating the infantile fixation, he is *'sensitized'* to even ordinarily trivial behaviour on the part of the analyst and reacts to it with massive affect, i.e., with a psychical anaphylactic reaction.*

In ordinary analysis, however, the recognizable triviality of the occasion conjoined with a prompt analysis of its significance usually prevents a 'second fixation'† occurring. Now since the final stage of analysis is agreed to be arrived at through the analytic dissolution of the 'transference neurosis' anything in the nature of a 'second fixation' must surely constitute a difficulty.

* [This is a crucial point in Glover's critique of Ferenczi. Here a generation gap may be detected. Ferenczi belonged to the first generation of pioneers, and to him, getting the analysand to free associate with the concomitant affect was the crucial issue. For Glover the full flowering of the transference and its transformation into a tranference neurosis is the decisive point in psychoanalytic technique. This idea was stated in one of Freud's papers (1914c), but here it appears for the first time as a central point in psychoanalytic technique. Anaphylactic reaction is the medical term for increased pathological sensitivity to infection or a foreign substance.]

† [The term "second fixation" was coined by Glover. It would be the psychic equivalent to an iatrogenic disease in medicine, an illness produced by the physician. Glover here raises the problem of an unanalyzable parameter (see Eissler 1953), that can cause a secondary fixation.]

The answer of the 'activist' to this criticism is in effect that he is throwing a sprat to catch a mackerel, that the most important repetition is wanting, being more or less actively satisfied elsewhere, and in such cases, and in such cases only, the durability of the transference can be put to the hazard. If he fails, and the analysis is broken off, he is in no worse case than the protagonist of passive methods who has merely attained stalemate.

This is still open to the counter that it is unnecessary to make a rule of involving the direct authority of the imago in such situations, and that repeated analysis of the gain through illness, of the gain through larval acts, or of the gain from indulging character-traits, can be made to focus the patient's attention on the *performance* or *non-performance* of such traits or acts. In so far as this focusing is arbitrarily determined it is an active step, but it avoids the necessity of the physician, so to speak, entering the arena clad in the mantle of the imago.*

As a matter of fact, although Ferenczi frequently mentions the danger of losing a patient inherent in the application of active technique, at only one point does he mention the opposite risk; speaking of influencing the outside life of patients incapable of coming to a decision, he says: 'Here the physician should be aware that he is no longer behaving as a psycho-analyst, that indeed his interference may cause positive difficulty as regards duration of treatment, e.g. an unwished-for strengthening of the transference-relationship.'

Repetition Phenomena. We know from Freud that the transference is in itself a repetition phenomenon, and that the greater the resistance the more does repetition replace memory-work. The main fight then is to prevent repetition obtaining motor discharge and to use the transference when serviceable as a playground in which the patient is given almost complete freedom to expand. This leads to the establishment of an artificial illness, the 'transference-neurosis', a provisional state having the characteristic of real experience. But the interpretation of this experience does not immediately overcome resistance; the patient must be allowed time to work through the compulsion. At this stage, says Freud, 'the physician can only wait and permit a course which can neither be avoided nor indeed hastened' (Freud 1914c). He summarizes the position later (1920) by saying that this transference neurosis must be allowed as little repetition as possible, but notes that the relationship between memory and reproduction varies in every case. The patient as a rule cannot be spared this part of the treatment, part of his forgotten existence must be re-experienced. It would seem then that the conditions under which varying degrees of play can be allowed to this repetition-compulsion ought to be accurately studied before any conclusions can be drawn as to the point at which active interference might be permissible.

We are now familiar with the general economic function of the biological

---

* [Glover means behaving like the parent that forbids masturbation.]

repetition-compulsion in binding traumatic stimuli, and so in working through traumata. There is, however, a natural tendency to regard transference phenomena (involving, as they do, relations with an imago) as *in themselves* the *complete* representation of this economic function.

The extension of libido to the object by means of primary identification, the ultimate mode of object-choice and the vicissitudes which this choice undergoes provide a series of situations during the repetition of which the analyst plays a repertory part. The rôle is mainly that of an object, but even where narcissistic choice* has prevailed developmentally over the anaclitic type and the analyst is made to play from time to time the part of subject by identification, the situation in both cases represents an extension of libido from ego to object. Since this series of situations has developed gradually from early stages of primary narcissism,† it is small wonder that the subject-object polarity should occupy the foreground of the analytic picture, and that the part played by narcissistic libido in repetition should tend to be minimized. Repetition can, however, make use of the analytic technique itself for the working through of auto-erotic vicissitudes, i.e. unconnected with the object, or more correctly, connected with the self as object. This represents the primary narcissistic stage in the modification of instinct before the impulse is turned towards the object (1915d). Now, although both auto-erotic and subject-object activities might be included under the common heading of ontogenetic vicissitudes, or individual modifications, of the compulsion to repeat, they are clearly distinguishable from each other as regards amenability to transference influence. It is, of course, true that auto-erotic manifestations are capable of influence through the transference in two ways: first, that historically the subject was induced to abandon conscious manifestations either through the direct influence of the object, or by the influence of the object

* [In the paper, "On Narcissism: An Introduction" (1914b), Freud differentiated the two paths by which a love object may be chosen. One was on the model of the self (as one was, is, or wanted to be). In this case it would be a narcissistic choice. Or, on the model of the supporting mother. This constitutes love based on anaclitic choice. Glover now discusses at considerable length the transformation of narcissistic libido into object libido. He thus moves to the level of metapsychology. These problems are only indirectly relevant to the controversy with Ferenczi. It was in the tradition of the times to move freely from technique to metapsychology.]

† [The term "primary narcissism" was used by Freud to designate narcissistic libido that was never converted into object libido. It was distinguished from secondary narcissism, which resulted from the withdrawal of libido from disappointing objects and converted into narcissistic libido. Since Spitz (1945) has demonstrated from observations of newborn babies that very early in life the baby is dependent on mother's love, the term primary narcissism, as Freud designated it, lost its meaning. In current psychoanalytic usage, the term primary narcissism has not been abandoned, but it now connotes a stage of development where self and objects are not yet separated. See Jacobson's (1964) review of Freud's ideas on primary narcissism, pp. 3–7. For justification for the retention of the term see p. 115; see also the discussion of transference in the editors' introduction on p. 17. That narcissistic transference is a special kind of transference, is discussed by Kohut (1971) and Kernberg (1975).]

indirectly as introjected ego-ideal; secondly, that where auto-erotic manifestations are regressively activated, the regression has taken a path which is still associatively linked to the object. It retraces the steps taken in the first limitation of auto-erotic impulses under object influence, and a situation arises which is somewhat loosely analogous to that of regressive hate which really continues love at the anal level.*

But whilst this degree of modification of auto-erotic impulses through the object exists, we know that many of the component-impulses continue from the primary stage to the point of serving the interests of genital primacy without direct modification (Freud 1915d); indeed, that they continue to serve pleasure interests apart from object-choice and genital primacy, just as narcissism runs a course apart from the contributions made by narcissism to object-choice. In this sense, then, they differ from the sadism and exhibitionism 'pairs' by being unmodified, and are autonomic by permission of the pleasure-principle;† repetitions are therefore found not in the transference situation, but in the patient's own aberrations in following the analytic rule, his traits and mannerisms, i.e. not in his relations to the analyst, but to the technique of the analysis. Again, however much the ego may be influenced by the object or by the ego-ideal, it is arguable that the abandonment of narcissistic enjoyment may, under certain conditions, such as ego-sensitiveness, or perhaps a time factor, constitute in itself a trauma comparable with and even stronger than the traumata which lead later to the abandonment of the parental Œdipus relation (or which are caused by this abandonment). I am indebted here to a suggestion of Mrs. Isaacs‡ in reference to suckling, that there

---

* The various tissue changes induced under hypnosis might be brought forward in support of the complete accessibility of narcissistic libido to object influence. Without going deeply into theoretical consideration of this point (on which much light is shed by Ferenczi in his paper on hysterical materialization-phenomena), it may be said that such alterations presuppose not only a strong transference capacity (and therefore strong object-modification of subject-impulse), but, as Ferenczi suggests, an advanced state of modification in which the body is "genitalized." The induction of such changes, themselves in the nature of a transference "conversion," does not preclude the co-existence of a stream of auto-erotic activity more or less inaccessible to transference influence.

[Whether to regard the aggression concomitant to the psychosexual phases as an independent drive or as belonging to the early libidinal phases of development has been a problem in psychoanalytic theory, which originally assumed that the pregenital phases are ambivalent and the genital phase alone is postambivalent (see Abraham 1924a). This distinction became more difficult after the formulation of the dual instinct theory, when the aggressive drive was conceptualized as being independent from the libido. For an attempt to reconcile these difficulties, see Gero 1962.]

† It seems probable that even in the case of modified 'component' pairs, especially the exhibitionism-scoptophilia pair, the primary narcissistic stage has still continuous gratification throughout life. This is less capable of direct proof owing to the fact that in the second stage of modification, namely, the turning of the impulse from the object against the self, a pseudo-narcissistic phase is attained. In the case of the erotogenic zones the continuance of primary organ-pleasure apart from any modification is more easily demonstrable.

‡ [Susan Isaacs, a well-known British child analyst and educator.]

may be an optimum psychic duration of this process, curtailing or lengthening of which may prove a trauma in itself. Here then would be an additional source of 'pain,' likely to be worked through by auto-erotic repetition and less amenable to the transference.

Now, although these ontogenetic or individual modifications of the compulsion to repeat comprise the larger part of analytic repetition phenomena, and even so with 'subject-object' repetitions forming, as it were, a screen behind which auto-erotic repetitions are more difficult to distinguish, we cannot afford to jettison entirely what might be called the phylogenetic aspects of the compulsion. These will consist mainly of two manifestations: first, the primary economic biological function of any organism to use repetition as a 'binding' mechanism (cf. Freud 1920); and secondly, the repetitions of racial traumata not yet racially worked through.* Some hint of the latter is given in the so-called archaic reactions, as in some female types of castration reaction, and in the incompleteness with which in certain archaic types active impulses have undergone passive changes; of course, the idea of psychic phylogeny is in keeping with this assumption.

At this point we reach the delicate question of the hereditary factors operating on ego-development, not only phylogenetically, but individually. In particular one would have to consider whether the history of racial libido-development can work or has wrought any permanent change in ego-structure, and secondly how far, in any individual, permanent ego-injury can be wrought by massive libido disorder.

However this may be, a point arises in the consideration of active technique calling for careful decision; how far, that is, repetitions should be merely interpreted, or, assuming that they may be actively interfered with, what interval should be allowed for working through? This problem, interestingly enough, is not necessarily solved even when acting is converted into memory work, since the function of repetition may still be operative auto-erotically when ontogenetic libido-fixation has been loosened, and is probably in any case a permanent factor in the sense of biological function. Not only so; the conversion of acting into memory-work may reach a stage in working back at which early experiences, e.g. primal scenes, etc., cease to be capable of direct reproduction in adult recollection, and may quite conceivably be only capable of reproduction as repetitions. Perhaps the best example of this class of experience would be the cumulative engrams connected with the gratification of the oral libido.

The question of determining the optimum amount of repetition in analysis is obviously one requiring the nicest judgment† a rather outstanding

---

* [Glover here shares Freud's belief in the collective unconscious where racial traumata are subject to the repetition compulsion. Freud believed in the inherence of acquired characteristics. These ideas are no longer acceptable to geneticists.]

†[Glover touches here upon a crucial point in psychoanalytic technique. Some

example of the difficulty exists in cases such as* one treated consecutively by two analysts for over two years, in which daily one-half of the time was spent in the working through of rage affect. Of an opposite type are cases of extreme transference passion where insight is obscured by greater or lesser degrees of projection; here some limitation of analytical repetition is called for if the ultimate success of the analysis is not to be jeopardized. Again, where the technique is adapted by the patient to satisfy mainly urethral, anal and onanistic impulses, the procedure must vary greatly; some hint as to the proper procedure might probably be gained by estimating the amount of modification such impulses seem to have undergone in the history of the individual. Where the larval formations are in the nature of regressions, or where they are adapted to the expression of guilt consciousness or object-defiance, it is probable that active prohibition can be employed effectively. Again, where the larval expression in 'association' form of anal or urethral activities plays into the hands of the latent exhibitionist, active interference will probably have fruitful result. On the other hand, one might go so far as to say that, where narcissistic fixation is strong or where the links originally binding the ego to the object have been weakly forged, active technique is bound to fail in that the transference does not hold the key to the situation. It is possible that in the cases described by Ferenczi regression or guilt factors largely determined the persistence of larval traits, but it is clear from a consideration of the second stage of his active technique (when, after interpretation, the newly encouraged or demanded activity is prohibited) that there is a danger of allowing too short a repetition interval to elapse.

Ætiological Factors. Here, again, a decision would depend on numerous factors, of which the condition of falling ill would seem to be the most important. We know from Freud (1912c) that, apart from that evolution of illness represented in the series privation, introversion and phantasy investment, regression, conflict and compromise-formation in the symptom, a second type exists which falls ill in an attempt to fulfil the demands of reality, i.e. not because of a privation imposed by the outer world, but because, in an attempt to exchange from an older gratification to a later sanctioned gratification, the patient wrecks himself against inner difficulties. An exaggeration of this latter type is seen in the third type, where, owing to developmental inhibitions, the patient turns ill as soon as he passes childhood and has, outside childhood, never reached a normal phase of health. The fourth type exists where, at certain ages and for certain biological reasons, the libido is suddenly increased, and consequently a relative privation occurs. Of course,

---

patients need to repeat a great deal without being open to insight. How much of these repetitions the analyst should encourage, or even tolerate, and when he should curb the repetition compulsion by interpretation, has remained an important problem in psychoanalytic technique.]

* K. Abraham; personal communication.

none of these types are pure, but it would seem that only in the first and last (where absolute or relative libido frustration occurs) is the application of active technique at all promising, and that in the second and third types (where ego-development is faulty) active interference, if any, should be more of the pedagogic type. The criterion is in the last event the condition of the ego. A similar condition is seen in the analysis of young people and of those of rather advanced years. Hug-Hellmuth (1921) shows how in the adolescent the technique is altered in an active direction, but more as a strengthening under educational guidance. As regards Ferenczi's type of activity, she thinks that the setting of tasks to children, especially those with inferiority-feeling, is certainly indicated in the later stages, but she is none too enthusiastic, and says later, 'A careful avoidance of *direct* prohibition is more important, and taking counsel with the child is better than both'.

Again, Abraham (1919a) has shown us that in advanced years cases are not necessarily refractory to psycho-analysis, that the age of the neurosis is of more importance than the age of the patient. He definitely alters his technique, however, by treating such cases more like children, encouraging more and explaining more, and often providing stimuli by spontaneous reference to previous work.

Alterations in the usual passive technique such as those of Hug-Hellmuth and Abraham, based as they were on mature consideration of empirical data, go far to confirm the suggestion that no active step should be taken in the usual analysis until something definite is known of the patient's ego-structure. To say this is, in one sense, merely to repeat one of Ferenczi's pre-requisites, viz. a serviceable transference; but, on the other hand, it is important to distinguish the disorders of the whole ego induced by libido disturbance from more serious permanent impairments of ego-function. Finally, the possibility that a neurosis may itself be a kind of defensive screen protecting underlying ego-disorder adds a degree of urgency to the suggestion.

A Special Difficulty. During the theoretical consideration of the transference situation it was suggested that one of the dangers of applying active technique was the production of a 'second fixation', in that the analyst's injunctions would lend colour in reality to the unconscious identifications of the patient. A practical instance of this, by no means uncommon in routine analysis, may give rise to especial difficulty, and justifies being singled out for emphasis. It is found in those persons who see in the analytical situation a substitute for coitus, where the bearing of the analyst is summed up by the patient in terms of sexual aggression and is interpreted in accordance with heterosexual or homosexual identifications.

The converse situation, in which the physician himself regards the analytic situation in terms of aggression, finds an interesting and, from the point of view of active therapy, a telling illustration in a paper delivered at the Berlin Congress (1922), where von Hattingberg (1924) considered the significance of the analytical situation itself, paying meticulous attention to the relation of physician and patient, and the state of aggression represented by the supine position of the latter relative to the analyst. It might be argued, of course, that the deeper one carries the analysis either preliminary to or as the result of active technique, the less likelihood there is of such confusion. This, however, would scarcely apply in the case of masochistic impulses which are so deeply rooted. The use of orders and prohibitions with their avowed intention of causing 'pain' is surely calculated to play into the hands of the masochist and possibly strengthen the guilt feeling, which, as Freud has pointed out, is responsible for so many of the cases which remain refractory after a long and seemingly complete analysis (1923b).* In such cases active therapy would defeat its own ends by providing another displacement in place of the one attacked.

It might be added here that, although Ferenczi has wisely emphasized the inherent dangers of the method, and the risks of failure, he has not yet published a detailed account of the mechanisms leading to failure; this would have been a valuable supplement to a most valuable contribution.

Summary. The application of active technique tends to increase the difficulties of transference solution by inducing a 'second fixation', especially where the patient exhibits strong masochistic trends. It affects transference repetitions (involving object-choice), and these have to be distinguished from auto-erotic vicissitudes of the repetition-function. The latter, especially those adapted to unmodified narcissistic gratification, are less accessible to transference influence. Further, there are various phylogenetic manifestations of the compulsion to repeat which operate functionally or in response to ego-defect. Hence the determining of an optimum period for 'working through' must vary widely. The valuation of developmental or secondary injuries (of whatever source) to ego-structure is an essential preliminary to the tentative application of active technique.†

---

* [The reference is to the negative therapeutic reaction; see editors' introduction, pp. 33–34.]

† [A section on "Other Active Methods" has been omitted here.]

Part Five

Freud's Case Histories
in the Light of
Later Knowledge

CHAPTER 10

MARK KANZER

THE TRANSFERENCE NEUROSIS OF THE RAT MAN

INTRODUCTORY NOTE

To read Freud's case history, "Notes upon a Case of Obsessional Neurosis" (1909b) and then to follow up this reading with Kanzer's analysis of the "Rat Man" should be an important as well as delightful psychoanalytic learning experience. Kanzer's style is lucid. He is an excellent clinician, and there is much to learn from his critique of Freud's case. One should bear in mind that the two were written forty years apart.

Unlike Freud, Kanzer notices transference manifestations from the very first hour. For Kanzer, the patient is never in the role of the narrator of his own life story only; he also communicates with the analyst by sending out "transference feelers." At all times more than an anamnesis is involved, and Freud is a respondent not just a chronicler. This dichotomy is discussed in Part One of this volume. Kanzer's emphasis on communication as distinct from anamnesis can be followed in his other writings (see Kanzer 1955 and 1966).

Whether Kanzer is justified in entitling the paper "The Transference Neurosis of the Rat Man" (as opposed to "Transference Manifestations") is open to question. The bulk of the case history consists of the first seven sessions, and one would not ordinarily expect a transference neurosis to set in quite so early (see editors' introduction, p. 28).

Originally published in *The Psychoanalytic Quarterly* 21 (1952):181–189. Reprinted by permission.

WHAT HAS BECOME familiarly known as the 'Rat Man' is the classical 'Notes Upon a Case of Obsessional Neurosis' (1909b) described by Freud, which represents an early phase of psychoanalytic theory and technique. As Kris (1951) points out, it reflects the 'conspicuous intellectual indoctrination' of patients which prevailed at the time, and the little emphasis on reliving in the transference which analysis was later to acquire. Freud nevertheless was stressing even in this paper that transference is the effective therapeutic agent; interestingly, however, from the standpoint of the evolution of analytic thinking, he did not then clearly apprehend the transference significance of many of the exchanges between the Rat Man and himself. In reconstructing this stage of analytic technique, it appears that much of the intellectual indoctrination then considered necessary and compatible with the 'mirror role' of the analyst was actually, on an unconscious level at least, a recognition of the resistances and a more or less active intervention which modified the patient's attitude toward the physician. The Rat Man contains remarkable material for a study of the intuitive processes by which Freud explored the minds of his patients, as well as of the clinical experiences that determined the direction analytic formulations were to take.

In introducing us to his methodology, Freud cited Alfred Adler, 'formerly an analyst', as having drawn attention to the peculiar importance of the very first communications made by patients. He then confirms this observation by giving the evidence of homosexual object choice in the Rat Man's initial remarks. Freud did not, however, draw from this the apparent inferences with respect to the developing transference; moreover, at that time homosexuality was not connected with ego and superego functioning. Thus, for example, the introductory words of the Rat Man referred to a friend to whom he always used to go when tormented by some criminal impulse, to see whether this man would regard him as a culprit. The friend, however, would give him moral support by assuring him that he was certainly of excellent character and merely in the habit of taking a dark view of himself. That this introductory communication embodied the motivation for seeking treatment, and made clear the need to appease and yet deceive the superego—thereby offering an unmistakable focal point for detecting and dealing with resistance —is substantiated by the subsequent course of the analysis.

The remainder of the first analytic session seems also, in retrospect, to raise some doubts as to the strictness with which the patient conformed to his 'pledge' to follow the rule of free association. He is reported to have given a detailed history of his early childhood sexual experiences, a circumstance surely not unrelated to the fact that he had read The Psychopathology of Everyday Life and had selected Freud as his therapist for this very reason. These memories were likewise not devoid of interest from the standpoint of formulating the initial dynamics of the transference and the resistances. They abounded in voyeuristic fantasies which were coupled with fears of being observed. There were recollections from childhood of misgivings that the patient's parents could read his thoughts, an idea that must have had immediate pertinence to the analysis, and which reached a climax in his confession of the obsessive thought that his father, on understanding his son's secret fantasies, would die. Presumably, the battle lines with the analyst were thus being drawn.

The next interview, with the famous narrative of the encounter between the patient and the sadistic army officer who precipitated the patient's neurosis, was no mere anamnestic account, as it was considered to be, but was already a flowering of the transference. The story of the Rat Man, it will be recalled, concerned a neurotic military man who became violently agitated when a mess hall conversation turned to a sadistic punishment practiced in the East, wherein a pot containing rats was turned upside down over the buttocks of criminals. Shortly thereafter, the patient had developed obsessional doubts over the details of payment on a package that was brought to him by the same officer who had described this exotic torture with apparent relish.

During the recital of these events the patient became so perturbed that he frequently had to break off and rise from the couch. The analyst sought to come to his assistance by supplying details that the analysand could not verbalize: 'I went on to say', Freud records, 'that I would do all I could . . . to guess the full meaning of any hints he gave me'. This guessing game apparently opened the way to an amusing and apparently unsuspected bit of acting out. When the young man approached the crucial details of the rat punishment, he was able to draw Freud into a dialogue that was actually a reproduction of the proceedings described. As he falteringly told how the rodents were applied to the buttocks, he rose again from the couch exclaiming with signs of horror and resistance, 'They bored their way in . . .', but could proceed no further. At this point Freud intervened and completed the unspoken thought by suggesting, correctly, that the rats had found their way into the anus.

Actually, the analyst was being seduced into the role not only of the cruel officer, who told the story, but also of the rats which invaded the victim's body. The rules of the analysis clearly lent themselves to interpretation of the unconscious as a forcible violation of the patient's mind—a point already foreshadowed in the preceding session by recollections of concern as to whether his parents had once been able to read his thoughts: '. . . I made him pledge himself to submit to the one and only condition of the treatment', Freud noted, '—namely, to say everything that came into his head, even if it was unpleasant . . .'. The analysand, even as he described the rat punishment, pleaded to be released from this vow: 'I assured him that I myself had no taste whatever for cruelty and certainly had no desire to torment him', Freud added with apparent feeling for the subtle accusation thus brought against him, 'but that naturally I could not grant him something which was beyond my power. . . . The overcoming of resistances was a law of treatment and on no consideration could it be dispensed with.'

The patient, with the typical cunning of the obsessive neurotic, managed to twist the analytic rule into an instrument suited to his own purposes; moreover, the analyst was persuaded to condone a violation of the pledge by permitting the patient to arise from the couch and actively joining in the violation by revealing the contents of his own mind rather than discovering those of the patient—namely by himself speaking the magic words which were the equivalent of the action: 'into the anus'.

Freud observed: 'At all the more important moments while he was telling his story his face took on a very strange, composite expression. I could only interpret it as one of *horror at pleasure of his own of which he himself was unaware.*' The full significance of this pleasure might have been gleaned from the fact that '. . . the patient behaved as though he were dazed and bewildered. . . . He repeatedly addressed me as "Captain" . . .'. This Freud explained (completely overlooking the alternative possibility) as '. . . probably because at the beginning of the hour I had told him that I myself was not fond of cruelty like Captain M . . .'.

At the next session, the patient was still in conflict over a vow—this time, with regard to an obsessive ritual he had concocted for payment on a package delivered to him by the captain. This had precipitated his flight from the army encampment and refuge in Vienna, where he had sought relief from his guilt, first from a friend and then from Freud. His account of this affair during the session arouses the suspicion that again this was no mere anamnesis but had definite transference implications. Now that the analyst had taken the place of the cruel captain, was the patient planning anew to break a pledge (to undergo treatment), flee from Vienna and return to camp? He did, at any rate, tell Freud during this hour that he had originally sought a physician only to receive from him a certificate which would have enabled him to return to the Army and carry out the terms of his obsession. Such shuttling back and forth between persons and places is characteristic of this type of neurosis. Freud comments: 'Many months later, when his resistance was at its height, he once more felt a temptation to travel to P—— after all, to look up Lieutenant A. and to go through the farce of returning him the money'.

The succeeding interview presumably saw the continuance of the inner debate as to the ability of the patient to rely upon and confide in his physician. Fears of his own aggression and of retaliatory hostility preoccupied him. He reminisced about the sudden death of his father and his own ensuing guilt: his parent had passed away as the young man lay resting for an hour (death wishes toward Freud?); nevertheless, he had never quite accepted the reality of his father's death, and on hearing some witticism, would find himself thinking, 'I must tell father that' (the analytic rule?). When he walked into a room, he would expect to find his father in it (analyst's office?); at other times, however, he became so depressed by self-reproaches about his father's death that only the reassurances of a friend that he was not guilty were helpful—a circumstance which leads back to the patient's very first communications and his need to seek out persons (including Freud) who would attest their confidence in him. The friend or analyst, as unconscious surrogate fathers, could reassure him of his innocence in no more convincing way than to demonstrate, by their existence, that he had not killed them.

Freud responded to this with a lengthy theoretical discourse on the relationship between idea and affect in the neuroses, in order to persuade the patient that his sense of guilt must indeed have had some valid unconscious justification and to induce him to search within himself for the explanation of his self-reproaches. Thus analysis was sharply differentiated from the reassuring techniques which the young man had evolved for himself and which would have been gratified by most other forms of therapy. Nevertheless, reassurance was given indirectly by Freud, for when the patient inquired as to the value of discovering his hidden motives, he was told that his troublesome feelings would probably be dissipated in this way.

This implied absolution was followed by the confession, vague and tentative, of misdemeanors in childhood. Freud at once seized upon this and assured him at great length that precisely the incidents which occurred in early life were of the greatest importance and that the patient would discover for himself the laws of the unconscious. (In what sense this was meant is not quite clear, since the Rat Man was already acquainted with Freud's writings.) This intellectual explanation seems to have been accompanied by some signs of approval and satisfaction on the part of the analyst; moreover, the patient was again invited, in a friendly way, to prove his capacity for self-analysis. Actually, the analysand next reacted with suspicion; he wished to know whether the procedure thus advocated could really undo such long-standing ideas. Thereupon he was assured that Freud had formed a good opinion of him—a statement to which he reacted with 'visible pleasure'. In short, far more than theoretical expositions were employed by the therapist during this interview.

In any case, the resistance seems to have been lessened by these exchanges. At the beginning of the next session the patient renewed his earlier statements that he believed his parents could read his thoughts (a tribute to the deftness of the analyst?), and bolstered his courage at last to confess a childhood fantasy concerning his father's death. This recollection could now be conceded, since Freud had already guessed and shown tolerance of this idea. The latter was not yet satisfied, however, and after some further prodding and reassurance by means of 'theoretical explanations', further confessions were forthcoming.

Freud comments in a note that 'It is never the aim of discussions like this to create conviction. They are only intended to bring the repressed complexes into consciousness, to set the conflicts going in the field of conscious mental activity, and to facilitate the emergence of fresh material from the unconscious.' The Rat Man reacted to Freud's expositions with the now well-known tendency of the obsessive to draw the therapist into further intellectual explanations in the course of which the fundamental rule is increasingly reversed. In this instance, however, Freud neatly avoided the trap set for him by the patient's questions and remarked that surely the latter must have some answers prepared and need but follow the trend of his own thoughts to discover them.

There followed a chain of associations from, first, matters that might not be communicated to the father, to, second, envy of a younger brother, and, third, recollections of an incident in which he had enticed this brother to look into the barrel of a gun, whereupon he had pulled the trigger. The interpretation of the transference here may be made both actively and passively: in the tacit struggle over penetration of each other's minds, the patient both intended and feared aggression in his relation with the physician; the 'rat punishment' had thoroughly infiltrated the unconscious significance of the fundamental rule and the analytic task was to expose and dislodge it.

Freud pursued his course in this situation by insisting that the recollection of the incident with the brother was merely a cover for hostile intentions toward the father; in this way, the problem of concealed aggression was kept in the foreground.* The analyst's theoretical comments constituted a sharp probing instrument; the patient himself was, in the imagery of his own uncon-

---

* [The relevant passage in Freud's case report reads as follows: "We were very fond of each other at the same time, and were inseparable; but I was plainly filled with jealousy, as he was the stronger and better-looking of the two, and consequently the favourite. . . . We both had toy guns of the usual make. I loaded mine with the ramrod and told him that if he looked up the barrel he would see something. Then, while he was looking in, I pulled the trigger. He was hit on the forehead and not hurt; but I had meant to hurt him very much indeed. Afterwards I was quite beside myself, and threw myself on the ground and asked myself how ever I could have done such a thing. But I *did* do it [Freud speaking]—I took the opportunity of urging my case. If he had preserved the recollection of an action so foreign to him as this, he could not, I maintained, deny the possibility of something similar, which he had not forgotten entirely, having happened at a still earlier age in relation to his father" (pp. 184–185).

The passage is of interest from another point of view. Freud deflected prematurely, as we understand it today, the aggression from the brother to the father. This was done in the service of psychoanalytic theory. The brother was seen by Freud merely as a substitute for the Oedipal father and not as a rival in his own right. It is likely that if Freud had not intervened, the associations themselves would have gone toward sibling rivalry and death wishes for the brother. Eventually, the working through of this sibling rivalry would have lessened the Rat Man's guilt toward the brother, and only later would he have been ready to face the unconscious Oedipal hostility, i.e., the hostility toward the father. But Freud's commitment to the Oedipal hostility as the kernel of the neurosis did not permit him to follow the trend suggested by the patient. The short cut necessitated a reconstruction that could only constitute an intellectual acceptance, an obedience to Freud's theoretical position. Kanzer pursues the sexual meaning of the incident, whereas to me, the shortcutting of the guilt feelings toward the brother seems as important.]

scious, pressed increasingly into the plight of a cornered rat. At this juncture, he lamented his own 'cowardliness', but Freud mitigated his plight by telling him that he ought not consider himself responsible for the residue of infantile dispositions within himself. In this way, the patient managed to convert into analytic terms his habitual disposition to persuade a respected friend to assure him that he was not really a criminal despite his reprehensible impulses.

Unfortunately Freud concluded his formal presentation of the case at this point in order himself to discuss the theoretical aspects of obsessional neuroses, into which the Rat Man gave him such unprecedented insight. Some further references, however, enable us to glean a few details of the subsequent analysis. The memory of injuring the brother after persuading him to look into the gun barrel has unmistakable sexual implications which are probably illuminated by a strange ritual which the patient had at one time practiced. During his student days, and after the death of his father, he had developed the habit of interrupting his studies between twelve and one at night, and opening the door as though someone were standing outside. Then he would take out his penis and contemplate it in the looking glass.

Freud supposed, in explanation, that the patient was expressing ambivalence toward his father, seeking to please him by his diligence in studying late at night, but at the same time affronting him by his sex play. Certainly this incident suggests some transposition of the earlier one with the brother, who was lured under false pretenses into looking at the gun (penis). Transference implications also may be discerned in the one-hour interval during which the patient alternated between conforming to and defying his father's wishes (the fundamental rule); the opening of the door to confront the father's ghost which, as we have previously seen, had settled itself on the figure of the analyst; the struggle to control his exhibitionism (the urge to confess to the therapist, with ultimate sexual aims), and even in the role of the mirror (mirror = analyst?).

Evidence accumulated to justify Freud in declaring that 'it was only over the painful road of transference that [the patient] was able to reach conviction' of the truth of the theoretical postulates. He dreamed that he saw his analyst's daughter standing before him with two patches of dung instead of eyes, which Freud interpreted as meaning that he would marry the girl not for her beautiful eyes but for her money. (Various other possibilities arise if we consider the dream figure as the Rat Man himself.) Supplementary fantasies came in the form of visions of Freud as a wealthy and powerful man whose interest in the young man arose out of his desire to have him as a son-in-law. A testing of this hypothesis took place in the form of acting out: 'How could a gentleman like you, sir', he asked, 'let yourself be abused by a low good-for-nothing wretch like me? You ought to turn me out.' The occasion was then seized for violating and reversing the analytic rule by rising from the couch and striding up and down, watching the analyst and averring that he feared attack for his impudence—a situation that presented an advanced counterpart of the first session in which Freud was tacitly lured into enacting the part of the cruel captain and of the rat penetrating the anus. The transference elements were now more clearly to be discerned.

In retrospect, we may say that at this stage in the development of psychoanalysis, there was not yet a full appreciation of the extent to which memories of the past were represented by or constituted actual reflections of attitudes in the present. Reconstructions of former events were entered into in preference to a dynamic analysis of the immediate transference, a danger which Freud had recently come to see in the case of Dora, but whose application he had not yet entirely grasped. The footnotes on the Rat Man do indeed discuss the fact that 'childhood memories' are distorted and consolidated with the events of later years; much remained, however, to be worked out with respect to the resulting implications.

The predilection for reconstructing the past also played a large part, as may be noted in the Rat Man's case, in disposing to theoretical indoctrination of the patient and to the subsequent need to find supplementary means of providing him with emotional conviction. Transference interpretations, focused on the immediate affect (aggression toward and distrust of the analyst) are more apt to touch the affective core of the resistance, and fit more frequently with the rule that resistance shall be interpreted before content. Nevertheless, in retrospect, it may be seen with what skill and intuition Freud's theoretical explanations took cognizance of and dealt with the transference.

CHAPTER 11

ELIZABETH ZETZEL

ADDITIONAL NOTES UPON A CASE OF OBSESSIONAL NEUROSIS: FREUD 1909

INTRODUCTORY NOTE

When Zetzel wrote this paper she had available Freud's notes on the Rat Man. They were published in 1955, when Volume 10 of the Standard Edition *appeared. They were, therefore, not available to Kanzer. But the differences are more profound, and the two papers illustrate how two analysts can approach a clinical case in different ways.*

IT IS A GREAT HONOUR and an even greater responsibility to open this first Scientific Session for the 24th International Psycho-Analytical Congress. The patient I am discussing is not only well known—he was the subject of the

Originally published in *International Journal of Psycho-Analysis* 47 (1966):123–129. Reprinted under the title "An Obsessional Neurotic: Freud's Rat Man," in Elizabeth R. Zetzel, *The Capacity for Emotional Growth* (New York: International Universities Press, 1970), chap. 13. © Elizabeth R. Zetzel 1970. Reprinted by permission of International Universities Press. This paper was read at the Boston Psychoanalytic Society, May 26, 1965, and at the Twenty-fourth International Psycho-Analytic Congress, Amsterdam, July 1965.

first presentation at the first international meeting of psychoanalysts ever held, in April 1908. On this basis alone it seems appropriate that a Congress which plans to devote a significant part of its programme to a contemporary review of the obsessional neurosis and its psycho-analytic treatment should begin with a re-examination of the first and possibly most famous obsessional patient discussed in detail by Sigmund Freud.

It was my intention when I first accepted this assignment to base my discussion primarily on the 1909 report published in Freud's *Collected Papers*. Fortunately, however, I decided to re-read the case history in the *Standard Edition*. I was surprised and excited by the discovery I made—namely, the unique salvage of Freud's daily notes covering the first four months of this analysis. These informal notes, as Strachey suggests, permit us to identify ourselves with Freud's continuous scrutiny of the material presented by his patient; with his awareness of areas in which the patient's conflicts may have impinged on his own; and with his concurrent reflections as to the possible significance of this analysis for more general understanding of the obsessional neurosis. His frank allusions, finally, to his own participation serve as salutary reminders of the degree to which the papers in which Freud recommended coldness, neutrality, and mirror-like detachment were based on an implicit differentiation between the analyst's position *vis-à-vis* the transference neurosis and the man's warm and spontaneous participation in the one-to-one doctor-patient relationship which is an indispensable feature of the analytic situation.

The 1909 publication stands in its own right as one of the classics of psycho-analytic literature. It provided concrete empirical material to demonstrate the continued impact of early instinctual life in determining the content and nature of adult symptomatology. It defined and elaborated in relation to this patient's thought processes most of the mechanisms which characterize the obsessional neurosis: reaction formation, indecision, isolation, undoing, intellectualization, and magical thinking. Despite the emphasis given to oedipal content in Freud's explicit interpretations, the anal sadistic implications of the patient's basic conflict were also clearly recognized. The regressive reemergence of unconscious, unresolved conflicts in both symptom formation and transference analysis were convincingly demonstrated. Last, but by no means least, Freud's repeated reference to the patient's positive attributes highlights one of the major criteria for analysability—namely, availability of the healthy, intact part of the patient's personality as one partner in the analytic situation.

In his Introduction to the familiar 'Notes,' Freud described this patient's obsessional neurosis as one of moderate severity. This evaluation was made after the patient's analysis had been successfully completed. His symptoms, however, as described in the early phases of treatment, had at times been extremely disabling. The possibility must therefore be entertained that Freud's evaluation was determined by his implicit sensitivity to the conceptual distinction he was to make only two years later in 'The Two Principles of Mental Functioning' between pathology determined by developmental failure of the ego and symptomatology attributable to instinctual fixation and/or regression.*

Freud also acknowledged in the same Introduction his drastic curtailment of this case and its treatment. The original notes, which suggest that the Rat Man, like Irma,† moved in circles which impinged on Freud's social life, support one reason for many of the omissions—protection of the patient's anonymity. Would Freud, however, have saved these notes if all his omissions had served their purpose? Surely this would have been an excellent reason for destroying them. He himself hinted at another explanation:

> I must confess that I have not yet succeeded in completely penetrating the complicated texture of a *severe* case of obsessional neurosis. . . . An obsessional neurosis is in itself not an easy thing to understand—much less so than a case of hysteria.

Only four years later, in his paper on the 'Predisposition to Obsessional Neurosis', Freud himself indicated the degree to which both obsessional and hysterical symptomatology might prove highly deceptive. The criteria for analysability are not determined by the content or the severity of the presenting symptoms. As Knight (1953) has demonstrated, obsessional thinking and compulsive behaviour may serve as bulwarks, however unsatisfactory, behind which psychotic disorder remains partially hidden. It may be suggested, therefore, that the obsessional patients whom Freud described as either severe or

---

* ["Formulations on the Two Principles of Mental Functioning" (Freud 1911c). The relevant passage reads: "The supercession of the pleasure principle, by the reality principle, with all the psychical consequences involved, which is here schematically condensed into a single sentence, is not in fact accomplished all at once; nor does it take place simultaneously all along the line. For while this development is going on in the ego-instincts, the sexual instincts become detached from them in a very significant way. The sexual instincts behave auto-erotically at first; they obtain their satisfaction in the subject's own body and therefore do not find themselves in the situation of frustration which was what necessitated the institution of the reality principle; and when, later on, the process of finding an object begins, it is soon interrupted by the long period of latency, which delays sexual development until puberty. These two factors—auto-erotism and the latency period—have as their result that the sexual instinct is held up in its psychical development and remains far longer under the dominance of the pleasure principle, from which in many people it is never able to withdraw.

"In consequence of these conditions, a closer connection arises, on the one hand, between the sexual instinct and phantasy and, on the other hand, between the ego-instincts and the activities of consciousness" (p. 222).

The difference between pathology determined by developmental failure and pathology resulting from intrapsychic conflict which causes the ego to regress to pregenital fixations is gaining in importance in recent psychoanalytic literature. See particularly Anna Freud 1972.]

† [Irma, Freud's patient, the subject of the first dream to be interpreted in *The Interpretation of Dreams*.]

not yet understood may have differed from the Rat Man in respect of basic ego functions. The Wolf Man obviously comes to mind as a case in point.

Kanzer [1952; see this volume, Chapter 10] has emphasized the acting out element in this patient's behaviour during critical stages of his analysis. I would propose, rather, that this same behaviour which after all occurred before the sanctification of the couch illustrated in dramatic form those ego attributes which are prerequisite to the crucial therapeutic split between fantasy or transference and reality or therapeutic alliance. The Rat Man, it will be recalled, remained at all times aware of and disturbed by the ego alien negative transference fantasies which determined his behaviour. Despite, in addition, his intellectual defences, his tendency to isolate, and his use of denial, the Rat Man demonstrated in his dreams, his fantasies, and his associations the capacity for instinctual regression which is a necessary concomitant of an analysable transference neurosis.

The original publication, successful though it was in demonstrating both the form and the content of obsessional symptoms, remained difficult and obscure in respect of some of Freud's theoretical speculations. This is particularly striking in his efforts to account for the inexorable either-or which characterized the alternating feelings of love and hate which the Rat Man directed towards his father and his lady:

> The conflicts of feeling in our patient which we have here enumerated separately were not independent of each other but were bound together in pairs. His hatred of his lady was inevitably coupled with his attachment to his father and inversely, his hatred of his father with his attachment to his lady. But, (and this is the statement I wish to emphasize), the two conflicts of feeling which result from this simplification . . . namely, the opposition between his relation to his father and to his lady, and the contradiction between his love and his hatred within each of these relations—had nothing to do with each other either in their content or in their origin.

Freud's reference to an inexorable 'either-or' and his sharp differentiation between the dichotomy masculine/feminine and the conflict between love and hate within individual object relations are worthy of comment. It might be suggested that his discussion of these problems might well be compared with his own references to the obscure and puzzling features of those dream elements which impinge most closely on problems of decisive and crucial importance. Freud himself presented a hypothesis for the occurrence of 'such a strange state of affairs' which might aptly be cited by a follower of Klein or Fairbairn:*

> At a very early age, somewhere in the prehistoric period of infancy, the two opposites should have been split apart.

* [W. H. D. Fairbairn, a Scottish psychoanalyst, whose book *An Object-Relations Theory of the Personality* was published by Basic Books in 1954. Although strictly speaking not a follower of Melanie Klein, his orientation is close to that of the English school.]

What is implicit in this discussion is the distinction which could not have been made in 1909 between substantial failure to integrate perceptions and emotions initially experienced as mutually exclusive—e.g. pain and pleasure, love and hate, activity and passivity, omnipotence and helplessness, and regressive impairment during neurotic symptom formation of fusions and integrations which had previously been established. Recognition and substantial mastery of the conflict between love and hate which Freud described as 'a strange state of affairs' is familiar to us today as one of the crucial developmental tasks integral to healthy self-object differentiation and early ego identification. The developmental achievements which determine at least one of the criteria for analysability concern just this specific area. The individual who, like the Rat Man, is capable of maintaining a real object relationship despite the emergence of conflicting negative feelings has been able, at whatever sacrifice, to recognize and tolerate concurrent feelings of love and hate towards one and the same object. His love, moreover, though perhaps by a narrow margin, has been substantially successful in achieving what might truly be described as a pyrrhic victory. In his case, this was shown by the crippling inhibitions, the shifting doubt, and the smoke cloud of compulsions which characterized his severe but nevertheless analysable obsessional neurosis. Despite the inexorable either-or which characterized the alternation between love and hate in his neurotic symptom formation and transference neurosis, the Rat Man proved capable of tolerating considerable ambivalence in the analytic situation.

It is not surprising that Freud's brilliant speculations touched on the phase of psychic development in which this capacity is first initiated, namely, what was then the prehistoric period of early infancy. Neither the importance of early object relations nor their possible relevance to the analytic situation had been envisaged in 1909. Freud's then current theory of object relations was indicated by a long footnote in which he stressed the overriding importance of early autoerotism and instinctual gratification. It is well known, however, that the stages of ego development and significant object relations which intervene between autoerotic activity preceding self-object differentiation and the capacity for adult heterosexual object love still remain one of the most difficult and controversial areas in psycho-analytic theory.

Freud's reconstruction of the Rat Man's early development was inevitably based on his approach to instinctual impulses and autoerotic gratification. The father was seen as an important real object—one who interfered with or threatened his son's instinctual impulses. Early object love, either pregenital or genital, was given relatively little attention. The patient's mother, for example, was only mentioned in six brief, essentially unrevealing, statements. Although, in addition, Freud acknowledged the possible importance of the death of the patient's older sister, he was led to the reconstruction that its primary significance related to the patient's subsequent conviction that 'you die if you masturbate'.

In striking contrast with the 1909 publication, there are more than forty references to a highly ambivalent mother-son relationship in the original clinical notes. Freud published his initial consultation almost verbatim with one significant omission: 'After I had told him my terms, he said he must consult his mother.' The patient, it will be recalled, was 29 years old at that time. On 18 October he reported that he had not taken over his inheritance from his father but had left it with his mother who allowed him a small amount of pocket money. Mention of his mother was relatively scanty during the first weeks of his analysis. As the analytic situation became more secure, however, there is evidence to suggest an increasingly positive identification with Freud, who thus noted with some pleasure on 8 December:

> He has stood up *manfully* against his mother's lamentation over his having spent thirty florins of pocket money during the last month instead of sixteen.

On 19 December his negative feelings about his mother became manifest and intense:

> He gets everything that is bad in his nature from his mother's side. He hands over all his money to his mother because he does not want to have anything from her.

These and many other references to financial problems, cleanliness and dirt, hostile fantasies and the reaction formations against them clearly point to a major area of instinctual fixation.

There are, in addition, notes which suggest that Freud perceived, although he did not conceptualize, the type of mother-child relationship and ego identification characteristic of many future obsessional characters and obsessional neurotics. On 21 December Freud says:

> He has been identifying himself with his mother in his behaviour and treatment transference. . . . It seems likely that he is also identifying himself with his mother in his criticisms of his father and is thus continuing the differences between his parents within himself.

Could we have on the one hand a better description of the process later to be defined as introjection? Is there anywhere in our literature a more precise account of the mechanism Anna Freud was to describe as 'identification with the aggressor'? Longitudinal observations of young children have in recent years demonstrated the significance of this defensive identification as one important precursor of the harsh superego of the future obsessional.

It must be noted, however, that this highly ambivalent relationship with his mother was not expressed in the opening phases of the Rat Man's analysis. It only emerged after the patient had mastered some of his ambivalence and established a positive therapeutic alliance with his analyst, Freud. The fact that he could do so raises questions as to how far unresolved ambivalence and significant identification with the aggressor had characterized this patient's initial relationship with his mother. Alternative hypotheses which might help us to understand his positive attributes may be suggested: First, that an essentially positive infantile mother-child relationship had been threatened or impaired by the birth of a younger brother when he was 18 months old. Second, that he had turned during his second and third years to a sister enough older to have enjoyed a maternal role. Third, that his pre-oedipal relationship with a father who emerges as an essentially warm and loving parent had been predominantly positive. An essentially normal, although partially displaced, oedipal triangle, may thus have emerged before the onset of this sister's fatal illness. Both his severe childhood neurosis and his adult predisposition to an obsessional illness might in this case be attributed to certain regressive responses to trauma rather than the continuation into adult life of initial developmental failure.

The original notes provide many hints of the importance of this relationship to both children. Katherine's attachment to the patient is indicated by her statement: 'On my soul, if you die I shall kill myself.' The patient reports a few recollections of Katherine's incipient illness. He remembers someone carrying her to bed; he remembers that she had for a long time complained of feeling tired. 'Once when Dr. P. was examining her he turned pale.' He also remembers asking: 'Where is Katherine?', and described his father sitting in a chair weeping. His famous—but not subjectively remembered—outburst of rage almost certainly occurred during the course of Katherine's fatal illness. In this affective storm the little boy attacked his father, calling him a 'towel', 'lamp', and 'plate'. Was this choice of inanimate objects determined, as Freud suggests, by the patient's lack of a wider vocabulary? Was it indicative of a direct death wish towards an oedipal rival? The separation from and impending loss of an important early object must also be considered. The outburst could have represented on the one hand his desperate longing for his sister. The terms of abuse might then have an additional meaning—reproach to a loved but devalued father for his withdrawal, unhappiness, and inability to help or console the anguished child. Not only the sister, but also the father failed to meet the child's need for love and support.

I have elsewhere [Zetzel 1964] related similar acute affective responses to separation to the recognition and tolerance of depression as an ego state. Relative developmental failures in this specific area represent one important determinant of the ego defences which are predominant in the obsessional neurotic. Bornstein [1949] related the childhood neurosis of the future obsessional Frankie to his attempts to ward off depression during a period of separation from his mother. Neither the Rat Man nor Frankie appear to have acknowledged or demonstrated overt depression. The context within which the

Rat Man's outburst of rage occurred suggests, however, that Katherine's illness and death may well have mobilized regressive defences against the re-emergence of depressive anxiety and related feelings of helplessness. This may have involved subsequent reinforcement of the defences characteristic of the earlier anal sadistic developmental period—(e.g. magical thinking, reaction formation, isolation, and intellectualization). It may also have led to substantial retreat from the triangular relations integral to the genital oedipal situation to the more primitive one to one relations of an earlier period. The inexorable either-or which characterized his adult neurosis may thus have represented a revival in adult life of this earlier regressive response to trauma. The reported memory of childhood fears lest his parents could read his thoughts suggests, in addition, threatened impairment of self-object differentiation and the use of projection as a mechanism of defence.

The early development of individuals who become healthy or analysable adults is characterized by the pregenital achievement of genuine one-to-one relations with both parents. In such circumstances the oedipal conflict can emerge and develop without sacrifice of sustained object relationships. Substantial developmental failure in the capacity for such object relations, although it may not preclude incestuous oedipal fantasies, usually retains an all or nothing quality which impairs individual capacity for a positive therapeutic alliance. Such developmental failure should be differentiated as far as possible from regressive responses to traumatic experience which may sometimes present misleadingly similar symptomatology.

The loss of an incestuous object at a time when the attachment is intense may have prolonged long-term after-effects. Insofar as the child experiences the loss as a punishment for his sexual wishes, his inhibitions, guilt, and ambivalence would, as Freud indicated, be considerably increased. In addition, however, the ability to recognize and work through later bereavements may be seriously impaired as a result of early loss. Denial, a defence which gradually diminished during healthy maturation, may be substantially reinforced. The Rat Man's continued use of denial in adult life was conspicuous in his striking inability genuinely to recognize, grieve, or accept the finality of his father's death. He failed, for example, to mention the fact that his father had been dead for nearly ten years when he first recounted the story of the rat punishment. Other episodes underlined his persistent feeling that his dead father might walk into the room. He frequently thought about him as though he were still alive. Although many of his fantasies in the area of sexual activity were overtly hostile, an undertone of sustained positive feeling is clearly apparent. The patient would have welcomed his father's return. His positive object relationship with his father appears to have been *at least* as important as the hostile oedipal rivalry which was stressed in the 1909 publication.

That Freud recognized the importance of the Rat Man's denial is indicated by his note of 21 December:

> I pointed out to him that this attempt to deny the reality of his father's death was the whole basis of his neurosis.

Freud is here obviously referring to his adult neurosis. I would like to suggest that a parallel but much earlier denial in respect of his sister Katherine had at least equal importance for his predisposition. This, I believe, also determined his attachment in adult life to a young woman, Gisela, in whom he found a suitable replacement for his dead sister. From the published notes and daily record we get a picture of Gisela as (i) a first cousin; (ii) possibly too old for him (her age is not mentioned); (iii) almost certainly sterile, a fact which made her resemble a prepuberty little girl; and (iv) a woman who was subject to frequent serious and disabling periods of ill health. In addition, the fact that this cousin who was herself highly ambivalent may also have been 'abused' by her stepfather, and was at least as disturbed in respect of her psychosexual life as the patient, suggests that her own personality loaned itself to a relationship characterized by many infantile features.

There is a wealth of material in the original notes to support the hypothesis that the Rat Man's persistent attachment to his ailing cousin represented an overdetermined, necessarily ambivalent effort to revive his sister as he last recalled her, namely, as an increasingly tired little girl who was finally carried away to the room in which she was to die. Recovery of this lost object entailed sacrifice—that is, substantial renouncement of libidinal wishes. On 27 October he dreamt, in this context, that another sister was very ill. A friend told him: 'You can only save your sister by renouncing all sexual pleasure.' His cousin was not only sterile; she also suffered periods of illness during which it may be assumed that sexual interest was prohibited. During one such illness 'when his affection and sympathy were at their greatest, she was lying on a sofa and he suddenly thought: "May she lie like this forever!"' Although the hostile death wishes which Freud deduced from this incident are not to be excluded, underlying fear of loss must also be acknowledged. Insofar as Gisela represented Katherine, her illness may have been experienced as imminent death in the Rat Man's repressed unconscious.

It may be suggested that just as Katherine's death had precipitated a childhood regression, his father's death, before he had reached full maturity, not only impeded adaptive utilization of a developmental second chance, but also undermined the precarious adjustment so far maintained. Neither loss, however, led to irreversible ego regression, as shown by his capacity to tolerate a difficult analytic situation. Some of the difficulties, it may be suggested, may have derived from his regressive wish to reestablish his passive pre-oedipal father-son relationship. Such wishes would inevitably conflict with oedipal rivalry and the unconscious search for the lost heterosexual object.

These passive wishes may well have been an important factor in his intolerance of the couch and the defensively overdetermined negative transference material.

The sustained positive undertone of the definitive father-son relationship may go far to account for both the qualities which Freud admired and for the stability of his alliance in the analytic situation. His periods of greatest symptomatic distress related to information and/or fantasy which devaluated Freud or his father. The significance of such devaluation as a death wish leading to guilt about oedipal rivalry is not to be minimized. The positive wish for a strong father as an ego ideal and an object for identification appears to have been at least equally important.

The analytic situation is a one-to-one relationship which draws on the strengths and reveals the weaknesses of the initial mastery of ambivalence in an essentially passive situation. The early mother-child relationship has been mentioned by many analysts. For example, Gitelson (1962) made explicit reference to the importance of the analyst's diatrophic responses during the opening phases of clinical psycho-analysis. Greenacre has referred to the 'matrix' of transference. I have myself, particularly in the paper (Zetzel, 1965) published in honour of Hartmann's 70th birthday, attempted to delineate the parallels and the differences between the parents' empathic responses to a young child and the analyst's intuitive responses to his patient's affective needs.

It may be suggested that this patient's continued unresolved ambivalence towards his mother might have made him vulnerable to ego regression in an uncommunicative analytic situation. Freud's spontaneous responses, however, as reported during the first few months of the Rat Man's analysis, appear to have differed considerably from his later theoretical models—his communications were not limited to interpretation of the transference neurosis. He acknowledged his patient's anxiety and took him into his confidence. He praised and encouraged him. He corrected realistic misinformation and explained the analytic reasons why he could not allow the patient to withhold names. Despite, moreover, the somewhat intellectualized terms in which some of his verbalizations were phrased, the underlying atmosphere appears to have been one of mutual respect and considerable understanding. If, therefore, my hypothesis in respect of the patient's early feelings towards his father is correct, his therapeutic alliance may have derived from this positive one-to-one relationship.

The original notes reveal Freud's comfort in correcting realistic information and in other spontaneous interactions which might be regarded as subject to question today. His subsequent recommendation of neutrality may have represented recognition of the fact that such interventions may sometimes prove unfortunate. Other patients may have responded less favorably to the type of activity which Freud showed in this analysis. The Rat Man's responses nevertheless illustrate a point which cannot be too strongly emphasized in our understanding of clinical psycho-analysis. A good analytic situation, although it may temporarily be distorted or modified, will not be undermined by occasional defects from traditional technique* on the part of the analyst. If a good analytic situation has not been achieved, technically correct interpretations will have little, if any, therapeutic value.

I will give two brief examples to illustrate Freud's technique in this analysis. Someone had told the patient that a distant Hungarian relative of Freud's had been a criminal. The patient only reported this gossip after a painful struggle. Freud laughingly relieved the patient's anxiety, saying that he had no relatives in Budapest. Two days later the patient reported a more significant realistic reinforcement of his active negative transference neurosis. His sister had once remarked that Freud's brother Alex would be the right husband for the patient's lady. The patient's fear lest Freud had designs on him as a husband for his daughter was compounded by the fantasy that the patient's lady would be taken over by Freud's brother. The familiar hostile transference material was thus doubly determined. This example illustrates on the one hand the realistic reasons for certain omissions. The patient's ability to report this disturbing gossip, suggests, however, that his response to the first correction had been helpful rather than harmful.

The second illustration is both startling and unusual. Freud's notes for 28 December commence as follows: 'He was hungry and was fed.' Direct responses to oral demands have sometimes been mentioned as a concomitant to the treatment of psychotic patients. The indications for such procedures relate to contemporary developmental theory in relation to the genesis of psychosis. As already noted, however, Freud's 1909 understanding of psychic development placed little explicit emphasis on early maternal functions. It is highly improbable, therefore, that he regarded his action as a therapeutic manoeuvre. Just as the correction of misinformation, however, had been followed on the earlier occasion by further revelations, it is noteworthy that the patient felt free to reject in words the gratification which he had partially accepted in fact. During the same hour he referred to his need to diet in order to lose weight. Within the next few days he verbalized with greater freedom the identification with his mother as an aggressor which I have already noted. He mentioned, in addition, that he had left the herring which had been offered untouched because he 'disliked herring intensely'. These responses suggest that an intervention which must be defined as unanalytic had not impeded the progress of this patient's treatment. The fact that he could reveal with increasing clarity his hostility to his mother, his comfort in rejecting part of the meal,

---

* [The phrase "occasional defections from traditional technique" is unfortunately vague for a scientific discussion. For a more rigorous approach to the problem of deviation from the standard technique, compare Eissler 1953.]

verbalizing in this context certain criticisms of Freud, confirms the positive therapeutic alliance which he had achieved by the end of the year.

Neither the published report nor the original notes permit us fully to understand the significance of his symptomatic recovery. The former demonstrated both a positive therapeutic alliance and the emergence of an analysable transference neurosis. The original notes permit us to reconsider certain aspects of his childhood neurosis and adult predisposition within the context of contemporary theory. In summary, I would suggest that the little boy who grew up to be the Rat Man might not have developed a serious neurosis were it not for the impact of a significant loss sustained at the height of the infantile neurosis. His relatively brief psycho-analysis appears at the very least to have helped him to retrieve the developmental achievements which had thus been undermined in early childhood. His positive identification with a father surrogate, Freud, may have been the central factor which impelled him towards greater mastery of unresolved intrapsychic conflict. The underlying vulnerability in his relationship with his mother may have remained a potential Achilles' heel. He may, however, have become a well integrated, somewhat obsessional character instead of a decompensated obsessional neurotic. We have no final note as to his definitive heterosexual achievement. It is nevertheless evident that Freud acted as an ally rather than a hostile menace in the patient's efforts to reintegrate genital potency and heterosexual object love. Whether and how far symptomatic remission would have enabled him, had he survived World War I, to reach and sustain his full potentiality must always remain an open question. Freud's willingness, however, to let him try his wings once his serious symptoms had disappeared is relevant to the indications for interruption or termination of psycho-analysis. This patient might have been caught in an interminable analysis if theoretical considerations had taken precedence over the demands of reality.*

Like Freud's publication, my discussion has focussed mainly on the patient's early development and the opening stages of his therapeutic analysis. The original notes have provided not only new information about this patient's pathology; they have also enriched our understanding of the more positive qualities to which Freud made several references. The patient's delightful sense of humour and his capacity for imaginative fantasy not only indicate the maturity of certain ego functions; they also illustrate his capacity to regress in the service of the ego. One of his fantasies suggests that his

---

* [Zetzel's evaluation of Freud's technique seems excessively deferential. It is one thing to appreciate Freud's work in the historical perspective, and quite another to hold it up as a model in 1966. The very fact that the role of the mother was neither recognized nor analyzed makes it impossible to call this therapy a psychoanalysis. It would seem more correct to say that the Rat Man had a valuable psychotherapeutic experience, with Freud as an auxiliary superego supplying an important corrective experience, but not a psychoanalysis. Since the Rat Man was killed during World War I, his subsequent development could not be evaluated.]

claims in respect of the power of his thoughts and wishes were not entirely without foundation. He reported an encounter with a little girl of 12 whom he had seen on the stairs of Freud's house. Whether or not his conviction was correct, he perceived this child as a daughter of the house, and Anna Freud was 12 years old during the last months of 1907. The possibility must be entertained that the Rat Man's interpretation of his own fantasy was prophetic. The child represented the new and young science, psycho-analysis. Although Miss Freud does not wish any formal celebration in this year of her 70th birthday, I would like to conclude with an almost verbatim quotation of the spontaneous tribute suggested by the Rat Man himself:

> It was the child who solved the problem with gay superiority—with smiling virtuosity she has indeed exposed many of the disguises which determine both the predisposition to and the treatment of an obsessional neurosis.

CHAPTER 12

ERIK H. ERIKSON

REALITY AND ACTUALITY: AN ADDRESS

An Excerpt

INTRODUCTORY NOTE

From a larger paper devoted to the meaning of actuality, a section dealing with Freud's patient Dora has been excerpted (Freud 1905a). Erikson has enriched psychoanalysis with a number of new concepts. Of these, ego identity, identity diffusion, basic trust, fidelity, and generativity have gained general currency (see Erikson 1959). In this paper he uses the Dora case to illustrate yet another concept, actuality. His approach throws new light on a failure in communication that took place between Freud and his young patient. For a further discussion of Dora see K. Lewin 1973–1974 and Marcus 1974.

WHEN WE USE Freud's work for the elucidation of what we are groping to say, it is for one very practical reason: all of us know the material by heart. Beyond this, we always find in Freud's writings parenthetical data worthy of the attention of generations to come. We must assume, of course, that Freud selected and disguised the clinical data he published, thus rendering reinterpretations hazardous. Yet, the repeated study of Freud's case reports strengthens the impression that we are dealing with creations of a high degree

Excerpted from "Reality and Actuality: An Address." Originally published in *Journal of the American Psychoanalytic Association* 10 (1962):454–457, 459–461. © 1962 by the American Psychoanalytic Association. Reprinted by permission of International Universities Press.

of psychological relevance and equivalence even in matters of peripheral concern to the author. Freud concludes his report on the treatment of Dora with an admission as frank as it is rare in professional publications: "I do not know what kind of help she wanted from me" (1905a).

Dora, you will remember, had interrupted the treatment after only three months, but had come back a year later (she was twenty years old then), "to finish her story and to ask for help once more." But what she told him then did not please Freud. She had in the interval confronted her family with certain irresponsible acts previously denied by them and she had forced them to admit their pretenses and their secrets. Freud considered this forced confrontation an act of revenge not compatible with the kind of insight which he had tried to convey to the patient through the analysis of her symptoms. The interview convinced him that "she was not in earnest over her request" for more help, and he sent her away. His displeasure he expressed in the assurance—apparently not solicited by the patient—that he was willing "to forgive her for having deprived [him] of the satisfaction of affording her a far more radical cure for her troubles" (1905a). Since Dora was intelligent, however, the judgment that she was "not in earnest" suggested insincerity on her part. And, indeed, Felix Deutsch who was consulted by Dora in her late middle age gives an unfavorable picture of her fully developed character—as unfavorable as may be seen in clinical annals (1957). Yet, in Freud's original description of the girl, Dora appeared "in the first bloom of youth—a girl of intelligent and engaging looks." If "an alteration in her character" indeed became one of the permanent features of her illness, one cannot help feeling that Dora was, as it were, confirmed in such change by the discontinuance of her treatment.

The description of Freud's fragmentary work with Dora has become the classical analysis of the structure and the genesis of a hysteria. It is clear from his description that Freud's original way of working and reporting was determined by his first professional identity as a physiological investigator: his clinical method was conceived as an analogy to clean and exact laboratory work. It focused on the "intimate structure of a neurotic disorder"—a structure which was really a reconstruction of its genesis and a search for the energies, the "quantities of excitation," which had been "transmuted" into the presenting symptoms, according to the dominant physicalistic configurations of his era (Erikson 1956a).

As to the unbearable excitations "transmuted" into Dora's symptoms, may it suffice to remind you of the two traumatic sexual approaches made to the girl by a Mr. K., a married man who kissed her once when she was fourteen under circumstances indicating that he had set the scene for a more thorough seduction; and who propositioned her quite unequivocally at an outing by an Alpine lake when she was sixteen. She had rebuked the man; but the sensations, affects, and ideas aroused on these two occasions were translated into the symptom language of hysteria, which was then decoded by Freud. But how clinically alive and concrete is his question as to what more, or what else, Dora had a right to expect of him. He could not see, Freud relates, how it could have helped her if he "had acted a part . . . and shown a warm personal interest in her." No patient's demands, then, were to make him dissimulate his integrity as an investigator and his commitment to the genetic kind of truth: they were *his* criteria of the respect due to a patient. But if in the patient's inability to live up to his kind of truth Freud primarily saw repressed instinctual strivings at work, he certainly also noted that Dora, too, was in search of some kind of truth. He was puzzled by the fact that the patient was "almost beside herself at the idea of its being supposed that she had merely fancied" the conditions which had made her sick; and that she kept "anxiously trying to make sure whether I was being quite straightforward with her." Let us remember here that Dora's father had asked Freud "to bring her to reason." Freud was to make his daughter let go of the subject of her seduction by Mr. K. The father had good reason for this wish, for Mr. K.'s wife was his own mistress, and he seemed willing to ignore Mr. K.'s indiscretions if he only remained unchallenged in his own. It was, therefore, highly inconvenient that Dora should insist on becoming morbid over her role as an object of erotic barter.

I wonder how many of us can follow today without protest Freud's assertion that a healthy girl of fourteen would, under such circumstances, have considered Mr. K.'s advances "neither tactless nor offensive." The nature and severity of Dora's pathological reaction make her the classical hysteric of her case history; but her motivation for falling ill, and her lack of motivation for getting well, today seem to call for developmental considerations. Let me pursue some of these.

Freud's report indicates that Dora was concerned with the historical truth as known to others, while her doctor insisted on the genetic truth behind her own symptoms. At the same time she wanted her doctor to be "truthful" in the therapeutic relation, that is, to keep faith with her on her terms rather than on those of her father or seducer. That her doctor did keep faith with her in terms of his investigative ethos she probably appreciated up to a point; after all, she did come back. But why then confront him with the fact that she had confronted her parents with the historical truth?

This act may impress some of us even today as "acting out." With Freud, we may predict that the patient would gain a permanent relief from her symptoms only by an ever better understanding of her own unconscious, an understanding which would eventually permit her to adjust to "outer reality," meaning to what cannot be helped. Strictly speaking, however, we could expect such utilization of insight only from a "mature ego," and Dora's neurosis was rooted in the crisis of adolescence.* The question arises whether today we would consider the patient's emphasis on the historical truth a mere matter of resistance to the genetic one; or whether we would discern in it also an adaptive pattern genuine for her stage of life, and challenged by her circumstances.

To return once more to Dora: if fidelity is a central concern of young adulthood, then her case appears to be a classical example of fatefully perverted fidelity. A glance back at her history will remind us that her family had exposed her to multiple sexual *infidelity*, while all concerned—father and mother, Mr. K. and Mrs. K.—tried to compensate for all their pervading *perfidy* by making Dora their *confidante*, each burdening her (not without her perverse provocation, to be sure) with half-truths which were clearly unmanageable for an adolescent. It is interesting to note that the middle-aged Dora, according to Felix Deutsch's report, was still obsessed with infidelities—her father's, her husband's, and her son's; and she still turned everybody against everybody else (1957). But lest it appear that I agree with those Victorian critics to whom Dora seemed only a case illustrating typical Viennese and sexual infidelity, I must add that other and equally malignant forms of fidelity-frustration pervade late adolescent case histories in other societies and periods.

If fidelity, then, emerges against the background of diverse historical perspectives, *identity*—as I had an opportunity to report to you a few years ago—must prove itself against sometimes confusing role demands (Erikson, 1956b). As a *woman*, Dora did not have a chance. A vital identity fragment in her young life was that of the *woman intellectual* which had been encouraged by her father's delight in her precocious intelligence, but discouraged by her brother's superior example as favored by the times: she was absorbed in such evening education as was then accessible to a young woman of her class. The negative identity of the *"déclassée" woman* (so prominent in her era) she tried to ward off with her sickness: remember that Mr. K., at the lake, had tried to seduce her with the same argument which, as she happened to know, had previously been successful with a domestic. She may well have sought in Mrs. K., whom Freud recognized primarily as an object of Dora's ambivalent homosexual love, that mentor who helps the young to overcome unusable identifications with the parent of the same sex: Dora read books with Mrs. K. and took care of her children. But, alas, there was no escape from her mother's *"housewife's psychosis,"* which Dora blended with her own then fully acquired *patient identity*. Felix Deutsch reports that the middle-aged Dora, "chatting in a flirtatious manner . . . forgot about her sickness . . . displaying

---

* For a consideration of prolonged adolescence, see P. Blos, "Prolonged Adolescence: The Formation of a Syndrome and Its Therapeutic Implications," *Amer. J. Orthopsychiat.* 24 (1954):733–742.

great pride in having been written up as a famous case" (1957). To be a famous, if uncured, patient had become for this woman one lasting positive identity element; in this she kept faith with Freud. We know today that if patienthood is permitted to become a young patient's most meaningful circumstance, his identity formation may seize on it as a central and lasting theme (Erikson 1956b).

This brings us, finally, to the question of the developmental aspects of the therapeutic relationship itself. At the time, Freud was becoming aware of the singular power of transference and he pursued this in his evidence. Today we know that this most elemental tie always is complemented by the patient's relation to the analyst as a "new person."* Young patients in particular appoint and invest the therapist with the role of mentor, although he may strenuously resist expressing what he stands for. This does not obligate him, of course, to "play a part," as Freud so firmly refused to do. True mentorship, far from being a showy form of emotional sympathy, is part of a discipline of outlook and method. But the psychotherapist must recognize what role he is, in fact, playing in what we are here trying to circumscribe as the actuality of a young person.

In summary, it is probable that Dora needed to act as she did not only in order to vent the childish rage of one victimized but also in order to set straight the historical past so that she could envisage a sexual and social future of her choice; call infidelities by their name before she could commit herself to her own kind of fidelity; and establish the coordinates of her identity as a young woman of her class and time, before she could utilize more insight into her inner realities.

Beyond the case of Dora, however, we face here a problem of general therapeutic urgency: some mixture of *"acting out"* and of *age-specific action* is to be expected of any patient of whatever age; and all patients reach a point in treatment when the recovering ego may need to test its untrained or long-inhibited wings of action. In the analysis of children, we honor this to some extent; but in some excessively prolonged treatments of patients of all ages, we sometimes miss that critical moment while remaining adamant in our pursuit of totally cleansing the patient of all "resistance to reality." Is it not possible that such habitual persistence obscures from us much of the ego's relation to action, and this under the very conditions which would make observation possible on clinical homeground?

You may wonder whether Dora's dreams, the focus of Freud's analytic attention, support the emphasis which I am adding here to his conclusions. A comprehensive answer to this question would call for a discussion of the representation of ego interests in dreams. As an example, I can propose only

---

* This has been most forthrightly formulated by Loewald [1960] who anticipates much of my argument about the role of reality testing within the actuality of the therapeutic relationship.

most briefly that in Dora's first dream the *house* and the *jewel case*, besides being symbols of the female body and its contents, represent the adolescent quandary: if there is a fire in "our house" (that is, in our family), then what "valuables" (that is, values) shall be saved first? And indeed, Freud's interpretation, although psychosexual and oedipal in emphasis, assigns to the father standing by the girl's bed not the role of a wished-for seducer, but that of a hoped-for protector of his daughter's inviolacy.

Part Six

Psychoanalytic Technique in the Twenties

CHAPTER 13

KARL LANDAUER

"PASSIVE" TECHNIQUE

On the Analysis of

Narcissistic Illnesses

INTRODUCTORY NOTE

Landauer's article appears here in translation for the first time. It is rich in new ideas, a pioneering article, anticipating developments as yet to come. In part, Landauer is responding to Ferenczi's "Active Technique" by an emphasis on his own passive technique. However, beyond the dispute with Ferenczi, Landauer's article forms an indispensable link to the understanding of the place of Wilhelm Reich in the history of psychoanalytic technique. Because the article had not been translated, it is not mentioned in the general textbooks, such as Fenichel (1945), Glover (1955), Greenson (1967), and Menninger (1958). For the same reason, Reich's indebtedness to Landauer has gone unnoticed.

PSYCHOANALYSIS aims at making what is unconscious conscious and to fit into reality these strivings which seek nothing but pleasure.* The uncon-

Originally entitled " 'Passive' Technik: Zur Analyse narzisstischer Erkrankungen," *Internationale Zeitschrift für Psychoanalyse* 10 (1924):415–422. Abstracted by Edward George Glover in *International Journal of Psycho-Analysis* 6 (1925):467–468. Reprinted by permission.
    * [This is an interesting definition of the aim of psychoanalytic therapy: to make the unconscious conscious and to bring to the unconscious strivings that had been made conscious under the domination of the reality principle. The second aim has often been neglected.]

scious is reached by way of the patient's free associations as well as his
actions, his entire conduct, that is to say, his communications in words on the
one hand, on the other in representations.* Since the physician within the
four walls of the treatment room is the analysand's only living object, he
becomes his opposite player [*Gegenspieler*] in this representation. The major
part of the material appears in this form of "transference onto the physician."
It is his task as early as in the *first indications* contained in the representations
to recognize the instinctual forces and to render them conscious to the patient.

The procedure is lengthy and demands expenditure of time, not only
from the one who is seeking help, but above all from the physician; this
means, it demands love and patience; he has to force back curiosity and his
urge to be active just as he has to renounce quick successes which would
gratify his own narcissism. This is not easy and the wish to shorten the
process is natural. One is only too ready to step out of one's reserve and,
forgetting that it is a question of repressed impulses that are being kept from
the patient's consciousness by countercurrents, to impose actively upon him
one's own wishes, *one's own* associations, *one's own* self. However, in this
way one never succeeds in removing the cause of illness, the patient's *individual* repressions. In this way the analyst is never able to undo the patient's *very
own* regressions nor to help him adapt to *his own* reality.

However, there seems to be one course of action that shortens the analysis without relinquishing the direction indicated by analytic theory. An observation shows the way, an unintentional observation made frequently by
every analyst in consequence of a technical error; it happens often that the
first indications of impulses of transference—in practice usually negative object relations and projections—are overlooked by the physician who is at-
tuned to communication in words. Not perceived by the physician,
unconscious to the patient these impulses grow stronger, become a resistance
to analysis, even an insurmountable obstacle. Thus: *Impulses that are made
conscious lose their strength; those that remain unconscious and are dammed
up, become powerful.* What if we were consistently to take advantage in
therapy of this familiar phenomenon?

An example: analysis has shown us that in *depressions*, positive object
relations are very loose, whereas negative ones, above all the sadistic ones,
are given vent to in narcissistic ways. The love-relations, which quickly establish a vague transference and—when discussed and rendered conscious—
became ever more shadowy, are superficial. The impulses of hatred growing
stronger repel the external world with ever greater intensity and gather ever
more tightly around the ego as an object, often finally destroying it. Now, one
can proceed in the following way: immediately, upon their first indications,

* [Representations is the translation of the German *Darstellungen.* Later in the
article, it will become clear that what Landauer means by representation is transference
feelings.]

one takes up impulses of hate and death-wishes, makes them, as well as their narcissistic transformations, conscious to their full extent and has worked them through; while this is happening the patient's positive wishes concerning the object (the analyst) are ignored and under that protective cover allowed to be dammed up for months. (In my work, it was usually for three to five months.)*

The negative transference can be placed in the foreground in the following way: one begins treatment explaining to the patient that there exists an unconscious cause of his illness and by conveying to him the fundamental rule of psychoanalysis (1913a). Then the analyst continues: "What we are to fathom—as stated before—is unconscious to *you* and unknown to *me*. You will, therefore, not be able to communicate it in verbal form, just as I am unable to tell it to you. However, experience has shown us that impulses within the analysand, without being conscious or becoming so, are able to express themselves through actions. These are representative in character. Instead of saying: hate lives in me, you will hate. Instead of saying: I love, you will love. Since I am the only person present in this room when your impulses are emerging, it is I to whom, for the most part, falls the role of opposite player. Hence you are going to hate me, love me, feel compassion for me and fear me. That, then, will constitute the information that you hate someone, love someone, and so on. Whatever you are going to do during the course of treatment, we shall evaluate as such a communication by way of representation."†

When after a short time the familiar silence sets in and for instance, the patient's mind is occupied with his situation, there are two questions that should be asked: "What are you concealing?" and "Why are you silent?" The first leads in most instances, to positive currents, the second to negative ones.

Using this technique I have brought to a happy conclusion, after four to six months, all my analyses of depressions in recent years; they were so-called endogenous depressions, melancholias, as well as hypochondriases. A chief reason for this success may be that the denial of love in which the patient is kept, meets his masochism halfway.

Schizophrenic illness is another area suitable for the application of the passive technique. Temporarily neglecting object-transference, be it positive or negative, one can go directly at identification and projection. It is especially in the case of psychoses with auditory hallucinations that use will be made of

---

* [Landauer, here, is in basic agreement with Reich. Since Reich's paper, "The Technique of Interpretation and the Resistance Analysis," appeared in 1927 (Reich 1927b), Landauer should be given priority for the insistence that the analysis of negative attitudes should precede the analysis of positive feelings.]

† [Such anticipatory statements about a transference situation that has not yet developed were not customary. Landauer suggests that these be explained at the same time as the basic rule of psychoanalysis is conveyed to the analysand.]

the device of expressing oneself in "hallucinatory terms," which means: speaking of patient and physician in the third person or impersonally rather than in the first and second person. Furthermore, it means giving expression to the compulsive character of the process. For instance, a patient who remains silent when addressed directly, still gives an answer to the question: "Must one hear voices?" or "What does it think within him?" Trifles of this sort should not be underestimated—where, in analysis are there "trifles" anyway?—Rather, it is a question of a very complex undertaking: by those around the patient as well as by his own reality-testing, it is constantly being impressed upon him that the tendencies that are expressed in his hallucinations must be dissimulated.* What we are doing is to raise up into reality these instinctual drives that had been forced away into the realm of the unreal and phantastic; we are doing so by inscribing ourselves—the real external world—into irreality of his phantasy-world.† We grant the patient the pleasure-gain of removing ourselves, by way of identification, the hated object, while we allow him to strengthen the weak positive object-transference under the protective cover of its being ignored. Otherwise, he would defend himself against the positive object transference by way of projection, clinically expressed by blocking. In this technique—briefly stated—instead of working with "transference," we are working with "carrying in" [*Eintragung*].‡

In this way I succeeded in curing, within six and nine months respectively, two illnesses giving the impression of hebephrenias, as well as favorably influencing several very severe chronic institutional cases. For instance, a catatonic, who had been lying in bed almost mute for twenty-six years, was induced, within three months, to chat, get out of bed and work. Within this period, his fear of contact was also removed, a fear that had prevented him from shaking hands with anyone, of touching door-handles, and the like. A woman patient who had abstained from food for five years and had to be tube fed, ate by herself after only two months of treatment. Up to now, I have tested my technique almost exclusively in the most severe cases; however, even in these, a great deal—at least in social respects—was improved within a short period; among those who were cured there was at least one quite serious case.§

---

* As one of my patients puts it: "One must not hear voices."

† [The technique of treatment of schizophrenics advocated here, in 1924, has much in common with Wexler's approach (1951). There, too, the therapist attempts to help the schizophrenic by speaking to him with a harshness that corresponds to the schizophrenic's own superego.]

‡ Of a positive as well as of a negative sort.

[The play on words is not translatable. The German word for transference is *über-tragung*, literally to carry over. By contrast, Landauer coined the term to "carrying in."]

§ A state of babbling imbecility in a 40-year-old man (which had begun with a depression following an hallucinatory excitement) had continued unchanged for fourteen years.

Certain forms of *perversion* and *impotence* admit of a similar procedure. In *obsessional neurosis*, too, one is often able to strengthen the lukewarm, positive transference by means of apparent disregard; however, one has to go after the negative impulses and often projections. It is quite safe to tune the patient's attachment to the physician to the negative transference—at least on a conscious level. Under these conditions, it is the more certain that the positive transference, remaining undiscussed, will grow and will be better equipped to make possible a final transference onto an adequate real love-object.*

It is, of course, difficult to select the moment at which the patient is to be informed of the fact that there exist positive love-relations in the transference onto the physician, and of the extent to which they are present. I think, Freud has already stated what is essential here (1913a, p. 98): we ought to discuss transference only when it turns into resistance, not for instance, when now and then there is a brief pause in the patient's talking; or when he concerns himself with our room-furnishings, and the like; only at the point when spiteful rebellion makes its appearance. But even then it is generally a question of a projection, behind which is concealed the positive transference.

One might hesitate to use this technique because one doubts whether the necessary detachment of the transference can be successfully achieved when one has caused it to become so strong [lit. massed it together]. Fortunately, experience has disclosed that the fear is unwarranted. Rather, it is precisely the positive transference which has been ignored for sometime and has grown strong in the unconscious which shifts (once it has been made conscious) more easily and more completely onto a new object that is in keeping with reality in contradistinction to the longing of love [*Liebesdrang*] which is brought to light bit by bit, again and again.

However, it has to be admitted that my experiences, even though ac-cumulated in more than ten years of work, still are those of one person only. I would therefore not dare as yet to appear as early as today before the public with this passive technique, were it not for the fact that just now technique occupies the center of discussion. Within the framework of this discussion, the title of my remarks might be interpreted as polemics against the "active tech-nique" advocated by *Ferenczi* and *Rank*. [See this volume, Part Four.] In truth, however, it is to be understood essentially as a protest against "wild

---

* [Reich acknowledged his indebtedness in the following paragraph:

"To my knowledge, it was Landauer who first pointed out that every interpretation of a transferred emotional attitude first decreases its intensity and increases that of the opposite tendency. The goal of analytic therapy is that of crystallizing out the genital object libido, of liberating it from repression and from its admixture with narcissistic, pregenital and destructive impulses. From this it follows that one must, as long as possible, interpret only, or predominantly, the expressions of a narcissistic and negative transference while letting the signs of a beginning love impulse develop without interference until it is concentrated, without ambivalence, in the transference" (Reich 1933, p. 126).]

psychoanalysis" such as it was—and still is—practised and not only by pseudo-analysts.

Technique is a most personal matter. Therefore, I may be permitted to relate the genesis of my method within the context of my contribution to the debate on technique; this is the only way for me to show clearly the goals at which it aims. My first analyses were attempts to treat psychoses along the lines of the "early Zurich school": translations from psychotic language into normal speech. This was work exclusively on the part of the physician; if he was lucky enough to have found an interpretation, he would fling it in the patient's teeth; a descriptive performance. One could not help noticing that, with this procedure, almost regularly the patient's condition grew worse; blocking, stupors, confusional hallucinatory states would appear. Freud* taught us how to understand their dynamics and the laws they follow. For the purpose of avoiding these pitfalls in technique, even the classical technique which was unsuccessful with schizophrenias had to be modified.

Four important impressions pointed the way:

1. The fact that blocking could be delayed by way of avoidance of interpretative activity. Passivity in this respect, reserve, became the precept or command.

2. The lucky chance that I was able to gain an insight into the mechanism of a spontaneous recovery in a case of stupor (Landauer 1914). Here the eminent importance of identification became evident. (In this case, projection was added to identification) (cf. Freud's Schreber case, 1911b). The task presented itself now of placing these mechanisms of narcissistic object-choice into the service of the cure.

3. The experience that an analysis failed owing to a negative transference that had manifested itself as early as in the second session but had at first been overlooked by me. As a result, the problem was set of coping as quickly as possible with the negative transference while allowing the positive transference to remain undiscussed and thus to grow.

4. Lastly, I had the good fortune of having very soon as a patient an actress who was inexhaustible in performing symptomatic actions. My interest was therefore focussed on her "representations"—the term suggested itself in connection with the analysand's profession. A simple statement of the fact of the patient's performances, or as the case might be, my describing them as exactly as possible, invariably had a more convincing effect than all my work of interpreting. The patient was active, while I was merely a passive spectator. She presented before me the entire process of recovery as an exciting play. The whole train of associations was transformed into symptomatic action. The communication in words was complemented by an at least equivalent communication by way of representation in the sense of Freud's "Psychopathology of Everyday Life."

---

* First during the discussion of my paper presented to the Vienna Psycho-Analytic Society in 1913 [Nunberg and Federn 1962–1975, vol. 4, pp. 218–220].

As to these leading trends, probably all psychoanalysts are in agreement. The contrast between the various ways of proceeding becomes apparent only in specific examples, more clearly, it seems, in that of setting a date for the termination of an analysis [cf. Ferenczi's 1927 paper on this subject, this volume, Chapter 16]. In most cases, I can do without it. If, however, it does become necessary, it develops in a way that I should like to demonstrate briefly by way of a case of psychic impotence.

In the case of a thirty-four-year-old man a "deficiency" of libido had been present for twelve years; consequently, despite a strong conscious desire for intercourse, it never occurred. The condition had begun with an *impotentia coeundi* in relation to an employee of the firm where he was working. A short time before, he had engaged in pleasurable sexual relations for months with a girl who lived in the same house, but had allowed the influence of an older spirited friend and boarding house mate to destroy her value for him; this friend would often declare that women do not feel deep sexual pleasure; the patient then had been seduced by him to practicing mutual masturbation. Analysis disclosed that this experience was a new edition [*Übermalung*—a painting over of . . .] of an earlier traumatic scene, in which he, as a thirteen year old had harbored designs on his sister's nursemaid but gave them up when he observed his older brother about to have intercourse with the same girl, his penis erect and appearing fabulously large to the boy. It was for this ideal penis that his attitude of intense ambivalence was meant, an attitude that manifested itself above all, in his feeling of inferiority with regard to the real membrum (to his actual membrum) whose smallness he traced back to the sin of masturbation. From here, paths were leading over similar scenes to the revelation of the most varied unconscious component instincts and object-attachments. Finally, it was even possible to render conscious the primal scene in which he had offered himself in vain to his mother as a substitute for the father who was absent. All this took up about a half year, the following months bringing only minor supplements; the analysis that had penetrated very deep (back to the second year of life), had "become flat." The patient stated this.

When he made another statement to this effect a few weeks later, I drew his attention to the "communication by way of representation" commenting that, in spite of the fact that in his opinion the analysis had already disclosed the essential matters, he had taken no action. This gave occasion for tracing the causes of his behavior ("passivity,"* clinging to the analyst, therefore: positive transference, pleasure in anxiety, and so on) back to the primal fixation in "pre-existence" and ending in the *patient's himself* setting the date for the termination of his treatment for six weeks later. This date was analyzed with the result that it was moved ahead four weeks. There followed ten

---

* The expression "passivity" covers a very complicated confluence of instincts. I do not consider it a primal phenomenon (as Jung does inertia).

days of just hanging around without decisive preparation. Four days before the expiration of the time set intercourse occurred and was repeated with pleasure the next day. Analysis of these recent events followed. Punctually, the patient took his leave feeling that he was completely cured.

I have chosen this example because, at first glance, it may seem as if I had proceeded in a most active manner. In fact, however, I abstained from giving any order, limiting myself solely to having the patient bring his associations to his passive conduct. The result was his renouncing the numerous pleasure-gains derived from passivity and abandoning himself to those obtainable from activity. Had I myself taken on the active role in setting the date for the termination of the analysis, I would have forced the patient to repeat his passive behavior in transference. This was the reason at that particular point, responsible for my reserve. In addition, I was influenced by the observation that analyses for which a time limit is set from the start, invariably do not end well.* Moreover, I am sure that a comment made by Freud to this effect had been suggestive for me. The essential point, however, was the general conviction gained from my analyses that the patient has to cure himself just as he had made himself sick. *Natura sanat, medicus curat:* it is the physician's only task to tend the healing tendencies within the patient.† Fitting into reality comes from the patient himself.

Indeed, the analyst does even less: the patient comes to him and, driven by his suffering, he acts out whatever affects him in front of the mirror—the physician—which reflects his motions. And the patient compares ever anew his topical image with his ego ideal and adjusts them to each other until he is able to step forth in life.

However, I have with fervor stood up for the passivity of my technique and have above even endeavored to defend it from the reproach of activity which might be leveled against it. This calls to mind that the analysand, speaking about a forbidden impulse, often says to the physician: "Now you will think that I have this impulse." It may then well be that notwithstanding my passive conduct, I am active in Ferenczi's and Rank's sense. And this not only in that significant silence and commanding *laissez-faire* often urge on more effectively than numerous words and busyness, but that by these means —and most successfully in this way—we "force" (cf. Ferenczi 1924a [this volume, Chapter 8]) recollections, dreams, phantasies, representations. In fact, it is especially in depressions and schizophrenias, as I have explained above, that the passive technique allows the forbidden impulses to emerge. So my suggestions in a certain respect can be perceived as specific advice relating to an active method, and reserve, conscious passivity becomes utmost activity.

---

* With the exception of analyses during school vacations with children and adolescents.

† [Latin for "Nature makes healthy; the physician only cures." This is a variant on the pious statement, God heals, the physician only dresses the wounds.]

CHAPTER 14

HERMAN NUNBERG

THE WILL TO RECOVERY

INTRODUCTORY NOTE

In contrast to other psychotherapeutic schools, Freudian psychoanalysis has put less stress on the will to recovery. The reasons for this seeming indifference are explained by Nunberg in this essay. The professed conscious reasons are not those of the unconscious, and it is the unconscious needs that often determine the fate of treatment. In clinical work, it is often of great importance for the therapist to discover as soon as possible the unconscious needs that brought the patient to treatment, as well as those that oppose treatment. With this knowledge he has a better chance of preventing a premature termination.

In structural terms the aims of the ego are different from those of the superego or the id, and in addition to these intersystemic conflicts, there may be intrasystemic ones. For example, the superego may demand that the patient stay in treatment and another part of the same superego may demand that he accomplish the cure by himself.

By later standards, Nunberg in 1925 is not yet a skillful therapist. Like Freud in the treatment of Dora, he confronts his patients with unconscious wishes of considerable depth in the first few sessions, possibly before the transference has been established. As a result, they often flee from treatment. This paper still belongs to id psychology and Nunberg writes as if the unconscious

Originally entitled "Über den Genesungswunsch," *International Journal of Psycho-Analysis* 7 (1926):64–78. Reprinted in Herman Nunberg, *Practice and Theory of Psychoanalysis*, Vol. 1, Nervous and Mental Disease Monographs, No. 74 (New York: Coolidge Foundation, 1948), pp. 75–88. Reissued by International Universities Press, 1961. Reprinted by permission. This paper was read before the Vienna Psychoanalytic Society, March 26, 1924.

wish is the only one that matters. The reader who wishes to enter into the world of the psychoanalytic pioneers will find this paper fascinating. Beginning psychotherapists will learn from this paper—even in light of later developments in technique—how to avoid many pitfalls.

THE QUESTION why the neurotic patient desires to get well and comes for treatment is not as paradoxical as it appears at a superficial glance. We know that neurosis is an indication of unsuccessful repression, which gives rise to symptoms, and that in his symptoms every patient tries to attain pleasure in some disguised form, if only symbolically. So it is not obvious why, in spite of all the resistances of the ego, the neurotic should forthwith renounce this pleasure.

It would be possible to answer the question by saying that it is the suffering accompanying illness (that is, unpleasure) which of itself rouses the endeavor to get well. But when we reflect that suffering in itself may be a source of pleasure we attach less importance to the pain of illness as the sole motive in the desire to get well. And if, further, we consider that the ego is passive and simply carries out the will of the id and the superego,* we are bound to look carefully for unconscious motives for the will to health.

We can never calculate the probable duration and success of analytic treatment from the conscious wish to get well. Every psychoanalyst knows that those neurotics who are so impatient that they can hardly wait for the beginning of treatment are not the easiest ones to treat. But it is amazing how obstinately these very patients cling to analysis, in spite of the enormous resistances with which they oppose treatment from the very beginning. We might perhaps account for this phenomenon by the transference. But these patients generally begin the treatment with a negative transference, which surely would be likely to alienate them from the analyst, rather than attach them to him. Moreover, some patients have not even had time at the beginning to form a transference at all; for quite a long while they adopt a waiting attitude, that of tranquil observation—and yet they go on with the treatment.

How then are we to explain the contradiction that in spite of the constantly operative repression, the patient readily understands the initial principles of psychoanalytic treatment and often in the very first hour confides the most intimate matters of his life to the physician, who is a total stranger?

At this point I should like to draw attention to a phenomenon of a general character. We take it as a matter of course that neurotics bring resistance to psychoanalytic treatment. But we forget that the majority of these patients like going to physicians, visit one after another, pour out to each the lamentable history of their sufferings and, further, can immediately

* [The ego the reader encounters here is the ego described by Freud in *The Ego and the Id* (1923b), an ego entirely at the service of the id. See editors' introduction, p. 36.]

give a reason for their illness, generally some frightful event that has befallen them. I remember an obsessional neurotic who, though he knew nothing about psychoanalysis, came for the first time with a written history of his illness, which showed so much insight that during all the rest of the treatment it needed only to be enlarged and completed. We know, too, that many patients produce the most valuable material in the very first sittings and that, if we do not let ourselves be confused by the resistances that arise during the analysis but keep hold of the first communications, we arrive most quickly at our goal. Sometimes, too, it happens that a patient who at first cannot make up his mind to analysis keeps on coming back, sometimes for years, and is obviously trying to communicate the history of his illness bit by bit unobserved. Such patients give the impression of being under a compulsion that drives them to a physician again and again, and when we consider also the multitude of people who do not believe themselves to be ill but constantly force themselves upon the analyst, quite obviously with the intention of confiding to him intimate matters in their lives, our impression that there is a compulsion to self-revelation is only confirmed, apart from the fact of its being borne out by innumerable parapraxes. Indeed, we need no further proof of this than the existence of the practice of confession in the Catholic Church.*

Since the tendency to self-revelation leads in extreme instances to the exposure of various strata of the unconscious it coincides in a certain sense with another tendency, which makes its appearance in symptoms—I mean that of abrogating repression†—the two tendencies that meet one another halfway and reinforce each other. If, in addition, they combine with the subject's desire to get rid of his illness they impel him to have recourse to treatment, as we shall presently see. So it is not at all surprising if we discover in the analysis that the conscious desire to get well is made up of unconscious motives. For, if we do not rely on the patient's conscious statements but go beyond them we soon see that there is always a "misunderstanding": the physician and patient are speaking at cross-purposes, for by mental "health"‡ the two mean totally different things.

The first time I was most clearly conscious of this speaking at cross-purposes was when I had a schizophrenic patient under observation. The academic view is that in the psychoses (paranoia, schizophrenia, melancholia, etc.) the subject has no insight into his illness. To my very great surprise,

---

* One of the motives of the tendency to self-revelation is easy to recognize: not only confession but the behavior of many patients in analysis shows that self-revelation relieves the sense of guilt up to a certain point.

† The tendency towards abrogation of the repression is theoretically explained by the fact that all unconscious impulses have a progressive tendency, which manifests itself in the endeavor to get control of the system Cs. and of motility.

‡ I do not wish here to enter into a discussion of the concept "health." We shall see that in some circumstances "health" also is to be construed as a "reaction" or "symptom."

however, I have noticed that these patients (especially schizophrenics) at times display a marked striving towards recovery and therefore no doubt have a sense of being ill. I had not to wait long for the analytic explanation of the will to health in these patients. It soon became clear that their wish to get well was overdetermined; it arose out of several motives which in the deepest stratum of the unconscious were merged into a single one.

On the surface there was generally a desire to overcome sensations of weakness and distress which originated in previous feelings of hypochondriac anxiety. At a deeper level this desire went back to infantile tendencies belonging to the period in childhood when the child feels an impulsion to busy occupation and has delusions of his own grandeur—the period of omnipotence and magic. The desire culminated in the single endeavor to return into the womb and be re-born from oneself (cf. Nunberg 1920, 1921 [see also Rank 1924]).

Although I do not agree with Rank's view that intra-uterine and rebirth-fantasies play the same part in all neuroses, nevertheless I think that the desire for recovery does always originate in the instinctual life of early infancy. At the same time there is invariably a misunderstanding in those who undergo analysis, not excluding those who are analyzed in order to learn the technique; the analysand expects from psychoanalysis something other than it can give.

Perhaps the best illustration of the fact that physician and patient mean two different things by "cure" is a case communicated by Ferenczi to the Vienna Psychoanalytic Society. The patient's object was to have his nose cured by psychoanalysis, while he was really suffering from an affection of the penis.

I had experience with a similar case some years ago. A patient imagined that there was something wrong with her teeth, although they were perfectly sound. At the bottom of this symptom was a marked unconscious cannibalistic tendency and a powerful castration complex. She was quite right in seeking to be cured by mental therapy, but she was not clear about the motives that impelled her to undergo it. Her conscious wish was: "I want to have sound teeth", but the content of the unconscious wish was: "I want to have a penis". It was the physician's task to bring the unconscious into her consciousness.

The motives that impel neurotics towards recovery are as manifold as the motives of their illness. Of course, first and foremost, the perfectly conscious unpleasure of the illness may send the patient to the physician; the clearest instance of this is in neurotic anxiety. But when we recollect that neurotic anxiety is a manifestation of a disturbance of libido and that at the beginning of the treatment the patient generally loses his anxiety for a time owing to the binding of the libido in the transference* to the physician, the

* Later he gets rid of the symptom through analysis.

typical representative of the parent-imagos, the helper of humanity who is endowed with every mysterious quality, the instinctual element becomes unmistakable.

All kinds of psychical impotence show in a perhaps still more striking way that the wish to get well is activated by unconscious motives. For instance, we have the case of a man of thirty-four, who fell in love with a married woman, the mother of six children and the wife of a friend of his. With her he was impotent and at the same time became so with other women. He came to analysis with the desire to become potent with this particular woman; he did not mind about any other. He wanted to separate her from her husband, whose death he desired, and he had dealings with fortune-tellers who, to please him, naturally prophesied the fulfilment of his wishes. The patient, a cultivated and in other respects an intelligent man, could not perceive the folly of the situation and expected from the treatment the realization of his infantile wishes. Thus his unconscious endeavor was to remain infantile, but it was the duty of the physician to free him from his infantile fixations.*

As a rule impotent men expect from the treatment not average, normal potency, but (and in this I confirm what Rank [1923] has observed) nearly always hyperpotency. Not infrequently it happens during analysis that when part of the castration-complex has been overcome these patients suddenly develop hyperpotency. This pleases them, and they are proud of their genital capacity and their ability to satisfy their women. At the same time they enjoy a narcissistic satisfaction through identification with the genital, which has now become efficient. Though they themselves remain sexually (genitally) unsatisfied they regard themselves as well and wish to break off the treatment.† After a short time the old condition reasserts itself, and then the second part of the treatment begins.

Perhaps a few short examples will show what ideas patients have about the health which they believe to be worth striving after and what is the ego-ideal which hovers before them when they think of it.

One patient had the feeling that he was turned, back to front and upside down, as if he were made up of two people. He thought that one of these

---

* [We assume that as a mother of six children, the woman stood for the patient's mother. The transfer of the incest taboo made the patient impotent. The patient comes to treatment hoping that the father, in this case the therapist, will allow the son to possess the mother. This wish is symbolically expressed by the patient becoming potent in the course of the treatment. An ego analyst interpreting this situation would stress that the woman is not the mother of the patient, that the relationship is not an incestuous one, and the analyst is not the father, and therefore can neither give nor deny his blessing to the union. Nunberg, on the other hand, sees the whole problem from the point of view of the superego and asks the patient to give up his infantile fixation. It is unlikely that the patient will comply.]

† [The wish for hyperpotency is the expression of phallic exhibitionistic wishes. Such patients have not reached the genital phase (see Abraham 1924a).]

people looked forward and the other backward. He was afraid to walk in the street, for he thought that his toes peeped out of his heels, and so he was afraid of tripping over his own feet. When he spoke he always had to take hold of the top of his head to convince himself where his head and his face were, and so forth. He pictured that if he were cured the man in him who looked forward would disappear. This symptom was overdetermined,* its deepest significance was an identification with the mother, who was embodied in him in the person looking backward. He therefore expected of the treatment that it would enable him to be completely absorbed in the mother and change him into an avowed homosexual, his former attitude representing inversion under a disguise.

A girl of seventeen-and-a-half, an obsessional neurotic who suffered from brooding mania, was exceedingly refractory. She imagined that she was oppressed by her parents and all other grown-ups, and that when the cure was completed she would know everything, be able to solve all problems, to produce anything from any material whatsoever, and no longer be oppressed by grown-ups but always able to carry out her own will without being influenced by anyone. At the beginning of the treatment she could of course not be convinced that her hopes could not be fulfilled. She wanted to remain as she was, only without suffering.

The following case is an instance of how deeply the conscious desire to get well may be intermingled with unconscious impulses tending in exactly the opposite direction. An impotent patient was violently eager to begin analysis for he was afraid that his wife would leave him if he did not soon get well, and without her, he said, he could not live. Naturally he brought to the treatment enormous resistances. Even at the first consultation he asked me if it would not be better for him at once to leave his wife for a time, but I insisted that for the present everything must remain as it was. Nevertheless, at the next sittings he constantly recurred to this question, and fantasies came to light in which he wished not to leave his wife temporarily but to separate from her for good. After we had found out that his present total impotence represented undischarged feelings of revenge in reaction to a certain experience, dreams and fantasies emerged in which he left his wife and went back to his home, to his mother. Thus his conscious wish to become potent ran counter to an unconscious desire to return to his mother, who apparently meant to him "health".

The two following dreams will perhaps show how deeply rooted in early infantile tendencies are such desires to get well and how they seek in them their fulfillment.

A girl whose illness caused her much suffering dreamt that she was at her home in the house where she had first lived as a child and that she was with

---

* [The term overdetermination belongs to Freud's earliest formulations. Freud was impressed by the fact that hysterical symptoms were determined by many events and therefore overdetermined; see also Waelder 1936.]

some companions of her own age, whom she had known at about the time of puberty. She told me that it was they who first enlightened her on sexual matters, and especially one of these girls, who had more freedom than the others because her parents were dead. This led to an infantile recollection of her third or fourth year when she was seduced by a nurse. This dream was the translation of a thought which the patient, who was feeling particularly ill at the time, had uttered aloud before going to sleep: "How happy I should be if I could be a child again".

Still more instructive is the dream of another patient. This was a woman who came for treatment on account of frigidity. At the second sitting she produced a transference dream as follows. She was with me and I had many women patients who made love to me. I chose her, however, and kissed her on the lips, but I was younger and better looking in the dream than in reality. Then she was on a big ship that had to pass from a large expanse of water into a little stream, and it depended on the steersman up above whether the ship would get through the narrow passage without stranding on the rocky shore which projected into the water. As she was getting out she saw a beautiful palace and an old woman with a basket.

In the very first hour she told me that her frigidity caused her no distress but that she came for treatment for another reason. After she and her husband were divorced she had "used up" several men in a comparatively short time and she was afraid that, if it went on, in a few years she would end on the streets. In order to save herself from this fate she came for analysis. Thus her reasons were moral ones, and she hoped that when the cure was finished she would be a different person.*

On the day before the dream she was with some people who knew something about psychoanalysis. Someone remarked that in psychoanalytic treatment two things had to be done: first, the transference had to be established and second, it had to be resolved, and that this was the more difficult task. She replied jokingly that she would manage, for she would marry the physician at the end of the treatment.

The fact that in the dream I chose her from amongst many other women was founded on a peculiar circumstance. In real life she always had the unfortunate experience that any man whom she loved turned away from her and attached himself to one of her friends. This was what had happened with the husband whom she had divorced. She was the youngest but one in a family of several children and had always had the impression that her mother loved her the least and that she was the Cinderella of the family. When in later years she found out that her mother had not suckled her, and that she was a "bottle-baby", she could scarcely contain herself.

The kiss on the lips corresponded to a recollection. As a young girl, the

* [Unlike the previous cases reported by Nunberg, this patient comes to treatment under the pressure of the superego.]

first time she was kissed by a man she hastened to her mother and asked her if a girl could have a child because a man had kissed her. At the next day's analysis she produced a second, much more important, memory. Her favorite game as a child had been to play at "mother and daughter" with her youngest sister. The patient acted the daughter and the sister the mother. The game consisted in the patient's lying on a sofa, shutting her eyes and pretending to be asleep. After a time the younger sister ("the mother") came and waked the "daughter" (the patient) with a kiss on her lips. (Compare the situation on the sofa in analysis!)

Thus the patient's infantile fantasies were used to represent her wish to be preferred by me to other women (children), to be made pregnant and wake to a new life. The same wish is even more clearly expressed in the second part of the dream (water). This part again begins with an oedipus-dream, passing by association into a memory of a childish game with a brother. But it ends with a symbolic birth-dream* (old woman—basket—water—palace). By association the old woman represents at once a midwife and the mother. The child in the basket is identified with the dreamer. Unfortunately, motives of discretion prevent my communicating the rest of the material.

This dream may be regarded as programmatic of the treatment. The conscious wish to become a new person through psychoanalysis and to marry the physician is in the unconscious represented as a return into the mother's body and rebirth. The heterosexual object-choice is merely superficial. The dream begins with incest but ends in a deeper stratum of the mind in union with the mother. But the whole meaning of the intra-uterine and birth-fantasies can be grasped only when we know the unconscious meaning of the childish games. The conscious wish to get well stirs up an unconscious fantasy and leads to an infantile situation of gratification. Consciously the patient wished to lose her frigidity with men but, unconsciously, to retain it.†

This case also throws a certain light on the relation between the will to health and the transference. We shall see this more clearly in the next case.

A woman came for treatment on account of hysterical hypochondriacal symptoms. At the third sitting I first of all called forth resistances by a clumsily worded question as to why she came for treatment. She looked at me

---

*[Here, Nunberg interprets the dream not through the associations of the patient but directly through symbols. Anna Freud has warned against such an interpretation; See editors' introduction p. 39, also Bergmann 1968. Rank's influence is still great. Lewin (1950) offered an alternate interpretation to these womb fantasies in what he called "the oral triad."]

† [Nunberg is not sufficiently explicit. The frigidity could be due to a fixation on an earlier oral level, or due to unconscious hostility toward men, or due to the prohibitions of the superego; all three may be involved. The treatment had just begun. It is too early for Nunberg to guess at the meaning of her frigidity. However, such leaps were common in the early years of psychoanalysis.]

in amazement and answered in an offended manner that naturally she came to me to be cured. After a fortnight I found out something different. A friend of hers, whom I was treating, was at that time suffering very much owing to a violent transference and had confided to the patient the story of her unreciprocated love for me. The latter was indignant at my having rebuffed her friend, whom she taunted with not having been able to win me, adding that that would never happen to her—if she wanted she could in any circumstances cause her love to be reciprocated. Although she had already determined to go for treatment, she was always hesitating, and only her indignation toward me and the desire to show her friend how these things could be managed better hastened her decision to begin the treatment. At this point we soon discovered that she distrusted all other women and suspected that they felt nothing but envy and jealousy. She was very strongly fixated to her father; till his death, which took place when she was eleven years old, she had slept in his bed and had completely ousted her mother. She blamed the latter for his death and was reconciled to her only after her own marriage. In analysis, from the very first day she formed a transference of as passionate and uncomfortable a kind as had her friend, whom she had so vehemently upbraided on that account.

Thus the wish to get well, which is nourished from the unconscious, makes use of the transference in order to attain in the present the infantile instinctual aims. Hence the transference is mobilized by the wish to get well and replaces it to a greater or less degree during the treatment. This is, of course, according to circumstances important for the duration and success of the treatment. In this patient it failed, because the desire to get well gave place altogether to the transference.*

We can understand, therefore, that treatment encounters insuperable difficulties in cases where either the will to recovery is altogether lacking from the outset, or where the conscious wish to get well and the unconscious tendencies cannot be brought into line, and finally in every case in which the will to health is wholly replaced (in very passive natures) by transference. I can illustrate this by two cases. In the first a homosexual broke off analysis when it became plain that what he expected from the treatment was to get back a lover whom he had lost. Another patient, who was abused, tormented and humiliated by his wife and yet could not leave her, wished that analysis, which he underwent submissively, should bring him to the point of resolving to get a separation. When, however, it was revealed that he was excessively passive and had strong fantasies of being beaten and bound, he ceased to

---

* [An overstrong transference would not today be accepted as a reason for the failure of a treatment. On the contrary, the stronger the transference, the more does it fuel the treatment process (see editor's comments, this volume, pp. 27–29). It is, however, possible that the patient could not withstand the deprivation of the gratifications of her wishes. Because of the intimacy she had with her father, she may well belong to the type described by Jacobson (1959) as "the exceptions."]

come for treatment. Unconsciously he did not in the least desire to be free from his wife. He had been driven to analysis by a mistress with whom he was impotent. Consciously he wished to become normal, but unconsciously he desired to remain a masochist.*

That the will to recovery may in itself lead to recovery even without transference is proved not only by many cases of spontaneous cure in schizophrenia, melancholia, etc., but above all by those neurotics who get well without any medical help at all. Since these patients, like others who are never treated psychically, are not easily accessible to observation, it is difficult to form any ideas about their motives for recovering.†

* [Undertaking psychoanalysis at the wish of someone else is usually an unfavorable sign, but it is nevertheless possible that the sequence and timing of the interpretations resulted in a premature confrontation between analyst and patient. The patients who could neither cope nor repress the unwelcome insight left treatment. The analyst failed to establish a therapeutic alliance with the healthy ego against the masochistic wishes.]

† The following section of this paper has been omitted.

CHAPTER 15

HERMAN NUNBERG

PROBLEMS OF THERAPY

INTRODUCTORY NOTE

This article contains one of the most forceful statements as to what psycho-analysis can achieve and how it goes about achieving it. Written within the era of Freud's basic reformulations (editors' introduction, p. 36), the paper shows how Nunberg assimilated the new knowledge and transformed it into a theory of technique. Within the historical context, two of Nunberg's formulations are of particular interest. He differs sharply with Wilhelm Reich on the importance of positive transference in the initial phase of treatment. He approaches treatment through the modification of the superego. He is, therefore, not an ego psychoanalyst. (See editors' introduction, pp. 37–38.)

THE NEED for a technique of psychoanalysis, that is, for precise directions on how mentally ill persons are to be treated, is self-evident. This need has always existed; however, in the course of time it increased to such an extent that it produced a series of proceedings which, allegedly, had only to be followed conscientiously in order to achieve certain success. The failure bound to result from such procedures is based, I think, on the one-sidedness, and in a sense, preconceived opinion, with which the patient is approached. It is certainly tempting to overestimate a real or presumed discovery and to make use of it immediately in therapy.

Originally entitled "Probleme der Therapie," *Internationale Zeitschrift für Psycho-analyse* 14 (1928):441–457. Reprinted in Herman Nunberg, *Practice and Theory of Psychoanalysis*, Vol. 1, Nervous and Mental Disease Monographs, No. 74 (New York: Coolidge Foundation, 1948), pp. 105–118. Reissued by International Universities Press, 1961. Reprinted by permission.

Jung was most impressed by the mythological formations, and now believes that the human mind is a myth. Adler found feelings of inferiority and he believes that man consists exclusively of feelings of inferiority [Rank 1924; Alexander 1927; Reik 1925]. Rank saw in neurosis a reaction to birth anxiety. Reik and Alexander are of the opinion that punishment is the most important motive force for neurosis. Reich bases his theory on the character changes appearing in the reaction-formations of the compulsive neurosis, believes that character consists of resistances only, and wants to unroll the entire analysis from a single so-called character trait.

Aside from the fact that all of these so-called discoveries are not new and that they had long before found their proper place in the psychoanalytic system, it is impossible to set in motion an apparatus as complicated as the human psyche from one single point.

If one wants to discuss practical advice for treatment, i.e. the technique of psychoanalysis, one first has to try to take account of the theoretical bases of therapy. Therefore I do not intend here to discuss technique, but rather shall attempt to clarify the changes that occur in the human being by means of the present method of treatment, as recommended by Freud.

Before undertaking a description of the process of recovery one must in the first place be clear about the nature of the illness. Psychic illness is by no means a simple process. We distinguish between a primary and a secondary process of illness. The *primary* process, thus the core of the neurosis, is formed by anxiety in relation to an instinctual danger and by the neurotic conflict closely connected with it. The conflict may be either cause or result of the anxiety, according to stratification.* The *secondary* process of illness is formed by the symptoms in the broadest meaning of the word; the ego indeed introduces all the complicated processes of repression and other defense measures in order to withdraw from the danger situation, to *avoid the unpleasure and to ban the anxiety*. And the entire symptomatology results from this defensive struggle. But at the same time the success of the defensive struggle means to the ego a solution of the neurotic conflict. The final result of this attempt at a solution is inhibition and modification of the course followed by the instincts, i.e. the barring of the ideational content from consciousness and of the affectivity from motor discharge (cf. Freud 1926a).

Consequently one might assume that the first therapeutic task is to help the instincts to discharge and to procure access to consciousness for them. Aside from the anxiety felt by the ego at such a direct attempt, there are other reasons for which this fails in most cases. The longer the neurosis lasts, the more the personality loses contact with reality. There are cases, indeed, in which the slightest contact with the external world results in an even more violent rejection of the world. The neurotic is *over-sensitive* to external and

* [Nunberg, at this point, had already assimilated the basic point made by Freud (1926a). See editors' introduction, p. 36.]

internal stimuli. The patient becomes *asocial* through his illness, he is self-sufficient, and he satisfies his needs not through changes of the outside world but through auto-plastic changes of his own organization. Certain parts of the ego are split off in the neurosis; the ego becomes more primitive in its reactions, especially more infantile in relation to danger situations. It clings to infantile reactions and considers as dangerous stimuli that are no longer so; on the other hand, the actual conditions that demand adequate adaptation to reality have thoroughly changed since the first fixations took place. Hence there is little prospect of success in the attempt to influence this *split-off ego*. Aside from this estrangement from reality, there is another difficulty that stems from the ego. The ego has repressed the instinctual demands, and erected various dams against them in the shape of reaction-formations. The ego does not affirm, but rejects the strivings of the id, and thus is deprived of its synthetic function, of its part as mediator between the contrasting impulses within the id on the one hand, and between these impulses and the outside world on the other. Even more: by virtue of their continuous nature the instincts cannot be banned by carrying out a defense measure just one time. Rather, the ego has to expand defensive forces constantly in order to maintain the first repressions. We call this permanent expenditure *resistance*; it manifests itself as a protector of repression. If one is clear about this nature of the resistances one will not consider it likely that an ego that expends so much energy for protection from instinct danger, an ego that has prepared itself for the rejection of instinctual life and opposes any weakening of this rejection, is going to abandon these resistances without a struggle. Certainly the power of holding out is not the same in each of these "resistances". Some of them increase the capacity of the ego to oppose the external influence and veil the connections to the point of utter obscurity, others yield more easily. However, it is impossible to state a priori how obstinate the expected resistance is going to be. One might assume that those resistances that develop last in the course of the neurosis, as for instance, the secondary gain from illness, will be destroyed most easily through direct influence. But not even that is always true. An excellent example is the accident-neurosis. Yet in other forms of illness also, such as obsessional neurosis, it may take the patient a very long time to give up his secondarily erected ideal of a hypermoral person; and in paranoia a similar narcissistic resistance, such as pride in the intellectual achievement of the delusional system, is altogether unassailable. The actual situation is that certain resistances may be influenced directly at one time, and at another time only by being undermined from the depths. Resistance due to the sense of guilt, i.e. that which develops under the influence of the superego might, in the same way, appear easily assailable.* But the feeling of guilt has an unconscious root, and in the form of the unconscious need for punishment

* [Nunberg here reiterates the five groups of resistances Freud had outlined in 1926*a*, p. 160. See also editors' introduction, p. 37.]

it energetically opposes any external influence. This powerful resistance not only serves the superego, but in its deepest layer also gratifies the instinctual needs; thus it forms an insurmountable obstacle against any attempt to attack from the surface the behaviour which it determines.

Direct access to the third ego resistance, *the resistance due to repression*, is bound to be more difficult: indeed, it stands at the cradle of the process of illness, introduces the defense, and is in most cases the actual cause for all later resistances which have the task of reinforcing it and of insuring its success.

This enumeration does not exhaust the series of resistances. Still another resistance appears during treatment, the "resistance due to transference" which results from the psychoanalytic situation. Thus it does not seldom happen, for example, that the "love" of a woman patient for her male analyst is conscious to her, but she cannot admit it to herself for reasons of self-esteem (narcissistic protection). Or else the instinct, which is under the pressure of defense, is revived in analysis, comes to be related to the person of the analyst, and under the influence of the repetition-compulsion is expressed through acting out rather than through a memory; in which case, as in the feeling of guilt, an ego tendency combines with an id tendency to form a joint resistance. Moreover, since this resistance serves to gratify a repressed instinctual impulse, it will try to maintain itself under all circumstances.

The true repetition compulsion, as a resistance of the unconscious, is independent of any transference. Its power is shown in the attraction of the unconscious models for the course of the repressed instincts. This resistance unconsciously repeats the defense; apparently it takes its course automatically. Since we know, moreover, that the path from the outside world to the repressed id must lead through the ego, direct access to the symptoms by way of the unconscious id is even less conceivable.

Thus we see that it is just as impossible directly to influence the resistances through the id as through the ego.

These resistances are thoroughly familiar to us. Some of them represent in a sense a narcissistic protection. But it is too simple to consider them as an integral part of character, as Reich does, and to give them the designation "narcissistic shield".* Character is more than a conglomerate of resistances.

The neurosis develops from the rift between the strivings of the id and the demands of the ego. The ego always desires to resolve this conflict in some way or other by the introduction of repressions. At first, however, the defense measures of the ego prove to be too weak, and fail. Finally, the ego gains mastery and represses the instincts, *thereby harming them and in turn being harmed*. The therapeutic task may therefore, in principle, consist of making

---

* [It is customary to speak of narcissistic armor rather than narcissistic shield. Terms borrowed from medieval forms of warfare have already become "dead metaphors" and are used by psychoanalysts rather loosely.]

peace between the two parts of the personality, the ego and the id, in the sense that the instincts no longer continue to lead a separate existence excluded from the organization of the ego, and the ego recovers its synthetic power. *To influence the ego as well as the id thus becomes the goal of the therapeutic task.*

Considering that the neurotic person surrounds himself with so many protective measures it seems a problem why he seeks treatment at all, and especially psychoanalytic treatment, for he must suspect that his various resistances will be fought. Certainly it is a fact that, in spite of what has been said before, these resistances can sometimes be attacked directly for a short time by taking the patient by surprise. But one pays a heavy toll for this, for the patient learns in analysis to camouflage his resistances even better. The small opening soon closes. The surgeon, too, has to prepare as broad a field of operation as possible in order to be able to work with technical precision.

However, the problem of therapy is not so hopeless. It is a fact that psychoanalysis shows successes. How then, is treatment set in motion, and recovery eventually accomplished?

In most organic illnesses there are certain tendencies toward recovery that are present either from the very beginning or that arise with the onset of the illness as an automatic reaction. It is therefore the physician's foremost task to create favorable conditions for the normal course of the process of illness by a reinforcement of the "natural" recuperative forces. What then, are the natural tendencies towards cure on which the analyst may count?

We cannot treat a person in a twilight state, nor a severe melancholic, a catatonic, and so on; in short, we cannot treat persons who have *completely* lost contact with their surroundings. It is a necessary pre-condition for treatment that a part of the personality has remained intact, i.e., there must be basic possibility for mutual understanding: the most primitive function of the ego, the capacity of perception and expression, must have been preserved. We cannot influence a person who is deaf or mute. Indeed the word is the most important expedient in analysis. Social relations must not have been completely dissolved. Briefly, some free object libido is essential for any treatment. This, however, is not all that is necessary. Insight into one's illness must also be present, that is to say, a feeling of rejection or strangeness in relation to the neurotic symptoms. Furthermore, there must be an incentive causing the patient to seek treatment. This incentive is provided by the neurotic suffering from which the patient is to be liberated by treatment, and thus is a wish for recovery. It owes its origin to the secondary unpleasure brought on by the illness. However, this wish is not so simple, for it is released by two motives that seem mutually exclusive.

The first motive is the following: every suffering and sick human being is helpless. In his helplessness he is inclined, like a child or a primitive person, to overestimate the power of the one who promises help. He will perhaps look upon the physician as a magician, in the broadest meaning of the word. Indeed, every physician utilizes this "superstition" more or less consciously. The psychoanalyst does not behave any differently when he promises the patient at the introduction of the treatment to cure him, under the condition, it is true, that the patient follow certain rules of the treatment.* Incidentally, the importance of the physician's personality for the favorable course of the process of healing becomes apparent right here. If a favorable relationship has been established by the helpless and superstitious patient to his psychoanalyst, whom he has equipped with magic power, much has been gained for the beginning, because the unconscious part of the patient's personality now sides with the physician.† The patient thinks, indeed, that the psychoanalyst has nothing else to do but what he himself has done unconsciously for years, that is, to protect him by magic means (as for instance in compulsion neurosis) from the dangers presented by the instincts; thus, to liberate him from his suffering. The patient has outwitted the physician in a sense, but also himself: and yet, in doing so, he has found in his ego some points of contact with the physician. He can thus identify himself with the analyst in the endeavor to procure help for himself. In spite of all resistance he becomes the analyst's helper and begins collaborating with him. In most cases, however, this collaboration is at first misunderstood. The obsessive neurotic, for instance, considers the psychoanalytic rules as magic formulae which he has to comply with conscientiously to the letter. Even where this is not the case, psychoanalysis is sacred to most patients, and the analytic hour is an hour of devotion. Every patient has some kind of ceremonial with which he surrounds the analysis.

However these details may vary, if we appeal to the patient's intact ego by promising help, we meet with unconscious advances from a helpless ego looking for support, even though this support is of an entirely different kind from that which we are willing to give. At any rate, the patient's anxiety is thereby mitigated.

Just as the patient understands the term "help" in a different sense from the physician, that is, he wants to borrow magic powers from the physician for the fight against the unconscious instinct dangers, so the wish for recovery contains still another motive, which likewise leads to a misunderstanding, as I have tried to demonstrate in another context [Nunberg 1926, this volume, Chapter 14]. By "health", too, the patient means something different from what the physician does, namely, the *gratification* of all kinds of desires,

---

* [Nunberg here differs sharply with Ferenczi, who was strongly opposed to promises (1928, this volume Chapter 17).]

† [Here Nunberg differs from Reich, who mistrusts the positive transference in the beginning of the analysis.]

impulses, expectations, hopes, and so on. To mention just one example, the impotent patient expects of the cure not merely normal adequate potency but a hyper-potency corresponding to the over-compensation for his castration complex. Thus the wish for recovery contains, in the unconscious, two contrasting roots, one that emanates from the ego and hopes to gain control over the instincts, and the other coming from the id, which hopes for gratification of the instincts. Although so self-contradictory, the wish for recovery has one advantage indispensable for analysis: it begins to establish the transference and even becomes its mainstay. And out of the positive transference, out of love for the analyst, the patient starts collaborating with him—which at first means fighting off the resistances.

What gratification does he find in the transference? What means does he use to set the work in motion? In favorable cases the patient gains gratification from realization of the motives contained in the wish for recovery. He fulfills conscientiously his first task of giving free associations. Analysis becomes rich in content right in the beginning and steers toward its goal without much difficulty. No patient, however, obeys the basic rule exactly, and most patients submit to it only in form. The pleasure derived from speaking becomes evident from the manner in which the speech is brought forth. In analysis the word regains its ancient magic power.* The patient believes he delights and dazzles the analyst by his words. This way of speaking veils the old resistances. The patient is able to keep his secret in spite of his apparent obedience to the fundamental psychoanalytic rule. The first gratification derived from the treatment is thus a narcissistic one: to fascinate through speech and not to yield secrets. Not speaking may have exactly the same double meaning. The second direct gratification is derived from the fact that the analyst gives his attention to the patient, listens to him, and is occupied with his troubles. The patient concludes from this that it is love that the analyst feels for him. Disappointment comes later. The necessity to become absorbed in himself, to look into himself, offers a *similar narcissistic gratification.* The same is true of the *intellectual achievement* if at an early stage the patient succeeds in discovering psychological connections. This gratification increases with the progress of the analysis as soon as the patient has gained intellectual interest in it. All of these gratifications may become initial resistances as well, but certainly they are not the only early resistances encountered, nor does each patient raise the same difficulties in the beginning.

* [Strictly speaking, this need not happen. In the analysis, the analysand will frequently experience certain words as having magic power (slang words often belong to this category). If the patient really believes that he has recaptured magic powers, the analysis is in danger. Similarly, an analysand may temporarily believe that he can dazzle the psychoanalyst, but should he persist in this conviction, the analysis is in jeopardy. One notes here, that Nunberg is not an ego psychoanalyst; that is, his stress is on the unconscious and not on the relationship of the ego to the unconscious material.]

This introductory phase, which may vary in duration, does not promote the treatment. It is at best agreeable to the patient. However, something else is now added which, although often disagreeable to the conscious ego, yet brings relief.

We know that any process started in the Unconscious has a progressive tendency. It endeavors to reach the system Pcpt-Cs: to become conscious and to find affective and motor discharge. In neurosis, where direct instinct expansion is inhibited, i.e., repressed, the pressure from below, from the unconscious, is stronger a priori; *therefore the need for discharge is more powerful than it is with instinct energy that is not blocked.* Because of this the neurotic involuntarily unmasks his repressed Unconscious constantly. One may plainly speak of a *self-unmasking tendency* dominating every neurotic. If in addition the patient suffers from a strong sense of guilt, this tendency assumes the character of the *compulsive confession* (Reik), by which the unconscious need for punishment is gratified. But just as the sinner in the Catholic confession expects absolution and clemency in return for his confessions, so does the patient hope for the same from his analyst.

The relation to the analyst is deepened by all these circumstances and the *transference is established.* The weak and helpless ego submits to the analyst, finds support in him and allows itself to be guided by him in the struggle against the resistances. The transference now becomes the bearer of the will to recovery, and substitutes for it, and is placed in the services of the real psychoanalytic task.

In the beginning we stressed that the neurotic patient cannot be influenced in a direct way to give up his resistance. However, insofar as it is a question of helplessness and ungratified instinctual needs, of object hunger, the analyst is able to steal into the patient's ego and there start the breakdown of the resistances. The analyst is not only identified with the patient's magic ego, he is even raised to the ideal. He replaces the patient's ego-ideal. Since the analyst is surrounded by libido in the patient's ego, he neutralizes the severity of the superego. The superego becomes more indulgent and milder through the libidinal absorption of the analyst.* The ego need not fear the superego as much as before, just as it needs no longer fear the instinctual demands, because it is protected by the analyst on both fronts. Combining in his person the strivings of the patient's ego and id, the analyst is predestined, so to say, to mediate in the neurotic conflict and to reconcile the conflicting parts of the neurotic personality. Furthermore, since the analyst is benevolent toward the repressed instinctual components, the patient's ego successively abandons its resistances that are due to repression; the patient feels allied with the analyst, in accord with him, protected by him, and therefore need not fear

---

* [Nunberg comes close to the ideas expressed by Alexander in 1925, this volume, Chapter 6.]

danger situations which, moreover, have long ceased to be real. This explains also why the most severe states of anxiety frequently disappear right after analysis has begun.

To repeat briefly: the patient raises the analyst to his ideal, identifies with him after the fashion of the hypnotized, finds protection in him, and finally transfers the strivings of the id upon him. Under these circumstances influence from within can be achieved: the slow breakdown of the resistances due to repression may be undertaken.

My own experience has taught me that in cases where this relationship could not be established, treatment has had to be stopped sooner or later.

We know that the resistances due to repression show their power mainly in keeping from consciousness the preconscious ideational material. Although it is true that in most of the phobias or obsessive neuroses pathogenic ideas are largely contained in consciousness, they are nevertheless always substitutive ideas, quite removed from the original ones. Therefore the first task of every analysis is to obtain the repressed ideational material in the form of recollections. If the patient has reached the state where he feels protected by alliance with the analyst, he will drop his aversion to remembering and give up all resistances pertaining to it. Even more: he will give up not only the amnesias, as does the hysteric, but also the taboo against contact, as does the obsessive neurotic (for whom remembering is generally easier); and he will connect the remembered experiences in time and space. Connection is established between preconscious and unconscious memory material through recollecting, and thus the repressed object-presentation is enabled to reenter consciousness. Here it is recognized as no longer an actual menace and so is tolerated with a minimum development of anxiety. The real traumatic situation is then reproduced through as close a recollection as possible. A comparison of that situation with the present one shows its non-actual and unreal character. As a result, not only is less anxiety developed, but also can reality-testing be introduced. Since through the work of remembering the patient is brought back to an unpleasurable situation, which he now relives without anxiety, *he learns, moreover, to tolerate unpleasure.* Indeed, the oversensitivity to unpleasure and the closely connected anxiety-readiness, the incapability of tolerating major need-tensions, is the characteristic trait of the neurotic person.

Two factors facilitate the work of remembering. The object-relationship has been disturbed by repression. In obsessional neurosis there is even danger of the loss of the object. In order to evade this danger, the obsessional neurotic forms substitutive ideas. Analysis therefore supports a natural process, with this difference, however: it penetrates through the substitutive images to the primary ones, thus to the real objects.*

* [Nunberg refers to the analysis of the transference leading back to the original object, usually the parent.]

The second factor is of a more general nature and applies also to hysteria. We spoke of the progressive tendency of the Unconscious. This progression is interrupted by repression; the instincts are under high energy tension due to the blocking of the channels of discharge, and being continuous they cause incessant unrest for the psychic apparatus. This instinctual tension is the *second* factor that impels the unconscious strivings on towards cathecting the system Pcpt.-Cs: towards discharge through the act of becoming conscious and through affectivity. This tendency is disturbed by the repression, and the patient feels relieved when analysis offers him an opportunity for discharge, even though it is under resistances.

It can easily be understood that discharge in affectivity may bring relief. However it cannot without further consideration be understood why the act of becoming conscious should have a similar result. We know how uncomfortable it is when we try to remember something and cannot, and with what a sigh of relief we greet the sought-for recollection when it emerges. The relief felt by a patient when he succeeds in remembering an important part of forgotten material is even more striking. However it is noteworthy that the relief is felt but one time, that is, at the moment when the forgotten material reappears in consciousness as though illuminated by a searchlight. In order to understand this we have to realize the function of the apparatus of consciousness. Perception of external as well as of internal stimuli is the main task of this apparatus. It has no memory, that is, it cannot preserve memory-traces. An experience becomes conscious only when a preconscious idea combines with an unconscious one and enters consciousness by means of the special act of hyper-cathexis. Therefore the moment of becoming conscious is fleeting: consciousness cannot lastingly maintain its susceptibility to the same idea. Since the system Cs does not preserve traces of stimuli (recollections), since the ego gets relief through the act of becoming conscious, and since, furthermore, this act occurs but once (though it may be repeated indefinitely), it frees the psychic system of tensions. This relief may be understood in the following way: the psychic energy bound to the repressed ideas is discharged in the act of remembering. According to Freud, *energy thus freed is spent in the act of becoming conscious.* Indeed, it happens that a patient may forget the recollections brought to light through his analysis and yet stay well.*

The old theory of abreacting remains valid. However, we understand by it not merely the discharge of affects, but also the discharge involved in the act of becoming conscious. But we know that abreacting does not always eliminate the symptoms, although it gives momentary relief; something else must be added in order to complete the process of healing.

* [Some patients, and even some therapists, believe that as much as possible of an analysis should be remembered. Such therapists do not differentiate between ordinary forgetting and repression. Nunberg explains in metapsychological terms why a cure can take place independent of whether the patient remembers how it happened.]

Free association, which must finally lead to the emergence of repressed memories, never proceeds smoothly. It is almost a rule that resistances increase with the deepening of the analysis; it can be observed over and over again that at a certain point during the chain of associations a feeling of discomfort—often an uncanny feeling—of variable intensity, arises, and may increase to the poiint of anxiety. These feelings can be overcome only with the help of the analyst, but still the patient's *active collaboration* is absolutely necessary. We have stressed before that the patient turns his active interest to inner processes, inner experiences composed of memories, not only out of love for his analyst but also because he feels *protected* by him. This protection enables him to drop his fear of recollecting and to permit the other affects uniting the memories to take their course freely.* Thus, for love of the analyst and in the knowledge of protection by him, the ego starts to work actively toward lifting the memories from the unconscious; at the same time it causes discharge of the instincts through "abreacting" and thus brings about an unintentional gain, as it were.

We have to consider still another factor. The further the neurosis has developed, the more the patient loses contact with reality. He withdraws, and renounces real gratification. He beccmes more or less indifferent and passive toward the outside world.

Complete harmony reigns for some time between patient and analyst. The patient relies completely upon the analyst, he accepts all his interpretations, and if it were possible he would ask the analyst to do the work of recollecting for him. The time soon comes, however, when this harmony is disturbed. As mentioned before, the resistances become ever stronger the deeper the analysis penetrates, i.e., the more the original pathogenic situation is approached. Added to these difficulties, there is the moment of frustration that is bound to appear in the transference sooner or later. Most patients react to the frustration with slackening in their work, with spite, and with "acting out", i.e., reenacting previous experiences. The patient leaves a part of the active work to the analyst: to guess what the patient wants to express but is unable to say. As a rule this concerns fantasies of being loved. The omnipotence of the patient's own means of expression (which may be mute) and the physician's omnipotence (his magic) are tested to the utmost. In part, the analyst succeeds in unmasking these resistances; in part, it is impossible for him to guess what the patient hides. The conflict now is no longer an inner one, but one between the patient and the analyst and it reaches an apex. The analysis is threatened with failure, i.e., the patient is confronted with the choice between losing the analyst and his love, or resuming active work. The patient becomes fearful of this loss if the transference is sufficiently firm, i.e.,

* [By contrast to Reich, Nunberg shows once more that significant psychoanalytic work can be done in the initial phase of positive transference.]

if a minimum amount of object-libido has been loosened from its fixations and is again at his disposal.

Frequently in such cases the analysis takes a remarkable turn. When the analyst has given up all hope of a successful conclusion of the analysis and lost interest in the case, the patient suddenly begins to yield a wealth of material, that promises a speedy conclusion of the analysis. This reminds us of those situations where some patients bring interesting material only at the end of the hour in the hope of extending the time they spend with their analyst. The only way in which I can explain this behaviour is: the patient becomes aware that the analyst is losing interest in him and consequently develops fear of losing the analyst's love. To avoid this he tolerates the unpleasure springing from the frustration and from the reproduction of the pathogenic traumatic situation, submits to active collaboration, and raises the last repressed memories from the unconscious. The inertia of instinct life, which found its expression in the repetition compulsion, is now overcome by the activity of the ego. The fear of loss of love, which is mobilized object libido, is the motive power bringing about the transformation of the passivity of instinct life (in the shape of the repetition compulsion) into the activity of the ego.*

The activity of the ego serves not only to loosen the last fixations of the instincts and to create the best conditions for their abreacting, but also to facilitate the reality-testing. This testing is prepared for in the analysis in that it is there proved, through conscious recollection, that the infantile strivings are psychic and historical formations corresponding to nothing in present reality. The transference likewise offers increasing opportunity for learning to distinguish between psychic and external reality. In this connection I should like to mention one example that is very instructive. When after many months analysis had succeeded in uncovering an important part of a patient's attachment to her father, she started the following hour with a question concerning my teeth. Puzzled, I tried to obtain some information about this question. The patient related that the day before, when taking leave of me, she had been struck by the appearance of my teeth. Until then she had been convinced that my front teeth had spaces between them. Now she noticed that this was not the case. She could not understand her error until suddenly, while she was discussing this illusion with me, she recalled that her father's teeth had spaces between them.†

It is almost self-evident that the greater precision of reality-testing acquired through analysis must lead to the abandonment of omnipotence and magic, just as it is evident that the recovered activity may bring about real changes in the external world which serve to create conditions for actual

---

* [This happened to Freud in the analysis of the Wolf Man. See editors' introduction, p. 25.]

† [Nunberg expanded this theme in a later publication; see Nunberg 1951.]

gratification of those instinctual needs that have just been released from repression.

Were this the final result, analysis at best would create a human being obedient to his instincts, capable of abreacting his erotic and destructive strivings on suitable objects. However, this is not the case; nor would it be possible in a civilized community. On the contrary, people after successful analysis are not only freer in their instinctual life but also better able to tolerate unpleasure and to control themselves; they are more able to endure instinct tensions, to sublimate, and to adapt to reality; and yet they do not fall ill from neurotic conflicts. Of course not even the best analysis can save them from actual conflicts.

We know that the ego is disorganized by the process of symptom-formation; the repressed part of instinct life is expelled from the ego-organization, subjected to the laws of the Unconscious, and inaccessible to the influence of the ego. The repressed part leads a separate existence. Briefly, in neurosis the ego has more or less lost its synthetic function.

Conscious thinking is a synthetic process which in neurosis is partly disturbed. Restitution of the synthetic function of the ego is introduced with the first act of recollecting. Indeed, remembering in analysis is accompanied by making connections and by uniting the repressed ideational elements with the actual ego, by mediating between the two, i.e., by assimilating what has been repressed. In this process analysis merely makes use of an already existing tendency of the ego: it is amazing with what tenacity even the most primitive person clings to his need for causality. The patient obeys the analyst's request to search for the hidden "causes" of his sufferings perhaps more readily and understandingly than he does any other request. The need for causality is in a sense gratified the very first day of analysis, and the last phases of analysis proceed entirely under this stimulus. Remembering then becomes the most important means of finding the "cause". To have found it means to unite what had been estranged, expelled, and repressed from the actual ego. It is through such assimilation that the continuity of the personality is restored and the real entrance of a psychic act into consciousness brought about. In some instances, such as in obsessional neurosis, remembering alone is not sufficient for entrance into consciousness. For this purpose seemingly heterogeneous elements have to be connected with each other. In analysis a process takes place similar to that in the spontaneous attempt at recovery in paranoia—though with different material and on a different level. Thus here too is manifested the power of Eros whose derivatives, in the form of desexualized libido of the ego, exercise their mediating and binding influence in the process of healing.

I intend to discuss this topic in greater detail on another occasion. At this time I should like only to indicate that it is at this point, that some other psychotherapeutic methods begin treatment—whether or not they are called "psychoanalysis". The basic difference between those methods and ours consists in the fact that in their case patients assimilate something foreign imposed upon them from outside, whereas our patients have to assimilate with painful self-conquest, what was *originally their own*.

The various resistances are gradually recognized during the psychoanalytic work and the ego rejects them as unsuitable ways of working. Simultaneous with the unmasking of resistances and the removal of repressions, major changes take place within the structure of the personality itself. The repressed libido is released and the representatives of the instincts are able to enter consciousness at any time and to discharge themselves in affects and actions. As the ego has absorbed the beloved and benign analyst, the severity of the superego is reduced. Since, furthermore, the ego has become freer of anxiety as a result of identification with the helping analyst, it admits and accepts the previously repressed instinctual impulses, and what has been repressed is once again absorbed by the ego. The ego recovers its synthetic function by regaining its capacity to mediate between the superego and the id on the one hand and between the id and objects of the external world on the other. It is now capable of establishing harmony between objects of the external world and the strivings of the id.

The description of the process of recovery is not yet ended. The ego, no longer having to spend energy in accomplishing the work of repression, becomes more adequate in its foremost task, that of reality-testing. Consequently, it becomes able to distinguish between real and psychic danger, between external and internal processes. Having gained ability to tolerate unpleasure, it also acquires better capacity to master non-ego-syntonic instinctual demands, i.e., those demands that are accompanied by external danger, and to direct them to other aims, as for instance, sublimation. In doing so it does not harm itself too much. It can procure gratification for other ego-syntonic instincts by useful changes of the external world. Through consideration of and regard for the objects of the external world the patient becomes more social and more resistant to instinct-tensions, i.e., to unpleasure.

Thus the psychoanalytic method of treatment makes use of "natural" recuperative forces which arise with the outbreak of the illness itself. They stem partly from the ego, partly from the id. The method frees the instincts from their fixations but at the same time aids the ego in its struggle against them.

In the ideal case the changes brought about by these recuperative forces involve the entire personality and are therefore as follows: *the energies of the id become more mobile, the superego more tolerant, the ego freer of anxiety and the synthetic function of the ego is restored.* Analysis is therefore *actually a synthesis.*

It should be emphasized that I do not say that the ego becomes free of anxiety, the id mobile, and the superego tolerant: I wish to indicate that analysis can bring about *relative* changes only; that *quantitative differences*, which at best can only be estimated, not measured, are involved here.

I am aware that I have not given a full account of the course of analysis here. That, however, was not my intention. I wanted to select from the almost inexhaustible wealth of problems only those that forced themselves upon my attention through my need for a better understanding of them. My presentation, while referring to an ideal case, is in general a composite of impressions gathered during many years of experience, which, however, I have been able to formulate only since Freud's latest works were published. In this description, I believe that I have come fairly close to the *actual* course of an analysis.

The implications for technique are self-evident: to permit the process of healing to go on undisturbed, but to make use of the natural and concomitant recuperative tendencies. Nevertheless, when the analysis reaches deeper levels it cannot be avoided that the resolution of one resistance is immediately followed by a new one, which arises as a reaction. Indeed, analysis takes an undulating course. In view of the present stage of our theoretic knowledge I should not venture to set up strict rules nor to give precise advice.

CHAPTER 16

SANDOR FERENCZI

THE PROBLEM OF THE TERMINATION OF THE ANALYSIS

Address to the Tenth International Psycho-Analytical Congress at Innsbruck, 3 September 1927

INTRODUCTORY NOTE

As long as psychoanalysts treated symptoms only, the question of termination did not arise. The analysis, if successful, ended when the symptoms disappeared. In the course of further developments, it became evident that one symptom could be exchanged for another, and the symptom could disappear during positive transference, and reappear when the transference became negative. Increasingly, as psychoanalysis became the analysis of character, the question at what point, and in what manner, to terminate, became acute.

Originally entitled "Das Problem der Beendigung der Analysen," *Internationale Zeitschrift für Psychoanalyse* 14 (1928):1–10. Reprinted in *Final Contributions to the Problems and Methods of Psychoanalysis*, ed. Michael Balint, trans. Eric Mosbacher et al. (New York: Basic Books, 1955), pp. 77–86. Reprinted by permission of Basic Books, Inc., Publishers.

Ferenczi's paper is the first of a long series of papers devoted to this subject. Freud's "Analysis Terminable and Interminable" (1937b) was, in part, a response to Ferenczi. In 1949, the British Psychoanalytic Society held a Symposium on the subject in which Annie Reich (1950) evaluated this paper. The question of termination remains a vexing problem in psychoanalytic technique.

LADIES AND GENTLEMEN,

Let me begin by referring to a case which some time ago caused me a great deal of concern. While dealing with a patient who, apart from certain neurotic difficulties, came to analysis chiefly because of certain abnormalities and peculiarities of his character, I suddenly discovered, incidentally after more than eight months, that he had been deceiving me the whole time in connexion with an important circumstance of a financial nature. At first this caused me the greatest embarrassment. The fundamental rule of analysis, on which the whole of our technique is built up, calls for the true and complete communication by the patient of all his ideas and associations. What, then, is one to do if the patient's pathological condition consists precisely in mendacity? Should one declare analysis to be unsuitable for dealing with such cases? I had no intention of admitting such poverty on the part of our scientific technique. Instead I continued with the work, and it was in fact investigation of the patient's mendacity which first enabled me to gain an understanding of some of his symptoms. During the analysis, before the lie was detected, he had on one occasion failed to appear for his appointment, and next day he failed even to mention the fact. When I taxed him with it, he obstinately denied his non-appearance, and, as it was certain that it was not I whose memory was at fault, I pressed him energetically to find out what had really happened. We both soon came to the conclusion that he had completely forgotten, not only his appointment with me, but the events of the whole day in question. He was able only partially to fill in the gaps in his memory, partly by questioning others. I do not propose to go into details of this incident, interesting though they are, and shall confine myself to mentioning that he had spent the forgotten day in a state of semi-drunkenness in various places of ill repute in the company of men and women of the lowest sort, all strangers to him.

He turned out to have had such memory disturbances before. When I obtained incontrovertible evidence of his conscious mendacity I was convinced that the split personality, at any rate in his case, was only the neurotic sign of his mendacity, a kind of indirect confession of his character-defect. So in this case the proof of the patient's lying turned out to be an advantageous event for his analysis.*

* I have no hesitation in generalizing this single observation, and interpreting all

It soon occurred to me that the problem of lying and simulation during analysis had been considered a number of times before. In an earlier paper I offered the suggestion that in infancy all hysterical symptoms were produced as conscious fictional structures. I recalled also that Freud used occasionally to tell us that it was prognostically a favourable sign, a sign of approaching cure, if the patient suddenly expressed the conviction that during the whole of his illness he had really been only shamming: it meant that in the light of his newly-acquired analytic insight into his unconscious drives he was no longer able to put himself back into the state of mind in which he had allowed his symptoms automatically to appear without the slightest intervention of his conscious self. A real abandonment of mendacity therefore appears to be at least a sign of the approaching end of the analysis.

We have in fact met the same set of circumstances before, though under a different name. What in the light of morality and of the reality principle we call a lie, in the case of an infant and in terms of pathology we call a fantasy. Our chief task in the treatment of a hysteria is essentially the seeking out of the automatically and unconsciously produced fantasy structure. During this process a large proportion of the symptoms disappear. We had come to the conclusion that the laying bare of the fantasy, which could be said to possess a special kind of reality of its own (Freud called it a psychical reality), was sufficient for a cure, and that it was a secondary matter from the point of view of the success of the analysis how much of the fantasy content corresponded to reality, i.e. physical reality or the recollections of such reality. My experience had taught me something else. I had become convinced that no case of hysteria could be regarded as cleared up so long as a reconstruction, in the sense of a rigid separation of reality and fantasy, had not been carried out. A person who admits the plausibility of the psycho-analytical interpretations he has been given but is not unquestionably convinced of their secure basis in fact preserves the right to take flight from certain disagreeable events into illness, i.e. into the world of fantasy; his analysis cannot be regarded as being terminated if by termination is meant a cure in the prophylactic sense. One might generalize and say that a neurotic cannot be regarded as cured if he has not given up pleasure in unconscious fantasy, i.e. unconscious mendacity. No bad way of ferreting out such fantasy nests is detecting the patient in one of those distortions of fact, however insignificant, which so often appear in the course of analysis. The patient's pride, his fear of losing the analyst's friendship by disclosing certain facts or feelings, invariably and without any excep-

---

cases of so-called split personality as symptoms of partially conscious insincerity which force those liable to it to manifest in turn only parts of their personality. In terms of metapsychology, one might say that these people have several super-egos which they have failed to fuse. Even scholars who do not *a priori* deny the possibility of 'many truths' about the same thing are probably people whose scientific morality has not developed into a unity.

tion whatever betray him into occasional suppression or distortion of facts. Observations of this nature have convinced me that calling on all our patients for full and complete free association from the outset represents an ideal which, so to speak, can be fulfilled only after the analysis has ended.* Associations which proceed from such distortions frequently lead back to similar but far more important infantile events, i.e. to times when the now automatic deception was conscious and deliberate.

We can confidently put down every infantile lie as an enforced lie and, as mendacity in later years is connected with infantile lying, perhaps there is something compulsive about every lie. This would be entirely logical. Frankness and honesty are certainly more comfortable than lying, the only inducement to which must be the threat of some greater unpleasure. What we describe by the fine names of ideal, ego-ideal, super-ego, owes its origin to the deliberate suppression of real instinctual urges, which thus have to be denied and repudiated, while the moral precepts and feelings imposed by education are paraded with exaggerated assiduity. Painful though it must be to students of ethics and moral theologians, we cannot avoid the conclusion that lying and morality are somewhat interconnected. To the child everything seems good that tastes good; he has to learn to think and feel that a good many things that taste good are bad, and to discover that the highest happiness and satisfaction lie in fulfilling precepts which involve difficult renunciations. In such circumstances it is not surprising—and our analyses demonstrate it beyond any possibility of doubt—that the two stages, that of original amorality and of subsequently acquired morality, are separated by a more or less long period of transition, in which all instinctual renunciation and all acceptance of unpleasure is distinctly associated with a feeling of untruth, i.e. hypocrisy.

From this point of view, if the analysis is to be a true reeducation, the whole process of the patient's character-formation, which was accompanied by the protective mechanism of instinctual repression, must be followed back to its instinctual foundations. The whole thing must, so to speak, become fluid again, so that out of the temporary chaos a new, better-adapted personality may arise under more favourable conditions. In other words, theoretically no symptom analysis can be regarded as ended unless it is a complete character analysis into the bargain. In practice, of course, many symptoms can be cured analytically without such radical changes being effected.

* [Ferenczi was apparently the first to recognize that completely free associations can take place only after all resistances have been overcome. In 1963, Eissler remarked: "In the psychoanalytic process 'saying everything' includes not only reporting every event past and present, every feeling, impulse, fantasy, but also that which is considered by the patient to be a lie, a falsification, unimportant, unnecessary. In order to reach the point of bringing all this material into analysis certain changes must take place in the patient. Strange as it may seem, to live up to this requirement is one of the most difficult tasks, and it is questionable whether anyone has ever lived up to it completely" (p. 198).]

Naïve souls, who do not know how instinctively human beings gravitate towards harmony and stability, will take fright at this, and wonder what happens to a man who loses his character in analysis. Are we able to guarantee delivery of a new character, like a new suit, in place of the old one which we have taken away? Might not the patient, divested of his old character, dash off in characterless nudity before the new clothing was ready? Freud has demonstrated how unjustified is this fear, and how psycho-analysis is automatically followed by synthesis. In reality the dissolution of the crystalline structure of a character is only a transition to a new, more appropriate structure; it is, in other words, a recrystallization. It is impossible to foresee in detail what the new suit will look like; the one thing that it is perhaps possible to say is that it will be a better fit, i.e. that it will be better adapted to its purpose.

Certain characteristics common to those who have been thoroughly analysed can, however, be mentioned. The far sharper severance between the world of fantasy and that of reality which is the result of analysis gives them an almost unlimited inner freedom and simultaneously a much surer grip in acting and making decisions; in other words it gives them more economic and more effective control.

In the few cases in which I have approached this ideal goal I have found myself compelled also to pay attention to certain external features of the patient's appearance or behaviour of a kind which previously we have often left unheeded. In my attempt to gain an understanding of the narcissistic peculiarities and mannerisms of sufferers from *tic* I observed how often relatively cured neurotics remained untouched by the analysis so far as these symptoms were concerned. A thoroughgoing character analysis cannot, of course, stop short at these peculiarities; in the last resort we must, so to speak, hold a mirror up to the patient, to enable him for the first time to become aware of the oddities in his behaviour and even in his personal appearance. Only those who have observed, as I have, how even people who have been cured by analysis continue surreptitiously to be smiled at by everybody because of their facial expressions or bodily posture or awkward mannerisms, etc., though they themselves have not the slightest suspicion that they are peculiar in any way, will shrink from regarding as a fearful but inevitable task imposed by a radical analysis the necessity of making those whom they primarily concern aware of these, so to speak, open secrets.* The analyst, who must, of course, always be tactful, must be particularly tactful in his handling of the acquisition by the patient of this particular kind of self-knowledge. I have made it a principle never to tell a patient directly about these

* This is the point at which psycho-analysis comes for the first time into practical contact with problems of physiognomy and physical constitution (as well as their derivatives, such as mimicry, graphology, etc.).

things; with the continuation of the analysis he must sooner or later become aware of them himself with our help.*

The phrase 'sooner or later' contains a hint of the importance of the time factor if an analysis is to be fully completed. The completion of an analysis is possible only if, so to speak, unlimited time is at one's disposal. I agree with those who think that the more unlimited it is, the greater are the chances of quick success. By this I mean, not so much the physical time at the patient's disposal, as his determination to persist for so long as may be necessary, irrespective of how long this may turn out to be. I do not wish to imply that there are not cases in which patients abundantly misuse this timelessness.

In the course of the time at our disposal, not only must the whole unconscious psychical material be lived through again in the form of memories and repetitions, but the third factor in analytic technique must also be applied. I refer to the factor of analytic working through. Freud has emphasized that this factor is as important as the other two, but its importance has not been properly appreciated. My present aim is to bring this working through, or the effort that we apply to it, into dynamic relation with the repression and the resistance, that is to say with a purely quantitative factor. Finding the pathogenic motivation and the conditions which determined the appearance of the symptom is a kind of qualitative analysis. This can be nearly complete without producing the expected therapeutic change. But further repetition of the same transference and resistance material, which has perhaps been gone through a countless number of times already, sometimes leads unexpectedly to an important advance which we can explain only as the successful effect of the factor of working through. Often enough the opposite takes place, i.e. after long working through, access to new memory material is gained which may herald the end of the analysis.

A really difficult but at the same time interesting task, which has in my opinion to be accomplished in every single case, is the gradual breaking-down of the resistances consisting in more or less conscious doubts about the dependability of the analyst. By this we mean his complete dependability in all circumstances, and in particular his unshakable good will towards the patient, to whatever extremes the latter may go in his words and his behaviour. One might actually speak of the patient's unconscious attempt consistently and in the greatest possible variety of ways to test the analyst's patience in this respect, and to test it, not just once, but over and over again. Patients sharply observe the physician's reaction, whether it takes the form of speech, silence, or gesture, and they often analyse it with great perspicuity. They detect the slightest sign of unconscious impulse in the latter, who has to submit to these

* [At this point, Ferenczi's view is in opposition to that of Wilhelm Reich, who used such confrontations to break through what he called "the character armor" of his patients (1933, this volume, Chapter 18).]

attempts at analysis with inexhaustible patience; this often makes superhuman demands upon him which, however, are invariably worth while. For if the patient fails to detect the analyst in any untruth or distortion, and comes gradually to realize that it is really possible to maintain objectivity in relation even to the naughtiest child, and if he fails to detect the slightest sign of unfounded superiority in the physician in spite of all his efforts to provoke signs of such a thing, if the patient is forced to admit that the physician willingly confesses the mistakes and inadvertences that he occasionally commits, it is not at all uncommon for the latter to reap as a reward of his labours a more or less rapid alteration in the patient's attitude. It seems to me exceedingly probable that, when patients do these things, they are attempting to reproduce situations in which non-understanding educators or relatives reacted to the child's so-called naughtiness with their own intense affectivity, thus forcing the child into a defiant attitude.

To stand firm against this general assault by the patient the analyst requires to have been fully and completely analysed himself. I mention this because it is often held to be sufficient if a candidate spends, say, a year gaining acquaintance with the principal mechanisms in his so-called training analysis. His further development is left to what he learns in the course of his own experience. I have often stated on previous occasions that in principle I can admit no difference between a therapeutic and a training analysis, and I now wish to supplement this by suggesting that, while every case undertaken for therapeutic purposes need not be carried to the depth we mean when we talk of a complete ending of the analysis, the analyst himself, on whom the fate of so many other people depends, must know and be in control of even the most recondite weaknesses of his own character; and this is impossible without a fully completed analysis.

The analysis of course shows that it was libidinous tendencies and not just the need for self-assertion or the desire for revenge that were the real factors in the patient's character formation and his often grotesquely disguised resistances. After the naughty, defiant child has fired off all the shots in his locker in vain, his concealed demands for love and tenderness come naïvely into the open. No analysis has been completed until most of the fore- and end-pleasure activities of sexuality, both in their normal and abnormal manifestations, have been emotionally lived through in the conscious fantasy. Every male patient must attain a feeling of equality in relation to the physician as a sign that he has overcome his fear of castration; every female patient, if her neurosis is to be regarded as fully disposed of, must have got rid of her masculinity complex and must emotionally accept without a trace of resentment the implications of her female role.* This analytic goal more or less

* [In 1937, Freud singled out these two points as limiting the effectiveness of the psychoanalytic cure (1937b).]

corresponds to that resuscitation of primeval innocence which Groddeck demands of his patients.* The difference between him and me is that he sets out for this goal straight from the symptoms, while I try to reach it by means of the 'orthodox' analytic technique, though at a slower pace. With the necessary patience the same result falls into our lap without any special pressure on our part.

Renunciation of pressure does not involve renunciation of those technical aids which I once referred to under the name of 'activity'. I still adhere to what I said on the subject at our Homburg congress. Perhaps no analysis can be ended until the patient, in agreement with our advice (which, however, must not contain anything in the nature of a command) † decides, in addition to free association, to consider some alterations in his way of life and behaviour, which then help in the discovery and eventual control of certain otherwise inaccessible nuclei of repression. In some cases bringing the analysis abruptly to an end may produce results, but this procedure is in principle to be rejected. While the pressure of an accidental external circumstance sometimes hastens the analysis, pressure imposed by the analyst often unnecessarily prolongs it. The proper ending of an analysis is when neither the physician nor the patient puts an end to it, but when it dies of exhaustion, so to speak, though even when this occurs the physician must be the more suspicious of the two and must think of the possibility that behind the patient's wish to take his departure some neurotic factor may still be concealed. A truly cured patient frees himself from analysis slowly but surely; so long as he wishes to come to analysis, he should continue to do so. To put it another way, one might say that the patient finally becomes convinced that he is continuing analysis only because he is treating it as a new but still a fantasy source of gratification, which in terms of reality yields him nothing. When he has slowly overcome his mourning over this discovery he inevitably looks round for other, more real sources of gratification. As Freud so long ago discovered, the whole neurotic period of his life appears to him in the light of analysis as one of pathological mourning, which he now seeks to displace into the transference situation; and the revelation of its true nature puts an end to the tendency to repeat it in the future. The renunciation of analysis is thus the final winding-up of the infantile situation of frustration which lay at the basis of the symptom-formation.

Another theoretically important experience in really completed analyses is the almost invariable occurrence of symptom transformation before the

* [Groddeck is best known for the fact that Freud took the term "id" from his *The Book of the It*, published in 1923. Groddeck believed that all illnesses, even those classified as organic, were psychologically determined. See also C. M. and S. Grossman's biography of Groddeck, 1965.]

† [The reader of Ferenczi's early papers, reprinted in this volume, will note that Ferenczi himself in his younger years was less careful in issuing commands, than he is here.]

end. We know from Freud that the symptomatology of neuroses is nearly always the result of a complicated psychical development. The obsessional patient, for instance, only gradually changes his emotions into obsessional behaviour and obsessional thinking. The hysteric may struggle for a long time with painful ideas of some kind before he succeeds in converting his conflicts into physical symptoms. The subsequent schizophrenic or paranoiac starts his pathological career as a victim of anxiety hysteria and succeeds often only after some hard work in finding a kind of pathological self-cure in exaggerated narcissism. We ought therefore not to be surprised if an obsessional begins producing symptoms of hysteria after a sufficient degree of loosening up and undermining of his intellectual obsessional system has taken place, and if the formerly so carefree sufferer from conversion hysteria, after analysis has made his physical symptoms inadequate, begins producing ideas and memories in place of the physical movements lacking in conscious content with which he previously expressed himself. It is thus a favourable sign if an obsessive neurotic begins to manifest hysterical emotion instead of affectless ideas, and if a hysteric temporarily produces obsessional ideas. It is, however, disagreeable if psychotic features make their appearance in the course of symptom transformation, but it would be mistaken to be excessively alarmed by them. I have seen cases in which there was no way to a final cure except through a temporary psychosis.

I have brought all these observations to your notice to-day in support of my conviction that analysis is not an endless process, but one which can be brought to a natural end with sufficient skill and patience on the analyst's part. If I am asked whether I can point to many such completed analyses, my answer must be no. But the sum-total of my experiences forces me to the conclusions that I have stated in this lecture. I am firmly convinced that, when we have learned sufficiently from our errors and mistakes, when we have gradually learned to take into account the weak points in our own personality, the number of fully analysed cases will increase.

CHAPTER 17

SANDOR FERENCZI

THE ELASTICITY OF

PSYCHO-ANALYTIC

TECHNIQUE

Lecture Given to the Hungarian

Psycho-Analytical Society, 1927

INTRODUCTORY NOTE

This article is of importance because of Ferenczi's admonition to future psy-choanalysts to refrain from promising good results or otherwise encouraging prospective analysands to undergo psychoanalysis. As he points out, analysis requires neither hope nor faith in its results, nor is it relevant for the success of the treatment whether the patient believes or is skeptical about psycho-analysis.

EFFORTS to make the technique which I use in my psycho-analyses available to others have frequently brought up the subject of psychological

Originally entitled "Die Elastizität der psychoanalytischen Technik," *Internationale Zeitschrift für Psychoanalyse* 14 (1928):197–209. Reprinted in *Final Contributions to the Problems and Methods of Psychoanalysis*, ed. Michael Balint, trans. Eric Mosbacher et al. (New York: Basic Books, 1955), pp. 87–101. Reprinted by permission of Basic Books, Inc., Publishers.

understanding. If it were true, as so many believe, that understanding of the mental processes in a third person depends on a special, inexplicable, and therefore untransferable faculty called knowledge of human nature, any effort to teach my technique to others would be hopeless. Fortunately it is otherwise. Since the publication of Freud's 'Recommendations' (1912b) on psychoanalytic technique, we have been in possession of the first foundations for a methodical investigation of the mind. Anyone who takes the trouble to follow the master's instructions will be enabled, even if he is not a psychological genius, to penetrate to unsuspected depths the mental life of others, whether sick or healthy.* Analysis of the parapraxes of everyday life, of dreams, and in particular of free associations, will put him in a position to understand a great deal about his fellows which was previously beyond the range of any but exceptional human beings. The human predilection for the marvellous will cause this transformation into a kind of craft of the art of understanding human nature to be received with disfavour. Artists and writers in particular appear to regard it as a kind of incursion into their domain and, after showing initial interest, summarily reject and turn their backs on it as an unattractive and mechanical technique. This antipathy causes us no surprise. Science is a process of progressive disillusionment; it displaces the mystical and the miraculous by universally valid and inevitable laws, the monotony and ineluctability of which provoke boredom and displeasure. However, it may serve as a partial consolation that among the practitioners of this craft, as of all others, artists will from time to time appear to whom we may look for progress and new perspectives.

From the practical point of view, however, it is an undeniable advance that analysis has gradually succeeded in putting tools for the more delicate investigation of the kind into the hands of the physician and student of only average gifts. A similar development occurred in regard to surgery; before the discovery of anaesthesia and asepsis it was the privilege of only a favoured few to exercise that 'healing art'; in the conditions of the time the art of working *cito, tuto, et jucunde*† was confined to them alone. True, artists in the technique of surgery still exist to-day, but the progress that has taken place enables all the thousands of physicians of average ability to exercise their useful and often life-saving activity.

A psychological technique had, of course, been developed outside the field of mental analysis; I refer to the methods of measurement used in psychological laboratories. This kind of technique is still in vogue, and for certain simple, practical purposes it may be sufficient. But analysis has much more

* [Ferenczi, like Freud, assumed that once the basic discoveries were made, it would take little skill to be a psychoanalyst. Freud compared psychoanalysis to a microscope. It took a genius to discover it, but any intelligent person can learn to use it. History has not justified his optimism.]

† [Latin for "quickly, safely, joyfully."]

far-reaching aims: the understanding of the topography, dynamics, and economy of the whole mental apparatus, undertaken without impressive laboratory apparatus, but with ever-increasing claims to scientific certainty and, above all, with incomparably greater results.

Nevertheless, there has been, and still is, a great deal in psycho-analytic technique which has created the impression that it involves a scarcely definable, individual factor. This has been chiefly due to the circumstance that in analysis the 'personal equation' has seemed to occupy a far more important place than we are called on to accept in other sciences. In his first essays on technique Freud himself left open the possibility that there was room in psycho-analysis for methods other than his own. But that expression of opinion dates from before the crystallizing out of the second fundamental rule of psycho-analysis, the rule by which anyone who wishes to undertake analysis must first be analysed himself. Since the establishment of that rule the importance of the personal element introduced by the analyst has more and more been dwindling away. Anyone who has been thoroughly analysed and has gained complete knowledge and control of the inevitable weaknesses and peculiarities of his own character will inevitably come to the same objective conclusions in the observation and treatment of the same psychological raw material, and will consequently adopt the same tactical and technical methods in dealing with it.* I have the definite impression that since the introduction of the second fundamental rule differences in analytic technique are tending to disappear.

If one attempts to weigh up the unresolved residue of this personal equation, and if one is in a position to see a large number of pupils and patients who have been analysed by others, and if, like me, one has to wrestle with the consequences of one's own earlier mistakes, one can claim the right to express a comprehensive opinion about the majority of these differences and mistakes. I have come to the conclusion that it is above all a question of psychological tact whether or when one should tell the patient some particular thing, when the material he has produced should be considered sufficient to draw conclusions, in what form these should be presented to the patient, how one should react to an unexpected or bewildering reaction on the patient's part, when one should keep silent and await further associations, and at what point the further maintenance of silence would result only in causing the patient useless suffering, etc. As you see, using the word 'tact' has enabled me only to reduce the uncertainty to a simple and appropriate formula. But what is 'tact'? The answer is not very difficult. It is the capacity for empathy. If, with the aid of the knowledge we have obtained from the dissection of many minds, but above all from the dissection of our own, we have succeeded in

* [Once again, history has not justified this optimism. On the contrary, different schools of psychoanalysis see the same psychological raw material—i.e., dreams, memories, transference reactions, etc.—in different lights and interpret it differently.]

forming a picture of possible or probable associations of the patient's of which he is still completely unaware, we, not having the patient's resistances to contend with, are able to conjecture, not only his withheld thoughts, but trends of his of which he is unconscious. At the same time, as we are continuously aware of the strength of the patient's resistance, we should not find it difficult to decide on the appropriateness or otherwise of telling him some particular thing or the form in which to put it. This empathy will protect us from unnecessarily stimulating the patient's resistance, or doing so at the wrong moment. It is not within the capacity of psycho-analysis entirely to spare the patient pain; indeed, one of the chief gains from psycho-analysis is the capacity to bear pain. But its tactless infliction by the analyst would only give the patient the unconsciously deeply desired opportunity of withdrawing himself from his influence.

All these precautions give the patient the impression of good will on the analyst's part, though the respect that the latter shows for the former's feelings derives solely from rational considerations. In my subsequent observations I shall try in a certain sense to justify the creation of this impression upon the patient. The essence of the matter is that there is no conflict between the tact which we are called upon to exercise and the moral obligation not to do to others what in the same circumstances we should not desire to have done to ourselves.

I hasten to add that the capacity to show this kind of 'good will' represents only one side of psycho-analytic understanding. Before the physician decides to tell the patient something, he must temporarily withdraw his libido from the latter, and weigh the situation coolly; he must in no circumstances allow himself to be guided by his feelings alone.

I now propose to give a few brief examples to illustrate this general point of view.

Analysis should be regarded as a process of fluid development unfolding itself before our eyes rather than as a structure with a design pre-imposed upon it by an architect. The analyst should therefore in no circumstances be betrayed into promising a prospective patient more than that, if he submits to the analytic process, he will end by knowing much more about himself, and that, if he persists to the end, he will be able to adapt himself to the inevitable difficulties of life more successfully and with a better distribution of his energies.* He can also be told that we know of no better and certainly of no more radical treatment of psychoneurotic and character difficulties; and we should certainly not conceal from him that there are other methods which hold out quicker and more definite prospects of a cure; and, if patients then

* [When Ferenczi opposes analyzing with a pre-imposed design, he is arguing against Reich; when he insists that no promises be given to the patient he is in sharp contrast to Nunberg's view in his 1928 paper (this volume, Chapter 15).]

reply that they have already submitted for years to treatment by suggestion, occupational therapy, or methods of strengthening the will, we may feel pleased. In other cases we may suggest that they try one of these much-promising methods before coming to us. But we cannot allow to pass the objection usually made by patients that they do not believe in our methods or theories. It must be explained to them at the outset that our technique makes no claim to be entitled to such unmerited confidence in advance, and that they need only believe in us if or when their experience of our methods gives them reason to do so. To another objection, namely that in this way we repudiate in advance any responsibility for a possible failure of the analysis and lay it squarely upon the patient and his impatience, we have no reply, and we have to leave it to the patient himself to decide whether or not to accept the risk of undertaking analysis under these difficult conditions. If these questions are not definitely settled in advance, one is putting into the hands of the patient's resistance a most dangerous weapon, which sooner or later he will not fail to use against us and the objective of the analysis. Even the most alarming question must not divert us from firmly establishing this basis for the analysis in advance. Many prospective patients ask with visible hostility whether the analysis may not last for two, three, five, or even ten years. We have to reply that it is possible. 'But a ten-year analysis would in practice be equivalent to a failure', we have to add. 'As we can never estimate in advance the magnitude of the difficulties to be overcome, we can promise you nothing, and can only say that in many cases the time needed is much shorter. But, as you probably believe that physicians like making favourable prognoses, and as you have probably heard, or will soon hear, many adverse opinions about the theory and technique of psycho-analysis, it will be best, from your point of view, if you will regard analysis as a bold experiment, which will cost you a great deal of toil, time, and money; and you must decide for yourself whether or not the amount of suffering which your difficulties are causing you is sufficient to make the experiment worth while in spite of all that. In any case, think it over carefully before beginning, because without the earnest intention to persist, even in spite of inevitable aggravations of your condition, the only result will be to add one more to the disappointments you have already suffered.'

I believe this preparation, which certainly errs on the side of pessimism, is certainly the better; in any case it is in accordance with the requirements of the 'empathy rule'. For, behind the prospective patient's often all-too-excessive display of faith in us, there is nearly always concealed a strong dose of distrust, which he is trying to shout down by his passionate demands on us for promises of a cure. A characteristic question that is often put to us by a prospective patient, after we have spent perhaps an hour trying to explain to him that we regard his case as suitable for analysis, is: 'But, doctor, do you really think that analysis would help me?' It would be a mistake to reply by simply saying yes. It is better to say that we do not believe in offering further assurances. Even if the prospective patient professes the most glowing opinion of analysis, this does not eliminate his concealed suspicion that the physician is, after all, a business man with something to sell. The patient's concealed incredulity is even more manifest in the question: 'But, doctor, don't you think that your methods might make me worse?' I generally reply to this with a counter-question. 'What is your occupation?' I ask. Suppose the answer is: 'I'm an architect.' 'Well, what would you say,' I reply, 'if you laid the plan for a new building before a client, and he asked you whether it wouldn't collapse?'* This generally puts an end to further demands for assurances, because it dawns on the patient that the practitioner of any craft is entitled to a certain amount of confidence in his own speciality, though that does not, of course, exclude disappointments.

Psycho-analysis is often reproached with being remarkably concerned with money matters. My own opinion is that it is far too little concerned with them. Even the most prosperous individual spends money on doctors most unwillingly. Something in us seems to make us regard medical aid, which in fact we all first received from our mothers in infancy, as something to which we are automatically entitled, and at the end of each month, when our patients are presented with their bill, their resistance is stimulated into producing all their concealed or unconscious hatred, mistrust, and suspicion over again. The most characteristic example of the contrast between conscious generosity and concealed resentment was given by the patient who opened the conversation by saying: 'Doctor, if you help me, I'll give you every penny I possess!' 'I shall be satisfied with thirty kronen an hour,' the physician replied. 'But isn't that rather excessive?' the patient unexpectedly remarked.

In the course of the analysis it is as well to keep one eye constantly open for unconscious expressions of rejection or disbelief and to bring them remorselessly into the open. It is only to be expected that the patient's resistance should leave unexploited no single opportunity for expressing these. Every patient without exception notices the smallest peculiarities in the analyst's behaviour, external appearance, or way of speaking, but without previous encouragement not one of them will tell him about them, though failure to do so constitutes a crude infringement of the primary rule of analysis. We therefore have no alternative but to detect ourselves from the patient's associations when we have offended his aesthetic feelings by an excessively loud sneeze or blowing of the nose, when he has taken offence at the shape of our face, or when he feels impelled to compare our appearance with that of others of more impressive physique.

I have on many other occasions tried to describe how the analyst must

* [Today we recognize that there are people whose ego structure is fragile and thus their fear of being analyzed may well be justified.]

accept for weeks on end the role of an Aunt Sally on whom the patient tries out all his aggressiveness and resentment. If we do not protect ourselves from this, but, on the contrary, encourage the only-too-hesitant patient at every opportunity that presents itself, sooner or later we shall reap the well-deserved reward of our patience in the form of the arrival of the positive transference. Any trace of irritation or offence on the part of the physician only prolongs the duration of the resistance period; if, however, the physician refrains from defending himself, the patient gradually gets tired of the one-sided battle and, when he has given full vent to his feelings, he cannot avoid confessing, though hesitantly, to the friendly feelings concealed behind his noisy defence; the result being deeper access into latent material, in particular into those infantile situations in which the foundation was laid (generally by non-understanding educators) for certain spiteful character traits.

Nothing is more harmful to the analysis than a schoolmasterish, or even an authoritative, attitude on the physician's part. Anything we say to the patient should be put to him in the form of a tentative suggestion and not of a confidently held opinion, not only to avoid irritating him, but because there is always the possibility that we may be mistaken. It is an old commercial custom to put 'E. & O.E.' ('errors and omissions excepted') at the bottom of every calculation, and every analytic statement should be put forward with the same qualification. Our confidence in our own theories should be only conditional, for in every case we may be presented with a resounding exception to the rule, or with the necessity of revising a hitherto accepted theory. I recall, for instance, an uneducated, apparently quite simple, patient who brought forward objections to an interpretation of mine, which it was my immediate impulse to reject; but on reflection not I, but the patient, turned out to be right, and the result of his intervention was a much better general understanding of the matter we were dealing with. Thus the analyst's modesty must be no studied pose, but a reflection of the limitations of our knowledge. Incidentally this may suggest the point from which, with the help of the lever of psychoanalysis, an alteration in the attitude of the doctor to the patient may be brought about. Compare our empathy rule with the lofty attitude to the patient generally adopted by the omniscient and omnipotent doctor.

Of course I do not mean that the analyst should be over-modest. He is fully justified in expecting that in the great majority of cases his interpretations, being based on experience, will sooner or later turn out to be correct, and that the patient will end by accepting the accumulation of evidence. But the analyst must wait patiently until the patient makes up his own mind; any impatience on the physician's part costs the patient time and money and the physician a great deal of work which he could very well spare.

A patient of mine once spoke of the 'elasticity of analytic technique', a phrase which I fully accept. The analyst, like an elastic band, must yield to the patient's pull, but without ceasing to pull in his own direction, so long as one position or the other has not been conclusively demonstrated to be untenable.

One must never be ashamed unreservedly to confess one's own mistakes. It must never be forgotten that analysis is no suggestive process, primarily dependent on the physician's reputation and infallibility. All that it calls for is confidence in the physician's frankness and honesty, which does not suffer from the frank confession of mistakes.

Analysis demands of the physician, not only a firm control of his own narcissism, but also a sharp watch on his emotional reactions of every kind. It used to be held that an excessive degree of 'antipathy' was an indication against undertaking an analysis, but deeper insight into the relationship has caused us to regard such a thing as unacceptable in principle, and to expect the analysed analyst's self-knowledge and self-control to be too strong for him to yield to such idiosyncrasies. Such 'anti-pathetic features' are in most cases only fore-structures, behind which quite different characteristics are concealed; dropping the patient in such cases would be merely leaving him in the lurch, because the unconscious aim of intolerable behaviour is often to be sent away. Knowledge of these things gives us the advantage of being able coolly to regard even the most unpleasant and repulsive person as a patient in need of help, and even enables us not to withhold our sympathy from him. The acquisition of this more than Christian humility is one of the hardest tasks of psycho-analytic practice, and striving to achieve it may incidentally lead us into the most terrible traps. I must once more emphasize that here too only real empathy helps; the patient's sharp wits will easily detect any pose.

One gradually becomes aware how immensely complicated the mental work demanded from the analyst is. He has to let the patient's free associations play upon him; simultaneously he lets his own fantasy get to work with the association material; from time to time he compares the new connexions that arise with earlier results of the analysis; and not for one moment must he relax the vigilance and criticism made necessary by his own subjective trends.

One might say that his mind swings continuously between empathy, self-observation, and making judgements. The latter emerge spontaneously from time to time as mental signals, which at first, of course, have to be assessed only as such; only after the accumulation of further evidence is one entitled to make an interpretation.

Above all, one must be sparing with interpretations, for one of the most important rules of analysis is to do no unnecessary talking; over-keenness in making interpretations is one of the infantile diseases of the analyst. When the patient's resistance has been analytically resolved, stages in the analysis are reached every now and then in which the patient does the work of interpretation practically unaided, or with only slight prompting from the analyst.

And now let us return for a moment to the subject of my much-praised and much-blamed 'activity'.* I believe I am at last in a position to give the details on timing for which I was rightly asked. You know, perhaps, that I was originally inclined to lay down certain rules of behaviour, in addition to free association, so soon as the resistance permitted such a burden. Experience later taught me that one should never order or forbid any changes of behaviour, but at most advise them, and that one should always be ready to withdraw one's advice if it turned out to be obstructive to the analysis or provocative of resistance. My original conviction that it was always the patient and never the physician who must be 'active' finally led me to the conclusion that we must content ourselves with interpreting the patient's concealed tendencies to action and supporting his feeble attempts to overcome the neurotic inhibitions to which he had hitherto been subject, without pressing or even advising him to take violent measures. If we are patient enough, the patient will himself sooner or later come up with the question whether he should risk making some effort, for example to defy a phobic avoidance. In such a case we shall certainly not withhold our consent and encouragement, and in this way we shall make all the progress expected from activity without upsetting the patient and falling out with him. In other words, it is the patient himself who must decide the timing of activity, or at any rate give unmistakable indications that the time is ripe for it. It remains true, of course, that such attempts by the patient produce changes of tension in his psychical system and thus prove to be a method of analytic technique in addition to the associations.

In another technical paper [1927, this volume, Chapter 16] I have already drawn attention to the importance of working through, but I dealt with it rather onesidedly, as a purely quantitative factor. I believe, however, that there is also a qualitative side to working through, and that the patient reconstruction of the mechanism of the symptom and character formation should be repeated again and again at every forward step in the analysis. Every important new insight gained calls for a revision of all the material previously produced, and may involve the collapse of some essential parts of what may have been thought to be a complete structure. The more subtle connexions between this qualitative working through and the quantitative factor (discharge of affect) must be left to a more detailed study of the dynamics of analytical technique.

A special form of this work of revision appears to occur, however, in every case; I mean the revision of the emotional experiences which happened in the course of the analyses. The analysis itself gradually becomes a piece of the patient's life-history, which he passes in review before bidding us farewell.

---

* See my papers on technique in *Further Contributions*.

In the course of this revision it is from a certain distance and with much greater objectivity that he looks at the experiences through which he went at the beginning of his acquaintanceship with us and the subsequent unravelling of the resistance and transference, which at the time seemed so immediate and important to him; and he then turns his attention away from analysis to the real tasks of life.

In conclusion I should like to hazard some remarks about the metapsychology of our technique.* It has often been said, by myself among others, that the process of recovery consists to a great extent of the patient's putting the analyst (his new father) in the place of the real father who occupies such a predominant place in his super-ego, and his then going on living with the analytic super-ego thus formed. I do not deny that such a process takes place in every case, and I agree that this substitution is capable of producing important therapeutic effects. But I should like to add that it is the business of a real character analysis to do away, at any rate temporarily, with any kind of super-ego, including that of the analyst. The patient should end by ridding himself of any emotional attachment that is independent of his own reason and his own libidinal tendencies. Only a complete dissolution of the super-ego can bring about a radical cure. Successes that consist in the substitution of one super-ego for another must be regarded as transference successes; they fail to attain the final aim of therapy, the dissolution of the transference [Alexander 1925, this volume, Chapter 6].

I should like to mention, as a problem that has not been considered, that of the metapsychology of the analyst's mental processes during analysis. His cathexes oscillate between identification (analytic object-love) on the one hand and self-control or intellectual activity on the other. During the long day's work he can never allow himself the pleasure of giving his narcissism and egoism free play in reality, and he can give free play to them in his fantasy only for brief moments. A strain of this kind scarcely occurs otherwise in life, and I do not doubt that sooner or later it will call for the creation of a special hygiene for the analyst.

Unanalysed ('wild') [Freud 1910c] analysts and incompletely cured patients are easily recognizable by the kind of 'compulsive analysing' from which they suffer; in contrast to the unhampered mobility of the libido which is the result of a complete analysis, which makes it possible to exercise analytic self-knowledge and self-control when necessary, but in no way hampers free enjoyment of life. The ideal result of a completed analysis is precisely that elasticity which analytic technique demands of the mental ther-

---

* By metapsychology we mean, of course, the sum-total of the ideas about the structure and dynamics of the psychical apparatus which our psychoanalytic experiences have caused us to adopt. See Freud's papers on metapsychology in his *Collected Papers*, Vol. 4. [Freud 1914b, 1915b, 1915c, 1917a, and 1917b. For a discussion of the significance of metapsychology for psychoanalytic technique see Part One of this volume.]

apist. This is one more argument for the necessity of the second fundamental rule of psycho-analysis.

Having regard to the great importance of any new technical recommendation, I could not make up my mind to publish this paper before submitting it to a colleague with a request for criticism.

'The title ["Elasticity"] is excellent,' he replied, 'and should be applied more widely, for Freud's technical recommendations were essentially negative. He regarded it as his most important task to emphasize what one should not do, to draw attention to all the temptations and pitfalls that stand in the way of analysis, and he left all the positive things that one should do to what you called "tact". The result was that the excessively docile did not notice the elasticity that is required and subjected themselves to Freud's "don't's" as if they were taboos. This is a situation that requires revision, without, of course, altering Freud's rules.

'True though what you say about tact is, the concessions you make to it seem to me to be questionable in the form in which you have put them. Those who have no tact will find in what you say a justification for arbitrary action, i.e. the intervention of the subjective element (i.e. the influence of their own unmastered complexes). What in reality we undertake is a weighing up— generally at the preconscious level—of the various reactions that we may expect from our intervention, and the most important aspect in this is quantitative assessment of the dynamic factors in the situation. Rules for making such an assessment can naturally not be given. The decisive factors are the analyst's experience and normality. But tact should be robbed of its mystical character.'

I entirely share my critic's view that these technical precepts of mine, like all previous ones, will inevitably be misused and misunderstood, in spite of the most extreme care taken in drafting them. There is no doubt that many—and not only beginners, but all who have a tendency to exaggeration —will seize on what I have said about the importance of empathy to lay the chief emphasis in their handling of patients on the subjective factor,* i.e. on intuition, and will disregard what I stated to be the all-important factor, the conscious assessment of the dynamic situation. Against such misinterpretations repeated warnings are obviously useless. I have even discovered that, in spite of the caution with which I put it forward—and my caution in the matter increases as time goes on—some analysts have used 'activity' as an excuse to indulge their tendency to impose on their patients entirely unanalytic rules, which sometimes border on sadism. I should therefore not be surprised if one day I heard my views on the patience required of an analyst

* [This was indeed Reik's point of view (1933, this volume, Chapter 26). It is interesting to note that Ferenczi doesn't mention either Alexander or Reik by name.]

used to justify a masochistic technique. Nevertheless the method which I follow and recommend—the method of elasticity—is not equivalent to non-resistance and surrender. True, we try to follow the patient in all his moods, but we never cease to hold firm to the viewpoint dictated to us by our analytic experience.

My principal aim in writing this paper was precisely to rob 'tact' of its mystical character. I agree, however, that I have only broached the subject and have by no means said the last word about it. I am, perhaps, a little more optimistic than my critic about the possibility of formulating positive advice on the assessment of certain typical dynamic conditions. For the rest, his belief that the analyst should be experienced and normal more or less corresponds to my own belief that the one dependable foundation for a satisfactory analytic technique is a complete analysis of the analyst himself. The processes of empathy and assessment will obviously take place, not in the unconscious, but at the pre-conscious level of the well-analysed analyst's mind.

Obviously under the influence of the above warnings, I feel it necessary to clarify one of the ideas put forward in this paper. I refer to my suggestion that a sufficiently deep character analysis must get rid of any kind of super-ego. An over-logical mind might interpret this as implying that my technique aimed at robbing people of all their ideals. In reality my objective was to destroy only that part of the super-ego which had become unconscious and was therefore beyond the range of influence. I have no sort of objection to the retention of a number of positive and negative models in the pre-conscious of the ordinary individual. In any case he will no longer have to obey his pre-conscious super-ego so slavishly as he had previously to obey his unconscious parent imago.

Part Seven

---

*The Controversy
Around
Wilhelm Reich's
Character Analysis*

CHAPTER 18

WILHELM REICH

THE TECHNIQUE OF

CHARACTER ANALYSIS

INTRODUCTORY NOTE

Wilhelm Reich, one of the most controversial figures in the history of psychoanalytic technique, is also one of the most interesting. Whether one agrees or disagrees with his method, there is no doubt that he was a highly original thinker. The chapter excerpted here deals with the core of his technique, that of Character Analysis. His point of view, as Hartmann (1951, p. 36) and Kris (1951, pp. 26–27) have pointed out, remained prestructural. Although he wrote after the publication of The Ego and the Id, *there is no evidence that he looked upon the neurotic as suffering from a conflict between superego, ego, and id. The term superego hardly appears in his writings. Instead, Reich saw character as a layering of resistances, where id impulses are defended against by ego impulses. As a teacher of a technical seminar he was afraid that an analysis would result in unending stalemates. He feared chaos, mistrusted intuition, and wanted to convert psychoanalytic technique into a procedure both exact and duplicable. His description of case material was masterful and he added to psychoanalytic technique a hitherto lacking awareness of the significance of how the analysand speaks, smiles, moves, or even lies on the couch.*

Originally entitled *Charakteranalyse: Technik und Grundlagen für studierende und praktizierende Analytiker* (Vienna: Selbsverlag d Verf [author's own publication], 1933). English translation entitled *Character-Analysis: Principles and Technique for Psychoanalysts in Practice and in Training*, trans. T. P. Wolfe (Rangeley, Me.: Orgone Institute Press, 1945), pp. 39–113, 119–143. Reprinted by permission of Gladys Meyers Wolfe. First presented at the Tenth International Psychoanalytic Congress, Innsbruck, 1927.

In the previous chapters, he delineated, without saying so explicitly, his points of departure from Freud. These may be summarized as follows:

1. *Many and perhaps most patients cannot free associate. It is, therefore, useless to demand that they follow the basic rule. New techniques have to be devised to make them analyzable.*

2. *Making the unconscious conscious releases only a very small quantity of energy. These are insufficient to bring about a characterological change, or a capacity to stay healthy after the analysis.*

3. *Only those patients who, in the course of their analysis, become orgastically potent can remain free from neurosis. The neurotic character is, therefore, the opposite of the genital character.*

4. *All neuroses are caused by the damming up of sexual energy. They, therefore, follow the model of Freud's "actual neurosis."*

5. *Every patient has the tendency to remain ill. The phrase, "you don't want to get well," should be eliminated from the analyst's vocabulary. Every unresolved stoppage in the analysis is the fault of the analyst.*

Reich left the psychoanalytic movement in 1934 and started his own form of therapy, which he called "orgonomy." His subsequent developments are therefore no longer of interest in the current context (see the biography in this volume, pp. 66–67).

1. / Introductory Review

OUR THERAPEUTIC METHOD is determined by the following basic theoretical concepts. The *topical** standpoint determines the technical principle that the unconscious has to be made conscious. The *dynamic* standpoint determines the rule that this has to take place not directly but by way of resistance analysis. The *economic* standpoint and the psychological structure determine the rule that the resistance analysis has to be carried out in a certain order according to the individual patient.†

As long as the topical process, the making conscious of the unconscious, was considered the only task of analytic technique, the formula that the unconscious manifestations should be interpreted *in the sequence in which they appeared* was correct. The dynamics of the analysis, that is, whether or not the making conscious also released the corresponding affect, whether the

* [The term "topical" is a mistranslation. What is meant is the "topographic" point of view.]

† [Reich here discusses three of the six categories of psychoanalytic metapsychology. See editors' introduction, pp. 6–8.]

analysis influenced the patient beyond a merely intellectual understanding, that was more or less left to chance. The inclusion of the dynamic* element, that is, the demand that the patient should not only remember things but also experience them, already complicated the simple formula that one had to "make the unconscious conscious." However, the dynamics of the analytic affect do not depend on the contents but on the resistances which the patient puts up against them and on the emotional experience in overcoming them. This makes the analytic task a vastly different one. From the topical stand-point, it is sufficient to bring into the patient's consciousness, one after the other, the manifest elements of the unconscious; in other words, the guiding line is the *content* of the material. If one also considers the dynamic factor one has to relinquish this guiding line in favor of another which comprehends the content of the material as well as the affects: that of the *successive resistances*. In doing so we meet, in most patients, with a difficulty which we have not yet mentioned.

2. / *Character Armor and Character Resistance*

(a)   *The inability to follow the fundamental rule*

Rarely are our patients immediately accessible to analysis, capable of following the fundamental rule and of really opening up to the analyst.† They cannot immediately have full confidence in a strange person; more impor-tantly, years of illness, constant influencing by a neurotic milieu, bad experi-ences with physicians, in brief, the whole secondary warping of the personality have created a situation unfavorable to analysis. The elimination of this difficulty would not be so hard were it not supported by the character of the patient which is part and parcel of his neurosis. It is a difficulty which has been termed "narcissistic barrier." There are, in principle, two ways of meet-ing this difficulty, in especial, the rebellion against the fundamental rule.

One, which seems the usual one, is a direct education to analysis by information, reassurance, admonition, talking-to, etc. That is, one attempts to educate the patient to analytic candor by the establishment of some sort of positive transference. This corresponds to the technique proposed by Nun-

* [Reich uses the word "dynamic" in an idiosyncratic way. The term "dynamic" in psychoanalytic usage merely implies bringing to light the opposing forces operating on the personality at the same time.]

† [Reich, like Ferenczi before him, was motivated to seek modifications of the psychoanalytic technique, because so many patients did not follow the fundamental rule.]

berg.* Experience shows, however, that this pedagogical method is very uncertain; it lacks the basis of analytic clarity and is exposed to the constant variations in the transference situation.

The other way is more complicated and as yet not applicable in all patients, but far more certain. It is that of *replacing the pedagogical measures by analytic interpretations*. Instead of inducing the patient into analysis by advice, admonitions and transference manoeuvres, one focuses one's attention on the actual behavior of the patient and its meaning: *why* he doubts, or is late, or talks in a haughty or confused fashion, or communicates only every other or third thought, why he criticizes the analysis or produces exceptionally much material or material from exceptional depths. If, for example, a patient talks in a haughty manner, in technical terms, one may try to convince him that this is not good for the progress of the analysis, that he better give it up and behave less haughtily, for the sake of the analysis. Or, one may relinquish all attempts at persuasion and wait until one understands why the patient behaves in this and no other way. One may then find that his behavior is an attempt to compensate his feeling of inferiority toward the analyst and may influence him by consistent interpretation of the meaning of his behavior. This procedure, in contrast to the first-mentioned, is in full accord with the principle of analysis.

This attempt to replace pedagogical and similar active measures seemingly necessitated by the characteristic behavior of the patient, by purely analytic interpretations led unexpectedly to the analysis of the *character*.

Certain clinical experiences make it necessary to distinguish, among the various resistances we meet, a certain group as *character resistances*. They get their specific stamp not from their content but from the patient's specific way of acting and reacting. The compulsive character develops specifically different resistances than does the hysterical character; the latter different resistances from the impulsive or neurasthenic character.† The *form* of the typical reactions which differ from character to character—though the contents may be the same—*is determined by infantile experiences just like the content of the symptoms or phantasies*.

---

* [Herman Nunberg (Reich gives no references). However, in a textbook published a year earlier Nunberg (1932) said: "When the positive transference sets in, the real collaboration with the analyst can begin. Thus, in spite of inner forces working against analysis, the patient starts his treatment when the conscious wish for recovery is supported by the unconscious id."]

† [Neurasthenic character denotes a functional neurosis based on irritability and fatigue-ability; see Freud "On the Grounds for Detaching a Particular Symptom from Neurasthenia Under Description of Anxiety Neurosis" (1895a). The concept and term were considered outmoded for many years in American psychiatry and in psychoanalysis but persisted in European psychiatry. The terms neurasthenic neurosis and asthenic character were reintroduced in the 1968 revision of the psychiatric nomenclature (see Diagnostic and Statistical Manual–II).]

(b)   *Whence the character resistances?*

Quite some time ago, Glover* worked on the problem of differentiating character neuroses from symptom neuroses. Alexander† also operated on the basis of this distinction. In my earlier writings, I also followed it. More exact comparison of the cases showed, however, that this distinction makes sense only insofar as there are neuroses with circumscribed symptoms and others without them; the former were called "symptom neuroses," the latter, "character neuroses." In the former, understandably, the symptoms are more obvious, in the latter the neurotic character traits. But, we must ask, are there symptoms without a neurotic reaction basis, in other words, without a neurotic character? The difference between the character neuroses and the symptom neuroses is only that in the latter the neurotic character also produced symptoms, that it became concentrated in them, as it were. If one recognizes the fact that the basis of a symptom neurosis is always a neurotic character, then it is clear that we shall have to deal with character-neurotic resistances in *every* analysis, that every analysis must be a character-analysis.

Another distinction which becomes immaterial from the standpoint of character-analysis is that between chronic neuroses, that is, neuroses which developed in childhood, and acute neuroses, which developed late. For the important thing is not whether the symptoms have made their appearance early or late. The important thing is that the neurotic character, the reaction basis for the symptom neurosis, was, in its essential traits, already formed at the period of the Oedipus phase. It is an old clinical experience that the boundary line which the patient draws between health and the outbreak of the disease becomes always obliterated during the analysis.

Since symptom formation does not serve as a distinguishing criterion we shall have to look for others. There is, first of all, insight into illness, and rationalization.

The lack of insight into illness is not an absolutely reliable but an essential sign of the character neurosis. The neurotic symptom is experienced as a foreign body and creates a feeling of being ill. The neurotic character trait, on the other hand, such as the exaggerated orderliness of the compulsive character or the anxious shyness of the hysterical character, are organically built into the personality. One may complain about being shy but does not feel ill for this reason. It is not until the characterological shyness turns into pathological blushing or the compulsion-neurotic orderliness into a compulsive ceremonial, that is, not until the neurotic character exacerbates symptomatically, that the person feels ill.

---

* [No reference given. However, from the content it appears that Reich was referring to Glover's article, "The Neurotic Character" (1925).]

† [No reference given. However, Reich probably had Alexander's "The Castration Complex in the Formation of Character" (1921) in mind.]

True enough, there are also symptoms for which there is no or only slight insight, things that are taken by the patient as bad habits or just peculiarities (chronic constipation, mild ejaculatio praecox, etc.). On the other hand, many character traits are often felt as illness, such as violent outbreaks of rage, tendency to lie, drink, waste money, etc. In spite of this, generally speaking, insight characterizes the neurotic symptom and its lack the neurotic character trait.

The second difference is that the symptom is never as thoroughly rationalized as the character. Neither a hysterical vomiting nor compulsive counting can be rationalized. The symptom appears meaningless, while the neurotic character is sufficiently rationalized not to appear meaningless or pathological. A reason is often given for neurotic character traits which would immediately be rejected as absurd if it were given for symptoms: "he just is that way." That implies that the individual was born that way, that this "happens to be" his character. Analysis shows this interpretation to be wrong; it shows that the character, for definite reasons, had to become that way and no different; that, in principle, it can be analyzed like the symptom and is alterable.

Occasionally, symptoms become part of the personality to such an extent that they resemble character traits. For example, a counting compulsion may appear only as part of general orderliness or a compulsive system only in terms of a compulsive work arrangement. Such modes of behavior are then considered as peculiarities rather than as signs of illness. So we can readily see that the concept of disease is an entirely fluid one, that there are all kinds of transitions from the symptom as an isolated foreign body over the neurotic character and the "bad habit" to rational action.

In comparison to the character trait, the symptom has a very simple construction with regard to its meaning and origin. True, the symptom also has a multiple determination; but the more deeply we penetrate into its determinations, the more we leave the realm of symptoms and the clearer becomes the characterological reaction basis. Thus one can arrive—theoretically —at the characterological reaction basis from any symptom. The symptom has its immediate determination in only a limited number of unconscious attitudes; hysterical vomiting, say, is based on a repressed fellatio phantasy or an oral wish for a child. Either expresses itself also characterologically, in a certain infantilism and maternal attitude. But the hysterical character which forms the basis of the symptom is determined by many—partly antagonistic —strivings and is expressed in a specific attitude or *way of being*. This is not as easy to dissect as the symptom; nevertheless, in principle it is, like the symptom, to be reduced to and understood from infantile strivings and experiences. While the symptom corresponds essentially to a single experience or striving, the character represents the specific way of being of an individual, an expression of his total past. For this reason, a symptom may develop suddenly while each individual character trait takes years to develop. In saying this we should not forget the fact that the symptom also could not have developed suddenly unless its characterological neurotic reaction basis had already been present.

The totality of the neurotic character traits makes itself felt in the analysis as a compact *defense mechanism* against our therapeutic endeavors. Analytic exploration of the development of this character "armor" shows that it also serves a definite economic purpose: on the one hand, it serves as a protection against the stimuli from the outer world, on the other hand against the inner libidinous strivings. The character armor can perform this task because libidinous and sadistic energies are consumed in the neurotic reaction formations, compensations and other neurotic attitudes. In the processes which form and maintain this armor, anxiety is constantly being bound up, in the same way as it is, according to Freud's description, in, say, compulsive symptoms. We shall have to say more later about the economy of character formation.

Since the neurotic character, in its economic function of a protecting armor, has established a certain *equilibrium*, albeit a neurotic one, the analysis presents a danger to this equilibrium. This is why the resistances which give the analysis of the individual case its specific imprint originate from this narcissistic protection mechanism. As we have seen, the mode of behavior is the result of the total development and as such can be analyzed and altered; thus it can also be the starting point for evolving the technique of character-analysis.

(c)   *The technique of analyzing the character resistance*

Apart from the dreams, associations, slips and other communications of the patients, their attitude, that is, *the manner* in which they relate their dreams, commit slips, produce their associations and make their communications, deserves special attention.* A patient who follows the fundamental rule from the beginning is a rare exception; it takes months of character-analytic work to make the patient halfway sufficiently honest in his communications. The manner in which the patient talks, in which he greets the analyst or looks at him, the way he lies on the couch, the inflection of the voice, the degree of conventional politeness, all these things are valuable criteria for judging the latent resistances against the fundamental rule, and understanding them makes it possible to alter or eliminate them by interpretation. The *how* of saying things is as important "material" for interpretation as is *what* the patient says. One often hears analysts complain that the analysis does not go

* The *form* of expression is far more important than the *ideational content*. Today, in penetrating to the decisively important infantile experiences, we make use of the form of expression *exclusively*. Not the ideational contents but the form of expression is what leads us to the biological reactions which form the basis of the psychic manifestations.

[This footnote was added in the 1945 edition. Compare Reich's view on this subject with Ferenczi's paper of 1919a (this volume, Chapter 5).]

well, that the patient does not produce any "material." By that is usually meant the content of associations and communications. But the manner in which the patient, say, keeps quiet, or his sterile repetitions, are also 'material" which can and must be put to use. There is hardly any situation in which the patient brings "no material"; it is our fault if we are unable to utilize the patient's behavior as "material."*

That the behavior and the form of the communications have analytic significance is nothing new. What I am going to talk about is the fact that these things present an avenue of approach to the analysis of the character in a very definite and almost perfect manner. Past failures with many cases of neurotic characters have taught us that in these cases the form of the communications is, at least in the beginning, always more important than their content. One only has to remember the latent resistances of the affect-lame, the "good," over-polite and ever-correct patients; those who always present a deceptive positive transference or who violently and stereotypically ask for love; those who make a game of the analysis; those who are always "armored," who smile inwardly about everything and everyone. One could continue this enumeration indefinitely; it is easy to see that a great deal of painstaking work will have to be done to master the innumerable individual technical problems.

For the purpose of orientation and of sketching the essential differences between character-analysis and symptom-analysis, let us assume two pairs of patients for comparison. Let us assume we have under treatment at the same time two men suffering from premature ejaculation; one is a passive-feminine, the other a phallic-aggressive character. Also, two women with an eating disturbance; one is a compulsive character, the other a hysteric.

Let us assume further that the premature ejaculation of both men has the same unconscious meaning: the fear of the paternal penis in the woman's vagina. In the analysis, both patients, on the basis of their castration anxiety which is the basis of the symptom, produce a negative father transference. Both hate the analyst (the father) because they see in him the enemy who frustrates their pleasure; both have the unconscious wish to do away with him. In this situation, the phallic-sadistic character will ward off the danger of castration by insults, depreciation and threats, while the passive-feminine character, in the same case, will become steadily more passive, submissive and friendly. In both patients, the character has become a resistance: one fends off the danger aggressively, the other tries to avoid it by a deceptive submission. It goes without saying that the character resistance of the passive-feminine patient is more dangerous because he works with hidden means: he produces a wealth of material, he remembers all kinds of infantile experiences, in short, he seems to cooperate splendidly. Actually, however, he

---

* [The emphasis on manner in addition to content is one of Reich's contributions to technique.]

camouflages a secret spitefulness and hatred; as long as he maintains this attitude he does not have the courage to show his real self.* If, now, one enters only upon *what* he produces, without paying attention to his way of behavior, then no analytic endeavor will change his condition. He may even remember the hatred of his father, but he will not *experience* it unless one interprets consistently the meaning of his deceptive attitude *before* beginning to interpret the deep meaning of his hatred of the father.

In the case of the second pair, let us assume that an acute positive transference has developed. The central content of this positive transference is, in either patient, the same as that of the symptom, namely, an oral fellatio phantasy. But although the positive transference has the same content in either case, the form of the transference resistance will be quite different: the hysterical patient will, say, show an *anxious* silence and a shy behavior; the compulsive character a *spiteful* silence or a cold, haughty behavior. In one case the positive transference is warded off by aggression, in the other by anxiety. And the form of this defense will always be the same in the same patient: the hysterical patient will always defend herself anxiously, the compulsive patient aggressively, no matter what unconscious content is on the point of breaking through. That is, *in one and the same patient, the character resistance remains always the same and only disappears with the very roots of the neurosis.*

In the character armor, the *narcissistic defense* finds its concrete chronic expression. In addition to the known resistances which are mobilized against every new piece of unconscious material, we have to recognize a constant factor of a *formal* nature which originates from the patient's character. Because of this origin, we call the constant formal resistance factor "character resistance."

In summary, the most important aspects of the character resistance are the following:

The character resistance expresses itself not in the content of the material, but in the formal aspects of the general behavior, the manner of talking, of the gait, facial expression and typical attitudes such as smiling, deriding, haughtiness, over-correctness, the *manner* of the politeness or of the aggression, etc.

What is specific of the character resistance is not *what* the patient says or does, but *how* he talks and acts, not *what* he gives away in a dream but *how* he censors, distorts, etc.

The character resistance remains the same in one and the same patient no matter what the material is against which it is directed. Different characters present the same material in a different manner. For example, a hysteric

* [In 1933 the term "self" was not part of the psychoanalytic vocabulary. Reich uses it in its commonsense meaning, not in the sense in which Hartmann (1950) differentiated between the ego and the self.]

patient will ward off the positive father transference in an anxious manner, the compulsive woman in an aggressive manner.

The character resistance, which expresses itself formally, can be understood as to its content and can be reduced to infantile experiences and instinctual drives just like the neurotic symptom.*

During analysis, the character of a patient soon becomes a resistance. That is, in ordinary life, the character plays the same role as in analysis: that of a psychic protection mechanism. The individual is "characterologically armored" against the outer world and against his unconscious drives.

Study of character formation reveals the fact that the character armor was formed in infancy for the same reasons and purposes which the character resistance serves in the analytic situation. The appearance in the analysis of the character as resistance reflects its infantile genesis. The situations which make the character resistance appear in the analysis are exact duplicates of those situations in infancy which set character formation into motion. For this reason, we find in the character resistance both a defensive function and a transference of infantile relationships with the outer world.

Economically speaking, the character in ordinary life and the character resistance in the analysis serve the same function, that of avoiding unpleasure, of establishing and maintaining a psychic equilibrium—neurotic though it may be—and finally, that of absorbing repressed energies. One of its cardinal functions is that of binding "free-floating" anxiety, or, in other words, that of absorbing dammed-up energy. Just as the historical, infantile element is present and active in the neurotic symptoms, so it is in the character. This is why a consistent dissolving of character resistances provides an infallible and immediate avenue of approach to the central infantile conflict.

What, then, follows from these facts for the technique of character-analysis? Are there essential differences between character-analysis and ordinary resistance analysis? There are. They are related to

(a) the selection of the sequence in which the material is interpreted;

(b) the technique of resistance interpretation itself.

As to (a): If we speak of "selection of material," we have to expect an important objection: some will say that any selection is at variance with basic psychoanalytic principles, that one should let oneself be guided by the patient, that with any kind of selection one runs the danger of following one's personal inclinations. To this we have to say that in this kind of selection it is not a matter of neglecting analytic material; it is merely a matter of *safeguarding a logical sequence* of interpretation which corresponds to the structure of the individual neurosis. All the material is finally interpreted; only, in any given situation this or that detail is more important than another. Incidentally, the analyst always makes selections anyhow, for he has already made a selection

---

* By the realization of this fact, the formal element becomes included in the sphere of psychoanalysis which, hitherto, was centered primarily on the content.

when he does not interpret a dream in the sequence in which it is presented but selects this or that detail for interpretation. One also has made a selection if one pays attention only to the content of the communications but not to their form. In other words, the very fact that the patient presents material of the most diverse kinds forces one to make a selection; what matters is only that one select *correctly* with regard to the given analytic situation.

In patients who, for character reasons, consistently fail to follow the fundamental rule, and generally where one deals with a character resistance, one will be forced *constantly to lift the character resistance out of the total material* and to dissolve it by the interpretation of its meaning. That does not mean, of course, that one neglects the rest of the material; on the contrary, every bit of material is valuable which gives us information about the meaning and origin of the disturbing character trait; one merely postpones the interpretation of what material does not have an immediate connection with the transference resistance until such time as the character resistance is understood and overcome at least in its essential features. I have already tried to show what are the dangers of giving deep-reaching interpretations in the presence of undissolved character resistances.

As to (b): We shall now turn to some special problems of character-analytic technique. First of all, we must point out a possible misunderstanding. We said that character-analysis begins with the emphasis on and the consistent analysis of the character resistance. It should be well understood that this does not mean that one asks the patient, say, not to be aggressive, not to deceive, not to talk in a confused manner, etc. Such procedure would be not only un-analytic but altogether sterile. The fact has to be emphasized again and again that what is described here as character-analysis has nothing to do with education, admonition, trying to make the patient behave differently, etc. In character-analysis, we ask ourself *why* the patient deceives, talks in a confused manner, why he is affect-blocked, etc.; we try to arouse the patient's interest in his character traits in order to be able, with his help, to explore analytically their origin and meaning. All we do is to lift the character trait which presents the cardinal resistance out of the level of the personality and to show the patient, if possible, the superficial connections between character and symptoms; it is left to him whether or not he will utilize his knowledge for an alteration of his character. In principle, the procedure is not different from the analysis of a symptom. What is added in character-analysis is merely that we isolate the character trait and confront the patient with it repeatedly until he begins to look at it objectively and to experience it like a painful symptom; thus, the character trait begins to be experienced as a foreign body which the patient wants to get rid of.

Surprisingly, this process brings about a change—although only a temporary one—in the personality. With progressing character-analysis, that impulse or trait automatically comes to the fore which had given rise to the character resistance in the transference. To go back to the illustration of the passive-feminine character: the more the patient achieves an objective attitude toward his tendency to passive submission, the more aggressive does he become. This is so because his passive-feminine attitude was essentially a reaction to repressed aggressive impulses. But with the aggression we also have a return of the infantile castration anxiety which in infancy had caused the change from aggressive to passive-feminine behavior. In this way the analysis of the character resistance leads directly to the center of the neurosis, the Oedipus complex.

One should not have any illusions, however. The isolation of such a character resistance and its analytic working-through usually takes many months of sustained effort and patient persistence. Once the breakthrough has succeeded, though, the analysis usually proceeds rapidly, with *emotionally* charged analytical experiences. If, on the other hand, one neglects such character resistances and instead simply follows the line of the material, interpreting everything in it, such resistances form a ballast which it is difficult if not impossible to remove. In that case, one gains more and more the impression that every interpretation of meaning was wasted, that the patient continues to doubt everything or only pretends to accept things, or that he meets everything with an inward smile. If the elimination of these resistances was not begun right in the beginning, they confront one with an insuperable obstacle in the later stages of the analysis, at a time when the most important interpretations of the Oedipus complex have already been given.

I have already tried to refute the objection that it is impossible to tackle resistances before one knows their *infantile* determination. The essential thing is first to see through the *present-day* meaning of the character resistance; this is usually possible without the infantile material. The latter is needed for the *dissolution* of the resistance. If at first one does no more than to show the patient the resistance and to interpret its present-day meaning, then the corresponding infantile material with the aid of which we can eliminate the resistance soon makes its appearance.

If we put so much emphasis on the analysis of the *mode* of behavior, this does not imply a neglect of the contents. We only add something that hitherto has been neglected. Experience shows that the analysis of character resistances has to assume first rank. This does not mean, of course, that one would only analyze character resistances up to a certain date and then begin with the interpretation of contents. The two phases—resistance analysis and analysis of early infantile experiences—overlap essentially; only in the beginning, we have a preponderance of character-analysis, that is, "education to analysis *by* analysis," while in the later stages the emphasis is on the contents and the infantile. This is, of course, no rigid rule but depends on the attitudes of the individual patient. In one patient, the interpretation of the infantile material will be begun earlier, in another later. It is a basic rule, however, not to give any deep-reaching interpretations—no matter how clear-cut the material—as long as the patient is not ready to assimilate them. Again, this is nothing new, but it seems that differences in analytic technique are largely determined by what one or the other analyst means by "ready for analytic interpretation." We also have to distinguish those contents which are part and parcel of the character resistance and others which belong to other spheres of experiencing. As a rule, the patient is in the beginning ready to take cognizance of the former, but not of the latter. Generally speaking, our character-analytic endeavors are nothing but an attempt to achieve the greatest possible security in the introduction of the analysis and in the interpretation of the infantile material. This leads us to the important task of studying and systematically describing the various forms of characterological transference resistances. If we understand them, the technique derives automatically from their structure.

(d)   *Derivation of the situational technique from the structure of the character resistance (interpretation technique of the defense)*

We now turn to the problem of how the situational technique of character-analysis can be derived from the structure of the character resistance in a patient who develops his resistances right in the beginning, the structure of which is, however, completely unintelligible at first. In the following case the character resistance had a very complicated structure; there were a great many coexistent and overlapping determinations. We shall try to describe the reasons which prompted me to begin the interpretation work with one aspect of the resistance and not with any other. Here also we will see that a consistent and logical interpretation of the defenses and of the mechanisms of the "armor" leads directly into the central infantile conflicts.

A Case of Manifest Inferiority Feelings

A man 30 years of age came to analysis because he "didn't get any fun out of life." He did not really think he was sick but, he said, he had heard about psychoanalysis and perhaps it would make things clearer to him. When asked about symptoms, he stated he did not have any. Later it was found that his potency was quite defective. He did not quite dare approach women, had sexual intercourse very infrequently, and then he suffered from premature ejaculation and intercourse left him unsatisfied. He had very little insight into his impotence. He had become reconciled to it; after all, he said, there were a lot of men who "didn't need that sort of thing."

His behavior immediately betrayed a severely inhibited individual. He spoke without looking at one, in a low voice, haltingly, and embarrassedly clearing his throat. At the same time, there was an obvious attempt to suppress his embarrassment and to appear courageous. Nevertheless, his whole appearance gave the impression of severe feelings of inferiority.

Having been informed of the fundamental rule, the patient began to talk hesitatingly and in a low voice. Among the first communications was the recollection of two "terrible" experiences. Once he had run over a woman with an automobile and she had died of her injuries. Another time, as a medical orderly during the war, he had had to do a tracheotomy. The bare recollection of these two experiences filled him with horror. In the course of the first few sessions he then talked, in the same monotonous, low and suppressed manner about his youth. Being next to the youngest of a number of children, he was relegated to an inferior place. His oldest brother, some twenty years his senior, was the parents' favorite; this brother had traveled a good deal, "knew the world," prided himself on his experiences and when he came home from one of his travels "the whole house pivoted around him." Although the content of his story made the envy of this brother and the hatred of him obvious enough, the patient, in response to a cautious query, denied ever having felt anything like that toward his brother. Then he talked about his mother, how good she had been to him and how she had died when he was 7 years of age. At this, he began to cry softly; he became ashamed of this and did not say anything for some time. It seemed clear that his mother had been the only person who had given him some love and attention and that her loss had been a severe shock to him. After her death, he had spent 5 years in the house of this brother. It was not the content but the tone of his story which revealed his enormous bitterness about the unfriendly, cold and domineering behavior of his brother. Then he related in a few brief sentences that now he had a friend who loved and admired him very much. After this, a continuous silence set in. A few days later he related a dream: *He saw himself in a foreign city with his friend; only, the face of his friend was different.* The fact that the patient had left his own city for the purpose of the analysis suggested that the man in the dream represented the analyst. This identification of the analyst with the friend might have been interpreted as a beginning positive transference. In view of the total situation, however, this would have been unwise. He himself recognized the analyst in the friend, but had nothing to add to this. Since he either kept silent or else expressed his doubts that *he* would be able to carry out the analysis, I told him that he had something against me but did not have the courage to come out with it. He denied this categorically, whereupon I told him that he also never had had the courage to express his inimical impulses toward his brother, not even to think them consciously; and that apparently he had established some kind of connection between his older brother and myself. This was true in itself, but I made the mistake of interpreting his resistance at too deep a level. Nor did the interpretation have any success; on the contrary, the inhibition became intensified. So

I waited a few days until I should be able to understand, from his behavior, the more important present-day meaning of his resistance. What was clear at this time was that there was a transference not only of the hatred of the brother but also a strong defense against a feminine attitude (*cf.* the dream about the friend). But an interpretation in this direction would have been inadvisable at this time. So I continued to point out that for some reason he defended himself against me and the analysis, that his whole being pointed to his being blocked against the analysis. To this he agreed by saying that, yes, that was the way he was generally in life, rigid, inaccessible and on the defensive. While I demonstrated to him his defense in every session, on every possible occasion, I was struck by the monotonous expression with which he uttered his complaints. Every session began with the same sentences: "I don't feel anything, the analysis doesn't have any influence on me, I don't see how I'll ever achieve it, nothing comes to my mind, the analysis doesn't have any influence on me," etc. I did not understand what he wanted to express with these complaints, and yet it was clear that here was the key to an understanding of his resistance.*

Here we have a good opportunity for studying the difference between the character-analytic and the active-suggestive education to analysis. I might have admonished him in a kindly way to tell me more about this and that; I might have been able thus to establish an artificial positive transference; but experience with other cases had shown me that one does not get far with such procedures. Since his whole behavior did not leave any room for doubt that he refuted the analysis in general and me in particular, I could simply stick to this interpretation and wait for further reactions. When, on one occasion, the talk reverted to the dream, he said the best proof for his not refuting me was that he identified me with his friend. I suggested to him that possibly he had expected me to love and admire him as much as his friend did; that he then was disappointed and very much resented my reserve. He had to admit that he had had such thoughts but that he had not dared to tell them to me. He then related how he always only *demanded* love and especially recognition, and that he had a very *defensive* attitude toward men with a particularly masculine appearance. He said he did not feel equal to such men, and in the relationship with his friend he had played the feminine part. Again there was material for interpreting his feminine transference but his total behavior warned against it. The situation was difficult, for the elements of his resistance which I already understood, the transference of hatred from his brother, and the narcissistic-feminine attitude toward his superiors, were strongly warded

* The explanation given here is insufficient, although it is psychologically correct. Today we know that such complaints are the immediate expression of a vegetative, that is, muscular armoring, The patient complains about affect-lameness because of a block in his plasmatic currents and sensations. The disturbance, then, is primarily of a *biophysical* nature. Vegetotherapy eliminates the block in motility not with psychological but with biophysical means. [This footnote was added in the 1945 edition.]

off; consequently, I had to be very careful or I might have provoked him into breaking off the analysis. In addition, he continued to complain in every session, in the same way, that the analysis did not touch him, etc.; this was something which I still did not understand after about four weeks of analysis, and yet, I felt that it was an essential and acutely active character resistance.

I fell ill and had to interrupt the analysis for two weeks. The patient sent me a bottle of brandy as a tonic. When I resumed the analysis he seemed to be glad. At the same time, he continued his old complaints and related that he was very much bothered by thoughts about death, that he constantly was afraid that something had happened to some member of his family; and that during my illness he had always been thinking that I might die. It was when this thought bothered him particularly badly one day that he had sent me the brandy. At this point, the temptation was great to interpret his repressed death wishes. The material for doing so was ample, but I felt that such an interpretation would be fruitless because it would bounce back from the wall of his complaints that "nothing touches me, the analysis has no influence on me." In the meantime, the secret double meaning of his complaint, "nothing touches me" ("*nichts dringt in mich ein*") had become clear; it was an expression of his most deeply repressed transference wish for anal intercourse. But would it have been justifiable to point out to him his homosexual love impulse—which, it is true, manifested itself clearly enough—while he, with his whole being, continued to protest against the analysis? First it had to become clear what was the meaning of his complaints about the uselessness of the analysis. True, I could have shown him that he was wrong in his complaints: he dreamed without interruption, the thoughts about death became more intense, and many other things went on in him. But I knew from experience that that would not have helped the situation. Furthermore, I felt distinctly the armor which stood between the unconscious material and the analysis, and had to assume that the existing resistance would not let any interpretation penetrate to the unconscious. For these reasons, I did no more than consistently to show him his attitude, interpreting it as the expression of a violent defense, and telling him that we had to wait until we understood this behavior. He understood already that the death thoughts on the occasion of my illness had not necessarily been the expression of a loving solicitude.

In the course of the next few weeks it became increasingly clear that his inferiority feeling connected with his feminine transference played a considerable role in his behavior and his complaints. Yet, the situation still did not seem ripe for interpretation; the meaning of his behavior was not sufficiently clear. To summarize the essential aspects of the solution as it was found later:

1. He desired recognition and love from me as from all men who appeared masculine to him. That he wanted love and had been disappointed by me had already been interpreted repeatedly, without success.

2. He had a definite attitude of envy and hatred toward me, transferred from his brother. This could, at this time, not be interpreted because the interpretation would have been wasted.
3. He defended himself against his feminine transference. This defense could not be interpreted without touching upon the warded-off femininity.
4. He felt inferior before me, because of his femininity. His eternal complaints could only be the expression of this feeling of inferiority.

Now I interpreted his inferiority feeling toward me. At first, this led nowhere, but after I had consistently held up his behavior to him for several days, he did bring some communications concerning his boundless envy, not of me, but other men of whom he also felt inferior. Now it suddenly occurred to me that his constant complaining could have only one meaning: "The analysis has no influence on me," that is, "It is no good," that is, "the analyst is inferior, is impotent, cannot achieve anything with me." *The complaints were in part a triumph over the analyst, in part a reproach to him.* I told him what I thought of his complaints. The result was astounding. Immediately he brought forth a wealth of examples which showed that he always acted this way when anybody tried to influence him. He could not tolerate the superiority of anybody and always tried to tear them down. He had always done the exact opposite of what any superior had asked him to do. There appeared a wealth of recollections of his spiteful and deprecatory behavior toward teachers.

Here, then, was his suppressed aggression, the most extreme manifestation of which thus far had been his death wishes. But soon the resistance reappeared in the same old form, there were the same complaints, the same reserve, the same silence. But now I knew that my discovery had greatly impressed him, which had *increased* his feminine attitude; this, of course, resulted in an intensified defense against the femininity. In analyzing the resistance, I started again from the inferiority feeling toward me; but now I deepened the interpretation by the statement that he did not only feel inferior but that, because of his inferiority, he felt himself in a female role toward me, which hurt his masculine pride.

Although previously the patient had presented ample material with regard to his feminine attitude toward masculine men and had had full insight for this fact, now he denied it all. This was a new problem. Why should he now refuse to admit what he had previously described himself? I told him that he felt so inferior toward me that he did not want to accept any explanation from me even if that implied his going back on himself. He realized this to be true and now talked about the relationship with his friend in some detail. He had actually played the feminine role and there often had been sexual intercourse between the legs. Now I was able to show him that his defensive attitude in the analysis was nothing but the struggle against the surrender to the analysis which, to his unconscious, was apparently linked up with the idea of surrendering to the analyst in a female fashion. This hurt his pride, and this was the reason for his stubborn resistance against the influence of the analysis. To this he reacted with a confirmatory dream: he lies on a sofa with the analyst, who kisses him. This clear dream provoked a new phase of resistance in the old form of complaints that the analysis did not touch him, that he was cold, etc. Again I interpreted the complaints as a depreciation of the analysis and a defense against surrendering to it. But at the same time I began to explain to him the economic meaning of this defense. I told him that from what he had told thus far about his infancy and adolescence it was obvious that he had closed himself up against all disappointments by the outer world and against the rough and cold treatment by his father, brother and teachers; that this seemed to have been his only salvation even if it demanded great sacrifices in happiness.

This interpretation seemed highly plausible to him and he soon produced memories of his attitude toward his teachers. He always felt they were cold and distant—a clear projection of his own attitude—and although he was aroused when they beat or scolded him he remained indifferent. In this connection he said that he often had wished I had been more severe. This wish did not seem to fit the situation at that time; only much later it became clear that he wished to put me and my prototypes, the teachers, in a bad light with his spite. For a few days the analysis proceeded smoothly, without any resistances; he now remembered that there had been a period in his childhood when he had been very wild and aggressive. At the same time he produced dreams with a strong feminine attitude toward me. I could only assume that the recollection of his aggression had mobilized the guilt feeling which now was expressed in the passive-feminine dreams. I avoided an analysis of these dreams not only because they had no immediate connection with the actual transference situation, but also because it seemed to me that he was not ready to understand the connection between his aggression and the dreams which expressed a guilt feeling.* Many analysts will consider this an arbitrary selection of material. Experience shows, however, that the best therapeutic effect is to be expected when an immediate connection is already established between the transference situation and the infantile material. I only ventured the assumption that, to judge from his recollections of his aggressive infantile behavior, he had at one time been quite different, the exact opposite of what he was today, and that the analysis would have to find out at what time and under what circumstances this change in his character had taken place. I told him that his present femininity probably was an avoidance of his aggressive masculinity. To this the patient did not react except by falling back into his old resistance of complaining that he could not achieve it, that the analysis did not touch him, etc.

* [The sparing use of dream interpretation while the character armor is not yet resolved is one of the characteristics of Reich's technique.]

I interpreted again his inferiority feeling and his recurrent attempt to prove the analysis, or the analyst, to be impotent; but now I also tried to work on the transference from the brother, pointing out that he had said that his brother always played the dominant role. Upon this he entered only with much hesitation, apparently because we were dealing with the central conflict of his infancy; he talked again about how much attention his mother had paid to his brother, without, however, mentioning any subjective attitude toward this. As was shown by a cautious approach to the question, the envy of his brother was completely repressed. Apparently, this envy was so closely associated with intense hatred that not even the envy was allowed to become conscious. The approach to this problem provoked a particularly violent resistance which continued for days in the form of his stereotyped complaints about his inability. Since the resistance did not budge it had to be assumed that here was a particularly acute rejection of the person of the analyst. I asked him again to talk quite freely and without fear about the analysis and, in particular, about the analyst, and to tell me what impression I had made on him on the occasion of the first meeting.* After much hesitation he said the analyst had appeared to him so masculine and brutal, like a man who is absolutely ruthless with women. So I asked him about his attitude toward men who gave an impression of being potent.

This was at the end of the fourth month of the analysis. Now for the first time that repressed attitude toward the brother broke through which had the closest connection with his most disturbing transference attitude, the envy of potency. With much affect he now remembered that he had always condemned his brother for always being after women, seducing them and bragging about it afterwards. He said I had immediately reminded him of his brother. I explained to him that obviously he saw in me his potent brother and that he could not open up to me because he condemned me and resented my assumed superiority just as he used to resent that of his brother; furthermore, it was plain now that the basis of his inferiority feeling was a feeling of impotence.

Then occurred what one always sees in a correctly and consistently carried-out analysis: *the central element of the character resistance rose to the surface.* All of a sudden he remembered that he had repeatedly compared his small penis with the big one of his brother and how he had envied his brother.

As might have been expected, a new wave of resistance occurred; again the complaint, "I can't do anything." Now I could go somewhat further in the interpretation and show him that he was acting out his impotence. His reaction to this was wholly unexpected. In connection with my interpretation of his distrust he said for the first time that he had never believed anyone, that he

---

* Since then I am in the habit of soon asking the patient to describe my person. This measure always proves useful for the elimination of blocked transference situations.

did not believe anything, and probably also not in the analysis. This was, of course, an important step ahead, but the connection of this statement with the analytic situation was not altogether clear. For two hours he talked about all the many disappointments which he had experienced and believed that they were a rational explanation of his distrust. Again the old resistance reappeared; as it was not clear what had precipitated it this time, I kept waiting. The old behavior continued for several days. I only interpreted again those elements of the resistance with which I was already well acquainted. Then, suddenly, a new element of the resistance appeared: he said he was *afraid of the analysis because it might rob him of his ideals.* Now the situation was clear again. He had transferred his castration anxiety from his brother to me. He was afraid of me. Of course, I did not touch upon his castration anxiety but proceeded again from his inferiority feeling and his impotence and asked him whether his high ideals did not make him feel superior and better than everybody else. He admitted this openly; more than that, he said that he was really better than all those who kept running after women and lived sexually like animals. He added, however, that this feeling was all too often disturbed by his feeling of impotence, and that apparently he had not become quite reconciled to his sexual weakness after all. Now I could show him the neurotic manner in which he tried to overcome his feeling of impotence: he was trying to recover a feeling of potency in the realm of ideals. I showed him the mechanism of compensation and pointed out again the resistances against the analysis which originated from his secret feeling of superiority. I told him that not only did he think himself secretly better and cleverer than others; it was for this very reason that he resisted the analysis. For if it succeeded, he would have taken recourse to the aid of somebody else and it would have vanquished his neurosis, the secret pleasure gain of which had just been unearthed. From the standpoint of the neurosis this would be a defeat which, furthermore, to his unconscious, would mean becoming a woman. In this way, by progressing from the ego and its defense mechanisms, I prepared the soil for an interpretation of the castration complex and of the feminine fixation.

The character-analysis had succeeded, then, in penetrating from his mode of behavior directly to the center of his neurosis, his castration anxiety, the envy of his brother because of his mother's favoritism, and the disappointment in his mother. What is important here is not that these unconscious elements rose to the surface; that often occurs spontaneously. What is important is the logical sequence and the close contact with the ego-defense and the transference in which they came up; further, that this took place without any urging, purely as the result of analytic interpretation of the behavior; further, that it took place with the corresponding affects. This is what constitutes a consistent character-analysis; it is a thorough working through of the conflicts assimilated by the ego.

In contrast, let us consider what probably would have happened without a consistent emphasis on the defenses. Right at the beginning, there was the possibility of interpreting the passive-homosexual attitude toward the brother, and the death wishes. Undoubtedly, dreams and associations would have provided further relevant material for interpretation. But without a previous systematic and detailed working through of his ego-defense, no interpretation would have affectively penetrated; the result would have been an intellectual knowledge of his passive desires alongside with a violent affective defense against them. The affects belonging to the passivity and the murderous impulses would have continued to remain in the defense function. The final result would have been a chaotic situation, the typical hopeless picture of an analysis rich in interpretations and poor in results.

A few months' patient and persistent work on his ego-defense,* particularly its form (complaints, manner of speech, etc.) raised the ego to that level which was necessary for the assimilation of the repressed, it loosened the affects and brought about their displacement in the direction of the repressed ideas. One cannot say, therefore, that in this case two different techniques would have been feasible; there was only one possibility if one was to alter the patient *dynamically.* I trust that this case makes clear the different concept of the application of theory to technique. The most important criterion of an orderly analysis is the giving of *few* interpretations which are to the point and consistent, instead of a great many which are unsystematic and do not take into consideration the dynamic and economic element. If one does not let oneself be led astray by the material, if, instead, one evaluates correctly its dynamic position and economic role, then one gets the material later, it is true, but more thoroughly and more charged with affect. The second criterion is a continuous connection between present-day situation and infantile situation. While in the beginning the various elements of the content coexist side by side without any order, this changes into a logical sequence of resistances and contents, a sequence determined by the dynamics and structure of the individual neurosis. With unsystematic interpretation, one has to make one new start after another, guessing rather than knowing one's way; in the case of character-analytic work on the resistances, on the other hand, the analytic process develops as if by itself. In the former case, the analysis will run smoothly in the beginning only to get progressively into more and more difficulties; in the latter case, the greatest difficulties are met in the first few weeks and months of the treatment, to give way progressively to smooth work even on the most deeply repressed material. The fate of every analysis depends on its introduction, that is, the correct or incorrect handling of the resistances. The third criterion, then, is that of tackling the case not in this or

---

* [Reich here does not speak of the defenses in the sense in which they are enumerated in Freud's Appendix A, "Repression and Defence," to "Inhibitions, Symptoms and Anxiety" (1926a), but only of ego defense. To Reich, the whole character armor is one defense.]

that spot which happens to be tangible but at the spot which hides the most essential ego-defense; and the systematic enlarging of the breach which has been made into the unconscious; and the working out of that infantile fixation which is affectively most important at any given time. A certain unconscious position which manifests itself in a dream or an association may have a central significance for the neurosis and yet may at any given time be quite unimportant with regard to its technical significance. In our patient, the feminine attitude toward the brother was of central pathogenic significance; yet in the first few months the technical problem was the fear of the loss of the compensation for the impotence by high ideals. The mistake which is usually made is that of attacking the central pathogenic point of the neurosis which commonly manifests itself somehow right at the beginning. What has to be attacked instead are the respective important present-day positions which, if worked on systematically, one after the other, lead *of necessity* to the central pathogenic situation. It is important, therefore, and in many cases decisive, *how, when* and from which side one proceeds toward the central point of the neurosis.

What we have described here as character-analysis fits without difficulty into Freud's theory of resistances, their formation and dissolution. We know that every resistance consists of an id-impulse which is warded off and an ego-impulse* which wards it off. Both impulses are unconscious. In principle, then, one would seem to be free to interpret first either the id-impulse or the ego-impulse. For example: If a homosexual resistance in the form of keeping silent appears right at the beginning of the analysis, one can approach the id-impulse by telling the patient that he is occupied with thoughts about loving the analyst or being loved by him; one has interpreted his positive transference, and if he does not take flight it will, at best, take a long time before he can come to terms with such a forbidden idea. The better way, then, is to approach first the *defense of the ego* which is more closely related to the conscious ego. One will tell the patient at first only that he is keeping silent because—*"for one reason or another,"* that is, without touching upon the id-impulse—he is defending himself against the analysis, presumably because it has become somehow dangerous to him. In the first case one has tackled the id aspect, in the latter case the ego aspect of the resistance, the defense.

Proceeding in this manner, we comprehend the negative transference in which every defense finally results, as well as the character, the armor of the ego. The superficial, more nearly conscious layer of *every* resistance must of necessity be a negative attitude toward the analyst, no matter whether the

* [Reich speaks of ego impulses rather than ego defenses. In psychoanalytic metapsychology, the ego is not conceptualized as having impulses. This is more than a slip of the pen. Reich believed in fact, that one id impulse is used to defend against another more dangerous one. For example, incest wishes he saw as defending against fear of the mother. He was not aware of the fact that the use of id wishes as a defense takes place only when the ordinary resources of the ego are insufficient (see Jacobson 1957, p. 62).]

warded-off id-impulse is hatred or love. The ego projects its defense against the id-impulse to the analyst who has become a dangerous enemy because, by his insistence on the fundamental rule, he has provoked id-impulses and has disturbed the neurotic equilibrium. In its defense, the ego makes use of very old forms of negative attitudes; it utilizes hate impulses from the id even if it is warding off love impulses.

If we adhere to the rule of tackling resistances from the ego side, we always dissolve, at the same time, a certain amount of negative transference, of hatred. This obviates the danger of overlooking the destructive tendencies which often are extremely well hidden; it also strengthens the positive transference. The patient also comprehends the ego interpretation more easily because it is more in accordance with conscious experience than the id interpretation; this makes him better prepared for the latter which follows at a later time.

The ego defense has always the same form, corresponding to the character of the patient, whatever the repressed id-impulse may be. Conversely, the same id-impulse is warded off in different ways in different individuals. If we interpret only the id-impulse, we leave the character untouched. If, on the other hand, we always approach the resistances from the defense, from the ego side, we include the neurotic character in the analysis. In the first case, we say immediately *what* the patient wards off. In the latter case, we first make clear to him *that* he wards off "something," then, *how* he does it, what are the means of defense (character-analysis); only at last, when the analysis of the resistance has progressed far enough, is he told—or finds out for himself— what it is he is warding off. On this long detour to the interpretation of the id-impulses, all corresponding attitudes of the ego have been analyzed. This obviates the danger that the patient learns something too early or that he remains affectless and without participation.

Analyses in which so much analytic attention is centered upon the attitudes take a more orderly and logical course while the theoretical research does not suffer in the least. One obtains the important infantile experiences later, it is true; but this is more than compensated for by the emotional aliveness with which the infantile material comes up *after* the analytic work on the character resistances.

On the other hand, we should not fail to mention certain unpleasant aspects of a consistent character-analysis. It is a far heavier burden for the patient; he suffers much more than when one leaves the character out of consideration. True, this has the advantage of a selective process: those who cannot stand it would not have achieved success anyhow, and it is better to find that out after a few months than after a few years. Experience shows that if the character resistance does not give way a satisfactory result cannot be expected. The overcoming of the character resistance does *not* mean that the character is altered; that, of course, is possible only after the analysis of its infantile sources. It only means that the patient has gained an objective view of his character and an analytic interest in it; once this has been achieved a favorable progress of the analysis is probable.

(e) The loosening of the narcissistic protection apparatus

As we said before, the essential difference between the analysis of a symptom and that of a neurotic character trait consists in the fact that the symptom is, from the beginning, isolated and objectively looked at while the character trait has to be continually pointed out so that the patient will attain the same attitude toward it as toward a symptom. Only rarely is this achieved easily. Most patients have a very slight tendency to look at their character objectively. This is understandable because it is a matter of loosening the narcissistic protection mechanism, the freeing of the anxiety which is bound up in it.

A man of 25 came to analysis because of some minor symptoms and because he suffered from a disturbance in his work. He showed a free, self-confident behavior but often one had the impression that his demeanor was artificial and that he did not establish any genuine relationship with the person to whom he talked. There was something cold in his manner of talking, something vaguely ironical; often he would smile and one would not know whether it was a smile of embarrassment, of superiority or irony.

The analysis began with violent emotions and ample acting out. He cried when he talked about the death of his mother and cursed when he described the usual upbringing of children. The marriage of his parents had been very unhappy. His mother had been very strict with him, and with his siblings he had established some sort of relationship only in recent years. The way in which he kept talking intensified the original impression that neither his crying nor his cursing or any other emotion came out really fully and naturally. He himself said that all this was not really so bad after all, that he was smiling all the time about everything he was saying. After a few hours, he began to try to provoke the analyst. For example, he would, when the analyst had terminated the session, remain lying on the couch ostentatiously for a while, or would start a conversation afterwards. Once he asked me what I thought I would do if he should grab me by the throat. Two days later, he tried to frighten me by a sudden hand movement toward my head. I drew back instinctively and told him that the analysis asked of him only that he say everything, not that he do things. Another time he stroked my arm in parting. The deeper meaning of this behavior which could not be interpreted at this time was a budding homosexual transference manifesting itself sadistically. When, on a superficial level, I interpreted these actions as provocations, he smiled and closed up even more. The actions ceased as well as his communications; all that remained was the stereotyped smile. He began to keep silent. When I pointed out the defensive character of his behavior, he merely smiled again and, after some period of silence, repeated, obviously with the intention of making fun of me, the word "resistance." Thus the smiling and the making fun of me became the center of the analytic work.

The situation was difficult. Apart from the few general data about his childhood, I knew nothing about him. All one had to deal with, therefore, were his modes of behavior in the analysis. For some time, I simply waited to see what would be forthcoming, but his behavior remained the same for about two weeks. Then it occurred to me that the intensification of his smile had occurred at the time when I had warded off his aggressions. I tried to make him understand the meaning of his smile in this connection. I told him that no doubt his smile meant a great many things, but at the present it was a reaction to the cowardice I had shown by my instinctive drawing back. He said that may well be but that he would continue to smile. He talked about unimportant things, and made fun of the analysis, saying that he could not believe anything I was telling him. It became increasingly clear that his smile served as a protection against the analysis. This I told him repeatedly over several sessions but it was several weeks before a dream occurred which had reference to a machine which cut a long piece of brick material into individual bricks. The connection of this dream with the analytic situation was all the more unclear in that he did not produce any associations. Finally he said that, after all, the dream was very simple, it was obviously a matter of the castration complex, and—smiled. I told him that his irony was an attempt to disown the indication which the unconscious had given through the dream. Thereupon he produced a screen memory which proved of great importance for the further development of the analysis. He remembered that at the age of about five he once had "played horse" in the backyard at home. He had crawled around on all fours, letting his penis hang out of his pants. His mother caught him doing this and asked what on earth he was doing. To this he had reacted merely by smiling. Nothing more could be learned for the moment. Nevertheless, one thing had been learned: his smile was a bit of mother transference. When I told him that obviously he behaved in the analysis as he had behaved toward his mother, that his smile must have a definite meaning, he only smiled again and said that was all well and good but it did not seem plausible to him. For some days, there was the same smile and the same silence on his part, while I consistently interpreted his behavior as a defense against the analysis, pointing out that his smile was an attempt to overcome a secret fear of me. These interpretations also were warded off with his stereotyped smile. This also was consistently interpreted as a defense against my influence. I pointed out to him that apparently he was always smiling, not only in the analysis, whereupon he had to admit that this was his only possible way of getting through life. With that, he had unwillingly concurred with me. A few days later he came in smiling again and said: "Today you'll be pleased, Doctor. 'Bricks,' in my mother-tongue, means horse testicles. Swell, isn't it? So you see, it is the castration complex." I said that might or might not be true; that, in any case, as long as he maintained this defensive attitude, an analysis of the dreams was out of the question; that, no doubt, he would nullify every association and every interpretation with his smile. It should be said here that his smile was hardly visible; it was more a matter of a feeling and an attitude of making fun of things. I told him he need not be afraid of laughing about the analysis openly and loudly. From then on, he was much more frank in his irony. His association, in spite of its fun-making implication, was nevertheless very valuable for an understanding of the situation. It seemed highly probable that, as happens so often, he had conceived of the analysis in the sense of a danger of castration; at first he had warded off this danger with aggression and later with his smile. I returned to the aggressions in the beginning of the analysis and added the new interpretation that he had tried to test me with his provocations, that he wanted to see how far he could go, how far he could trust me. That, in other words, he had had a mistrust which was based on an infantile fear. This interpretation impressed him visibly. He was struck for a moment but quickly recovered and again began to disavow the analysis and my interpretations with his smiling. I remained consistent in my interpretations; I knew from different indications that I was on the right track and that I was about to undermine his ego defense. Nevertheless, he remained equally consistent in his smiling attitude for a number of sessions. I intensified my interpretations by linking them up more closely with the assumed infantile fear. I told him that he was afraid of the analysis because it would revive his infantile conflicts which he thought he had solved with his attitude of smiling but that he was wrong in this belief because his excitation at the time when he talked about his mother's death had been genuine after all. I ventured the assumption that his relationship with his mother had not been so simple; that he had not only feared and ridiculed but also loved her. Somewhat more serious than usually, he related details concerning the unkindness of his mother toward him; one time when he had misbehaved she even hurt his hand with a knife. True, he added, "Well, according to the book, this is again the castration complex, isn't it?" Nevertheless, something serious seemed to go on in him. While I continued to interpret the manifest and latent meaning of the smiling as it appeared in the analytic situation, further dreams occurred. Their manifest content was that of symbolical castration ideas. Finally he produced a dream in which there were horses, and another where a high tower arose from a fire truck. A huge column of water poured from the tower into a burning house. At this time, the patient suffered from occasional bedwetting. The connection between the "horse dreams" and his horse game he realized himself, although accompanied by smiling. More than that, he remembered that he had always been very much interested in the long penes of horses; he thought that in his infantile game he had imitated such a horse. He also used to find a great deal of pleasure in urinating. He did not remember whether as a child he used to wet his bed.

On another occasion of discussing the infantile meaning of his smile he thought that possibly his smile on the occasion of the horse game had not been derisive at all but an attempt to reconcile his mother, for fear that she might scold him for his game. In this way he came closer and closer to what I had now been interpreting for months from his behavior in the analysis. The smiling, then, had changed its function and meaning in the course of time: originally an *attempt at conciliation*, it had later become a *compensation of an inner fear*, and finally, it also served as a means of *feeling superior*. This explanation the patient found himself when in the course of several sessions he reconstructed the way which he had found out of his childhood misery. The meaning was: "Nothing can happen to me, I am proof against everything." It was in this last sense that the smile had become a defense in the analysis, as a protection against the reactivation of the old conflicts. The basic motive of this defense was an infantile fear. A dream which occurred at the end of the fifth month revealed the deepest layer of his fear, the fear of being left by his mother. The dream was the following: "I am riding in a car, with an unknown person, through a little town which is completely deserted and looks desolate. The houses are run down, the windowpanes smashed. Nobody is to be seen. It is as if death had ravaged the place. We come to a gate where I want to turn back. I say to my companion we should have another look. There is a man and a woman kneeling on the sidewalk, in mourning clothes. I approach them to ask them something. As I touch them on the shoulder they jump and I wake up, frightened." The most important association was that the town was similar to that in which he had lived until he was four years of age. The death of his mother and the infantile feeling of being left alone were clearly expressed. The companion was the analyst. For the first time, the patient took a dream completely seriously, without any smiling. The character resistance had been broken through and the connection with the infantile had been established. From then on, the analysis proceeded without any special difficulty, interrupted, of course, by the relapses into the old character resistance as they occur in every analysis.

It goes without saying that the difficulties were far greater than may appear from this brief synopsis. The whole resistance phase lasted almost six months, characterized by derision of the analysis for days and weeks on end. Without the necessary patience and the confidence in the efficacy of consistent interpretation of the character resistance, one often would have been inclined to give up.

Let us see whether the analytic insight into the mechanism of this case would justify some other technical procedure. Instead of putting the emphasis consistently on the mode of behavior, one might have thoroughly analyzed the patient's scarce dreams. Possibly he might have had associations which one could have interpreted. It may not be important that previous to the analysis the patient did not dream or forgot all his dreams and did not produce any dreams with a content relevant to the analytic situation until after the consistent interpretation of his behavior. One might object that the patient would have produced these dreams spontaneously anyhow; this cannot be argued because it cannot be proved one way or the other. At any rate, we have ample experience which teaches us that such a situation as presented by our patient can hardly be solved by passive waiting alone; if so, it happens by accident, without the analyst having the reins of the analysis in his hand. Let us assume, then, that we had interpreted his associations in connection with the castration complex, that is, tried to make him conscious of his fear of cutting or of being cut. *Perhaps* this would have finally also led to a success. But the very fact that we cannot be sure that it would have happened, that we must admit the accidental nature of the occurrence, forces us to refute such a technique which tries to circumvent an existing resistance as basically un-analytic. Such a technique would mean reverting to that stage of analysis where one did not bother about the resistances, because one did not know them, and where, consequently, one interpreted the meaning of the unconscious material directly. It is obvious from the case history that this would mean, at the same time, a neglect of the ego defenses.*

One might object again that while the technical handling of the case was entirely correct one did not understand my argument; that all this was self-evident and nothing new, that this was the way all analysts worked. True, the general principle is not new; it is nothing but the consistent application of resistance analysis. Many years of experience in the Technical Seminar showed, however, that analysts generally know and recognize the principles of resistance technique, while in practice they use essentially the old technique of the direct interpretation of the unconscious. This discrepancy between theoretical knowledge and practical action was the source of all the mistaken objections to the systematic attempts of the Vienna Seminar to develop the consistent application of theory to therapy. If they said that all this was trite and nothing new, they had their theoretical knowledge in mind; if they objected that it was all wrong and not "Freudian" analysis, they thought of their own practice, which, as we have said, was quite different.

A colleague once asked me what I would have done in the following case: For the past four weeks he had been treating a young man who kept consistently silent but was otherwise very nice and showed a very friendly behavior before and after the analytic session. The analyst had tried all kinds of things, had threatened to break off the analysis and finally, when even dream interpretation failed, had set a date for the termination of the analysis. The scarce dreams had been filled with sadistic murder. The analyst had told

* [Reich uses the plural ego defenses, but the sense is the same as in the footnote on p. 250.]

the patient that, after all, he should realize from his dreams that in his phantasy he was a murderer. But it did not help. The colleague was not satisfied with my statement that it was incorrect to interpret such deep material in the presence of an acute resistance, no matter how clearly the material might appear in a dream. He thought there was no other way. When I told him that, first of all, the silence should have been interpreted as a resistance, he said that could not be done, for there was no "material" available to do it with. Is not the behavior itself, the silence during the hour in contrast to the friendly attitude outside, "material" enough? Does not this situation show clearly the one thing at least, that the patient expresses, with his silence, a negative attitude or a defense? And that, to judge from his dreams, it is a matter of sadistic impulses which, by his over-friendly behavior, he tried to compensate and camouflage? Why does one dare to deduce certain unconscious processes from a slip such as a patient's forgetting some object in the consultation room, and why does one not dare to deduce the meaning of the situation from his behavior? Is the total behavior less conclusive material than a slip? All this did not seem plausible to my colleague; he continued to insist that the resistance could not be tackled because there was "no material." There could be no doubt that the interpretation of the murderous impulses was a technical error; it could only have the effect of frightening the patient and of putting him all the more on his guard.

The difficulties in the cases presented in the Seminar were of a very similar nature: It was always the same underestimation or the complete neglect of the behavior as interpretable material; again and again the attempt to remove the resistance from the id side instead of by analysis of the ego defense; and finally, almost always, the idea—which was used as an alibi—that the patient simply did not want to get well or that he was "all too narcissistic."

In principle, the loosening of the narcissistic defense is not different in other types than in the one described. If, say, a patient is always affectless and indifferent, no matter what material he may be presenting, then one is dealing with the dangerous affect-block. Unless one works on this before anything else one runs the danger of seeing all the material and all the interpretations go to waste and of seeing the patient become a good analytical theorist while otherwise he remains the same. Unless one prefers in such a case to give up the analysis because of "too strong narcissism" one can make an agreement with the patient to the effect that one will continue to confront him with his affect-lameness but that, of course, he can stop whenever he wants to. In the course of time—usually many months, in one case it took a year and a half—the patient begins to experience the continued pointing out of his affect-lameness and its reasons as painful, for in the meantime one has acquired sufficient means of undermining the protection against anxiety which the affect-lameness presents. Finally the patient rebels against the danger which threatens from the analysis, the danger of losing the protective psychic armor and of being confronted with his impulses, particularly with his aggression. This rebellion activates his aggressivity and before long the first emotional outburst in the sense of a negative transference occurs, in the form of an attack of hatred. That achieved, the road becomes clear. When the aggressive impulses make their appearance, the affect-block is breached and the patient becomes capable of being analyzed. The difficulty consists in bringing out the aggressivity.

The same is true when narcissistic patients express their character resistance in their way of talking; they will talk, for example, always in a haughty manner, in technical terms, always highly correctly or else confusedly. Such modes of talking form an impenetrable barrier and there is no real experiencing until one analyzes the mode of expression itself. Here also, the consistent interpretation of the behavior results in narcissistic indignation, for the patient does not like to be told that he talks so haughtily, or in technical terms, in order to camouflage his feeling of inferiority before himself and the analyst, or that he talks so confusedly because he wants to appear particularly clever and is unable to put his thoughts into simple words. In this manner, one makes an important breach in the neurotic character and creates an avenue of approach to the infantile origin of the character and the neurosis. Of course, it is insufficient to point out the nature of the resistance at one time or another; the more stubborn the resistance, the more consistently does it have to be interpreted. If the negative attitudes against the analyst which are thus provoked are analyzed at the same time the risk of the patient's breaking off the analysis is negligible.

The immediate effect of the analytic loosening of the character armor and the narcissistic protection mechanism is twofold: First, the loosening of the affects from their reactive anchoring and hiding places; second, the creation of an avenue of approach to the central infantile conflicts, the Oedipus complex and the castration anxiety. An enormous advantage of this procedure is that one not only reaches the infantile experiences as such, but that one analyzes them in the specific manner in which they have been assimilated by the ego. One sees again and again that one and the same piece of repressed material is of different dynamic importance according to the stage which has been reached in the loosening of the resistances. In many cases, the affect of the infantile experiences is absorbed in character defenses; with simple interpretation of the contents, therefore, one may be able to elicit the memories but not the corresponding affects. In such cases, interpretation of the infantile material without *previous* loosening of the affect energies which are absorbed in the character is a serious mistake. It is responsible, for example, for the hopelessly long and relatively useless analyses of compulsive characters.* If,

* The following case illustrates the decisive importance of the neglect of a mode of behavior. A compulsive character who had been in analysis for twelve years without any appreciable result and knew all about his infantile conflicts, such as his central father

on the other hand, one first frees the affects from the defense formations of the character, a new cathexis of the infantile impulses takes place automatically. If the line of character-analytic resistance interpretation is followed, remembering without affect is practically out of the question; the disturbance of the neurotic equilibrium which goes with the analysis of the character from the very beginning makes it practically impossible.

In other cases, the character has been built up as a solid protective wall against the experiencing of infantile anxiety and has served well in this function, although at the expense of much happiness. If such an individual comes to analysis because of some symptom, this protective wall serves equally well as character resistance and one realizes soon that nothing can be done unless this character armor which covers up and absorbs the infantile anxiety is destroyed. This is the case, for example, in "moral insanity" and in many manic, narcissistic-sadistic characters. In such cases one is often confronted with the difficult question whether the symptom justifies a deep-reaching character-analysis. For one must realize that the character-analytic destruction of the characterological compensation temporarily creates a condition which equals a breakdown of the personality. More than that, in many extreme cases such a breakdown is inevitable before a new, rational personality structure can develop. One may say, of course, that sooner or later the breakdown would have occurred anyhow, the development of the symptom being the first sign. Nevertheless, one will hesitate about undertaking an operation which involves so great a responsibility unless there is an urgent indication.

In this connection another fact must be mentioned: character-analysis creates in every case violent emotional outbursts and often dangerous situations, so that it is important always to be master of the situation, technically. For this reason, many analysts will refuse to use the method of character-analysis; in that case, they will have to relinquish the hope for success in a great many cases. A great many neuroses cannot be overcome by mild means. The means of character-analysis, the consistent emphasis on the character resistance and the persistent interpretation of its forms, ways and motives, are as potent as they are unpleasant for the patient. This has nothing to do with education; rather, it is a strict analytic principle. It is a good thing, however, to point out to the patient in the beginning the foreseeable difficulties and unpleasantness.

---

conflict, talked in the analysis in a peculiarly monotonous, sing-song intonation and kept wringing his hands. I asked him whether this behavior had ever been analyzed, which was not the case. One day it struck me that he talked as if he were praying, and I told him so. He then told me that as a child he had been forced by his father to go to the synagogue and to pray. He had prayed, but only under protest. In the same manner he had also prayed—for twelve long years—before the analyst: "Please, I'll do it if you ask me to, but only under protest." The uncovering of this seemingly incidental detail of his behavior opened the way to the analysis and led to the most strongly hidden affects.

(f) *On the optimal conditions for the analytic reduction
of the present-day material to the infantile*

Since the consistent interpretation of the behavior spontaneously opens
the way to the infantile sources of the neurosis, a new question arises: Are
there criteria to indicate *when* the reduction of the present-day modes of
behavior to their infantile prototypes should take place? This reduction, we
know, is one of the cardinal tasks of analysis, but this formulation is too
general to be applied in everyday practice. Should it be done as soon as the
first signs of the corresponding infantile material appear, or are there reasons
for postponing it until a certain later time? First of all it must be pointed out
that in many cases the purpose of the reduction—dissolution of the resistance
and elimination of the amnesia—is not fulfilled: either there is no more than
an intellectual understanding, or the reduction is refuted by doubts. This is
explained by the fact that—as is the case with the making conscious of
unconscious ideas—the topical process is complete only if combined with the
dynamic-affective process of the becoming conscious. This requires the fulfilment of two conditions: first, the main resistances must be at least loosened
up; second, the idea which is to become conscious—or, in the case of the
reduction, is fo enter a new association—must become charged with a certain
minimum of affect. Now, we know that the affects are usually split off from
the repressed ideas, and bound up in the acute transference conflicts and
resistances. If, now, one reduces the resistance to the infantile situation before
it has fully developed, as soon as there is only a trace of its infantile origin,
then one has not fully utilized its affective energies; one has interpreted the
content of the resistance without also having mobilized the corresponding
affect. That is, dynamic considerations make it necessary not to nip the resistance in the bud, but, on the contrary, to bring it to full development in the
transference situation. In the case of chronic, torpid character incrustations
there is no other way at all. Freud's rule that the patient has to be brought
from acting out to remembering, from the present day to the infantile, has to
be complemented by the further rule that *first* that which has become chroni-
cally rigid must be brought to new life in the actual transference situation, just
as chronic inflammations are treated by first changing them into acute ones.
With character resistances this is always necessary. In later stages of the
analysis, when one is certain of the patient's cooperation, it becomes less
necessary. One gains the impression that with many analysts the immediate
reduction of as yet completely immature transference situations is due to the
fear of strong and stormy transference resistances; this fits in with the fact
that—in spite of better theoretical knowledge—resistances are very often
considered something highly unwelcome and only disturbing. Hence the tendency to circumvent the resistance instead of bringing it to full development
and then treating it. One should not forget the fact that the neurosis itself is

contained in the resistance, that with the dissolution of every resistance we dissolve a piece of the neurosis.

There is another reason why it is necessary to bring the resistance to full development. Because of the complicated structure of each resistance, one comprehends all its determinations and meanings only gradually; the more completely one has comprehended a resistance situation, the more successful is its later interpretation. Also, the double nature of the resistance—present-day and historical—makes it necessary first to make fully conscious the forms of ego defense it contains; only after its present-day meaning has become clear should its infantile origin be interpreted. This is true of the cases who have already produced the infantile material necessary for an understanding of the resistance *which follows*. In the other, more numerous cases, the resistance must be brought to full development for no other reason than that otherwise one does not obtain enough infantile material.

The resistance technique, then, has two aspects: *First, the comprehension of the resistance from the present-day situation through interpretation of its present-day meaning; second, the dissolution of the resistance through association of the ensuing infantile material with the present-day material.* In this way, one can easily avoid the flight into the present-day as well as into the infantile, because equal attention is paid to both in the interpretation work. Thus the resistance turns from an impediment of the analysis into its most potent expedient.

(g)   *Character-analysis in the case of amply flowing material*

In cases where the character impedes the process of recollection from the beginning, there can be no doubt about the indication of character-analysis as the only legitimate way of introducing the analysis. But what about the cases whose character admits of the production of ample memory material in the beginning? Do they, also, require character-analysis as here described? This question could be answered in the negative if there were cases without a character armor. But since there are no such cases, since the narcissistic protection mechanism always turns into a character resistance—sooner or later, in varying intensity and depth—there is no fundamental difference between the cases. The practical difference, though, is this: In cases such as described above, the narcissistic protection mechanism is at the surface and appears as resistance immediately, while in other cases it is in deeper layers of the personality so that it does not strike one at first. But it is precisely these cases that are dangerous. In the former case one knows what one is up against. In the latter case, one often believes for a long period of time that the analysis proceeds satisfactorily, because the patient seems to accept everything very readily, shows prompt reactions to one's interpretations, and even improvements. But it is just in these patients that one experiences the worst disappointments. The analysis has been carried out, but the final success fails to materialize. One has shot all one's interpretations, one seems to have made completely conscious the primal scene and all infantile conflicts; finally the analysis bogs down in an empty, monotonous repetition of the old material, and the patient does not get well. Worse still, a transference success may deceive one as to the real state of affairs, and the patient may return with a full relapse soon after his discharge.

A wealth of bad experiences with such cases suggested as a rather self-evident conclusion that one had overlooked something. This oversight could not refer to the contents, for in that respect these analyses left little to be desired; it could only be an unrecognized latent resistance which nullified all therapeutic endeavor. It was soon found that these latent resistances consisted precisely in the great willingness of the patients, in the lack of manifest resistances. In comparing them with successful cases, one was struck by the fact that these analyses had shown a constantly even flow, never interrupted by violent emotional outbursts; more importantly, they had taken place in almost constant "positive" transference; rarely, if ever, had there been violent negative impulses toward the analyst. This does not mean that the hate impulses had not been analyzed; only, they did not appear in the transference, or they had been remembered without affect. The prototypes of these cases are the narcissistic affect-lame and the passive-feminine characters. The former show a luke-warm and even, the latter an exaggerated "positive" transference.

These cases had been considered "going well" because they procured infantile material, that is, again because of a one-sided overestimation of the contents of the material. Nevertheless, all through the analysis, the character had acted as a severe resistance in a form which remained hidden. Very often, such cases are considered incurable or at least extremely difficult to handle. Before I was familiar with the latent resistances of these cases, I used to agree with this judgment; since then, I can count them among my most gratifying cases.

The character-analytic introduction of such cases differs from others in that one does not interrupt the flow of communications and does not begin the analysis of the character resistance until such time as the flood of communications and the behavior itself has unequivocally become a resistance. The following case will illustrate this as it will again show how character-analysis leads of itself into the most deeply repressed infantile conflicts. We shall follow this analysis farther along than those previously described, in order to show the logical development of the neurosis in the transference resistances.

A Case of Passive-Feminine Character

Anamnesis

A 24-year-old bank employee came to analysis because of his anxiety states; these had set in a year previously on the occasion of his going to a hygiene exhibit. Even before that he had had *hypochondriac* fears: he thought he had a *hereditary* taint, he would go *crazy* and would *perish in a mental institution*. For these fears, he seemed to have rational grounds: his father had acquired syphilis and gonorrhea ten years previous to his marriage. The paternal grandfather also was supposed to have had syphilis. A paternal uncle was very nervous and suffered from insomnia. The maternal heredity was even more serious: the mother's father committed suicide, as did one of her brothers. A great-aunt was "mentally abnormal." The patient's mother was an anxious and nervous woman.

This double "heredity" (syphilis on the paternal, suicide and psychosis on the maternal side) made the case all the more interesting in that psycho-analysis—in contradistinction to orthodox psychiatry—considers heredity only one of many etiological factors. As we shall see, the patient's idea about his heredity had also an irrational basis. He was cured in spite of his heredity and did not relapse during a follow-up period of five years.

This presentation covers only the first seven months of the treatment which were taken up with the analysis of the character resistances. The last seven months were presented only very briefly because, from the standpoint of resistance and character-analysis, they presented little which would be of interest. What is to be presented here is chiefly the introduction of the treatment, the course of the resistance analysis, and the way it established the contact with the infantile material. We shall follow the red thread of the resistances and their analysis. In reality, of course, the analysis was not as simple as it may appear here.

The patient's anxiety attacks were accompanied by palpitations and a paralysis of all initiative. Even in the intervals between the attacks he was never free of a feeling of malaise. The anxiety attacks often occurred spontaneously but also were precipitated by his reading about mental diseases or suicides in the newspaper. In the course of the past year his working capacity had begun to suffer and he was afraid that he might be discharged because of inefficiency.

Sexually he was severely disturbed. Shortly before the visit to the hygiene exhibit, he had attempted coitus with a prostitute and had failed. He said that this had not bothered him particularly. There was very little conscious sexual desire: he said he did not suffer from his sexual abstinence. A few years earlier, he had succeeded in carrying out the sexual act, although he had suffered from a premature and pleasureless ejaculation.

Asked whether his anxiety states had not had any precursors, he related that already as a child he had been very apprehensive, and particularly during puberty he had been *afraid of world catastrophes*. Thus he was very much afraid when in 1910 the end of the world through a collision with a comet was predicted; he was surprised that his parents could talk about it so calmly. This "fear of catastrophe" gradually subsided, being completely replaced by his fear of the hereditary taint. Severe anxiety states he had had since childhood, although less frequently.

Apart from the hypochondriac idea of the hereditary taint, the anxiety states and the sexual weakness, there were no symptoms. Awareness of illness was at first present only with regard to the anxiety states which was the symptom which bothered him most. The idea of the hereditary taint was too well rationalized and the sexual weakness produced too little suffering to produce insight into their pathological character. Symptomatologically speaking, then, we were dealing with the hypochondriac form of anxiety hysteria with a particularly marked actual-neurotic core (stasis neurosis).*

The diagnosis was hysterical character with hypochondriac anxiety hysteria. The diagnosis "hysterical character" is based on the analytic findings concerning his fixations. Phenomenologically, he was a typical passive-feminine character: he was always over-friendly and humble; he kept apologizing for the most trifling things; on arriving and on leaving he made several deep bows. In addition, he was *awkward, shy and circumstantial.* If he was asked, for example, whether his hour could be changed, he did not simply say, Yes, but assured me at length that he was completely at my disposal, that he was agreeable to any change I wished to make, etc. When he asked for something, he would stroke the analyst's arm. When I first mentioned the possibility of a distrust of the analysis, he returned on the same day, highly perturbed, saying that he could not stand the thought of my thinking him distrustful; he asked repeatedly for forgiveness in case he should have said something that could have given me any such impression.

The development and analysis of the character resistance

The analysis developed according to the resistances which were determined by this kind of character, as follows:

After being told the fundamental rule, he talked rather fluently about his family and the hereditary taint. He asserted that he loved both his parents equally well but had more respect for his father whom he described as an energetic, clear-thinking person. The father had always *warned him against masturbation and extramarital sexual intercourse.* He had told him about his

* [As already indicated, to Reich all neuroses result from the damming up of libido and are, therefore, at their core "actual neuroses."]

own bad experiences, his syphilis and gonorrhea, of his relationships with women which had come to a bad end; all this with the intention of saving the patient from similar experiences. The father had never beaten him but had always gotten his way by telling him, "I'm not forcing you, I only advise you to . . ."; this, however, had been done very forcefully. The patient described the relationship with his father as very good and his father as his very best friend in whom he had the greatest confidence.

Soon he switched to an extensive description of the relationship with his mother. She was always very solicitous and kind. He was also very kind to her; on the other hand he let her wait on him hand and foot. She took care of his laundry, brought breakfast to his bed, sat beside him until he went to sleep, even now, she combed his hair, in a word, he led the life of a pampered mother's boy.

After six weeks, he was *close to becoming conscious of the wish for coitus.* Apart from this, he had been fully conscious of the tender relationship with his mother, in part he had known it even before the analysis: he had often thrown his mother on his bed to which she had reacted with "bright eyes and flushed cheeks." When she would come in in her nightgown to say good night to him, he would embrace her and press her against him. Though he always tried to emphasize the sexual excitation on the part of his mother— undoubtedly in order to give away less of his own intentions—he mentioned several times, parenthetically as it were, that he himself had definitely felt sexual excitation.

A very cautious attempt to make him understand the real significance of these things, however, led to a violent resistance: he could assure me, he said, that he felt exactly the same thing with other women. I had made this attempt by no means in order to interpret the incest phantasy to him but only in order to see whether I was correct in surmising that this straight advance of his in the direction of the historically important incest love was actually a manoeu- vre to divert attention from something that *at present* was much more impor- tant. The material about his mother was unequivocal; it really appeared as if he needed only one more step to arrive at the core. But something militated against the interpretation of this material: the content of his communications was in striking contrast to the content of his dreams and to his over-friendly behavior.

For this reason, I centered my attention more and more on his behavior and on his dream material. He produced no associations to his dreams. Dur- ing the session, he enthused about the analysis and the analyst, while outside he was very much concerned about his future and ruminated about his heredi- tary taint.

The content of the dreams was of a twofold nature: On the one hand, they also contained incest phantasies; what he did not express during the day he expressed in the manifest dream content. For example, in one dream he went after his mother with a knife, or crept through a hole before which his mother was standing. On the other hand, there was often some obscure *murder story*, the hereditary taint, a crime which somebody committed or *derisive remarks made by somebody*, or *distrust* expressed by somebody.

During the first 4 to 6 weeks of the analysis, we had obtained the following material: his statements regarding the relationship with his mother; his anxiety states and the heredity idea; his over-friendly, submissive behavior; his dreams, those which continued the incest phantasy and those of murder and distrust; and certain indications of a positive mother transference.

Confronted with the choice of interpreting his clear-cut incest material or to emphasize the signs of his distrust, I chose the latter. For there could be no doubt that here was a *latent resistance* which for many weeks did not become manifest because it consisted precisely in that the patient presented too much and was too little inhibited. As was shown later, it was also the first important transference resistance the specific form of which was determined by the patient's character. *He was deceiving*: by offering up all the material on his experiences, which was therapeutically useless, by his over-friendly behavior, by his many clear-cut dreams, by his seeming confidence in the analyst.* He tried *to please* the analyst, as he had tried to please his father all along, and for the same reason: because he was *afraid of him*. If this had been my first case of this nature I could not possibly have known that such behavior was a decisive and dangerous resistance. Previous experience in such cases had shown, however, that such patients are incapable of producing a manifest resistance, over periods of months or even years; and further, that they do not react therapeutically in the least to the interpretations which, prompted by the clear-cut material, one gives them. One cannot say, therefore, that in such cases one should wait until the transference resistance makes its appearance; it is, in fact, present from the very first moment in a fully developed, but typically *hidden* form.

Clearly, the presented heterosexual incest material could not really be material which had broken through from the depths. If one pays any attention to the actual function of the presented material one often finds that deeply repressed impulses are temporarily used for the purpose of warding off *other* contents, without any change in the state of repression taking place. This is a peculiar fact, not easily understood depth-psychologically. It is obvious from this fact, though, that the direct interpretation of such material is a definite mistake. Such interpretation not only has no therapeutic effect; more than

---

* [Reich's view of the patient as a deceiver is similar to Ferenczi's use of the term, "unconscious mendacity." He regarded all patients as deceivers. In 1936, Anna Freud said to this point; "In my opinion, we do our patients a great injustice if we describe these transferred defense-reactions as 'camouflage' or say that the patients are 'pulling the analyst's leg' or purposely deceiving them in some other way. . . . The patient is in fact candid when he gives expression to the impulse or affect in the only way still open to him, namely, in the distorted defensive measure" (1936, pp. 20–21).]

that, it interferes with the maturing of the respective repressed contents for later interpretation. Theoretically one might say that psychic contents appear in consciousness under two totally different conditions: either born by the affects which specifically belong to them, or born by extraneous interests. In the first case, it is the result of the inner pressure of dammed-up excitation, in the latter case it occurs in the service of defense. It is the same difference as that between freely flowing love and manifestations of love which serve to compensate for hatred, that is, reactive love.*

In our patient, the handling of the resistance was, of course, far more difficult than it is in the case of manifest resistances. The meaning of the resistance could not be deduced from the patient's communications, but it could be deduced from his behavior and from the seemingly incidental details of many of his dreams. From these it was evident that, for fear of rebelling against his father, he had camouflaged his spite and distrust by reactive love and had escaped anxiety by being submissive.

The first resistance interpretation was given on the fifth day on the occasion of the following dream:

> My handwriting is submitted to a graphologist for an opinion. His opinion was: "This man belongs in a mental institution." My mother is completely desperate. I want to commit suicide. Then I wake up.

To the graphologist, he associated Professor Freud. He added that the Professor had told him that analysis cured such diseases as his with "absolute certainty."† I called his attention to the following contradiction: since in the dream he was afraid of having to be committed to a mental institution, he apparently did not believe that the analysis would help him. This he could not see; he refused to accept the interpretation and kept insisting that he had the fullest confidence in the analysis.

Until the end of the second month he dreamt much, though little that would have lent itself to interpretation, and continued talking about his mother. I let him talk, without urging him on and without giving interpretations, being careful all the time not to miss any indication of distrust. After the first resistance interpretation, however, he had camouflaged his secret distrust even more thoroughly, until finally he produced the following dream:

> A crime, possibly *a murder, has been committed.* Somehow and against my will, I have been implicated in it. *I am afraid of discovery and punish-*

---

* [The term "reactive love" was based on Reich's assumption that all manifestations of positive transference at the beginning of treatment are reaction formations against the latent hostility. He does not operate with the concept of ambivalence, that gives equal value to both feelings.]

† [If the patient has indeed quoted Freud correctly, then Freud did not follow Ferenczi's (1928a) admonition that no promises be given about the outcome of an analysis (see this volume, Chapter 17).]

ment. One of my fellow employees, who impresses me with his courage and decision, is there. I am keenly aware of his superiority.

I emphasized only the fear of discovery and related it to the analytic situation, telling him that his whole attitude indicated that he was hiding something. As early as the following night, he had the following confirmatory dream:

> *A crime is going to be committed in our apartment.* It is night and I am on the dark stairs. I know that *my father* is in the apartment. I want to go to his aid but *I am afraid of falling into the hands of the enemies.* I want to call the police. I have a roll of paper with me which has all the details of the intended crime on it. I need *a disguise,* otherwise *the leader of the gang,* who has placed a lot of spies, will prevent me. I take a large cape and a false beard and leave the house, bent over like an old man. The leader of the gang stops me and asks one of his men to search me. He finds the roll of paper. I feel that I am going to be lost if he reads its contents. *I act as innocently as possible* and tell him that they are notes which don't mean anything. He says he'll have to have a look anyhow. There is a moment of painful tension, then, in desperation, a look for a weapon. I find a revolver in my pocket and fire it. The man has disappeared, and suddenly I feel myself very strong. The leader of the gang has changed into a woman. I am seized by a desire for this woman. I pick her up and carry her into the house. I am overcome by a pleasurable feeling, and wake up.

At the end of the dream, we have the whole incest motif before us, but earlier in the dream unmistakable allusions to the patient's masquerading in the analysis. I entered only upon the latter because the patient would have to give up his attitude of deceit before deeper interpretations could be given.* This time, however, I went a step further in the resistance interpretation. I told him that not only was he distrustful of the analysis; that, furthermore, by his behavior, he pretended the exact opposite. Upon this, the patient became highly excited, and through six sessions he produced three different hysterical actions:

1. He thrashed around with arms and legs, yelling: "Let me alone, don't come near me, I'm going to kill you, I'm going to squash you!" This action often changed into another:

2. He grabbed his throat and whined in a rattling voice: "Please let me alone, please, I'm not going to do anything any more!"

3. He behaved not like one who is violently attacked but like a girl who is sexually attacked: "Let me alone, let me alone." This, however, was said

---

* [In this dream, heterosexual impulses break through (the patient finds he has a revolver), but these are not acknowledged by the analyst, who is afraid of being deceived by the patient. One may well ask if the behavior of the patient after such a one-sided interpretation, was not partially, at least, a response to the analyst's persistent suspicion. It is difficult to know whether the patient was indeed so suspicious, as Reich claimed, or whether he became so as a result of Reich's suspicions of him. Here we see, again, how Reich avoids the interpretation of dreams at this stage of the analysis.]

without the rattling voice and, while during the action of the second type he pulled up his legs, he now spread them apart.

During these six days he was in a manifest resistance, and continued to talk about his hereditary taint, from time to time falling back into the actions just described. Peculiarly enough, as soon as the actions would cease he would continue to talk calmly as if nothing had happened. He only remarked, "Certainly something queer goes on in me, Doctor."*

* [Limitations of space unfortunately prevent the complete reprinting of Reich's thorough description of his work with a patient.]

CHAPTER 19

OTTO FENICHEL

PROBLEMS OF
PSYCHOANALYTIC
TECHNIQUE
An Excerpt

IT IS THE MERIT of his [Reich's] important papers to have added to the meaning of the rules: "Interpretation of resistance precedes interpretation of content," and "Analyze always from the surface." In order to attain the desired dynamic and economic alteration, he said, it is necessary to recognize and name not only what is fended off, but also the *fending force* itself, and this "ego analysis" must take place systematically, consistently, and in the end, historically. When a patient does not follow the basic rule of free association, the analyst must not, impatient with such unsuitable behavior, try to influence him pedagogically or punish him by depriving him of treatment; he must try to understand analytically why the patient behaves thus, and why he does it just in this manner. Reich made especially clear the "frozen character resistances," and, above all, the fact I have tried to show in these discussions that in many cases the thawing out at just these points is the indispensable prerequisite to any subsequent progress in analytic treatment, even when at these same points relatively fluid living conflicts between instinct and defense are concurrently observable.

Translation by David Brunswick, originally published in *Psychoanalytic Quarterly* 7 (1938):421–442, 8 (1939):57–87, 164–185, 303–324, 438–470. Reprinted in *Problems of Psychoanalytic Technique* (New York: Psychoanalytic Quarterly, 1941).

With Reich's *Charakteranalyse*, which brings all these conclusions together and supplements them with some new ones, we must thoroughly agree in its essentials. However, the objection must be made that the book gives way so extensively to some personal characteristics of its author, especially to his penchant for schematic simplification, that the work as a whole suffers. We wish therefore to qualify our essential agreement by two minor theoretical objections and by some others that are directed not against Reich's principles, but against the way he applies them. The two criticisms of theory refer to: (1) the insufficient consideration of "faulting" and of "spontaneous chaotic situations"; (2) the neglect of "the collection of material." Objections of the way his principles are applied are: (1) "Shattering of the patient's defensive armor" is sometimes accomplished too aggressively with Reich and should be regulated by better dosage. (2) When a patient's aggression is mobilized by an aggressive act of the analyst, this aggression is not properly speaking a "negative transference"; or rather, to the extent that it still is one, it loses its ability to be demonstrated as such. (3) Reich's preference for "crises," "eruptions" and theatrical emotions makes one suspicious of a "traumatophilia" that has its roots in a love of magic. (4) The "shattering of the defense armor" is masochistically enjoyed by many patients, and specific transferences can hide behind such enjoyment and escape discovery.

CHAPTER 20

RICHARD F. STERBA

CLINICAL AND THERAPEUTIC ASPECTS OF CHARACTER RESISTANCE

INTRODUCTORY NOTE

As a young psychoanalyst, Sterba was under Wilhelm Reich's influence when Reich was the leader of the technical seminar. Sterba was also an active participant in the development of psychoanalytic ego psychology. An important contribution of his (1934) at the beginning of the development of psychoanalytic ego psychology is reprinted in this volume, Chapter 25.

This paper, recollected in tranquillity twenty years after the controversy with Wilhelm Reich was over, is the most extensive reply to Reich's approach. The article contains not only a critique but also a summary of Reich's view.

The juxtaposition of the chapter by Wilhelm Reich, and Fenichel's and Sterba's responses to Reich's views, should enable the reader to decide independently how much of Wilhelm Reich's approach he considers useful.

THE TERM, character resistance, may be found earlier in psychoanalytic literature—though not in Freud's writings—but for those who have participated in the development of psychoanalytic literature since the early twenties, it is tied up with the therapeutic theory and technique of Wilhelm Reich. It is,

Originally published in *The Psychoanalytic Quarterly* 22 (1953):1–20. Reprinted by permission. This paper was read before the meeting of The New York Psychoanalytic Society, October 30, 1951.

therefore, not possible to discuss the concept of character resistance without reviewing and re-evaluating the therapeutic ideas of Wilhelm Reich. I refer, of course, to those theories which he evolved while he was still a psychoanalyst; that is, before his sexobiological and 'orgonic' phase. I must confess to a feeling of hesitancy in presenting Reich's ideas because, having lived through the era of his impact on the therapeutic thinking of his time and having struggled out of it, I am not altogether in a position to make a completely objective appraisal of their significance for the present-day psychoanalyst. But judging from the interest in Reich's concepts which persists—especially among students—perhaps my re-examination will be of more than merely historical value.

Reich first presented his therapeutic ideas and developed his theory of resistance and his characterology in two papers: the first, On Technique of Interpretation and Resistance Analysis, (subtitled, On the Lawful Development of Transference Neurosis), published in 1927, and the second, On Character Analysis, which appeared the following year. The whole structure of his therapeutic theory and technique he expanded in his book, Character Analysis, published in 1933 in German, the English translation appearing in 1945.

To make his renovation of analytic therapy impressive and still legitimate within the framework of Freudian technique, Reich emphasizes repeatedly that the decisive change in psychoanalytic therapy took place when the therapeutic emphasis shifted from the symptom to the resistance. To this we must reply that a state of affairs in which the therapeutic efforts of psychoanalysis were focused only on the symptoms is an artificial construction, based on a confusion of different phases of the development of psychotherapy by Sigmund Freud. Actually preoccupation solely with the symptom belongs to the preanalytic phase of psychodynamic therapy. Psychoanalytic therapy was born when the resistances were taken into consideration, that is, when the symptoms were recognized as a result of a conflict between instinctual drives and ego, and the resistances, therefore, had to be taken into consideration in therapy. What changed gradually was the way in which psychoanalytic technique dealt with the resistances, and in this development Wilhelm Reich participated during a certain phase until further progress, mainly inspired by Anna Freud, nullified and cast into discard the greater part of Reich's therapeutic suggestions, including the concept of 'character resistance' as he had created it. But the interest that this therapeutic technique holds for many even today makes a critical investigation of his concepts necessary. What I shall undertake to demonstrate is that the concept of 'character resistance', as Reich formulated it, has to be discarded as an artifact which owed its existence to the peculiarities of Wilhelm Reich's theory and technique of psychoanalytic therapy.

The main aim of Reich's therapeutic efforts was to make analytic therapy a systematic procedure. Any student of psychoanalysis will agree that the main difficulty which he encounters as a beginner is to know what to consider as important among the wealth of the material offered by the patient, what to choose as the focus of his interest, and on what to make the patient focus and objectivate in order finally to interpret it. It appeared as an enormous help to the neophyte in therapy when Reich established his schematic theory of the structure of the neurosis and the neurotic personality, which was based on his technique of systematic analysis of resistances. I entered the Vienna Psychoanalytic Institute as a student just at the time when Reich took over the technical seminar and thus acquired a forum for the propagation of his technical ideas. Largely because of the impact of his forceful personality, he created the impression that resistances were only dealt with in psychoanalytic therapy since Reich had appeared on the analytic scene. We students were very much impressed; his systematic technical approach seemed to be the answer to our main technical problem and the way out of our therapeutic confusion. I readily admit that the interest he created in therapeutic technique led to considerable clarification and that the discussions in his seminars provided for me the first orientation in the difficult field of psychoanalytic therapy.

I consider Reich's first technical rule still valid with a few qualifications. This rule is to make the first approach to any material to be interpreted from the side of the ego, not from the side of the id; in other words, the defense or resistance has to be dealt with before the unconscious content is told to the patient. This point is particularly emphasized by Reich, one could say overemphasized, concerning all transference situations. It is from the transference situations that Wilhelm Reich develops his theory of personality structure and his characterology. Reich regards all transference situations, particularly in the beginning—in fact all relationship or nonrelationship to the analyst—as the expression of resistance.* His reasoning is that analytic therapy disturbs the neurotic equilibrium which the patient has difficulty in maintaining anyway. As the disturber of the patient's intrapsychic armistice, as it were, the analyst necessarily becomes an intruder and frightening enemy. The patient, therefore, will react to the analysis either with open rebellion or, if this reaction is felt as too daring or too dangerous, his defense will go 'underground' and he will react with superficial obedience, but underneath he will build up latent 'secret' resistances which are even more 'dangerous' than open negative reactions and to which Reich, therefore, pays particular attention. For Reich initial transference situations are never repetitions of genuine object relationships, which may or may not be used by the need to resist the analytic process. For him transference and resistance are identical. The result of this equation of transference and resistance is that Reich is full of suspicion about every positive transference manifestation at the beginning of treatment.

* [This broad definition of resistance was not unique to Reich, but can be found in a paper of 1915 by Reik.]

He does not trust it, refuses to accept it at face value and seeks systematically to destroy it. He calls this procedure 'systematic resistance analysis'.

According to Reich the initial transference resistance expresses itself in a specific form which is characteristic of the patient's personality. The distinguishing and important feature of this resistance is not the content, which might be the same with different personalities, but the form in which it is expressed and is felt by the analyst. The defense of a female patient with a male analyst might take the form of masculine aggressiveness, or of superiority and coldness, or of suspicion and mistrust and other similar negative and aggressive features. Reich postulates that the form of the initial resistance against the analytic approach—the 'how' by which it expresses itself—is specific for the personality because this form is taken over and over again by any further resistance that the patient develops during the course of his analysis. Since this form of resistance is characteristic for the personality of the patient, Wilhelm Reich considers it the expression of the patient's character and therefore calls it 'character resistance'. The consistent analysis of this form of resistance Reich presents as his therapeutic innovation, and terms it 'character analysis'. He is particularly concerned with what he calls the 'secret' resistances of the personality, and repeatedly emphasizes how 'dangerous' they are. The danger of secret resistances almost seems to haunt him. He is of the opinion that secret resistances, if neglected, destroy all therapeutic efforts, and he believes that if the analyst interprets beyond them, that is, if he deals with recognizable material from a deeper layer than the one to which the secret resistance belongs, he creates what he calls a 'chaotic situation'. He maintains that analyses conducted in such disorderly fashion are ruined and have to be given up without therapeutic results because the damage done by unsystematic interpretation is irreparable.

One can recognize that Reich's theory of therapy is closely tied up with another systematizing concept of his, that of 'stratification'. For Reich the mental apparatus is structured in the form of layers, and the order of the layers has to be considered in therapy, particularly in the interpretative approach to resistances. In his case histories, in which he shows all his brilliance as a clinician, he demonstrates very clearly what he meant by a systematic approach to a character resistance. If, for example, such a resistance manifests itself as a silent smile which accompanies the patient's productions, Reich considers it useless and even dangerous to interpret any material produced while the analytic situation is under the influence of such a resistance. According to him the only correct technique is to interpret the resistance expressed in this smiling according to the layers of its significance systematically from the most superficial and genetically latest layer to the next one and in proper order down to its deepest significance. In his book he demonstrates the different resistive meanings of the smile of a patient and unmasks it first as an attempt at reconciliation, in the next layer as a compensation for anxiety, and in the third as the expression of a feeling of superiority. He is convinced that had he interpreted the second significance before the first, or the third before the second, or, even worse, the third before the first, the whole analysis would have been hopelessly disturbed and would have resulted in chaotic disorder. Since the analysis of the resistive smiling led to different defensive attitudes established against painful experiences in the patient's infantile past, and since these contain the patient's most characteristic attitudes toward unpleasure, he considers this resistance an expression of the patient's character. He claims that even if this resistance is unfolded and removed by systematic interpretation, it will return with the appearance of every further resistance and will have to be dealt with systematically again, since this typical form of resisting is the most essential constituent of the patient's character. He therefore calls it 'character resistance'.

It might be appropriate to add a few words about the further development of this concept of 'character resistance' into what Reich calls 'character analysis'. When Reich examines systematically the origin of the resistance form of a patient he considers this study a characterological one, since he is of the opinion that it expresses the patient's character, and he makes the study of such a resistance the basis of his characterology. For Reich character is something that establishes a 'typical', specific resistance in analysis which repeats itself always in the same form. From the fact that the attitude of defense characteristic of a person serves as a permanent resistance in analysis, Reich makes the bold conclusion that character itself is a resistance, an apparatus of protection against the outside world and against instinctual drives. Since Reich's characterology is based on technical experience in therapy one could call it 'therapeutic-technical'. His technique, in which he focuses all technical efforts on the resistance expressed in a specific attitude, he terms 'character analysis', since he maintains that this attitude of defense forms the central and most essential part of the character. He claims that his technique is the only correct one because it strives to undermine the neurosis in all directions from a firm stronghold.

The ego's resistive reaction against analysis, representing the 'character' of the patient, Reich tries to trace back to the infantile experiences which were responsible for the formation of the specific way, the 'how' in which the patient resists the analytic effort to uncover unconscious material. Reich tries to establish a metapsychology of the characteristic attitudes of protection. According to Reich character from the topological viewpoint is an apparatus of protection, from the dynamic viewpoint it is composed of frozen resistances against unconscious drives plus instinctual satisfactions obtained through the character attitudes themselves, and economically it serves the avoidance of unpleasure, for example anxiety, as well as the establishment of the neurotic equilibrium mainly due to the satisfaction of repressed infantile drives through the character attitude itself. These metapsychological relationships of defense

attitudes Reich illustrates in his book with excellent examples. The case histories are the best part of his book.

Since character and resistance are practically identical for Reich, he
arrives at strange concepts about character which he tries to illustrate by a
simile. To him character is an armor formed through chronic 'hardening' of
the ego. The meaning and purpose of this armor is protection from inner and
outer dangers. And from this comparison of character with an armor he
makes statements which might be appropriate for the armor of a medieval
knight, but hardly for a person's character.* For example: character, Reich
says, results in a definite limitation of the psychic mobility of the total personality since a person cannot move freely if he is encased in armor. Nevertheless, the armor has to be imagined as somewhat movable with a normal
person, for one has to imagine that it has gaps or openings through which the
emission and withdrawal of object cathexes can take place. Reich speaks of
the 'rigid' armor of the emotionally blocked person, or the 'prickly' armor of
the querulent and aggressive type, of 'armoring of the surface' with patients
blocked in their emotions or with the compulsive character, of 'armoring of
the depth' with the hysterical character.

He further considers the libidinous gratification derived from the specific
defense as the linkage between the different defensive attitudes organized in
the character resistance, as the putty that fills out the gaps, as the joints
between the different parts of the armor, and similar farfetched comparisons.
I have presented this thumbnail sketch of Reich's characterology to show that
his characterology is the end product of his therapeutic theory, of which the
concept of character resistance, the subject of this paper, is the most essential
part. I shall now undertake to evaluate Reich's therapeutic theory and
technique.

The first postulate against which we have to raise objections is Reich's
assumption that particularly the initial transference is exclusively the means
and expression of resistance. There is probably no better argument against
this notion than the following quotation from Freud:

> The part taken by resistance in the transference-love is unquestionable and
> very considerable. *But this love was not created by the resistance*;† the latter
> finds it ready to hand, exploits it and aggravates the manifestation of it. Nor
> is its genuineness impugned by the resistance . . . it is true that the trans-
> ference-love consists of new editions of old traces and that it repeats infantile
> reactions. But this is the essential character of every love. There is no love
> that does not reproduce infantile prototypes. The infantile conditioning factor
> in it is just what gives it its compulsive character which verges on the
> pathological. The transference-love has perhaps a degree less of freedom than
> the love which appears in ordinary life and is called normal; it displays its

* [Reich's use of the term "character armor" illustrates the dangers of a metaphor
being transformed into a model; see Max Block 1962.]
  † Italics added.

CLINICAL AND THERAPEUTIC ASPECTS OF CHARACTER RESISTANCE

dependence on the infantile pattern more clearly, is less adaptable and capable of modification, but that is all and that is nothing essential.

By what other signs can the genuineness of a love be recognized? By its power to achieve results, its capacity to accomplish its aim? In this respect the transference-love seems to give place to none; one has the impression that one could achieve anything by its means.

Let us resume, therefore: one has no right to dispute the 'genuine' nature of the love which makes its appearance in the course of analytic treatment. However lacking in normality it may seem to be, this quality is sufficiently explained when we remember that the condition of being in love in ordinary life outside analysis is also more like abnormal than normal mental phenomena. The transference-love is characterized, nevertheless, by certain features which ensure it a special position. In the first place, it is provoked by the analytic situation; second, it is greatly intensified by the resistance which dominates this situation; and third, it is to a high degree lacking in regard for reality, is less sensible, less concerned about consequences, more blind in its estimation of the person loved, than we are willing to admit of normal love. We should not forget, however, that it is precisely these departures from the norm that make up the essential element in the condition of being in love (Freud 1915a, p. 157).

Freud further gives the following advice concerning the handling of the initial transference: 'One must wait until the transference, which is the most delicate matter of all to deal with, comes to be employed as resistance' (Freud 1913a, p. 139). Here it is again clear that initial transference and resistance are not identical, but that the transference becomes sooner or later employed by the resistance.

It is one of Reich's basic errors that he denies the genuine character of positive transference, particularly in the beginning of the analysis.* Reich's technique of dealing with the transference seemingly is an outgrowth of his own suspicious character and the belligerent attitude that stems from it. This makes him imply 'secret' resistances even where genuine transference-love is established. Under the impact of his technique, which is conditioned by the mistrust in the patient's positive transference reactions and by the disbelief in the genuineness of initial and even later transference-love, the patient must necessarily feel unaccepted and constantly questioned as to the truthfulness of his positive feelings toward the analyst, so that he finally has to develop negative reactions out of his feeling of being frustrated and rejected. If these negative reactions finally manifest themselves in dreams or otherwise Reich is triumphant because it proves to him that the initial transference was not genuinely positive. His whole therapeutic approach is full of aggressiveness and belligerency. It is revealing to observe how regularly he uses comparisons from the battlefield, and the way in which he uses them. Again and again he compares the resistance with a dangerous and tricky enemy who has to be exterminated at all costs. The consistent analysis of the character resistance

* [The opposite point of view was taken by Nunberg (1928, this volume, Chapter 15).]

he calls the 'stronghold' from which the therapist has to 'undermine' the enemies' position. The patient is assumed constantly to deceive the analyst who has to be on guard against 'secret agents' of resistance all the time. At the time he published his book in 1933, his technique had reached a degree of aggression that he himself felt it dangerous for his patients, so that he found it necessary to warn them of the possibility that their egos might break down under the constant hammering at their character resistances. When he claimed at the beginning of his therapeutic crusade that his new technique does away with the directly aggressive approach to the patient's resistance and replaces it by analytic dissolution, we must reply that, on the contrary, the further development of his character therapy is much more sadistic and destructive than the analytic technique as it was developed before him. In his recent book, Listen, Little Man!, he releases all the fury of mockery, abuse, irony and sarcasm toward the suffering neurotic.

When one reads Reich's analytic papers one will recognize that he lacks insight and understanding of one of the basic characteristics in man's emotional life. He ignores the phenomenon of ambivalence. Freud designates as a characteristic of infantile instinctual life '. . . the fact that the contrasting pair of impulses are developed in almost the same manner, a situation which was happily designated by Bleuler by the term *ambivalence*' (Freud 1905b, p. 199). Reich refuses to acknowledge this basic fact of the instinctual life of man; for him ambivalence is not something inherent, an innate characteristic of certain instinctual manifestations, but is acquired in the course of development due to the frustration of instinctual needs [A. Freud 1936; cf. editors' introduction, p. 39]. According to Reich our drives originally have only positive and loving aims and attitudes (although strangely enough he denies their genuine reappearance in the transference, at least in the beginning of the analysis); hatred and destructiveness are the result of frustration by reality. This denial of instinctual ambivalence as well as of the ambivalence of feelings or attitudes toward objects is essential in Reich's theory of personality, for it is responsible for his theoretical construction of the personality in the form of layers. If negative signs appear among positive ones in the transference, Reich considers their simultaneous appearance not the expression of an ambivalent attitude but believes that either the negative ones are breaking through the positive surface from a deeper layer or vice versa. The order of the layers depends on the quantitative relationship between the manifestations with opposite signs in the sense that the superficial layer predominates in its manifestation. This layered structure of the personality goes very deep. From five to six layers are sometimes enumerated by Reich. They all are supposed to contribute to the character resistance and, according to Reich, the systematic removal of their contributions to the initial and constant resistance in orderly sequence is supposed to be the only correct analytic procedure in character analysis.

The auxiliary construction of the psyche in the form of layers is not Reich's original idea. Freud used the concept of superficial and deeper layers of the mind repeatedly and emphasized that we have to reach out for spatial relationships in order to gain some plastic concept of the working of the complicated mental apparatus. 'Upper and lower level' are often used in dreams even of persons not psychoanalytically trained to represent the conscious mind and the unconscious.

What we have to object to in Reich's theories is the concept of multiple layers that can and have to be peeled off systematically in therapy with careful avoidance of penetrating a deeper layer before all the others above it are removed. Freud himself made an attempt to demonstrate the relationship of mental contents, or better cathexes of presently actual and regressive formations and tendencies, with a comparison, and when reading it one obtains the impression that he wants to emphasize how incorrect a strict application of the concept of stratification is in connection with mental material. He tries to demonstrate this by a comparison:

> Now let us make the fantastic supposition that Rome were not a human dwelling-place, but a mental entity with just as long and varied a past history: that is, in which nothing once constructed had perished, and all the earlier stages of development had survived alongside the latest. This would mean that in Rome the palaces of the Caesars were still standing on the Palatine and the Septizonium of Septimius Severus was still towering to its old height; that the beautiful statues were still standing in the colonnade of the Castle of St. Angelo, as they were up to its siege by the Goths, and so on. But more still: where the Palazzo Caffarelli stands there would also be, without this being removed, the Temple of Jupiter Capitolinus, not merely in its latest form, moreover, as the Romans of the Caesars saw it, but also in its earliest shape, when it still wore an Etruscan design and was adorned with terracotta antefixae. Where the Colosseum stands now we could at the same time admire Nero's Golden House; on the Piazza of the Pantheon we should find not only the Pantheon of today as bequeathed to us by Hadrian, but on the same site also Agrippa's original edifice; indeed, the same ground would support the church of Santa Maria sopra Minerva and the old temple over which it was built. And the observer would need merely to shift the focus of his eyes, perhaps, or change his position, in order to call up a view of either the one or the other (Freud 1930, p. 70).

From this comparison we gain the impression that Freud definitely was opposed to the concept of simple layers because it misrepresents the spatial coincidence of mental contents as if they were temporal and spatial successions.

Reich's demand for systematic technique based upon the alleged stratification of the character resistance is unjustified since the rigid concept of multiple stratification itself is incorrect. This does not mean that interpretation should not be given in an orderly and organized fashion, in which the sequence—interpretation of the defense first, of the unconscious content second, or id part after ego contribution of the formation to be interpreted—is almost always valid. Reich particularly emphasizes the damaging effect of too early or too deep interpretation which is supposed to upset the whole therapeutic schedule to an irreparable degree. The careful training in psychoanalytic institutes nowadays prevents beginners from making the mistake of making shocking interpretation that could do damage, scare the patient, or severely disturb him. The 'what' and 'how' of interpretation is something that can be taught only to some degree. The rest has to be afforded by the therapist's gift and intuition. Systematization of interpretation impedes the flexibility of the analyst in acting according to the patient's needs and according to the make-up of his own personality.

In my own experience I have found that even the oversight or neglect of an initial 'secret' resistance—the most dangerous mistake in Reich's view— does not have the irreparable consequences that Reich ascribes to such a technical blunder. Some years ago I took in therapeutic analysis a physician who had read about analysis and had friends who were analyzed. In the second or third hour he released a flood of disconnected obscenities, acting, as I thought, according to a misconception of free association. Since his profuse profanity made the impression of being exaggerated in its content and quantity, I simply told him, 'Don't force yourself', and explained free association to him. He calmed down and from then on spoke in a much less excited manner. The analysis proceeded in orderly fashion. He was in analysis for two and a half years and improved greatly. Since there was little movement during the last six months in his analysis, we decided to terminate temporarily. He was pleased with the result of his analysis which had produced a decisive change in his personality. From a rather timid individual with relatively limited capacity in his profession, he became a very successful, steady physician who was a strong support for his family, whereas his wife had dominated him before and had been the main provider. There were some slight anxieties which we considered to be the scars left from his neurosis. After a year of all-round success he returned for some more analysis, because he felt something had been left undone. When he was on the couch again, he began with bitter reproaches about my attitude at the beginning of his analysis. When I had responded to his flood of four-letter words with, 'Don't force yourself', he reacted only to the 'don't'. It prevented him from the free expression of his aggression. He never dared to come out openly again with all the dirty words, accusations and reproaches which he had desired to hurl at me and which had been pent up from childhood when his parents applied the 'don't' to his aggressive behavior. In the short period during which I saw him the second time, he released all the resentment that he had accumulated on account of my first 'don't' and which he had not dared to express in his analysis. Only after the first termination, when he was not any more under the direct influence of the analyst, did he accumulate the courage to come back and 'tell me off' and thus complete the analytic work.

According to Reich, the initial mistake which built up a defense against free expression toward the analyst—a typical dangerous 'secret' resistance—should have rendered all further analytic work null and void. The layer of resistance below the surface should have destroyed the effect of all later interpretations of further material. That this was not so is clearly demonstrated by this brief case report. The analysis was able to progress and achieve satisfactory results despite the secret resistance.

Reich propagates the systematic sequence of interpretation—by which he almost always means interpretation of resistance—because the cardinal resistance, initiating the transference and shaping all subsequent resistance throughout the analysis, is for him identical with the character of the patient. And here our main objection to Reich's concept of 'character resistance' has to be raised. Reich's concept of 'character' is far too limited. It might be appropriate to say a few words about the definition of character as I understand it. Character, in my opinion, designates the features of personality which are more or less indelibly engraved upon it, which express themselves in our actions and reactions, and by which one personality structure can be differentiated from others. Since actions and reactions as they manifest themselves are the business of the ego, we are accustomed to attribute the character of the personality to the ego. For Reich, character would be only the expression of the resisting ego. But a short deliberation reveals that character is formed not only by the specific way of dealing with the multitude of stimuli which impinge on the ego from outside and from within, but that it is deeply rooted in the quantity of specific pressings upward from the id in the form of instinctual strivings. In this respect character is based on the organic substratum of our psychic personality, as was already pointed out by Freud in his study of Leonardo: 'The tendency to repression, as well as the ability to sublimate, must be traced back to the *organic basis of the character*,* upon which alone the psychic structure arises' (Freud 1910b, p. 136).

A third contributor to character formation is the superego with its specific, more or less rigid demands. All three provinces of the mental structure of the personality, therefore, contribute to the formation of character. A definition of character, then, would designate it as the sum total of specific reactions of the individual, determined by the interaction of the three provinces of the mind according to their inherited and acquired dynamic contents. Ego, id and superego have to be involved where character studies are attempted.

From all this it is obvious that the concept of character implies more than resistance in analysis. There is no doubt that Reich's concept of character as a dynamic formation which produces specific and constant resistances in analysis is far too narrow and one-sided. A technical-therapeutic charac-

* Italics added.

terology is very insufficient to explain the many facets of the personality, and character certainly consists of more than defenses.

Reich's 'character resistance' is outdated and hardly useful nowadays. However, some of Reich's contributions to the theory and technique of psychoanalytic therapy were of considerable value. His ideas and the forceful way in which he presented them led to the clarification of many concepts in therapy. In my opinion the most important result of the commotion he created is Anna Freud's book, The Ego and the Mechanisms of Defense. When I read Anna Freud's book again recently, after having worked my way through Wilhelm Reich's papers on technique, I experienced great relief. One feels the pressure of Reich's technique even while reading his papers, and particularly his book on character analysis. Anna Freud's book, when read after Reich's technical papers, produces a feeling of liberation. It is as if, after being hurled in a boat through rapids, one emerged in a calm, wide body of water where the mountains that narrowed the river have receded into the background and a relaxed survey of the open landscape is possible in many directions. Though this comparison may sound somewhat poetic, it helps to illustrate the fundamental difference in the basic attitude of the authors. Anna Freud's concept of character is much broader than Reich's. According to her, character is approximately the whole set of attitudes habitually adapted by an individual ego for the solution of the never-ending series of inner conflicts. Character, then, is the single ego's typical way of dealing with the conflict between the instinctual urges coming from the id, the dangers coming from the outside world, and the threats of the superego which represent the incorporation of a most important part of what was once the outside world, the parental authority. Every ego is characterized by the choice it makes among the instinctual urges which it seeks to satisfy due to its access to motility, by the determination of those instinctual urges which it rejects, and by the methods of defense which it uses against the powers threatening it from outside and from within.

It is significant that in broadening Reich's therapeutic bottleneck Anna Freud emphasizes the concept of defense in contrast to Reich's almost exclusive preoccupation with the concept 'resistance'. It is well known, after the original introduction of the term 'defense' into psychopathology, Sigmund Freud abandoned it for more than forty years because during this time he was investigating mainly one type of defense, repression. In Inhibition, Symptom and Anxiety, he reinstated the concept of defense and put repression back in its place as only one among many typical defenses against inner and outer dangers. The reinstatement of the term 'defense' paved the way, as it were, for Anna Freud's studies of the defense mechanisms. Freud also laid the groundwork for the psychodynamic understanding of the ego. But while Freud's ego analysis was concerned with the ego's dynamic structure, its composition, and the development of its organization, Anna Freud's ego analysis is a study of the ego's activity, or at least of one very important and constantly applied activity, namely that of defense. It is defense analysis. The importance of this step in the development of our science can hardly be overestimated. Only after the study of the ego's activity of defense was added did psychoanalysis become a well-rounded science of man's mind and its working.

It is my impression that the importance of this newest addition to our science has not been sufficiently recognized and that it has not yet penetrated the thinking and therapeutic technique of most analysts. It is easy to understand why this is so. We are still very much impressed, even fascinated, by the id contents which psychoanalysis enables us to discover. The working of the ego is so inconspicuous and silent that we are hardly aware of it. It is only necessary to recall the contrast between the experience of dreaming—which may occur in the most vivid images and with most violent emotions and always manifests itself as a sensual perception—and the forgetting of the dream which occurs so quietly and unnoticed. Without having become aware of any activity within ourselves, we simply recognize the fact that the dream is gone or that we are left with only a few meager memory fragments of an experience that, a few minutes before, had filled us with great intensity of perception and feeling. We notice the result of the ego's defensive activity—in the case of dreams the forgetting through repression—but we are completely unable to perceive this activity itself, and this applies to all unconscious defense activities of the ego. We never can catch them at work; we can only reconstruct them from the result. While one can listen with the 'third ear' to the utterances of the id, it needs a most refined instrument to register the workings of the ego defenses. It has been my observation that it is a most difficult task to teach students to pay attention to these mute and subterranean workings of the ego. Even the experienced analyst must constantly exercise self-discipline in order to remain aware of the ego's defense measures in therapy.

Perhaps it is carrying coals to Newcastle to emphasize here the significance and importance of Anna Freud's contribution to psychoanalysis. But judging from my experience as a teacher, from our scientific meetings and from the current literature I find too little real influence of Anna Freud's studies, although often lip service is paid to them. 'Mechanism of defense' is used glibly to indicate the advanced state of one's analytic thinking, and 'identification with the aggressor' is mentioned in order to display consideration of the ego. I believe it will require a great deal of time and effort on the part of training analysts to make Anna Freud's discoveries of the silent activities of the ego penetrate general analytic thinking and improve psychoanalytic technique so that it will consist of id-plus-ego analysis, applied alternatingly.

I cannot here present all the technical implications and modifications that arise from Anna Freud's studies. I can only contrast them to Reich's therapeutic ideas and his technique. The difference at first sight is not very conspicuous; both emphasize ego consideration. But on closer inspection this difference is very profound; in fact one has to consider it fundamental. Reich looks for resistances, suspects them in every transference manifestation and considers it the only proper technique to bring them relentlessly into the open and to follow them up with interpretations through all layers down to their roots. Reich's ego consideration is only apparent; it concerns itself only with the ego as it resists the analytic process. The ego for Reich is the enemy of analysis, and a deceiving and tricky enemy at that. His approach to the ego, therefore, is a hostile, aggressive one as we have demonstrated. Anna Freud, in contrast to Reich, remains an objective observer of the ego and is only concerned with the understanding of its functioning inside and outside the analytic situation. She objects explicitly to any suspicious inimical and pressing attitude toward the ego's resistance and defense in the analytic situation. She obviously has Wilhelm Reich's technique in mind when she states: 'In my opinion we do our patients a great injustice if we describe these transferred defense-reactions as "camouflage" or say that the patients are "pulling the analyst's leg" or purposely deceiving him in some other way. . . . The patient *is* in fact candid when he gives expression to the impulse or affect in the only way still open to him, namely, in the distorted defensive measure' (A. Freud 1936). Defense mechanisms which the ego learned to use against inner and outer dangers during its lifetime, particularly in childhood, the ego will find useful and will by necessity have to apply in the analytic situation. The therapeutic task is to notice them during the analytic process, to observe them or better to reconstruct their working from the result and to demonstrate to the patient their general application by the personality in the present and past; furthermore, to trace their genesis and their deepest motivation in the form of infantile anxieties; and finally to render their use unnecessary through analytic comparison between the present and past. Only if in this way ego analysis complements the analysis of the id, is psychoanalytic method a study and therapy of the total personality. Only then can we hope to be successful in our therapeutic approach to what Anna Freud calls the 'innumerable transformations, distortions and deformities of the ego which are in part the accompaniment of and in part substitutes of neurosis'.

Anna Freud says especially that she considers the term 'character analysis' not very appropriate for ego analysis and defense analysis. The term 'character resistance' is not used by her at all, and I think it is rightly omitted in her book. Though a specific defense reaction which serves as a resistance in analysis might be characterisitc for the patient, we have no justification to identify this defense reaction with the totality of reactions of a personality, which is properly called character. Such an identification is implied if we use the term 'character resistance' for a major defense mechanism of a patient in analysis. I therefore feel we are justified in abandoning the term 'character resistance' as inappropriate.

Part Eight

Psychoanalytic
Technique
in the Thirties:
Controversy and
Synthesis

CHAPTER 21

SANDOR FERENCZI

THE PRINCIPLE OF
RELAXATION AND
NEOCATHARSIS

INTRODUCTORY NOTE

In this essay Ferenczi advocates the principle of "indulgence" in opposition to Freud's emphasis on abstinence (see editors' introduction, p. 30). The reader will encounter here the last and, once again, controversial innovation—the principle of indulgence. Like Ferenczi's "Active Technique," his last suggestions were rejected by the majority of Freudian psychoanalysts, but they exerted a great influence on other schools of psychotherapy. Ferenczi is the originator of Franz Alexander's corrective emotional experience. He is also the originator of a technique of treatment much in vogue after World War II where certain prospective analysands were given a period of preparation so they could later withstand the rigors of a Freudian analysis. In practice, the preparatory period was usually entrusted to a psychoanalytically trained social worker or a psychiatrist trained in supportive therapy. When the prospec-

Originally entitled "Relaxationsprinzip und Neokatharsis," *Internationale Zeitschrift für Psychoanalyse* 16 (1930):149–164. English translation originally published in *International Journal of Psycho-Analysis* 11 (1930):428–443. Reprinted in *Final Contributions to the Problems and Methods of Psycho-Analysis,* ed. Michael Balint, trans. Eric Mosbacher et al. (New York: Basic Books, 1955), pp. 108–125. Reprinted by permission of Basic Books, Inc., Publishers. This paper is an enlarged version of a paper read at the Eleventh International Psycho-Analytic Congress, Oxford, August 1929, entitled "Progresses in Psycho-Analytic Technique."

tive analysand was judged to have matured, psychoanalysis proper was undertaken. Thus, the first therapist "indulged" the patient and the second conducted the analysis in a state of abstinence.

Ferenczi is aware that his new ideas arouse opposition, and like the early Freud, he anticipates and verbalizes their objections. At the time that Ferenczi wrote, the relationship between psychoanalytic technique and the structure of the ego was only dimly understood. The reaction of the psychoanalytic community would have been more favorable had Ferenczi confined his innovations to those whose ego structure was too weak to enter fully into the psychoanalytic pact. It was only after World War II (see Eissler 1953; Stone 1954; Anna Freud 1954b) that the connection between the structure of the ego and the variations that should be introduced into the psychoanalytic technique was explored. In current psychoanalytic language, what Ferenczi called "indulgences" are now conceptualized as parameters.

AT THE CONCLUSION of this essay many of you will very likely have the impression that I ought not to have called it 'Progress in Technique', seeing that what I say in it might be more fittingly termed retrogressive or reactionary. But I hope that this impression will soon be dispelled by the reflection that even a retrograde movement, if it be in the direction of an earlier tradition, undeservedly abandoned, may advance the truth, and I honestly think that in such a case it is not too paradoxical to put forward an accentuation of our past knowledge as an advance in science. Freud's psycho-analytical researches cover a vast field: they embrace not only the mental life of the individual, but group psychology and study of human civilization; recently also he has extended them to the ultimate conception of life and death. As he proceeded to develop a modest psychotherapeutic method into a complete system of psychology and philosophy, it was inevitable that the pioneer of psycho-analysis should concentrate now on this and now on that field of investigation, disregarding everything else for the time being. But of course the withdrawal from facts earlier arrived at by no means implied that he was abandoning or contradicting them. We, his disciples, however, are inclined to cling too literally to Freud's latest pronouncements, to proclaim the most recently discovered to be the sole truth and thus at times to fall into error

My own position in the psycho-analytical movement has made me a kind of cross between a pupil and a teacher, and perhaps this double role gives me the right and the ability to point out where we are tending to be one-sided, and, without foregoing what is good in the new teaching, to plead that justice shall be done to that which has proved its value in days past.

The technical method and the scientific theory of psycho-analysis are so closely and almost inextricably bound up with one another that I cannot in this paper confine myself to the purely technical side; I must review part of the contents of this scientific doctrine as well. In the earliest period of psycho-analysis, a period of which I will give as concise a summary as possible, there was no talk of any such division, and, even in the period immediately succeeding, the separation of technique and theory was purely artificial and was made solely for purposes of teaching.

I

A genial patient and her understanding physician shared in the discovery of the forerunner of psycho-analysis, namely, the cathartic treatment of hysteria. The patient found out for herself that certain of her symptoms disappeared when she succeeded in linking up fragments of what she said and did in an altered state of consciousness with forgotten impressions from her early life. Breuer's remarkable contribution to psychotherapy was this: not only did he pursue the method indicated by the patient, but he had faith in the *reality* of the memories which emerged, and did not, as was customary, dismiss them out of hand as the fantastic inventions of a mentally abnormal patient. We must admit that Breuer's capacity for belief had strict limitations. He could follow his patient only so long as her speech and behaviour did not overstep the bounds marked out by civilized society. Upon the first manifestations of uninhibited instinctual life he left not only the patient but the whole method in the lurch. Moreover, his theoretical deductions, otherwise extremely penetrating, were confined as far as possible to the purely intellectual aspect, or else, passing over everything in the realm of psychic emotion, they linked up directly with the physical.

Psychotherapy had to wait for a man of stronger calibre, who would not recoil from the instinctual and animal elements in the mental organization of civilized man; there is no need for me to name this pioneer. Freud's experience forced him relentlessly to the assumption that in every case of neurosis a *conditio sine qua non* is a sexual trauma. But when in certain cases the patient's statements proved incorrect, he too had to wrestle with the temptation to pronounce all the material they had produced untrustworthy and therefore unworthy of scientific consideration. Fortunately, Freud's intellectual acumen saved psycho-analysis from the imminent danger of being once more lost in oblivion. He perceived that, even though certain of the statements made by patients were untrue and not in accordance with reality, yet the psychic reality of their lying itself remained an incontestable fact. It is difficult to picture how much courage, how much vigorous and logical thinking, and how much self-mastery was necessary for him to be able to free his mind from disturbing affects and pronounce the deceptive unveracity of his patients to be hysterical fantasy, worthy as a psychic reality of further consideration and investigation.

Naturally the technique of psycho-analysis was coloured by these successive advances. The highly emotional relation between physician and patient, which resembled that in hypnotic suggestion, gradually cooled down to a kind of unending association-experiment; the process became mainly intellectual. They joined, as it were, their mental forces in the attempt to reconstruct the repressed causes of the illness from the disconnected fragments of the material acquired through the patient's associations. It was like filling in the spaces in an extremely complicated crossword puzzle. But disappointing therapeutic failures, which would assuredly have discouraged a weaker man, compelled Freud once more to restore in the relation between analyst and patient the affectivity which, as was now plain, had for a time been unduly neglected. However, it no longer took the form of influence by hypnosis and suggestion —an influence very hard to regulate, and one whose nature was not understood.* Rather more consideration and respect were accorded to the signs of *transference of affect* and of *affective resistance* which manifested themselves in the analytical relation.

This was, roughly speaking, the position of analytical technique and theory at the time when I first became an enthusiastic adherent of the new teaching. Curiously enough, the first impetus in that direction came to me through Jung's association-experiments. You must permit me in this paper to depict the development of the technique from the subjective standpoint of a single individual. It seems as though the fundamental biogenetic law applies to the intellectual evolution of the individual as of the race; probably there exists no firmly established science which does not, as a separate branch of knowledge, recapitulate the following phases: first, enlightenment, accompanied by exaggerated optimism, then the inevitable disappointment, and, finally, a reconciliation between the two affects. I really do not know whether I envy our younger colleagues the ease with which they enter into possession of that which earlier generations won by bitter struggles. Sometimes I feel that to receive a tradition, however valuable, ready-made, is not so good as achieving something for oneself.

I have a lively recollection of my first attempts at the beginning of my psycho-analytical career. I recall, for instance, the very first case I treated. The patient was a young fellow-physician whom I met in the street. Extremely pale and obviously struggling desperately for breath, he grasped my arm and implored me to help him. He was suffering, as he told me in gasps, from nervous asthma. He had tried every possible remedy, but without success. I took a hasty decision, led him to my consulting-room, got him to give me his

* [For a more extensive discussion of the transition from the cathartic method to psychoanalysis, see editors' introductory essay, p. 18, as well as Coltrera and Ross 1967.]

reactions to an association-test, and plunged into the analysis of his earlier life, with the help of this rapidly sown and harvested crop of associations. Sure enough, his memory pictures soon grouped themselves round a trauma in his early childhood. The episode was an operation for hydrocele.* He saw and felt with objective vividness how he was seized by the hospital attendants, how the chloroform-mask was put over his face, and how he tried with all his might to escape from the anaesthetic. He repeated the straining of the muscles, the sweat of anxiety, and the interrupted breathing which he must have experienced on this traumatic occasion. Then he opened his eyes, as though awaking from a dream, looked about him in wonder, embraced me triumphantly, and said he felt perfectly free from the attack.

I could describe many other 'cathartic' successes similar to this, at about this time. But I soon discovered that, in nearly all the cases where the symptoms were thus cured, the results were but transitory, and I, the physician, felt that I was myself being gradually cured of my exaggerated optimism. I tried by means of a deeper study of Freud's work and with the help of such personal counsel as I might seek from him to master the technique of association, resistance, and transference. I followed as exactly as possible the technical hints that he published during this period. I think I have already told elsewhere how, with the deepening of my psychological knowledge as I followed these technical rules, there was a steady decrease in the striking and rapid results I achieved. The earlier, cathartic therapy was gradually transformed into a kind of analytical re-education of the patient, which demanded more and more time. In my zeal (I was still a young man) I tried to think out means for shortening the period of analysis and producing more visible therapeutic results. By a greater generalization and emphasizing of the principle of frustration (to which Freud [1919] himself subscribed at the Congress at Budapest in 1918), and with the aid of artificially produced accentuations of tension ('active therapy'), I tried to induce a freer repetition of early traumatic experiences and to lead up to a better solution of them through analysis. You are doubtless aware that I myself, and others who followed me, sometimes let ourselves be carried away into exaggerations of this active technique. The worst of these was the measure suggested by Rank and, for a time, accepted by myself—the setting of a term to the analysis. I had sufficient insight to utter a timely warning against these exaggerations, and I threw myself into the analysis of the ego and of character-development, upon which in the meantime Freud had so successfully entered. The somewhat one-sided ego-analysis, in which too little attention was paid to the libido (formerly regarded as omnipotent), converted analytical treatment largely into a process designed to afford us the fullest possible insight into the topography, dynamics, and economy of symptom-formation, the distribution of energy

---

* [Hydrocele is a collection of serous fluid in a sacculated cavity, specifically in the lining of the testis.]

between the patient's id, ego, and super-ego being exactly traced out. But when I worked from this standpoint, I could not escape the impression that the relation between physician and patient was becoming far too much like that between teacher and pupil. I also became convinced that my patients were profoundly dissatisfied with me, though they did not dare to rebel openly against this didactic and pedantic attitude of the analyst. Accordingly, in one of my papers on technique I encouraged my colleagues to train their patients to a greater liberty and a freer expression in behaviour of their aggressive feelings towards the physician. At the same time I urged analysts to be more humble-minded in their attitude to their patients and to admit the mistakes they made, and I pleaded for a greater elasticity in technique [1928, see this volume, Chapter 17], even if it meant the sacrifice of some of our theories. These, as I pointed out, were not immutable, though they might be valuable instruments for a time. Finally, I was able to state that not only did my patients' analysis not suffer from the greater freedom accorded them, but, after all their aggressive impulses had exhausted their fury, positive transference and also much more positive results were achieved. So you must not be too surprised if, once more, I have to tell you of fresh steps forward, or, if you will have it so, *backward* in the path that I have followed. I am conscious that what I have to say is not at all likely to be popular with you. And I must admit that I am afraid it may win most unwelcome popularity amongst the true reactionaries. But do not forget what I said at the beginning about progress and retrogression; in my view a return to what was good in the teaching of the past most emphatically does not imply giving up the good and valuable contributions made by the more recent development of our science. Moreover, it would be presumptuous to imagine that any one of us is in a position to say the last word on the potentialities of the technique or theory of analysis. I, for one, have learnt humility through the many vicissitudes which I have just sketched. So I would not represent what I am about to say as in any way final. In fact, I think it very possible that in a greater or lesser degree it will be subject to various limitations as time goes on.

II

In the course of my practical analytical work, which extended over many years, I constantly found myself infringing one or another of Freud's injunctions in his 'Recommendations on Technique'. For instance, my attempt to adhere to the principle that patients must be in a lying position during analysis would at times be thwarted by their uncontrollable impulse to get up and walk about the room or speak to me face to face. Or again, difficulties in the real situation, and often the unconscious machinations of the patient, would leave me with no alternative but either to break off the analysis or to depart from the general rule and carry it on without remuneration. I did not hesitate to adopt the latter alternative—not without success. The principle that the patient should be analysed in his ordinary environment and should carry on his usual occupation, was very often impossible to enforce. In some severe cases I was even obliged to let patients stay in bed for days and weeks and to relieve them of the effort of coming to my house. The sudden breaking-off of the analysis at the end of the hour very often had the effect of a shock, and I would be forced to prolong the treatment until the reaction had spent itself; sometimes I had to devote two or more hours a day to a single patient;* often, if I would not or could not do this, my inflexibility produced a resistance which I felt to be excessive and a too literal repetition of traumatic incidents in the patient's childhood; it would then take a long time even partly to overcome the bad effects of this unconscious identification of his. One of the chief principles of analysis is that of frustration, and this certain of my colleagues, and, at times, I myself applied too strictly. Many patients suffering from obsessional neurosis saw through it directly and utilized it as a new and quite inexhaustible source of resistance-situations, until the physician finally decided to knock this weapon out of their hands by *indulgence* (*Nachgiebigkeit*).

I had the greatest conscientious scruples about all these infringements of a fundamental rule (and about many others which I cannot instance in detail here), until my mind was set at rest by the authoritative information that Freud's 'Recommendations' were really intended only as warnings for beginners and were designed to protect them from the most glaring blunders and failures; his precepts contained, however, hardly any positive instructions, and considerable scope was left for the exercise of the analyst's own judgement, provided that he was clear about the metapsychological consequences of his procedure.

Nevertheless, the exceptional cases have become so numerous that I feel impelled to propound another principle, not hitherto formulated, even if tacitly accepted. I mean the *principle of indulgence*, which must often be allowed to operate side by side with that of frustration. Subsequent reflection

---

* [What Ferenczi called indulgences are now called parameters (Eissler 1953). However, while Ferenczi could not establish any valid generalizations as to when indulgences should be allowed, Eissler can be specific. A parameter is any variation in the technique when the psychoanalyst does something other than interpret. These include telephone conversations, visits to hospitals, seeing members of the family, giving advice, guidance, and many more. A parameter can be introduced into an analysis under the following conditions: (1) Where the treatment process is in jeopardy unless this parameter is introduced; (2) The parameter should be used as little as possible; (3) Before treatment ends, the parameter must be self-eliminating; (4) The parameter should not be allowed to convert the transference into a permanent unanalyzable relationship (p. 111).]

has convinced me that my explanation of the way in which the active technique worked was really a very forced one: I attributed everything that happened to frustration, i.e. to a 'heightening of tension'. When I told a patient, whose habit it was to cross her legs, that she must not do so, I was actually creating a situation of libidinal frustration, which induced a heightening of tension and the mobilization of psychic material hitherto repressed. But when I suggested to the same patient that she should give up the noticeably stiff posture of all her muscles and allow herself more freedom and mobility, I was really not justified in speaking of a heightening of tension, simply because she found it difficult to relax from her rigid attitude. It is much more honest to confess that here I was making use of a totally different method which, in contrast to the heightening of tension, may safely be called *relaxation*. We must admit, therefore, that psycho-analysis employs two opposite methods: it produces heightening of tension by the frustration it imposes and relaxation by the freedom it allows.

But with this, as with every novelty, we soon find that it contains something very, very old—I had almost said, something commonplace. Are not both these principles inherent in the method of free association? On the one hand, the patient is compelled to confess disagreeable truths, but, on the other, he is permitted a freedom of speech and expression of his feelings such as is hardly possible in any other department of life. And long before psycho-analysis came into existence there were two elements in the training of children and of the masses: tenderness and love were accorded to them, and at the same time they were required to adapt themselves to painful reality by making hard renunciations.

If the International Psycho-Analytical Association were not so highly cultivated and self-disciplined an assembly, I should probably be interrupted at this point in my discourse by a general uproar and clamour. Such a thing has been known to happen even in the British House of Commons, usually so dignified, when a particularly infuriating speech has been made. 'What on earth do you really mean?' some of you would shout. 'We have scarcely reconciled ourselves to some extent to the principle of frustration, which you yourself carried to all lengths in your active technique, when you upset our laboriously placated scientific conscience by confronting us with a new and confusing principle, whose application will be highly embarrassing to us.' 'You talk of the dangers of excessive frustration,' another and no less shrill voice would chime in. 'What about the dangers of coddling patients? And, anyhow, can you give us any definite directions about how and when one or the other principle is to be applied?'

Softly, ladies and gentlemen! We are not yet advanced far enough to enter on these and similar details. My only object for the moment was to prove that, even though we may not admit it, we do actually work with these two principles. But perhaps I ought to consider certain objections which naturally arise in my own mind too. The fact that the analyst may be made uncomfortable by being confronted with new problems surely need not be seriously discussed!

To compose your minds I will say with all due emphasis that the attitude of objective reserve and scientific observation which Freud recommends to the physician remains, as ever, the most trustworthy and, at the beginning of an analysis, the only justifiable one, and that, ultimately, the decision as to which is the appropriate method must never be arrived at under the influence of affective factors but only as the result of intelligent reflection. My modest endeavours have for their object merely a plain definition of what has hitherto been vaguely described as the 'psychological atmosphere'. We cannot deny that it is possible for even the cool objectivity of the physician to take forms which cause unnecessary and avoidable difficulties to the patient, and there must be ways and means of making our attitude of friendly goodwill during the analysis intelligible to him without abandoning the analysis of transference-material or falling into the errors of those who treat neurotics, not analytically, i.e. with complete sincerity, but with a simulation of severity or love.

III

I expect that various questions and objections, some of them I admit, very awkward ones, have arisen in your minds. Before we discuss them, let me state the main argument which, in my view, justifies me in emphasizing the principle of relaxation side by side with that of frustration and of objectivity (which is a matter of course for the analyst). The soundness of any hypothesis or theory is tested by its theoretical and practical usefulness, i.e. by its heuristic value, and it is my experience that the acknowledgement of the relaxation-principle has produced results valuable for both theory and practice. In a number of cases in which the analysis had come to grief over the patient's apparently insoluble resistances, and a new analysis was attempted in which a change was made from the too rigid tactics of frustration hitherto employed, it was attended with much more substantial success. I am not speaking of patients who had failed to get well with other analysts and who gratified me, their new analyst, by taking a turn for the better (partly out of revenge for the old, perhaps). I am speaking of cases in which I myself, with the one-sided technique of frustration, had failed to get any further but, cn making a fresh attempt and allowing more relaxation, I had not nearly so long-drawn-out a struggle with interminable personal resistances, and it became possible for physician and patient to join forces in a less disrupted analysis of the repressed material, or, as I might say, to tackle the 'objective resistances'. On analysing the patient's former obstinacy and comparing it with the readiness to give way, which resulted from the method of relaxation, we found that the rigid and cool aloofness on the analyst's part was experienced by the patient as a continuation of his infantile struggle with the grown-ups' authority, and made him repeat the same reactions in character and in symptoms as formed the basis of the real neurosis. Hitherto my idea about terminating the treatment had been that one need not be afraid of these resistances and might even provoke them artificially; I hoped (and to some extent I was justified) that, when the patient's analytical insight had gradually closed to him all avenues of resistance, he would be cornered and obliged to take the only way left open, namely, that which led to health. Now I do not deny that every neurotic must inevitably suffer during analysis; theoretically it is self-evident that the patient must learn to endure the suffering which originally led to repression. The only question is whether sometimes we do not make him suffer more than is absolutely necessary. I decided on the phrase 'economy of suffering' to express what I have realized and am trying to convey—and I hope it is not far-fetched—namely, that the principles of frustration and indulgence should both govern our technique.

As you all know, we analysts do not attach great scientific importance to therapeutic effects in the sense of an increased feeling of well-being on the patient's part. Only if our method results not merely in his improvement but in a deeper insight into the process of his recovery may we speak of real progress in comparison with earlier methods of treatment. The extent to which patients improved when I employed this relaxation-therapy in addition to the older method was in many cases quite astonishing. In hysterics, obsessional neurotics, and even in neurotic characters the familiar attempts to reconstruct the past went forward as usual. But, after we had succeeded in a somewhat deeper manner than before in creating an atmosphere of confidence between physician and patient and in securing a fuller freedom of affect, hysterical physical symptoms would suddenly make their appearance, often for the first time in an analysis extending over years. These symptoms included paraesthesias and spasms, definitely localized, violent emotional movements, like miniature hysterical attacks, sudden alterations of the state of consciousness, slight vertigo and a clouding of consciousness often with subsequent amnesia for what had taken place.* Some patients actually begged me to tell them how they had behaved when in these states. It was easy to utilize these symptoms as fresh aids to reconstruction—as physical memory symbols, so to speak. But there was this difference—this time, the reconstructed past had much more of a feeling of *reality and concreteness* about it

* [The appearance of new symptoms in the course of an analysis would now be taken as a sign that the transference neurosis is developing (see editors' introduction, p. 29).]

than heretofore, approximated much more closely to an actual *recollection*, whereas till then the patients had spoken only of possibilities or, at most, of varying degrees of probability and had yearned in vain for memories. In certain cases these hysterical attacks actually assumed the character of *trances*, in which fragments of the past were relived and the physician was the only bridge left between the patients and reality. I was able to question them and received important information about dissociated parts of the personality. Without any such intention on my part and without my making the least attempt to induce a condition of this sort, unusual states of consciousness manifested themselves, which might also be termed autohypnotic. Willy-nilly, one was forced to compare them with the phenomena of the Breuer-Freud *catharsis*. I must confess that at first this was a disagreeable surprise, almost a shock, to me. Was it really worth while to make that enormous detour of analysis of associations and resistances, to unravel the maze of the elements of ego-psychology, and even to traverse the whole metapsychology in order to arrive at the good old 'friendly attitude' to the patient and the method of catharsis, long believed to have been discarded? But a little reflection soon set my mind at rest. There is all the difference in the world between this cathartic termination to a long psycho-analysis and the fragmentary eruptions of emotion and recollection which the primitive catharsis could provoke and which had only a temporary effect. The catharsis of which I am speaking is, like many dreams, only a confirmation from the unconscious, a sign that our toilsome analytical construction, our technique of dealing with resistance and transference, have finally succeeded in drawing near to the aetiological reality. There is little that the palaeocatharsis has in common with this *neocatharsis*. Nevertheless we must admit that here, once more, a circle has been completed. Psycho-analysis began as a cathartic measure against traumatic shocks, the effects of which had never spent themselves, and against pent-up affects; it then devoted itself to a deeper study of neurotic fantasies and the various defence mechanisms against them. Next, it concentrated rather on the personal affective relation between analyst and patient, being in the first twenty years mainly occupied with the manifestations of instinctual tendencies, and, later, with the reactions of the ego. The sudden emergence in modern psycho-analysis of portions of an earlier technique and theory should not dismay us; it merely reminds us that, so far, no single advance has been made in analysis which has had to be entirely discarded as useless, and that we must constantly be prepared to find new veins of gold in temporarily abandoned workings.

IV

What I am now about to say is really the logical sequel to what I have said already. The recollections which neocatharsis evoked or corroborated lent an added significance to the original traumatic factor in our aetiological equations. The precautions of the hysteric and the avoidance of the obsessional neurotic may, it is true, have their explanation in purely mental fantasy-formations; nevertheless the first impetus towards abnormal lines of development has always been thought to originate from real psychic traumas and conflicts with the environment—the invariable precursors of the formation of nosogenic mental forces, for instance, of conscience. Accordingly, no analysis can be regarded (at any rate in theory) as complete unless we have succeeded in penetrating to the traumatic material. This statement is based, as I said, on experience acquired in relaxation-therapy; if it be true, it adds considerably (from the theoretical as well as the practical standpoint) to the heuristic value of this modified technique. Having given due consideration to fantasy as a pathogenic factor, I have of late been forced more and more to deal with the pathogenic trauma itself. It became evident that this is far more rarely the result of a constitutional hypersensibility in children (causing them to react neurotically even to a commonplace and unavoidable painful experience) than of really improper, unintelligent, capricious, tactless, or actually cruel treatment. Hysterical fantasies do not lie when they tell us that parents and other adults do indeed go monstrous lengths in the passionate eroticism of their relation with children, while, on the other hand, when the innocent child responds to this half-unconscious play on the part of its elders the latter are inclined to think out severe punishments and threats which are altogether incomprehensible to him and have the shattering effects of a shock.* To-day I am returning to the view that, beside the great importance of the Oedipus complex in children, a deep significance must also be attached to the *repressed incestuous affection of adults, which masquerades as tenderness.* On the other hand, I am bound to confess that children themselves manifest a readiness to engage in genital eroticism more vehemently and far earlier than we used to suppose. Many of the perversions children practise probably indi-

* [Here Ferenczi anticipated important findings of psychoanalysis after World War II, namely, the significance of the actual relationship between parent and child for the development of the child neurosis. Broadly speaking, three stages in the development of the psychoanalytic attitude toward the origin of the neurosis can be distinguished: The first was Freud's belief that all neuroses originate in trauma. The second stage was the belief that neuroses are based on infantile wishes that were neither overcome nor effectively repressed. The third period begins with this paper. It sees the wishes as developing within a real matrix of relationship between parent and child. The wishes are important but so is the real attitude of the parent to the child. Ferenczi is, therefore, the originator of the interpersonal approach.]

cate not simply fixation to a pregenital level but regression from an *early genital level*. In many cases the trauma of punishment falls upon children in the midst of some erotic activity, and the result may be a permanent disturbance of what Reich calls 'orgastic potency'. But the premature forcing of genital sensations has a no less terrifying effect on children; what they really want, even in their sexual life, is simply play and tenderness, not the violent ebullition of passion.

Observation of cases treated by the neocathartic method gave further food for thought; one realized something of the psychic process in the traumatic primal repression and gained a glimpse into the nature of repression in general. *The first reaction to a shock seems to be always a transitory psychosis*, i.e. a turning away from reality. Sometimes this takes the form of negative hallucination (hysterical loss of consciousness—fainting or vertigo), often of an immediate positive hallucinatory compensation, which makes itself felt as an illusory pleasure. In every case of neurotic amnesia, and possibly also in the ordinary childhood-amnesia, it seems likely that a *psychotic splitting off* of a part of the personality occurs under the influence of shock. The dissociated part, however, lives on hidden, ceaselessly endeavouring to make itself felt, without finding any outlet except in neurotic symptoms. For this notion I am partly indebted to discoveries made by our colleague, Elisabeth Severn, which she personally communicated to me.

Sometimes, as I said, we achieve direct contact with the repressed part of the personality and persuade it to engage in what I might almost call an infantile conversation. Under the method of relaxation the hysterical physical symptoms have at times led us back to phases of development in which, since the organ of thought was not yet completely developed, physical memories alone were registered.

In conclusion, there is one more point I must mention, namely, that more importance than we hitherto supposed must be attached to the anxiety aroused by menstruation, the impression made by which has only been recently properly emphasized by C. D. Daly;* together with the threat of castration it is one of the most important traumatic factors.

Why should I weary you, in a discourse which is surely mainly concerned with technique, with this long and not even complete list of half-worked-out theoretical arguments? Certainly not in order that you may wholeheartedly espouse these views, about which I myself am not as yet quite clear. I am content if I have conveyed to you the impression that a proper evaluation of the long neglected traumatogenesis promises to be fruitful, not only for practical therapy but for the theory of our science.

---

* [The reference is to "Der Menstruationskomplex" (1928), which has not been translated. Two articles by this author dealing with the psychology of menstruation have appeared in English; see Daly 1935 and 1943.]

V

In a conversation with Anna Freud in which we discussed certain points in my technique she made the following pregnant remark: 'You really treat your patients as I treat the children whom I analyse.' I had to admit that she was right, and I would remind you that in my most recent publication, a short paper on the psychology of unwanted children who later become subjects for analysis, I stated that the real analysis of resistances must be prefaced by a kind of comforting preparatory treatment.* The relaxation-technique which I am suggesting to you assuredly obliterates even more completely the distinction between the analysis of children and that of adults—a distinction hitherto too sharply drawn. In making the two types of treatment more like one another I was undoubtedly influenced by what I saw of the work of Georg Groddeck, the courageous champion of the psycho-analysis of organic diseases, whom I consulted about an organic illness.† I felt that he was right in trying to encourage his patients to a childlike *naïveté*, and I saw the success thus achieved. But, for my own part, I have remained faithful to the well-tried analytical method of frustration as well, and I try to attain my aim by the tactful and understanding application of *both* forms of technique.

Now let me try to give a reassuring answer to the probable objections to these tactics. What motive will patients have to turn away from analysis to the hard reality of life if they can enjoy with the analyst the irresponsible freedom of childhood in a measure which is assuredly denied them in actuality? My answer is that even in analysis by the method of relaxation, as in child-analysis, conditions are such that performance does not outrun discretion. However great the relaxation, the analysis will not gratify the patient's actively aggressive and sexual wishes or many of their other exaggerated demands. There will be abundant opportunity to learn renunciation and adaptation. Our friendly and benevolent attitude may indeed satisfy that childlike part of the personality which hungers for tenderness, but not the part which has succeeded in escaping from the inhibitions in its development and becoming adult. For it is no mere poetic licence to compare the mind of the neurotic to a double malformation, something like the so-called *teratoma*‡ which harbours in a hidden part of its body fragments of a twin-being which

* [The reference is to Ferenczi's important paper. "The Unwelcome Child and the Death Instinct" (1929).]

† [Groddeck, the author of the *Book of the It*, believed that even organic illnesses have a psychological basis; see Grossman and Grossman 1965.]

‡ [Teratoma is a tumor composed of multiple tissues including tissues not usually found in the organ in which the tumor arises including the three embryonic germ layers. They are most commonly found in the ovary where they are usually benign and in the testes where they are usually malignant.]

has never developed. No reasonable person would refuse to surrender such a *teratoma* to the surgeon's knife, if the existence of the whole individual were threatened.

Another discovery that I made was that repressed hate often operates more strongly in the direction of fixation and arrest than openly confessed tenderness. I think I have never had this point more clearly put than by a patient whose confidence, after nearly two years of hard struggle with resistance, I won by the method of indulgence. 'Now I like you and now I can let you go,' was her first spontaneous remark on the emergence of a positive affective attitude towards me. I believe it was in analysis of the same patient that I was able to prove that relaxation lends itself particularly well to the conversion of the repetition-tendency into recollection. So long as she identified me with her hard-hearted parents, she incessantly repeated the reactions of defiance. But when I deprived her of all occasion for this attitude, she began to discriminate the present from the past and, after some hysterical outbreaks of emotions, to remember the psychic shocks of her childhood. *We see then that, while the similarity of the analytical to the infantile situation impels patients to repetition, the contrast between the two encourages recollection.**

I am of course conscious that this twofold method of frustration and indulgence requires from the analyst himself an even greater control than before of counter-transference and counter-resistance. It is no uncommon thing for even those teachers and parents who take their task seriously to be led by imperfectly mastered instincts into excess in either direction. Nothing is easier than to use the principle of frustration in one's relation with patients and children as a cloak for indulgence in one's own unconfessed sadistic inclinations. On the other hand, exaggerated forms and quantities of tenderness may subserve one's own, possibly unconscious, libidinal tendencies, rather than the ultimate good of the individual in one's care. These new and difficult conditions are an even stronger argument in support of the view I have often and urgently put forward, namely, that it is essential for the analyst himself to go through an analysis reaching to the very deepest depths and putting him into control of his own character-traits.

I can picture cases of neurosis—in fact I have often met with them—in which (possibly as a result of unusually profound traumas in infancy) the greater part of the personality becomes, as it were, a *teratoma*, the task of adaptation to reality being shouldered by the fragment of personality which has been spared. Such persons have actually remained almost entirely at the child-level, and for them the usual methods of analytical therapy are not

* [Ferenczi here emerges as the real originator of Alexander's corrective emotional experience; see Alexander's biographical sketch, pp. 51–52.

Now Ferenczi no longer maintains as he did in his collaborative work with Rank (1924) that reliving rather than remembering is all important.]

enough. *What such neurotics need is really to be adopted and to partake for the first time in their lives of the advantages of a normal nursery.* Possibly the analytic in-patient treatment recommended by Simmel* might be developed with special reference to these cases.

If even part of the relaxation-technique and the findings of neocatharsis should prove correct, it would mean that we should substantially enlarge our theoretical knowledge and the scope of our practical work. Modern psychoanalysis, by dint of laborious effort, can restore the interrupted harmony and adjust the abnormal distribution of energy amongst the intrapsychic forces, thus increasing the patient's capacity for achievement. But these forces are but the representatives of the *conflict originally waged between the individual and the outside world.* After reconstructing the evolution of the id, the ego, and super-ego many patients repeat in the neocathartic experience the primal battle with reality, and it may be that the transformation of this last repetition into recollection may provide a yet firmer basis for the subject's future existence. His situation may be compared with that of the playwright whom pressure of public opinion forces to convert the tragedy he has planned into a drama with a 'happy ending'. With this expression of optimism I will conclude.

* [The reference is to Tegelsee, Simmel's experimental hospital, conducted on psychoanalytic principles, near Berlin, which opened that year.]

CHAPTER 22

HERMAN NUNBERG

THE SYNTHETIC
FUNCTION OF THE EGO

INTRODUCTORY NOTE

*Few papers have been as frequently quoted as and have exerted an influence
comparable to "The Synthetic Function of the Ego." Nunberg's name is usually associated with this term. However he was not the originator of this
concept. Seven years before the publication of this essay Freud (1923b, p.
45) postulated that the ego uses desexualized—that is, sublimated—libido for
the task of uniting and binding. He suggested that the tendency to bind is
particularly characteristic of the ego. A year later (1924b, p. 167) he
reiterated that the function of the ego is to reconcile the claims of the three
psychic institutions. In 1926, Freud said:*

> *The ego is an organization. It is based on the maintenance of free intercourse
> and the possibility of reciprocal influence between all its parts. Its desexualized
> energy still shows traces of its origin in its impulsion to bind together and
> unify, and this necessity to synthesize grows stronger in proportion as the
> strength of the ego increases. It is therefore only natural that the ego should
> try to prevent symptoms from remaining isolated and alien by using every
> possible method to bind them to itself in one way or another, and to incor-
> porate them into its organization by means of those bonds. As we know, a*

Originally entitled "Die synthetische Funktion des Ich," *Internationale Zeitschrift
für Psychoanalyse* 16 (1930):301–318. English translation in *International Journal of
Psychoanalysis* 12 (1931):123–140. Reprinted in *Practice and Theory of Psychoanalysis*,
Nervous and Mental Disease Monographs, No. 74 (New York: Coolidge Foundation,
1948). Reissued by International Universities Press, 1961. Reprinted with permission. The
paper reprinted here (with some omissions) is based upon a paper read at the Eleventh
International Psychoanalytic Congress, Oxford, July 1929.

tendency of this kind is already operative in the very act of forming a symptom. A classical instance of this are those hysterical symptoms which have been shown to be a compromise between the need for satisfaction and the need for punishment. Such symptoms participate in the ego from the very beginning, since they fulfil a requirement of the super-ego, while on the other hand they represent positions occupied by the repressed and points at which an irruption has been made by it into the ego-organization (Freud 1926a, pp. 98–99).

Nunberg, however, goes beyond Freud and postulates the need for causality as a fundamental need of men. Science satisfies this need, but so does mythology, and the systematized delusion in paranoia can also be attributed to the synthetic functions of the ego.

With maturation, as the quote from Freud shows, the ego's capacities for synthesis grow. This need not always be an advantage. Children can tolerate contradictions better than adults. Pressed by the need for synthesis, the adult is driven to repress and thus run the risk of developing a neurosis, whereas the child, whose ego demands are less stringent, can remain free of neurosis (Blos, 1972).

ACCORDING to the hypothesis of Freud the ego is a part of the id, the surface part, which has undergone modification. In the id there are accumulated various trends which, when directed towards objects in the outside world, lead to a union between these and the subject, thereby bringing into existence a new living being. These libidinal trends are ascribed by us to Eros, in the Freudian sense of the term. Our daily experience teaches us that in the ego also there resides a force that similarly binds and unites, although it is of a somewhat different nature. For its task is to act as an intermediary between the inner and the outer worlds and to adjust the opposing elements within the personality. It achieves a certain agreement between the trends of the id and those of the ego, an agreement producing a harmonious cooperation of all the psychic forces.

The period when the psychic harmony is most complete is probably that of earliest infancy, when the id's every impulse finds direct fulfilment in the ego (ideal-ego).* This state must very soon suffer various disturbances, most likely when there first arises a tension due to some craving and when gratification does not ensue.

Later on the psychic harmony, which probably corresponds to the condition within the "ideal-ego", is disturbed by the development of the superego and the subject's increasing adaptation to reality. The more power these two factors acquire over the ego, the more energetically do they oppose that

* [The term "ideal ego" should not be confused with the "ego ideal." Nunberg's "ideal ego" is ideal only from the point of view of the id. It employs its energy to fulfill the wishes of the id.]

tendency which has hitherto held undisputed sway over it—the tendency, namely, to translate into instant action and to contrive gratification for every instinctual claim made by the id. Only selected ego-syntonic strivings are admitted to gratification: the rest are repudiated. If the ego were to indulge all the trends in the id, it would come into conflict either with reality or with the superego. If, on the other hand, it complied with all the requirements of the superego, or of reality, it would encounter vigorous resistance on the part of the id. The antagonism of the separate psychic institutions is, however, not permanent: as a rule a balance is struck between the opposing forces.

An unmistakable effort is made to cancel the differentiation between ego and id and to reunite and fuse the diverging psychic forces. The tendency constantly to bring about a reunion between the ego and the id or to preserve their unity never wholly dies out, though in individual cases it may suffer disturbance. In this self-sufficient unity the id finds in the ego the gratification of its narcissism.*

Up to the point at which the superego is established, the ego's task is a simple one: it has only to act as an intermediary between the inner and the outer world, between the id and reality. But once the superego is fully developed, the task becomes more complicated, for the ego is called into action on several fronts at once. (1) It reconciles the conflicting elements in the autonomous instincts within the id and allies them one with another so that there is unanimity of feeling, action and will. (The ego tolerates no contradiction.) (2) It brings the instinctual trends of the id into harmony with the requirements of reality. (3) It strikes a balance between the claims of the superego and of reality on the one hand and of the id on the other.

Ultimately, then, the harmonious cooperation of all the psychic forces is restored. The introduction of the superego does indeed make this cooperation more complicated, but it by no means destroys the psychic harmony.

In everyday life we readily recognize this part played by the ego as an intermediary and a binding force. It develops slowly and has many aspects. Its earliest and clearest manifestation is in the oedipus constellation, when the superego is being formed. The ego's method of defending itself from the dangers of the oedipus situation is that of assimilating (ideationally) the id's objects and also the instinctual trends relating to them; this it does by identification. The objects thus assimilated into the ego are held together by the bond of feelings and affects corresponding to these instinctual trends. Hence, through the process of identification, certain instincts and objects which are not consonant with the ego are not merely warded off; they are also united, modified, fused, divested of their specific element of danger and transformed

---

* [Nunberg assumes that the id has at its disposal a narcissism of its own that finds gratification in the ego. Today (following Hartmann 1950), one would define narcissism as the libidinal cathexis of the whole self.]

into a new psychic creation—the superego.* We see then that this new creation is a product of the ego and arises out of the assimilation of insupportable inner and outer stimuli. It is in this process of assimilation that we have the first and plainest manifestation of the ego's influence as an intermediary and binding force, that is, of its synthetic function. But the ego's capacity for synthesis manifests itself, during the formation of the superego, not only in its mediation between the inner and the outer world and its assimilation of the two, but also in the manner in which it unites, modifies and fuses the separate psychic elements within itself. The synthetic capacity of the ego manifests itself, then, as follows: it assimilates alien elements (both from within and from without), and it mediates between opposing elements and even reconciles opposites and sets mental productivity in train.

We cannot make any final pronouncement or any very far-reaching conjectures about the innermost nature of ego-synthesis. But even a superficial survey reveals a clear analogy to the id, to those of its components which strive to unite and to bind—in short, to Eros. And, since the ego is derived from the id, it is probably from this very source (Eros) that it acquires its binding and productive power.

The function of passing negative judgments, is, in Freud's [1923b, p. 239] view, derived from the instincts of destruction. Consequently we need have no hesitation in tracing to the libidinal instincts the passing of affirmative judgments. I hold that these instincts also constitute one of the roots of causal thinking. In this connection, however, it is not causal thinking per se which interests us, but the motive which impels us to such thinking—in a word: the need for causality.

Before discussing this in detail I want to lay special stress on the fact that by no means is it my view that the need for causality is derived exclusively from libidinal tendencies. Assuredly other factors come into play, such as, for instance, the mastering of reality. I am not here concerned with the principle of causality but with the need for causality in general.

In the structure of certain paranoid delusional systems we are impressed by the fact that when once an idea or sensation has emerged into consciousness, the subject clings to it and endeavors to rationalize it and to establish its causal relations. The seeking and finding of such relations, i.e., the connecting of two facts in such a way that the second is shown to be conditioned by the first, is certainly a preconscious psychic activity, which constantly manifests itself in everyday life, quite apart from schizophrenia. It is called "rationalization" when the causal relation discovered is a fictitious one, giving an illusion

* [Whereas Nunberg assumes that the ego creates its own superego, it is now customary to conceptualize the development of the ego, superego, and id out of an undifferentiated matrix (Hartmann 1950).]

of fact where none exists. In the genesis of delusions rationalization seems to play the same part as that played in other preconscious thought-processes by secondary elaboration, which reconciles such antitheses as are too abrupt and fills up gaps in our thinking. We encounter it again where the ego has not wholly succeeded in distorting the unconscious processes by repression. Where repression has been most unsuccessful, as for instance, in schizophrenic disturbances in which the repressed material manages to gain direct access to consciousness, rationalization is most plainly in evidence. Here things that are wholly irrelevant to one another are quite uncritically brought into causal relation because, as I have already said, the ego has a tendency to unify and to connect, and is obviously no longer able to tolerate excessive contradictions. But if the ego has undergone complete disintegration as a rule it no longer even essays the secondary elaboration of unconscious wishes and fantasies, or else, if the attempt is made, it is so clumsy that the meaning of the unconscious material is scarcely distorted at all and can be immediately recognized.

So the business of rationalization, as the psychic processes go on, is to establish a causal connection between certain antagonistic elements in our thinking, to fill up the gap in it and so to give it the semblance of being subject to necessity and reason.

The fact of rationalization seems to me less mysterious than the passionate eagerness with which all men (even the most primitive) seek after a "first cause", that is, the apparently primary need for causality. The more a man lacks the critical faculty the easier is it for him to discover a causal basis for his actions and thoughts. We can see this clearly when we observe children. In most forms of schizophrenia this need is so imperative that we can almost regard it as typical, and surely certain obsessional neuroses, in which the patient is compelled constantly to ask questions, are based on this same need.

We do not know what is the origin of the need for causality. But when we recollect that in children the compulsion to ask questions is derived from their desire to inquire into sexual matters, the path we must follow in order to reach some understanding of this problem seems clearly indicated.

One special form that the child's obsessional questioning takes, the inquiry about the origin of things, really refers to the genesis of human beings. The child's craving to know, manifested in this question, is the psychic representative of the infantile sexual instinct, and in particular of the instinct of reproduction which biologically does not come to maturity until puberty. Now these two instincts are united in Eros, which according to Freud represents the sum of all the life-instincts, whose aim it is to bind together and unite two separate living beings in order that from them there may proceed a new living being. The compulsion which man is under to inquire into the first cause of the world of phenomena—the need for causality—is accordingly the sublimated expression of the reproductive instinct of Eros. That which in the id appears as a tendency to unite and bind together two living beings manifests itself in the ego also as a tendency to unite and to bind—not objects, however, but thoughts, ideas and experiences. Thus, in the need for causality the binding (synthetic) tendency of Eros reveals itself in a sublimated form in the ego. It would seem that this need represents a very important principle—that of connection—in the psychic realm as a whole.

That traces of primary instinctual attitudes still lurk behind this need is evident in our everyday modes of speech. In the common parlance of the present day the cause of a phenomenon is often personified, and events are endowed with a causal significance which they can assume only by virtue of identification with mankind. To give only one of countless examples, we speak of "fruitful rain", though the rain in itself has no fertilizing property but can only, by means of the soil, produce certain changes in plant-life, so that the already fertilized plants are stimulated to a more abundant growth. If we examine the causal links in a schizophrenic's chain of thought we nearly always come upon personified (animistic) causes, and as a rule the last link is a rationalized explanation of the genesis of the world and of human being. The ego assumes a function of the id—namely, of Eros—and this function loses merely its sexual tinge through the transition from the one psychic system to the other.

Every piece of research has, after all, a practical purpose, and this fact is especially obvious in empirical science. The object of research is to fathom the laws of nature and so make man independent of them and able on his own account to create something new. The same principle, I think, governs creative art. One of my patients began, while she was having treatment, to study drawing and painting. She described her work as an artist as follows: "When I copy a subject (a living model), I feel as if with every line and stroke I were appropriating part of him by touch and that I do not really get to know and understand his nature until I absorb it into myself. When I have done that he belongs to me. Then, if I want to draw him from memory, I am completely independent of him as an actual object: I can reproduce that from within myself as often as I like." This artist learns to know another being by allowing her ego to absorb and assimilate it. The (originally alien) object then belongs to and is united with herself. Then, and only then, can she do creative work and develop from within herself the capacity for production. Her analysis showed that her artistic creative power, which ventured to reveal itself only after she began treatment, was motivated by unconscious fantasies, the content of which was the incorporation of her father's genital and the subsequent bearing of a child.

Scientific, artistic or social work is the ego's extension in a sublimated form of the reproductive efforts of the id, or, as we may say, of the creative faculty of Eros.

The ego's tendency to unite, to bind and to create, goes hand in hand with a tendency to simplify and to generalize. Once more it is in schizophrenics that this is most evident. In this disease, and especially in its paranoid forms, wholly irrelevant thoughts are connected, and events and experiences that are directly contradictory, are intermingled. We see clearly revealed the tendency to crystallize, out of the chaos in which these patients find themselves, a conception of life that shall have unity and that shall contain no contradiction. They think that by generalizing they can assemble under one roof, as it were, not only all their contradictory delusions but the experiences belonging to the inner and the outer worlds, and construct a novel "philosophy of life" adapted to themselves alone. Actually, it is the philosophy of a mind deranged—out of range of reality. As a rule, it turns out to be a cosmology, based on the problem of the genesis of man. This tendency of the ego to simplify and to generalize is yet another manifestation of its synthetic function. We can see at once that this function is subject to an economic principle, in deference to which the ego economizes expenditure of labor. For instance, when it fuses two opposites into a unity, it performs only one piece of work—that of fusion—instead of pouring out energy in various directions in a conflict of ambivalence.

Thus the synthetic function of the ego does not merely unify the whole personality; it simplifies and brings economy into the ego's mode of operation. Children and primitive men have not yet developed a unified ego. They are able to harbor contradictions not only of thought, but also of feeling and action. With further development the ego becomes more unified in its aims and endeavors; with the total disintegration of the personality the ego's synthetic function fails altogether. When, on the other hand, its stability is most gravely threatened but it yet retains a certain measure of constructive energy (as in the paranoid forms of schizophrenia), its synthetic operations are immeasurably extended: whatever has any access to it is indiscriminately connected and fused, the result being new psychic productions of the most bizarre character. This is an exaggeration and a distortion of the ego's synthetic function (cf. many philosophical systems). It seems that in conditions such as that of schizophrenic disintegration, where Eros is most seriously menaced by the loss of object-libido, the ego makes the greatest efforts at synthesis.

By virtue of its primary propensity for mediation and combination the ego has to undertake as one of its principal tasks the solution of conflicts between the different parts of the personality. The solution will take whatever form the ego finds most satisfactory for the adjustment of opposing factors; it may be sublimation, or change in character, or neurosis. If the intermediary fails altogether in its function, the result will be a psychosis, the subject either passing into a state of mental hebetude or falling prey to his uncontrolled instincts.

Like so much in the psychic realm, sublimation is the result of a struggle within the ego. The solution of the oedipus complex in childhood already implies sublimation, for with the formation of the superego the earliest moral (social) sense is developed. As an illustration of this I may quote the choice of a profession made by a certain woman. It is a good example because as a general rule the process by which sublimation has been effected is clearest when it has been difficult. This woman's father, himself a physician, was strongly opposed to her studying medicine, but at last she overcame his objections and was allowed to begin. At the end of her first year as a student she married. Although she was most enthusiastic about her medical work and had passed her examination with distinction, she now wanted to give up her training. But before this she had tried to persuade her husband, who was a student in another faculty, to take up medicine instead. Her motive for abandoning the profession she had chosen of her own free will is to be found in the following train of thought: If her husband were a physician, she would not have to become one too, because she felt that if he adopted her father's profession her relations with her father might still be vicariously preserved. The lines of her sublimation were those usually followed when the superego is set up: (1) the libidinal object and the instincts focussed on it were absorbed into the ego, and (2) by the processes of identification and of deflection of the instinct from its direct aim, the object was renounced. In the ego object and instinct were united, the former as a (now unconscious) idea, the latter as an instinct deflected from its primary aim, i.e., as psychic energy, desexualized libido, which in this patient found an outlet in study and later in professional work. When she found her object and could give direct expression in life to the instinct, she was ready to cancel the sublimation.*†

The ego then redirects libido towards the symptom, which it unites with itself and incorporates anew in its own organization. The symptom becomes once more an integral part of the ego, which derives pleasure from the union. In the struggle over the symptom it becomes impoverished in libido. Through

---

* [This example illustrates how difficult it is to formulate theory on the basis of a case presentation. For example, one could argue that the wish to be a physician represented for this analysand an identification with the father, but also a sublimation of her masculine wishes which entered into conflict with her feminine desires when she fell in love. She could therefore not be a physician and wife at the same time. Transforming the husband into a physician was an attempt to solve this problem (see Bergmann 1971).]

† [Section omitted.]

symbiosis* with the symptom it obtains narcissistic gratification or else escapes a narcissistic wound, and thus the deficit in the libidinal economy of the ego is made up. To give an illustration: a patient suffered from agoraphobia. As time went on, the anxiety reached such a pitch that he could no longer go to his place of business. He became incapable of work and his wife had to earn enough to support him. The symptom existed first and the incapacity for work came later, but he reconciled himself to the symptom by seeing in his wife's maintaining him a proof of her love—a sacrifice which she made because she loved him—and this afforded him narcissistic gratification. In obsessional neurosis and melancholia the patients render their illness tolerable by making it a source of such gratification. They feel, for instance, that they are particularly moral people; the paranoiac is specially proud of his intellectual achievement in the formation of his complicated delusions, and so forth.

It follows that in this narcissistic gratification the patient has yet another motive for clinging to the disease which has already developed. The greater his narcissism the more tenacious will be his hold. Perhaps this is one of the main reasons why the schizophrenic's delusions are so inaccessible to any influence from without. We see then that with the narcissistic gratification derived from the symptom—the epinosic gain†—the subject also attains a disguised instinctual gratification within the ego. The epinosic gain reinforces the ego's narcissism as follows: (a) in hysteria by establishing an object-relation, of whatever sort (in phobias by excessive attachment to an object), (b) in obsessional neurosis by the gratification of the superego, (c) in schizophrenia by over-emphasis on some of the functions of the ego (e.g., in paranoia that of thinking, and so forth).

As I have already said, the synthetic function of the ego does not manifest itself for the first time in the epinosic gain. It appears still earlier. In conversion-hysteria it displays itself in the character of the symptoms themselves, which is that of a compromise. We see it most perfectly, however, in those cases of obsessional neurosis in which the patient succeeds in intermingling prohibition and gratification and thus in unifying the striving of the id and the demands of the superego. To cite a single example: One of my patients had the compulsion, every time he urinated, to wipe his penis (following a special ceremonial) until finally erection took place.

In the epinosic gain the synthetic work of the ego is reinforced by a process of compensation. When the disease sets in, it is the synthetic faculty which gives the impetus to symptom-formation and which so governs the course of the illness that in extreme cases it enables the repressed instincts to

---

* [Nunberg uses the term "symbiosis" in its original meaning, not in the way in which it has become familiar through the work of Mahler (1968).]

† [An epinosic gain is a secondary gain. It is one of the five resistances enumerated by Freud in 1926.]

break through (obsessional neurosis). Once the symptom has been established, the synthetic faculty as a rule ceases to function, but indirectly, in the epinosic gain, it comes into action again and makes up the deficit of libido. The neurotic conflict on the one hand and, on the other, the menace of disintegration of the ego spur it on to increased efforts at synthesis. Thus it is very rare for the synthetic tendency to perish; only it is specially striking and assumes the most bizarre forms where the threat to Eros is most serious, as we have seen in the instance of schizophrenia. In grave psychotic conditions (confusional states and catatonic stupor) it is put out of action. In neurosis it merely undergoes a disturbance, either manifesting itself where normally it would not do so or striking out in wrong directions.

In all forms of defense, most strikingly in repression, some psychic act becomes unconscious. From the manner in which repression is lifted (in analytic treatment) and the unconscious material is restored to consciousness we can draw certain conclusions about the way psychic acts in general enter consciousness.

In the light of what has already been said about psychic productivity we shall view the final act of conscious thinking, the comprehension of general relations, the forming of concept, etc., as a synthetic act. In neurosis the process of thought is partially disturbed; owing to the breaking of the communication between the unconscious and the conscious—the work of repression—the act of thought cannot take place when repressed material is touched upon. In this connection I refer to the prohibition of thinking and the occasional suspension of thought in some obsessional neuroses, phenomena which occur as a direct extension of the prohibition of onanism and the struggle to break off the habit. The result of repression is that the subject has no immediate perception of his unconscious mental processes; as we know, in hysteria significant thoughts or memories of important experiences are forgotten. On the other hand, it is well known that in obsessional neurosis amnesias play a lesser part, for in them pathogenic processes can be recollected. In schizophrenics there is generally immediate "consciousness" of inner processes that in neurotics can be brought into consciousness only by laborious psychoanalytic work. In obsessional neurosis there is no connection between the various pathogenic complexes of ideas and thoughts; in schizophrenia there is lacking, in addition, a connection between these complexes and the personality as a whole. It is true that schizophrenics and obsessional neurotics have perception of their ideas and thoughts, but these do not possess the "quality of consciousness". Thus, perception by itself does not constitute consciousness of a psychic process. If we succeed in getting rid of the amnesias in hysteria and connecting the pathogenic experiences in obsessional neurosis, they become conscious. Repression involves a breaking of the connection between the psychic systems; when the repression is removed this connection is reestablished and the capacity to connect is reborn. For running parallel to the process of recollection in analysis is the discovery of connection—the uniting and reconciling of the repressed ideational elements and the actual ego: the assimilation of the repressed. In all this, analysis simply utilizes an already existing tendency in the ego: the need for causality; the wish is aroused in patients to find out the 'cause' of their illness. We have derived the need for causality from Eros, that striving of the id which expresses itself most strikingly in the ego's tendency to unite and connect. Influenced by this tendency the patient discovers intimate relations (of which he has hitherto been unaware) between different experiences, memories, thoughts and fantasies, which he first connects with one another and then with the actual ego. Now the thoughts, experiences, impulses and fantasies of neurotics are detached from the ego, to which, by the process of defense, they have become alien. Hence, if we get rid of the amnesias, relate the memories, experiences and thoughts to one another, convert the latent dream-thoughts into the manifest dream-content and so forth, what we do is to connect and reconcile that which is alien to the ego—that which has been ejected from the ego-organization—and the intact ego. In short, we are effecting a synthesis.

As I have already said, it is not sufficient for the act of consciousness that a connection should be established between the systems Ucs. and Pcs., i.e., between the ideas of objects and of words representing them. So long as the system Cs. remains unreceptive to preconscious material, the process continues to be below the threshold of consciousness. We can readily observe that, as repression is progressively broken down during analytic treatment, the perceptual ego becomes accessible to the preconscious derivatives of the unconscious. The system Pcpt.-Cs. apparently becomes hypercathected, and it is through this alone that free intercourse between all three systems becomes possible, that is to say, connection, union, reconciliation and adjustment of opposites can take place amongst the psychic trends themselves and between them and the ego. The real becoming conscious is, as we see, the final act in a very complicated process; it is at the same time a manifestation of the synthetic function of the ego. Let me once more lay stress on the fact that where that function is specially called into play, a process takes place which is exactly opposed to that of repression. For repression depends on the ego's synthetic capacities being temporarily inadequate. Ultimately, then, the process of cure becomes a process of assimilation of those psychic trends which the defense-mechanisms have rendered alien to the ego and in this way it seems to ensure the continuity of the personality.

Generally speaking, the process that takes place in psychoanalytic treatment of the neuroses is something similar to that which occurs in the spontaneous attempts at recovery made by patients suffering from the different forms of schizophrenia. Here the most heterogeneous psychic elements are combined and often fused with impressions from the outside world to form new structures as, for instance, in delusional systems. Of course the synthesis in psychosis is applied to quite other material than that upon which it is employed in neurosis—material which is either wholly unrelated, or in very loose relation, to the true unconscious ideas of objects. For in neurosis the act of becoming conscious is preceded by the connecting of the preconscious ideas of words with the unconscious idea of the things which they represent. In schizophrenia this connection is lacking: the preconscious ideas of words are subject to elaboration by that primary process which dominates the unconscious in general, and they then receive a hypercathexis of psychic energy from the ego and are united with it. For the ego they play the part of actual things, although they leave no material substratum. Nevertheless, this phase of the illness represents an attempt to regain the lost world of reality—thus, an attempt at cure (Freud). And in actual fact the reconstruction of this world, though it be but in fantasy, corresponds to a spontaneous cure. This "cure" takes place under the influence not only of the direct libidinal strivings of the id after the lost objects, but also under that of the ego in its striving after synthesis. In the psychoanalytic cure of neuroses, too, we have seen the synthetic process at work. Thus cooperation with the analyst achieves what takes place spontaneously in psychosis.

Whatever the details of the process may be there is no doubt that in the final phase of cure of the neuroses there is a fresh manifestation of the power of Eros, whose derivatives even in the desexualized libido of the ego carry on their work of mediation and union.

It is probable that the other psychotherapeutic methods, including those which call themselves "psycho-analytic" without being so, attack the illness from this point also. But the essential difference between all these methods and our own is that in the former, patients have to assimilate something forced upon them from without, whereas in psychoanalysis by a process of painful self-mastery they have to admit into their ego and unite with it that which is a fundamental part of their own nature. This is probably one reason why many patients may be cured even though their analysis is not fully completed, provided that they accept and acknowledge as their own, repressed material whose existence it has been possible to infer in their analysis, although it has not actually been remembered.*

I think that the most important point to realize is that but in the very gravest cases of psychoses the synthetic faculty of the ego does not altogether cease to function: It merely goes off on false tracks. Analysis brings it back on to the right lines by enabling the ego to harmonize the strivings of the id with the requirement of the superego on the one hand and, on the other, with reality (the objects in the outside world). In other words, at the end of the

---

* [Nunberg here asserts his belief in the therapeutic values of constructions; see editors' introduction, p. 24. For a more recent view of the role of childhood memories, see Kris 1956b.]

treatment the ego-syntonic impulses are admitted to action and to conscious-ness, while those which are not consonant with the ego are restrained by it, transmuted, and not utilized in the production of neurotic symptoms but displaced into the realm of intellectual productivity—i.e., they are sublimated. At the end of an analysis that has been correctly carried through, the ego's synthesis will have been automatically corrected without the analyst's having consciously aimed at this result. Hence analysis is, properly speaking, a "synthesis".

To sum up: the ego has command not only of destructive tendencies but also of capacities for construction and synthesis, which extend over the whole field of psychic activity and impel man to the harmonious unification of all his strivings and to simplification and productivity in the broadest sense of the word. Under the influence of this synthetic faculty he creates that which is socially valuable, in science, art, etc., as well as that which is of no social value in illness.

CHAPTER 23

EDWARD GLOVER

THE THERAPEUTIC EFFECT OF INEXACT INTERPRETATION

A Contribution to the Theory of Suggestion

INTRODUCTORY NOTE

In this paper, without naming Freud, Glover takes issue with one of Freud's most fundamental ideas, the harmlessness of an incorrect construction. To take a position opposite to Freud's in the field of technique is a rare phenomenon. In 1937, Freud said: ". . . if nothing further develops we may conclude that we have made a mistake and we shall admit as much to the patient at some suitable opportunity without sacrificing any of our authority. Such an opportunity will arise when some new material has come to light which allows us to make a better construction and so to correct our error. In this way, the false construction drops out, as if it had never been made; and, indeed, we often get an impression as though, to borrow the words of Polonius, our bait of falsehood had taken a carp of truth. The danger of our leading a patient astray by suggestion, by persuading him to accept things

Originally published in *International Journal* of *Psycho-Analysis* 12 (1931):397–411. Reprinted in Edward Glover, *The Technique of Psycho-Analysis* (New York: International Universities Press, 1955). Reprinted by permission of Mary W. Crosbie and the Institute of Psycho-Analysis.

which we ourselves believe, but which he ought not to, has certainly been enormously exaggerated" (Freud 1937a, pp. 261–262; see also editor's foot-note, p. 151).

Freud took no cognizance of Glover's argument, although Glover's paper was published six years earlier. The distinction between the inexact and the incomplete interpretation, even though it is not always easy to make, is an important distinction for the psychotherapist to keep in mind. The defensive use that a patient can make of an inexact interpretation goes a long way to explain the successes of many schools of psychotherapy. The concept of inexact interpretation led Glover to see the relationship between psychoanaly-sis and other psychotherapies, particularly those that rely on suggestion in a new light.

The paper is argued with the same sharp logic as Glover's previous one published in this volume (1924, Chapter 9). Here, as in the previous publica-tion, Glover is concerned with the special difficulties posed by the narcissistic neurosis. He is the first to compare the impact of an analysis that interprets primarily the repressed sexual or libidinal wishes of the analysand with the impact of an analysis that stresses predominantly the derivative of the aggres-sive drive. He thus participates indirectly in the controversy with Wilhelm Reich. Of particular significance is Glover's conclusion that any analysis that continuously analyzes only one aspect of a neurotic conflict must inevitably lead to the same consequences as an inexact interpretation.

PSYCHO-ANALYTIC interest in theories of cure is naturally directed for the most part to the curative processes occurring in analytic treatment: the thera-peutic effect of other methods is, nowadays at any rate, more a matter of general psychological interest. In earlier times, of course, it was necessary to pay special attention to the theoretical significance of non-analytic psycho-therapy. Statements were frequently bandied about that psycho-analysis was nothing more than camouflaged suggestion: moreover, the fact that analytic method was based on experiences derived from situations of rapport between physician and patient, for example, in hypnosis, made some theoretical differ-entiation desirable. Most discussions of the 'resolution of transference' can be regarded as contributions to this problem, affording a rough but serviceable distinction between analytic and other therapeutic methods. And the special studies of Freud on group psychology (1921), Ferenczi on transference (1909), Ernest Jones on suggestion and auto-suggestion (1923), Abraham on Couéism (1926) and an unfinished study by Radó on the processes of cure (1925), have given a broader theoretical basis to this differentiation.

Nevertheless we are periodically stimulated to reconsider the relations between different forms of psychotherapy, more particularly when any ad-vance is made in analytic knowledge. When such advances occur we are bound to ask ourselves, 'what happened to our cases before we were in a position to turn this fresh knowledge to advantage?' Admittedly we would not be under this obligation had we not previously used terms such as 'cure', 'thorough analysis', etc., etc. But for many years now we have been in the habit of speaking in such terms and therefore cannot avoid this periodic searching of heart.

One possible answer is that the additional information does not affect therapeutic procedure at all; that, like M. Jourdain, we have been talking 'prose' all the time. This certainly applies to a great deal of recent work on super-ego analysis, anxiety and guilt. It is true we have been able to subdivide resistances into super-ego resistances, ego resistances and Id resistance. But we always endeavoured to reduce such resistances, even when we had no special labels to attach to them. On the other hand when we consider the actual content of repression, it is clear that the discovery of fresh phantasy systems set us a problem in the theory of healing. It might be stated as follows: what is the effect of inexact as compared with apparently exact interpretation? If we agree that accuracy of interpretation amongst other factors contributes towards a cure, and if we agree that fresh phantasy systems are discovered from time to time, what are we to make of the cures that were effected before these systems were discovered?

An obvious difficulty in dealing with this problem is the fact that we have no adequate and binding definitions of terms. Take for example standards of 'cure'. It may be that the standards have varied: that in former times the criterion was more exclusively a symptomatic one: that as our knowledge has increased our standards of cure have become higher or broader or more exacting. For example the application of analysis to character processes has certainly increased the stringency of therapeutic standards: whether it has given rise to fantastic criteria remains to be seen. In any case it is generally agreed that a distinction between analytic and non-analytic therapeutic processes cannot be solely or immediately established by reference to symptomatic changes.

Then as to the significance of phantasy systems, it might be suggested that presentation content is not in itself primarily pathogenic: that the history of the affect only is important in illness, hence that the value of fresh discoveries of phantasy content lies solely in providing more convenient or rapid access to affective reactions. The objection to this view is that it leaves the door open to complete interpretative distortion or glossing over of repressed content: moreover it would deprive us of a valuable distinction between psycho-analytical interpretation and pseudo-analytical suggestion.

Incidentally a somewhat cynical view would hint that fresh discoveries are not necessarily or invariably accurate, or indeed fresh. One is bound to recall here the rapidity with which some analysts were able to discover 'birth traumas' in all their patients for some time after Rank first published his book on the *Trauma of Birth*, and before his theory was officially exploded [1924; see also this volume, Chapter 6]. A less cynical view is that many new phantasy systems or elaborations of known systems are mainly repetitive in nature; repeating some central interest in varying idiom, the idiom being determined by stages of libido development and ego reaction. According to this view repetitions assist displacement and are therefore protective: the greater the number of systems we discover the more effectively we can prevent defensive displacement. We could then say that in the old days affective disturbances were worked through under a handicap (viz.: lack of knowledge of the variations of phantasy), but that they were nevertheless worked out.

The next view has some resemblances to the last but brings us closer to an impasse. It is that pathogenic disturbances are bound by fixation and repression to certain specific systems, but that these can be lightened by regression (displacements backwards) to earlier non-specific systems (Rückphantasieren) or again by distribution, i.e., forward displacement to later and more complicated systems of phantasy. Even then we could say that legitimate cures were effected in former times although under a handicap. But if anyone cared to claim that particular neuroses were defences against a specific set of unconscious phantasies, related to a specific stage of fixation and that unless these were directly released from repression no complete cure could be expected, we would be compelled to consider very carefully how cure came about in the days before these phantasies were discovered.

Obviously if such a claim were made, the first step in investigation would be to estimate the part played in previous cures by repression. This is always the unknown quantity in analyses. It does not require any close consideration to see that the rapid disappearance of symptoms which one occasionally observes in the opening phase of an analysis (e.g. in the first two or three months) is due partly to transference factors, but in the main to an increase in the effectiveness of repression. This efficiency reaches its height at one of two points; first when the amount of free anxiety or guilt has been reduced,* and second when the transference-neurosis threatens to bring out deep anxiety or guilt together with their covering layer of repressed hate. One is apt to forget, however, that the same factors can operate in a more unobtrusive way and take effect at a much later date in analysis. In this case the gradual disturbance of deep guilt is undoubtedly the exciting cause of increased repression. According to this view cures effected in the absence of knowledge of specific phantasy systems would be due to a general redressing of the balance of conflict by true analytic means, bringing in its train increased effectiveness of repression.

If we accept this view we can afford to neglect the practical significance

---

* [This is an interesting point not usually argued; namely, that reduction in anxiety or guilt need not "free the libido" for analytic work but rather enables the ego to repress better.]

of inexact interpretations. It will be agreed of course that in the hypothetical case we are considering, many of the interpretations would be inexact in that they did not uncover the specific phantasy system, although they might have uncovered systems of a related type with some symbolic content in common. Nevertheless, we are scarcely justified in neglecting the theoretical significance of inexact interpretations. After all, if we remember that neuroses are spontaneous attempts at self-healing, it seems probable that the mental apparatus turns at any rate some inexact interpretations to advantage, in the sense of substitution products. If we study the element of displacement as illustrated in phobias and obsessions, we are justified in describing the state of affairs by saying that the patient unconsciously formulates and consciously lives up to an inexact interpretation of the source of anxiety. It seems plausible, therefore, that another factor is operative in the cure of cases where specific phantasy systems are unknown; viz. that the patient seizes upon the inexact interpretation and converts it into a displacement-substitute. This substitute is not by any means so glaringly inappropriate as the one he has chosen himself during symptom-formation and yet sufficiently remote from the real source of anxiety to assist in fixing charges that have in any case been considerably reduced by other and more accurate analytical work. It used to be said that inexact interpretations do not matter very much, that if they do no good at any rate they do no great damage, that they glide harmlessly off the patient's mind. In a narrow symptomatic sense it does not seem a justifiable assumption. It is probable that there is a type of inexact interpretation which, depending on an optimum degree of psychic remoteness from the true source of anxiety, may bring about improvement in the symptomatic sense at the cost of refractoriness to deeper analysis.* A glaringly inaccurate interpretation is probably without effect unless backed by strong transference authority, but a slightly inexact interpretation may increase our difficulties. Some confirmation of this can be obtained by studying the spontaneous interpretations offered us by patients. These are often extremely accurate in reference to *some* aspect of their phantasy activity, more particularly when the interpretation is truly intuitive, i.e. is not stimulated by intellectual understanding or previous analytic experience. But it will be found that except in psychotic cases, the interpretation offered is not at the moment the true interpretation. Test this by appearing to acquiesce in the patient's view and in nine out of ten cases of neurosis the patient will proceed to treat you with the indifference born of relief from immediate anxiety. The moral is of course that, unless one is sure of one's ground, it is better to remain silent.†

The subject is one that could be expanded indefinitely, but I will con-

---

* [This is the closest Glover comes to alluding to Freud's view on this subject.]

† [For an excellent example of an inexact interpretation and its effect see Kris (1951, p. 23; Schmideberg 1938), where different interpretations of the same behavior were given by two analysts belonging to different schools.]

clude its purely analytic aspect here by giving a brief illustration. If we recall the familiar intrauterine phantasies which have been variously interpreted from being indications of birth traumas to being representations of pre-latency genital incest-wishes; or the phantasies of attacking the father or his penis in the mother's womb or vagina to which special attention was drawn by Abraham; or again the more 'abdominal' womb phantasies to which Melanie Klein has attached a specific meaning and significance, it will be seen that we have ample material to illustrate the problem under discussion. I would add only one comment by way of valuation. *It is that in the absence of definite evidence indicating specific fixation at some stage or another, the more universally such phantasies are found, the greater difficulty we have in establishing their value in any one case.* In other words the greater difficulty we have in establishing the neurotic option. In terms of a recent discussion* of precipitating factors in neurosis, we cannot speak of a specific qualitative factor in a precipitation series of events until by the uncovering of repression we have proved not only that the same factor existed in the predisposing series, but also that it was pathogenic.

Before leaving this aspect of the subject, and in order to prevent mis-understanding, it would be well to establish some distinction between an 'inexact' and an 'incomplete' interpretation. It is obvious that in the course of uncovering a deep layer of repressed phantasy, a great number of preliminary interpretations are made, in many cases indeed cannot be avoided. To take a simple example: it is common experience that in the analysis of unconscious homosexual phantasies built up on an anal organisation, much preliminary work has to be done at a genital level of phantasy. Even when genital anxieties are relieved and some headway has been made with the more primitive organisation, patients can be observed to reanimate their genital anxieties periodically. The anal system has for the moment become too strongly charged. In such a case the preliminary interpretations of genital phantasy would be perfectly accurate and legitimate, but in the pathogenic sense incomplete and indirect. If, however, no attempt were made to uncover anal phantasies and if genital phantasies alone were interpreted, the interpretation would be inexact. If subsequently in the course of analysing anal phantasies genital systems were recathected, and a genital interpretation alone were given, such an interpretation would be not only incomplete but inexact.†

A similar situation arises with sadistic components of an anal-sadistic system. Preliminary interpretation of the anal component would be incomplete: it would not be inexact unless the sadistic element were permanently

---

* "The Significance of Precipitating Factors in Neurotic Disorder," a symposium held by the British Psycho-Analytical Society, May 6, 1931.

† [For a step-by-step discussion of how an incomplete interpretation can be completed, see Loewenstein 1951, p. 4; see also editors' introduction, p. 11.]

neglected. This particular example is worthy of careful consideration: it brings out another point in the comparison of analytic results obtained in recent times with those obtained in earlier years. In the analyses of obsessional neuroses, it can be observed that when sadistic components are causing resistance, the resistance frequently takes the form of an exaggeration of seemingly erotic phantasy and ceremonial. And the patient is only too glad to accept an interpretation in terms of libidinal phantasy. The same applies to the defence of erotic components by a layer of sadistic phantasy. Now the whole trend of modern psycho-analytic therapy is in the direction of interpreting sadistic systems and guilt reactions.* We are bound, therefore, to consider whether some of the earlier symptomatic successes were not due to the fact that by putting the stress on libidinal factors and only slightly on sadistic factors, the patient was freed from anxiety but left with unresolved (repressed) sadistic systems. It would be interesting to compare the earlier results of analysis of transference and narcissistic neuroses respectively with those obtained in recent times. If the view I have presented is valid, one would expect to find that in former times the results in the narcissistic neuroses were comparatively barren, and the symptomatic results in the transference-neuroses more rapid and dramatic. As against this one would expect to find better results from the modern treatment of narcissistic neuroses and less rapid (if ultimately more radical) results in the transference-neuroses. The deep examination of guilt layers might be expected to postpone alleviation in cases where the maladaptation lay more patently in the libidinal organisation.†

One more comment on 'incomplete' interpretation. Apart from the degree of thoroughness in uncovering phantasy, in interpretation is never complete until the immediate defensive reactions following on the interpreta-

*[What Glover designates as the modern trend started with Freud's formulation of the dual instinct theory (1920), and the structural theory, which emphasized the role of the superego and guilt feelings (1923b); see editors' introduction, pp. 34–35.]

† If a companion paper were written 'on the exacerbating effect of inexact interpretation', it would doubtless be concerned mainly with the result of partial interpretation of sadistic phantasy. A common result of disturbing guilt systems without adequate interpretation is that the patient breaks off in a negative transference. Even if his anxiety symptoms have disappeared he may depart with increased inferiority feeling, a sure sign of activated guilt. Short of this dramatic termination, there are many other indications of active resistance following inexact interpretation. During the discussion of this paper, Miss Searl drew attention to a common source of resistance or stagnation during analysis. It is the interpretation of an Id system in terms of a super-ego system or vice versa. This observation is certainly sound. It can be demonstrated experimentally with ease during the analysis of obsessional cases. In the early stages of ceremonial formation the protective or cancelling ('undoing') system is dictated by the super-ego. Sooner or later this is infiltrated with repressed libidinal and sadistic (Id) elements. Continuance of the 'Super-ego' interpretation is then 'inexact' and if persisted in brings the analysis to a standstill.

[As indicated in the introductory statement, this is one of Glover's most significant postulates. Any one-sided line of interpretation becomes, in due course, an inexact interpretation.]

tion are subjected to investigation. The same applies to an interpretation in terms of 'guilt' or 'anxiety'; the latter is incomplete until the phantasy system associated with the particular affect is traced. The tracing process may lead one through a transference repetition to the infantile nucleus or through the infantile nucleus to a transference repetition (Glover 1955).

Turning now to the non-analytical aspect of the problem, there are one or two points worthy of consideration. The psycho-analyst has never called in question the symptomatic alleviation that can be produced by suggestive methods either of the simple transference type or of the pseudo-analytical type, i.e. suggestions based on some degree of interpretative appreciation. He has of course queried the permanence of results or speculated as to the price paid for them in general happiness or adaptability or emotional freedom. But he could not very well question the occurrence of such alleviations; in his own consultative practice the analyst has many occasions of observing the thera-peutic benefit derived from one or more interviews. Even in this brief space he is able to observe the same factors at work which have been described above. Patients get better after consultation either because they have relieved them-selves of trigger charges of anxiety and guilt, or because they have been frightened off unconsciously by the possibility of being analysed or because in the course of consultation the physician has made some fairly accurate ex-planations which are nevertheless sufficiently inexact to meet the patient's need.

Strictly speaking this observation is not an analytical one, but taken in conjunction with the earlier discussion of the effect of inexact interpretation in actual analysis, it seems to justify some reconsideration of current theory of suggestion. One is tempted to short-circuit the process by stating outright that whatever psycho-therapeutic process is not purely analytical must, in the long run, have something in common with the processes of symptom-formation. Unless we analyse the content of the mind and uncover the mental mech-anisms dealing with this content together with its appropriate affect, we automatically range ourselves on the side of mental defence. When therefore an individual's mental defence mechanisms have weakened and he goes to a non-analytical psychotherapeutist to have his symptoms (i.e. subsidiary de-fences) treated, the physician is bound to follow some procedure calculated to supplement the secondary defence (or symptomatic) system. He must employ a tertiary defence system.

Theoretical considerations apart, it would seem reasonable to commence by scrutinising the actual technique employed in suggestion. This can be done most conveniently by using a common standard of assessment, to wit, the amount of psychological truth disclosed to the patient. Or, to reverse the standard, suggestive procedure can be classified in accordance with the amount of deflection from psychological truth, or by the means adopted to deflect attention.

Using these standards it would no doubt be possible to produce an elaborate subdivision of methods, but there is no great advantage to be obtained by so doing. It will be sufficient for our purpose to contrast a few types of suggestive procedure, using analytical objectivity as the common measure. The most extreme form of deviation from objectivity is not generally regarded as a suggestive method at all. Yet there is no doubt that it belongs to suggestive procedure and produces very definite results. It is the method of 'neglect' combined with 'counter-stimulation' employed by the general practitioner or consultant (Glover 1929). The psychological truth is not even brushed aside; it is completely ignored. Nevertheless stimulated no doubt by intuitive understanding of counter-irritations and attractions, the practitioner recommends his patient to embark on activities outside his customary routine. He advises a change of place (holiday) or of bodily habit (recreation, sport, etc.) or of mental activity (light reading, music-hall, etc.). The tendencies here are quite patent. The physician unwittingly tries to reinforce the mechanism of repression (neglect) and quite definitely invokes a system of counter-charge, or anticathexis. His advice to go for a holiday or play golf or attend concerts is therefore an incitement to substitute (symptom-) formation. And on the whole it is a symptom of the obsessional type. The patient must do or think something new (obsessional ceremonial or thought), or take up some counter-attraction (anticathexis, cancellation, undoing, expiation). This counter-charge system no doubt contributes to the success of the general manœuvre but the repression element is important. The physician encourages the patient by demonstrating his own capacity for repression. He says in effect, 'You see, I am blind; I don't know what is the matter with you; go and be likewise'.

The next group, though officially recognised, does not differ very greatly from the unofficial type. It includes the formal methods of suggestion or hypnotic suggestion. Here again the tendency is in complete opposition to the analytical truth; but the repression aspect is not so strongly represented. The suggestionist admits that he knows something of his patient's condition but either commands or begs the patient to neglect it (auxiliary to repression). The patient can and will get better, is in fact better and so on. To make up for the inherent weakness of the auxiliary system, the suggestionist goes through various procedures (suggestions or recommendations) that are again of an obsessional type. Interest has to be transferred to 'something else' more or less antithetical in nature to the pathogenic interest; and of course in hypnotic procedure there are always remainders of magical systems (gestures and phrases).

A third group is distinguished by the fact that a certain amount of use is made of psychological truth or analytic understanding. Explanations varying in detail and accuracy are put before the patient or expounded to him. This is followed by direct or indirect suggestion. By exhortation or persuasion or implication the patient is led to believe that he is now or ought now to be relieved of his symptoms. Auxiliary suggestions of an antithetical type may or may not be added. Although varying in detail, all these procedures can be included under one heading, viz. *pseudo-analytical suggestion*. And as a matter of fact, although the view has aroused much resentment, analysts have made so bold as to describe all pseudo-Freudian analysis as essentially pseudo-analytic suggestion. The only difference they can see is that no open suggestive recommendations are made in the second or third stage of the procedure. As, however, the negative transference is not analysed at all, and very little of the positive, a state of rapport exists which avoids the necessity for open recommendation. Despite this, and presumably to make assurance doubly sure, a good deal of oblique ethical or moral or rationalistic influence is exerted.

There is one feature in common to all these methods; they are all backed by strong transference authority, which means that by sharing the guilt with the suggestionist and by borrowing strength from the suggestionist's super-ego, a new substitution product is accepted by the patient's ego. The new 'therapeutic symptom construction' has become, for the time, ego-syntonic.*

At this point the critic of psycho-analysis who for reasons of his own is anxious to prove that psycho-analysis is itself only another form of suggestion, may argue as follows: if in former times analysts did not completely uncover unconscious content, then surely the analytic successes of earlier days must have been due in part to an element of suggestion in the affective sense as distinct from the verbal sense. It may be remembered that the old accusation levelled against psycho-analysis was that analytic interpretations were disguised suggestions of the 'verbal' or ideoplastic order. At the risk of being tedious the following points must be made clear. Analysis has always sought to resolve as completely as possible the affective analytic bond, both positive and negative. It has always pushed its interpretations to the existing maximum of objective understanding. It is certainly possible that the factor of repression (always an unknown quantity) has dealt with psychic constructions that were incompletely interpreted, but analysis has always striven its utmost to loosen the bonds of repression. It is equally possible that when interpretation has been incomplete some displacement systems are left to function as substitutes or anticathexes; nevertheless analysis has always endeavoured to head off all known protective displacements. In short, it has never sought to maintain a transference as an ultimate therapeutic agent; it has never offered less than the known psychological truth; it has never sided with the mechanisms of repression, displacement or rationalisation. Having made its own position clear, psycho-analysis offers no counterattack to the criticism. It offers instead a theory of suggestion. It is prepared to agree that the criticism might be valid for bad analysis or faulty analysis or pseudo-analysis. It adds, however, that

---

* I have omitted here any detailed description of the dynamic and topographic changes involved by the processes of suggestion. These have been exhaustively described by Ernest Jones in the papers already quoted.

bad analysis may conceivably be good suggestion, although in certain instances it has some misgivings even on this point. For example, it has always been poor analysis to stir up repressed sadistic content and then, without analysing the guilt reactions fully, to remove the props of displacement. And it has probably always been good suggestion to offer new or reinforced displacement substitutes and to buttress what tendencies to withdraw cathexis are capable of conscious support. It is conceivably bad suggestion or more accurately bad pseudo-analytic suggestion to disturb deep layers of guilt. Presumably a good deal of the success of ethical suggestion and sidetracking is due not only to the fact that the patient's sadistic reactions are given an extra coating of rationalisation, but to the fact that the sidetracking activities recommended act as obsessional 'cancellings' of unconscious sadistic formations.*

In addition to these two factors of repression and substitution there is a third fundamental factor to be considered. A great deal of information has now been collected from various analytical sources to show that at bottom mental function is and continues to be valued in terms of concrete experience. There has of course always been some academic interest in the relation of perceptual to conceptual systems, but the contributions of psycho-analysis to this subject have been so detailed and original that it is for all practical purposes a psycho-analytical preserve. For the unconscious a thought is a substance, a word is a deed, a deed is a thought. The complicated variations which psycho-analysis has discovered within this general system depend on the fact that in the upper layers of the unconscious (if we may use this loose topographical term) the substance is regarded as having different origin, properties and qualities. Put systematically, the nature of the substance depends upon the system of libidinal and aggressive interest in vogue during the formation of the particular layer of psychic organisation.

During the primacy of oral interest and aggression, all the world's a breast and all that's in it good or bad milk. During the predominance of excretory interest and anal mental organisation, all the world's a belly. During infantile genital phases, the world at one time is a genital cloaca, at another a phallus. The overlappings and interdependence of these main systems give rise to the multiplicity and variety of phantasy formations. One element is, however, common to all phases, and therefore is represented in all variations of phantasy. This is the element of aggression direct or inverted. So all the substances in the world are benign or malignant, creative or destructive, good or bad.

Psycho-analysts have shown over and over again that, given the slightest relaxation of mental vigilance, the mind is openly spoken of as a bodily organ. The mind is the mouth; talk is urine or flatus, an ideal is fertile and procrea-

---

* In a personal communication Mrs. Riviere has emphasized the importance of sadistic factors in any assessment of analytic or suggestive method.

tive. Our patients are 'big with thought' and tell us so when off guard. This has been demonstrated with considerable detail in the analysis of transference phantasies. An interpretation is welcome or resented (feared) as a phallus. Analysts are reproached for speaking and for keeping silent. Their comments are hailed as sadistic attacks; their silences as periods of relentless deprivation. In short, analysis is unconsciously regarded as the old situation of the infant in or *versus* the world. An interpretation is a substance, good or bad milk, good or bad faeces or urine (or baby, or phallus). It is the supreme parent's substance, friendly or hostile; or it is the infant's substance, returning in a friendly or malignant form, after a friendly or hostile sojourn in the world.*

As I have pointed out elsewhere (1930) this innate tendency of the mind is a perpetual stumbling-block to objectivity not only on the patient's part but on the part of the analyst. It must be constantly measured and allowed for in all stages of analysis. This measurement and uncovering is the essence of transference interpretation. In both transference and projection forms it plays a large part in the fear of analysis which is universally observed. Only the other day a patient with intuitive understanding of symbolism, but without any direct, or indirect orientation in analytic procedure expressed the following views during the first stage of analysis: words are really urine and the stream of urine is an attacking instrument: associations may be either unfriendly or friendly urine: interpretation is generally friendly urine, except on days when erotic and sadistic phantasies are important: when the associations are bad the urine is bad; when the interpretation is bad the analyst is putting bad urine into the patient: the patient must get it out or as the case may be the analyst must take it out. Prognostically speaking, the situation in this case was not very good, but the material was entirely spontaneous.

As has been remarked this innate tendency of the mind is a perpetual stumbling-block to analysis. But what is a stumbling-block to analysis may be a key-stone to suggestion. At any rate part of a key structure. From the earliest times some appreciation of the significance of 'substance' has crept into theories of suggestion; it is to be seen in the old belief in a 'magnetic

---

* [This statement deserves amplification. The communication between analyst and analysand takes place on two levels. One is on the level of secondary processes, where concepts are clarified, connections established, and ideas exchanged. At the same time, there is also a communication on the level of the primary processes, where the interpretation may stand for milk, urine, or the penetrating phallus (Freud 1911c). This duality is present in all communications that are emotionally meaningful. But in psychoanalysis, the primary processes aspect is made conscious and subject to analysis. When thus analyzed, it makes possible the integration of the interpretation into the ego and other deeper levels of wishes can become activated. When one reads this paragraph of Glover's one gets the impression that he deplores the fact that the primary processes interfere, rather than welcoming them as supplying the motor force for the deepening of the analysis.]

fluid' and in the quite modern 'implantation' theories of Bernheim and others (ideoplasty). And it seems plausible that these, in their time apparently scientific explanations, are remote derivatives from a more primitive 'concrete' ideology such as is to be studied in the animistic systems of primitives, the delusional systems of paranoiacs and (given analytical investigation) the transference systems of neurotics. Janet, it will be remembered, regarded the 'somnambulistic passion' or craving as comparable with the craving of drug-addicts; and Ernest Jones has pointed out the relation of this to psycho-analytic ideas concerning the significance of alcohol (Abraham). Discredited or inadequate theories of suggestion thus come into their own in an unexpected fashion. They give us one more hint of the nature of hypnotic and suggestive rapport. And they give us some hint of the therapeutic limits of pseudo-analytic suggestion. The essential substance, symbolised by words or other medium of communication, must be a friendly curative substance. It must be capable of filling a dangerous space in the patient's (body)mind, it must be able to expel gently the dangerous substances in the patient's (body)-mind, or at the least it must be able to neutralise them. In the process of neutralising guilt, it must not awaken anxiety. The hysteric, for example, must not be made psychically pregnant in the course of psychic laparotomy.* So the pseudo-analytical suggestionist does well to alleviate anxieties before administering his suggestive opiate for guilt. And he should steer clear of analysing sadism. The general practitioner sets him a good example in his unofficial and unwitting system of suggestion. As we have seen, the latter not only weighs in on the side of repression and inculcates policies of obsessional anticathexis, but he caters for the patient's fundamental core of paranoia. He doesn't know what is wrong with his patient's mind but he knows, or thinks he knows, what is wrong with his patient's intestinal system. And he uses cathartic drugs or gentle laxatives to drive out the poison, following them up with friendly tonics and invigorating haematinics. In this way he deals with the paranoidal and dangerous omnipotence systems of his patient, without bringing the mind into the matter at all. The suggestionist who openly endeavours to deal with mind through mind should remember that in the last resort he must base his suggestive interferences on a system of 'friendly paranoia'.† Here again the difference between suggestion and true analysis becomes apparent. Analysis must at all times uncover this deepest mental system: the suggestionist with an eye on his patient's anxiety reactions must invariably exploit it.

Conclusion

There are many other factors in the operation of suggestion, concerning which analysis has had or will have much to say. But for the present purpose

---

* [Laparotomy is any surgical opening of the abdomen.]
† [The striking term "friendly paranoia" was coined by Glover.]

it is unnecessary to go into greater detail. Examination of the effect of inexact interpretation in analysis focuses our attention on the possibility that what is for us an incomplete interpretation is for the patient a suitable displacement. By virtue of the fact that the analyst has given the interpretation, it can operate as an ego-syntonic displacement system (substitution-product, symptom). Applying this to the study of methods of suggestion, we see that suggestion technique varies in accordance with the emphasis placed on various defensive mechanisms. All methods depend on the mechanism of repression, but as regards auxiliaries to repression there are quite definite variations in method. In general, non-analytical types of suggestion, by virtue of their complete opposition to the psychological truth and the stress they put on modifications of conduct and thought, might be regarded as 'obsessional systems of suggestion'. Pseudo-analytical types, although nearer the truth, are yet sufficiently remote to operate by focusing energy on a displacement, and in this respect might be called 'hysterical suggestions of a phobiac order'. But the most original and in a sense daring technician, who seldom gets credit for being an expert in suggestion, is the general practitioner or consultant. Intuitively he attempts to deal at once with the patient's superficial anxiety layers and his deepest guilt layers. He is unwittingly a pure 'hysterical suggestionist' in the sense that he plumps for repression and tacitly offers his own repressions (ignorance) as a model; but by his use of drugs he shows intuitive appreciation of the deeper cores of guilt which, under other circumstances, give rise to paranoia. And he plays the rôle of the 'friendly persecutor'. He is in this respect the lineal descendant of the first magical pharmacologists.

These conclusions do not pretend to be original. It has long been held that hypnotic manifestations represent an induced hysteria, and similar suggestions have been made by Radó for the abreaction phenomena of catharsis. Abraham considered that states of auto-suggestion were induced obsessional systems and of course the induction or development of a transference-'neurosis' during analysis is regarded as an integral part of the process. Current types of pseudo-analytical suggestion have not received the same amount of attention. And since they are being employed more and more frequently in psycho-therapeutic circles, it is high time to give them some more definite status. In the sense of displacement, the system they endeavour to exploit is a phobia system. For the treatment to be successful, the patient must develop an ego-syntonic phobia. One might regard this form of suggestion as a kind of homeopathy.* The suggestionist plays the patient at his own game of symptom-formation.

---

* [Homeopathy is a system of medical treatment that uses small quantities of drugs to cure a disease that can be produced by larger quantities of the same drug.]

CHAPTER 24

JAMES STRACHEY

THE NATURE OF THE THERAPEUTIC ACTION OF PSYCHO-ANALYSIS

INTRODUCTORY NOTE

Psychoanalytic ego psychologists object to Strachey's emphasis on the transference interpretation as the only mutative one. Although one-sided, this point of view is expressed so forcibly that the article can be reckoned among the most influential in the history of psychoanalytic technique. The reader will note also that, historically speaking, Strachey is indebted to Alexander, Rank, and, in his approach, he is a follower of Melanie Klein.

Introductory

IT WAS as a therapeutic procedure that psycho-analysis originated. It is in the main as a therapeutic agency that it exists to-day. We may well be surprised, therefore, at the relatively small proportion of psycho-analytical litera-

Originally published in *International Journal of Psycho-Analysis* 15 (1934):127–159. Reprinted in *Psychoanalytic Clinical Interpretation*, ed. Louis Paul (New York: Free Press, 1963). Copyright Angela Richards and the Institute of Psycho-Analysis. Reprinted by permission. Portions of this paper were read at a meeting of the British Psycho-Analytical Society, June 13, 1933.

ture which has been concerned with the mechanisms by which its therapeutic effects are achieved. A very considerable quantity of data have been accumulated in the course of the last thirty or forty years which throw light upon the nature and workings of the human mind; perceptible progress has been made in the task of classifying and subsuming such data into a body of generalized hypotheses or scientific laws. But there has been a remarkable hesitation in applying these findings in any great detail to the therapeutic process itself. I cannot help feeling that this hesitation has been responsible for the fact that so many discussions upon the practical details of analytic technique seem to leave us at cross-purposes and at an inconclusive end.* How, for instance, can we expect to agree upon the vexed question of whether and when we should give a 'deep interpretation', while we have no clear idea of what we *mean* by a 'deep interpretation', while, indeed, we have no exactly formulated view of the concept of 'interpretation' itself, no precise knowledge of what 'interpretation' is and what effect it has upon our patients? We should gain much, I think, from a clearer grasp of problems such as this. If we could arrive at a more detailed understanding of the workings of the therapeutic process we should be less prone to those occasional feelings of utter disorientation which few analysts are fortunate enough to escape; and the analytic movement itself might be less at the mercy of proposals for abrupt alterations in the ordinary technical procedure—proposals which derive much of their strength from the prevailing uncertainty as to the exact nature of the analytic therapy. My present paper is a tentative attack upon this problem; and even though it should turn out that its very doubtful conclusions cannot be maintained, I shall be satisfied if I have drawn attention to the urgency of the problem itself. I am most anxious, however, to make it clear that what follows is not a practical discussion upon psycho-analytic technique. Its immediate bearings are merely theoretical. I have taken as my raw material the various sorts of procedures which (in spite of very considerable individual deviations) would be generally regarded as within the limits of 'orthodox' psycho-analysis and the various sorts of effects which observation shows that the application of such procedures tends to bring about; I have set up a hypothesis which endeavours to explain more or less coherently why these particular procedures bring about these particular effects; and I have tried to show that, if my hypothesis about the nature of the therapeutic action of psycho-analysis is valid, certain implications follow from it which might perhaps serve as criteria in forming a judgment of the probable effectiveness of any particular type of procedure.

---

* [A historical explanation for this bewildering state of affairs has been offered in the editors' introduction pp. 27, 35.]

Retrospect

It will be objected, no doubt, that I have exaggerated the novelty of my topic.* 'After all', it will be said, 'we *do* understand and have long understood the main principles that govern the therapeutic action of analysis'. And to this, of course, I entirely agree; indeed I propose to begin what I have to say by summarizing as shortly as possible the accepted views upon the subject. For this purpose I must go back to the period between the years 1912 and 1917 during which Freud gave us the greater part of what he has written directly on the therapeutic side of psycho-analysis, namely the series of papers on technique and the twenty-seventh and twenty-eighth chapters of the *Introductory Lectures* [Freud 1916–1917, pp. 431–477; these chapters are entitled "Transference" and "Analytical Therapy"].

'Resistance Analysis'

This period was characterized by the systematic application of the method known as 'resistance analysis'. The method in question was by no means a new one even at that time, and it was based upon ideas which had long been implicit in analytical theory, and in particular upon one of the earliest of Freud's views of the function of neurotic symptoms. According to that view (which was derived essentially from the study of hysteria) the function of the neurotic symptom was to defend the patient's personality against an unconscious trend of thought that was unacceptable to it, while at the same time gratifying the trend up to a ctrtain point. It seemed to follow, therefore, that if the analyst were to investigate and discover the unconscious trend and make the patient aware of it—if he were to make what was unconscious conscious—the whole *raison d'être* of the symptom would cease and it must automatically disappear. Two difficulties arose, however. In the first place some part of the patient's mind was found to raise obstacles to the process, to offer resistance to the analyst when he tried to discover the unconscious trend; and it was easy to conclude that this was the same part of the patient's mind as had originally repudiated the unconscious trend and had thus necessitated the creation of the symptom. But, in the second place, even when this obstacle seemed to be surmounted, even when the analyst had

---

* I have not attempted to compile a full bibliography of the subject, though a number of the more important contributions to it are referred to in the following pages.

succeeded in guessing or deducing the nature of the unconscious trend, had drawn the patient's attention to it and had apparently made him fully aware of it—even then it would often happen that the symptom persisted unshaken. The realization of these difficulties led to important results both theoretically and practically.* *Theoretically*, it became evident that there were two senses in which a patient could become conscious of an unconscious trend; he could be made aware of it by the analyst in some intellectual sense without becoming 'really' conscious of it. To make this state of things more intelligible, Freud devised a kind of pictorial allegory. He imagined the mind as a kind of map. The original objectionable trend was pictured as being located in one region of this map and the newly discovered information about it, communicated to the patient by the analyst, in another. It was only if these two impressions could be 'brought together' (whatever exactly that might mean) that the unconscious trend would be 'really' made conscious. What prevented this from happening was a force within the patient, a barrier—once again, evidently, the same 'resistance' which had opposed the analyst's attempts at investigating the unconscious trend and which had contributed to the original production of the symptom. The removal of this resistance was the essential preliminary to the patient's becoming 'really' conscious of the unconscious trend. And it was at this point that the *practical* lesson emerged: as analysts our main task is not so much to investigate the objectionable unconscious trend as to get rid of the patient's resistance to it.

But how are we to set about this task of demolishing the resistance? Once again by the same process of investigation and explanation which we have already applied to the unconscious trend. But this time we are not faced by such difficulties as before, for the forces that are keeping up the repression, although they are to some extent unconscious, do not belong to the unconscious in the systematic sense; they are a part of the patient's ego, which is cooperating with us, and are thus more accessible. Nevertheless the existing state of equilibrium will not be upset, the ego will not be induced to do the work of re-adjustment that is required of it, unless we are able by our analytic procedure to mobilize some fresh force upon our side.

What forces can we count upon? The patient's will to recovery, in the first place, which led him to embark upon the analysis. And, again, a number of intellectual considerations which we can bring to his notice. We can make him understand the structure of his symptom and the motives for his repudiation of the objectionable trend. We can point out the fact that these motives are out-of-date and no longer valid; that they may have been reasonable when he was a baby, but are no longer so now that he is grown up. And finally we can insist that his original solution of the difficulty has only led to illness, while the new one that we propose holds out a prospect of health. Such

---

* [Strachey does not state how or under what circumstances this realization had taken place; see editors' introduction, pp. 29, 37.]

motives as these may play a part in inducing the patient to abandon his resistances; nevertheless it is from an entirely different quarter that the decisive factor emerges. This factor, I need hardly say, is the transference. And I must now recall, very briefly, the main ideas held by Freud on that subject during the period with which I am dealing.

Transference

I should like to remark first that, although from very early times Freud had called attention to the fact that transference manifested itself in two ways—negatively as well as positively, a good deal less was said or known about the negative transference than about the positive. This of course corresponds to the circumstance that interest in the destructive and aggressive impulses in general is only a comparatively recent development. Transference was regarded predominantly as a *libidinal* phenomenon. It was suggested that in everyone there existed a certain number of unsatisfied libidinal impulses, and that whenever some new person came upon the scene these impulses were ready to attach themselves to him. This was the account of transference as a universal phenomenon. In neurotics, owing to the abnormally large quantities of unattached libido present in them, the tendency to transference would be correspondingly greater; and the peculiar circumstances of the analytic situation would further increase it. It was evidently the existence of these feelings of love, thrown by the patient upon the analyst, that provided the necessary extra force to induce his ego to give up its resistances, undo the repressions and adopt a fresh solution of its ancient problems.* This instrument, without which no therapeutic result could be obtained, was at once seen to be no stranger; it was in fact the familiar power of suggestion, which had ostensibly been abandoned long before. Now however it was being employed in a very different way, in fact in a contrary direction. In pre-analytic days it had aimed at bringing about an increase in the degree of repression; now it was used to overcome the resistance of the ego, that is to say, to allow the repression to be removed.

But the situation became more and more complicated as more facts about transference came to light. In the first place, the feelings transferred turned out to be of various sorts; besides the loving ones there were the hostile

* [In 1912, Freud said: "If one's need for love is not entirely satisfied by reality, he is bound to approach every new person whom he meets with libidinal anticipatory ideas; and it is highly probable that both portions of his libido, the portion that is capable of becoming conscious, as well as the unconscious one, have a share in forming that attitude" (1912a, p. 100).]

ones, which were naturally far from assisting the analyst's efforts. But, even apart from the hostile transference, the libidinal feelings themselves fell into two groups: friendly and affectionate feelings which were capable of being conscious, and purely erotic ones which had usually to remain unconscious. And these latter feelings, when they became too powerful, stirred up the repressive forces of the ego and thus increased its resistances instead of diminishing them, and in fact produced a state of things that was not easily distinguishable from a negative transference.* And beyond all this there arose the whole question of the lack of permanence of all suggestive treatments. Did not the existence of the transference threaten to leave the analytic patient in the same unending dependence upon the analyst?

All of these difficulties were got over by the discovery that the transference itself could be analysed. Its analysis, indeed, was soon found to be the most important part of the whole treatment. It was possible to make conscious its roots in the repressed unconscious just as it was possible to make conscious any other repressed material—that is, by inducing the ego to abandon its resistances—and there was nothing self-contradictory in the fact that the force used for resolving the transference was the transference itself. And once it had been made conscious, its unmanageable, infantile, permanent characteristics disappeared; what was left was like any other 'real' human relationship. But the necessity for constantly analysing the transference became still more apparent from another discovery. It was found that as work proceeded the transference tended, as it were, to eat up the entire analysis. More and more of the patient's libido became concentrated upon his relation to the analyst, the patient's original symptoms were drained of their cathexis, and there appeared instead an artificial neurosis to which Freud gave the name of the 'transference neurosis'. The original conflicts, which had led to the onset of neurosis, began to be re-enacted in the relation to the analyst. Now this unexpected event is far from being the misfortune that at first sight it might seem to be. In fact it gives us our great opportunity. Instead of having to deal as best we may with conflicts of the remote past, which are concerned with dead circumstances and mummified personalities, and whose outcome is already determined, we find ourselves involved in an actual and immediate situation, in which we and the patient are the principal characters and the development of which is to some extent at least under our control. But if we bring it about that in this revivified transference conflict the patient chooses a new solution instead of the old one, a solution in which the primitive and unadaptable method of repression is replaced by behaviour more in contact with reality, then, even after his detachment from the analysis, he will never be able to fall back into his former neurosis. The solution of the transference

---

* [This is an important point. In a negative transference, the resistances come from the id. In a positive transference, the resistances come from the ego. It is the ego of the patient that is afraid that the love offering will be rejected.]

conflict implies the simultaneous solution of the infantile conflict of which it is a new edition. 'The change', says Freud in his *Introductory Lectures*, 'is made possible by alterations in the ego occurring as a consequence of the analyst's suggestions. At the expense of the unconscious the ego becomes wider by the work of interpretation which brings the unconscious material into conscious-ness; through education it becomes reconciled to the libido and is made willing to grant it a certain degree of satisfaction; and its horror of the claims of its libido is lessened by the new capacity it acquires to expend a certain amount of the libido in sublimation. The more nearly the course of the treatment corresponds with this ideal description the greater will be the suc-cess of the psycho-analytic therapy', (Freud 1933, p. 381). I quote these words of Freud's to make it quite clear that at the time he wrote them he held that the ultimate factor in the therapeutic action of psycho-analysis was sug-gestion on the part of the analyst acting upon the patient's ego in such a way as to make it more tolerant of the libidinal trends.

The Super-Ego

In the years that have passed since he wrote this passage Freud has produced extremely little that bears directly on the subject; and that little goes to show that he has not altered his views of the main principles involved. Indeed, in the additional lectures which were published last year, he explicitly states that he has nothing to add to the theoretical discussion upon therapy given in the original lectures fifteen years earlier (Freud 1933, p. 194). At the same time there has in the interval been a considerable further develop-ment of his theoretical opinions, and especially in the region of ego-psychol-ogy. He has, in particular, formulated the concept of the super-ego. The re-statement in super-ego terms of the principles of therapeutics which he laid down in the period of resistance analysis may not involve many changes. But it is reasonable to expect that information about the super-ego will be of special interest from our point of view; and in two ways. In the first place, it would at first sight seem highly probable that the super-ego should play an important part, direct or indirect, in the setting-up and maintaining of the repressions and resistances the demolition of which has been the chief aim of analysis. And this is confirmed by an examination of the classification of the various kinds of resistance made by Freud in *Hemmung Symptom und Angst* (1926a, pp. 117–118). Of the five sorts of resistance there mentioned it is true that only one is attributed to the direct intervention of the super-ego, but two of the ego-resistances—the repression-resistance and the transference-resistance

—although actually originating from the ego, are as a rule set up by it out of fear of the super-ego. It seems likely enough therefore that when Freud wrote the words which I have just quoted, to the effect that the favourable change in the patient 'is made possible by alterations in the ego' he was thinking, in part at all events, of that portion of the ego which he subsequently separated off into the super-ego. Quite apart from this, moreover, in another of Freud's more recent works, the *Group Psychology* (1921), there are passages which suggest a different point—namely, that it may be largely through the patient's super-ego that the analyst is able to influence him. These passages occur in the course of his discussion on the nature of hypnosis and suggestion (Freud 1933, p. 77). He definitely rejects Bernheim's view that all hypnotic phenomena are traceable to the factor of suggestion, and adopts the alternative theory that suggestion is a partial manifestation of the state of hypnosis. The state of hypnosis, again, is found in certain respects to resemble the state of being in love. There is 'the same humble subjection, the same compliance, the same absence of criticism towards the hypnotist as towards the loved object'; in particular, there can be no doubt that the hypnotist, like the loved object, 'has stepped into the place of the subject's ego-ideal'. Now since suggestion is a partial form of hypnosis and since the analyst brings about his changes in the patient's attitude by means of suggestion, it seems to follow that the analyst owes his effectiveness, at all events in some respects, to his having stepped into the place of the patient's super-ego. Thus there are two convergent lines of argument which point to the patient's super-ego as occupying a key position in analytic therapy: it is a part of the patient's mind in which a favourable alteration would be likely to lead to general improvement, and it is a part of the patient's mind which is especially subject to the analyst's influence.

Such plausible notions as these were followed up almost immediately after the super-ego made its first *début*.* They were developed by Ernest Jones, for instance, in his paper on 'The Nature of Auto-Suggestion' (1923). Soon afterwards† Alexander launched his theory that the principal aim of all psycho-analytic therapy must be the complete demolition of the super-ego and the assumption of its functions by the ego. According to his account, the treatment falls into two phases. In the first phase the functions of the patient's super-ego are handed over to the analyst, and in the second phase they are passed back again to the patient, but this time to his ego. The super-ego, according to this view of Alexander's (though he explicitly limits his use of the word to the *unconscious* parts of the ego-ideal), is a portion of the mental

---

* In Freud's paper read before the Berlin Congress in 1922, subsequently expanded into *The Ego and the Id* (1923b, pp. 34–36).

[In contemporary psychoanalytic theory the ego ideal is differentiated from the superego as a separate psychic structure; in the early 1920s the two terms were still interchangeable.]

† At the Salzburg Congress in 1924 [Alexander 1925, this volume, Chapter 6].

apparatus which is essentially primitive, out of date and out of touch with reality, which is incapable of adapting itself, and which operates automatically, with the monotonous uniformity of a reflex. Any useful functions that it performs can be carried out by the ego, and there is therefore nothing to be done with it but to scrap it. This wholesale attack upon the super-ego seems to be of questionable validity. It seems probable that its abolition, even if that were practical politics, would involve the abolition of a large number of highly desirable mental activities. But the idea that the analyst temporarily takes over the functions of the patient's super-ego during the treatment and by so doing in some way alters it agrees with the tentative remarks which I have already made.

So, too, do some passages in a paper by Radó upon 'The Economic Principle in Psycho-Analytic Technique'.* The second part of this paper, which was to have dealt with psycho-analysis, has unfortunately never been published; but the first one, on hypnotism and catharsis,† contains much that is of interest. It includes a theory that the hypnotic subject introjects the hypnotist in the form of what Radó calls a 'parasitic super-ego', which draws off the energy and takes over the functions of the subject's original super-ego. One feature of the situation brought out by Radó is the unstable and temporary nature of this whole arrangement. If, for instance, the hypnotist gives a command which is too much in opposition to the subject's original super-ego, the parasite is promptly extruded. And, in any case, when the state of hypnosis comes to an end, the sway of the parasitic super-ego also terminates and the original super-ego resumes its functions.

However debatable may be the details of Radó's description, it not only emphasizes once again the notion of the super-ego as the fulcrum of psychotherapy, but it draws attention to the important distinction between the effects of hypnosis and analysis in the matter of permanence. Hypnosis acts essentially in a temporary way, and Radó's theory of the parasitic super-ego, which does not really replace the original one but merely throws it out of action, gives a very good picture of its apparent workings. Analysis, on the other hand, in so far as it seeks to affect the patient's super-ego, aims at something much more far-reaching and permanent—namely, at an integral change in the nature of the patient's super-ego itself.‡ Some even more recent developments in psycho-analytic theory give a hint, so it seems to me, of the kind of lines along which a clearer understanding of the question may perhaps be reached.

* Also first read at Salzburg in 1924 [Radó 1925].
† *Int. J. Psycho-Anal.* 6 (1925); in a revised form in German, *Internationale Zeitschrift für Psychoanalyse* 12 (1926) [Radó 1925].
‡ This hypothesis seems to imply a contradiction of some authoritative pronouncements, according to which the structure of the super-ego is finally laid down and fixed at a very early age. Thus Freud appears in several passages to hold that the super-ego (or at all events its central core) is formed once and for all at the period at which the child emerges from its Œdipus complex. (See, for instance, *The Ego and the Id*, pp. 68–69)

Introjection and Projection

This latest growth of theory has been very much occupied with the destructive impulses and has brought them for the first time into the centre of interest; and attention has at the same time been concentrated on the correlated problems of guilt and anxiety. What I have in mind especially are the ideas upon the formation of the super-ego recently developed by Melanie Klein and the importance which she attributes to the processes of introjection and projection in the development of the personality. I will re-state what I believe to be her views in an exceedingly schematic outline (see Klein 1932, passim, esp. chaps. 8–9). The individual, she holds, is perpetually introjecting and projecting the objects of its id-impulses, and the character of the introjected objects depends on the character of the id-impulses directed towards the external objects. Thus, for instance, during the stage of a child's libidinal development in which it is dominated by feelings of oral aggression, its feelings towards its external object will be orally aggressive; it will then introject the object, and the introjected object will now act (in the manner of a super-ego) in an orally aggressive way towards the child's ego. The next event will be the projection of this orally aggressive introjected object back on to the external object, which will now in its turn appear to be orally aggressive. The fact of the external object being thus felt as dangerous and destructive once more causes the id-impulses to adopt an even more aggressive and destructive attitude towards the object in self-defence. A vicious circle is thus established. This process seeks to account for the extreme severity of the super-ego in small children, as well as for their unreasonable fear of outside objects. In the course of the development of the normal individual, his libido eventually reaches the genital stage, at which the positive impulses predominate. His attitude towards his external objects will thus become more friendly, and accordingly his introjected object (or super-ego) will become less severe and his ego's contact with reality will be less distorted. In the case of the

---

[Freud 1923b, pp. 34–36]. So, too, Melanie Klein speaks of the development of the super-ego 'ceasing' and of its formation 'having reached completion' at the onset of the latency period (*The Psycho-Analysis of Children*, pp. 250 and 252) [Klein 1932], though in many other passages (e.g. p. 369) she implies that the super-ego can be altered at a later age under analysis. I do not know how far the contradiction is a real one. My theory does not in the least dispute the fact that in the normal course of events the super-ego becomes fixed at an early age and subsequently remains essentially unaltered. Indeed, it is a part of my view that in practice nothing except the process of psycho-analysis *can* alter it. It is of course a familiar fact that in many respects the analytic situation re-constitutes an infantile condition in the patient, so that the fact of being analysed may, as it were, throw the patient's super-ego once more into the melting-pot. Or, again, perhaps it is another mark of the non-adult nature of the neurotic that his super-ego remains in a malleable state.

neurotic, however, for various reasons—whether on account of frustration or of an incapacity of the ego to tolerate id-impulses, or of an inherent excess of the destructive components—development to the genital stage does not occur, but the individual remains fixated at a pre-genital level. His ego is thus left exposed to the pressure of a savage id on the one hand and a correspondingly savage super-ego on the other, and the vicious circle I have just described is perpetuated.

The Neurotic Vicious Circle

I should like to suggest that the hypothesis which I have stated in this bald fashion may be useful in helping us to form a picture not only of the mechanism of a *neurosis* but also of the mechanism of its *cure*. There is, after all, nothing new in regarding a neurosis as essentially an obstacle or deflecting force in the path of normal development; nor is there anything new in the belief that psycho-analysis (owing to the peculiarities of the analytic situation) is able to remove the obstacle and so allow the normal development to proceed. I am only trying to make our conceptions a little more precise by supposing that the pathological obstacle to the neurotic individual's further growth is in the nature of a vicious circle of the kind I have described. If a breach could somehow or other be made in the vicious circle, the processes of development would proceed upon their normal course. If, for instance, the patient could be made less frightened of his super-ego or introjected object, he would project less terrifying imagos on to the outer object and would therefore have less need to feel hostility towards it; the object which he then introjected would in turn be less savage in its pressure upon the id-impulses, which would be able to lose something of their primitive ferocity. In short, a *benign* circle would be set up instead of the vicious one, and ultimately the patient's libidinal development would proceed to the genital level, when, as in the case of a normal adult, his super-ego will be comparatively mild and his ego will have a relatively undistorted contact with reality.*

But at what point in the vicious circle is the breach to be made and how is it actually to be effected? It is obvious that to alter the character of a person's super-ego is easier said than done. Nevertheless, the quotations that I have already made from earlier discussions of the subject strongly suggest that

---

* A similar view has often been suggested by Melanie Klein. See, for instance, *The Psycho-Analysis of Children*, p. 369 [1932]. It has been developed more explicitly and at greater length by Melitta Schmideberg: "Zur Psychoanalyse asozialer Kinder und Jugendlicher," *Internationale Zeitschrift für Psychoanalyse* 18 (1932).

the super-ego will be found to play an important part in the solution of our problem. Before we go further, however, it will be necessary to consider a little more closely the nature of what is described as the analytic situation. The relation between the two persons concerned in it is a highly complex one, and for our present purposes I am going to isolate two elements in it. In the first place, the patient in analysis tends to centre the whole of his id-impulses upon the analyst. I shall not comment further upon this fact or its implications, since they are so immensely familiar. I will only emphasize their vital importance to all that follows and proceed at once to the second element of the analytic situation which I wish to isolate. The patient in analysis tends to accept the analyst in some way or other as a substitute for his own super-ego. I propose at this point to imitate with a slight difference the convenient phrase which was used by Radó in his account of hypnosis and to say that in analysis the patient tends to make the analyst into an 'auxiliary super-ego'. This phrase and the relation described by it evidently require some explanation.

The Analyst as 'Auxiliary Super-Ego'

When a neurotic patient meets a new object in ordinary life, according to our underlying hypothesis he will tend to project on to it his introjected archaic objects and the new object will become to that extent a phantasy object. It is to be presumed that his introjected objects are more or less separated out into two groups, which function as a 'good' introjected object (or mild super-ego) and a 'bad' introjected object (or harsh super-ego). According to the degree to which his ego maintains contacts with reality, the 'good' introjected object will be projected on to benevolent real outside objects and the 'bad' one on to malignant real outside objects. Since, however, he is by hypothesis neurotic, the 'bad' introjected object will predominate, and will tend to be projected more than the 'good' one; and there will further be a tendency, even where to begin with the 'good' object was projected, for the 'bad' one after a time to take its place. Consequently, it will be true to say that in general the neurotic's phantasy objects in the outer world will be predominantly dangerous and hostile. Moreover, since even his 'good' introjected objects will be 'good' according to an archaic and infantile standard, and will be to some extent maintained simply for the purpose of counteracting the 'bad' objects, even his 'good' phantasy objects in the outer world will be very much out of touch with reality. Going back now to the moment when our neurotic patient meets a new object in real life and supposing (as will be the more usual case) that he projects his 'bad' introjected object on to it—the phantasy external object will then seem to him to be dangerous; he will be frightened of it and, to defend himself against it, will become more angry. Thus when he introjects this new object in turn, it will merely be adding one more terrifying imago to those he has already introjected. The new introjected imago will in fact simply be a duplicate of the original archaic ones, and his super-ego will remain almost exactly as it was. The same will be also true *mutatis mutandis* where he begins by projecting his 'good' introjected object on to the new external object he has met with. No doubt, as a result, there will be a slight strengthening of his kind super-ego at the expense of his harsh one, and to that extent his condition will be improved. But there will be no *qualitative* change in his super-ego, for the new 'good' object introjected will only be a duplicate of an archaic original and will only re-inforce the archaic 'good' super-ego already present.

The effect when this neurotic patient comes in contact with a new object *in analysis* is from the first moment to create a different situation. His super-ego is in any case neither homogeneous nor well-organised; the account we have given of it hitherto has been over-simplified and schematic. Actually the introjected imagos which go to make it up are derived from a variety of different stages of his history and function to some extent independently. Now, owing to the peculiarities of the analytic circumstances and of the analyst's behaviour, the introjected imago of the analyst tends in part to be rather definitely separated off from the rest of the patient's super-ego. (This, of course, presupposes a certain degree of contact with reality on his part. Here we have one of the fundamental criteria of accessibility to analytic treatment; another, which we have already implicitly noticed, is the patient's ability to attach his id-impulses to the analyst.) This separation between the imago of the introjected analyst and the rest of the patient's super-ego becomes evident at quite an early stage of the treatment; for instance in connection with the fundamental rule of free association. The new bit of super-ego tells the patient that he is allowed to say anything that may come into his head. This works satisfactorily for a little; but soon there comes a conflict between the new bit and the rest, for the original super-ego says: 'You must *not* say this, for, if you do, you will be using an obscene word or betraying so-and-so's confidences'. The separation off of the new bit—what I have called the 'auxiliary' super-ego—tends to persist for the very reason that it usually operates in a different direction from the rest of the super-ego. And this is true not only of the 'harsh' super-ego but also of the 'mild' one. For, though the auxiliary super-ego is in fact kindly, it is not kindly in the same archaic way as the patient's introjected 'good' imagos. The most important characteristic of the auxiliary super-ego is that its advice to the ego is consistently based upon *real* and *contemporary* considerations and this in itself serves to differentiate it from the greater part of the original super-ego.

In spite of this, however, the situation is extremely insecure. There is a constant tendency for the whole distinction to break down. The patient is liable at any moment to project his terrifying imago on to the analyst just as though he were anyone else he might have met in the course of his life. If this happens, the introjected imago of the analyst will be wholly incorporated into the rest of the patient's harsh super-ego, and the auxiliary super-ego will disappear. And even when the *content* of the auxiliary super-ego's advice is realised as being different from or contrary to that of the original super-ego, very often its *quality* will be felt as being the same. For instance, the patient may feel that the analyst has said to him: 'If you don't say whatever comes into your head, I shall give you a good hiding', or, 'If you don't become conscious of this piece of the unconscious I shall turn you out of the room'. Nevertheless, labile though it is, and limited as is its authority, this peculiar relation between the analyst and the patient's ego seems to put into the analyst's grasp his main instrument in assisting the development of the therapeutic process. What is this main weapon in the analyst's armoury? Its name springs at once to our lips. The weapon is, of course, interpretation. And here we reach the core of the problem that I want to discuss in the present paper.

Interpretation

What, then, *is* interpretation? and how does it work? Extremely little seems to be known about it, but this does not prevent an almost universal belief in its remarkable efficacy as a weapon: interpretation has, it must be confessed, many of the qualities of a *magic* weapon. It is, of course, felt as such by many patients. Some of them spend hours at a time in providing interpretations of their own—often ingenious, illuminating, correct. Others, again, derive a direct libidinal gratification from being given interpretations and may even develop something parallel to a drug-addiction to them. In non-analytical circles interpretation is usually either scoffed at as something ludicrous, or dreaded as a frightful danger. This last attitude is shared, I think, more than is often realized, by a certain number of analysts. This was particularly revealed by the reactions shown in many quarters when the idea of giving interpretations to small children was first mooted by Melanie Klein. But I believe it would be true in general to say that analysts are inclined to feel interpretation as something extremely powerful whether for good or ill. I am speaking now of our *feelings* about interpretation as distinguished from our reasoned beliefs. And there might seem to be a good many grounds for thinking that our feelings on the subject tend to distort our beliefs. At all events, many of these beliefs seem superficially to be contradictory; and the contradictions do not always spring from different schools of thought, but are apparently sometimes held simultaneously by one individual. Thus, we are told that if we interpret too soon or too rashly, we run the risk of losing a patient; that unless we interpret promptly and deeply we run the risk of losing a patient; that interpretation may give rise to intolerable and unmanageable outbreaks of anxiety by 'liberating' it; that interpretation is the only way of enabling a patient to cope with an unmanageable outbreak of anxiety by 'resolving' it; that interpretations must always refer to material on the very point of emerging into consciousness; that the most useful interpretations are really deep ones; 'Be cautious with your interpretations!' says one voice; 'When in doubt, interpret!' says another. Nevertheless, although there is evidently a good deal of confusion in all of this, I do not think these views are necessarily incompatible; the various pieces of advice may turn out to refer to different circumstances and different cases and to imply different uses of the word 'interpretation'.

For the word is evidently used in more than one sense. It is, after all, perhaps only a synonym for the old phrase we have already come across— 'making what is unconscious conscious', and it shares all of that phrase's ambiguities. For in one sense, if you give a German–English dictionary to someone who knows no German, you will be giving him a collection of interpretations, and this, I think, is the kind of sense in which the nature of interpretation has been discussed in a recent paper by Bernfeld.* Such descriptive interpretations have evidently no relevance to our present topic, and I shall proceed without more ado to define as clearly as I can one particular sort of interpretation, which seems to me to be actually the ultimate instrument of psycho-analytic therapy and to which for convenience I shall give the name of 'mutative' interpretation.

I shall first of all give a schematized outline of what I understand by a mutative interpretation, leaving the details to be filled in afterwards; and, with a view to clarity of exposition, I shall take as an instance the interpretation of a hostile impulse. By virtue of his power (his strictly limited power) as auxiliary super-ego, the analyst gives permission for a certain small quantity of the patient's id-energy (in our instance, in the form of an aggressive impulse) to become conscious.† Since the analyst is also, from the nature of things, the *object* of the patient's id-impulses, the quantity of these impulses which is now released into consciousness will become consciously directed

---

* "Der Begriff der Deutung in der Psychoanalyse," *Zeitschrift für angewandte Psychologie* 42 (1932). A critical summary of this by Gerö will be found in *Imago* 19 (1933).

† I am making no attempt at describing the process in correct metapsychological terms. For instance, in Freud's view, the antithesis between conscious and unconscious is not, strictly speaking, applicable to instinctual impulses themselves, but only to the ideas which represent them in the mind ("Unconscious," *Collected Papers*, Vol. 4, p. 109) [1915b, p. 177]. Nevertheless, for the sake of simplicity, I speak throughout this paper of 'making id-impulses conscious'.

towards the analyst. This is the critical point. If all goes well, the patient's ego will become aware of the contrast between the aggressive character of his feelings and the real nature of the analyst, who does not behave like the patient's 'good' or 'bad' archaic objects. The patient, that is to say, will become aware of a distinction between his archaic phantasy object and the real external object. The interpretation has now become a mutative one, since it has produced a breach in the neurotic vicious circle. For the patient, having become aware of the lack of aggressiveness in the real external object, will be able to diminish his own aggressiveness; the new object which he introjects will be less aggressive, and consequently the aggressiveness of his super-ego will also be diminished. As a further corollary to these events, and simultaneously with them, the patient will obtain access to the infantile material which is being re-experienced by him in his relation to the analyst.

Such is the general scheme of the mutative interpretation. You will notice that in my account the process appears to fall into two phases.* I am anxious not to pre-judge the question of whether these two phases are in temporal sequence or whether they may not really be two simultaneous aspects of a single event. But for descriptive purposes it is easier to deal with them as though they were successive. First, then, there is the phase in which the patient becomes conscious of a particular quantity of id-energy as being directed towards the analyst; and secondly there is the phase in which the patient becomes aware that this id-energy is directed towards an archaic phantasy object and not towards a real one.

The First Phase of Interpretation

The first phase of a mutative interpretation—that in which a portion of the patient's id-relation to the analyst is made conscious in virtue of the latter's position as auxiliary super-ego—is in itself complex. In the classical model of an interpretation, the patient will first be made aware of a state of tension in his ego, will next be made aware that there is a repressive factor at work (that his super-ego is threatening him with punishment), and will only then be made aware of the id-impulse which has stirred up the protests of his super-ego and so given rise to the anxiety in his ego. This is the classical

---

* [Strachey speaks of two phases in which the mutative interpretation takes effect. From the point of view of psychoanalytic ego psychology two different ego functions are involved here. In the first, the ego has to relax its vigil and let derivatives of the id or superego wishes become conscious. In the second, the ego has to reassert itself and employ its capacity to test reality. Some patients have difficulty in relaxing their control, while others, particularly the borderline patients, have difficulty in reasserting the reality testing.]

scheme. In actual practice, the analyst finds himself working from all three sides at once, or in irregular succession. At one moment a small portion of the patient's super-ego may be revealed to him in all its savagery, at another the shrinking defencelessness of his ego, at yet another his attention may be directed to the attempts which he is making at restitution—at compensating for his hostility; on some occasions a fraction of id-energy may even be directly encouraged to break its way through the last remains of an already weakened resistance. There is, however, one characteristic which all of these various operations have in common; they are essentially upon a small scale. For the mutative interpretation is inevitably governed by the principle of minimal doses. It is, I think, a commonly agreed clinical fact that alterations in a patient under analysis appear almost always to be extremely gradual: we are inclined to suspect sudden and large changes as an indication that suggestive rather than psycho-analytic processes are at work. The gradual nature of the changes brought about in psycho-analysis will be explained if, as I am suggesting, those changes are the result of the summation of an immense number of minute steps, each of which corresponds to a mutative interpretation. And the smallness of each step is in turn imposed by the very nature of the analytic situation. For each interpretation involves the release of a certain quantity of id-energy, and, as we shall see in a moment, if the quantity released is too large, the highly unstable state of equilibrium which enables the analyst to function as the patient's auxiliary super-ego is bound to be upset. The whole analytic situation will thus be imperilled, since it is only in virtue of the analyst's acting as auxiliary super-ego that these releases of id-energy can occur at all.

Let us examine in greater detail the effects which follow from the analyst attempting to bring too great a quantity of id-energy into the patient's consciousness all at once.* On the one hand, nothing whatever may happen, or on the other hand there may be an unmanageable result; but in neither event will a mutative interpretation have been effected. In the former case (in which there is apparently no effect) the analyst's power as auxiliary super-ego will not have been strong enough for the job he has set himself. But this again may be for two very different reasons. It may be that the id-impulses he was trying to bring out were not in fact sufficiently urgent at the moment: for, after all, the emergence of an id-impulse depends on two factors—not only on the permission of the super-ego, but also on the urgency (the degree of cathexis) of the id-impulse itself. This, then, may be one cause of an apparently negative response to an interpretation, and evidently a fairly harmless one. But the same apparent result may also be due to something else; in spite of the id-impulse being really urgent, the strength of the patient's own repressive forces (the degree of repression) may have been too great to allow his ego to listen

* Incidentally, it seems as though a *qualitative* factor may be concerned as well: that is, some *kinds* of id-impulses may be more repugnant to the ego than others.

to the persuasive voice of the auxiliary super-ego. Now here we have a situation dynamically identical with the next one we have to consider, though economically different. This next situation is one in which the patient accepts the interpretation, that is, allows the id-impulse into his consciousness, but is immediately overwhelmed with anxiety. This may show itself in a number of ways: for instance, the patient may produce a manifest anxiety-attack, or he may exhibit signs of 'real' anger with the analyst with complete lack of insight, or he may break off the analysis. In any of these cases the analytic situation will, for the moment at least, have broken down. The patient will be behaving just as the hypnotic subject behaves when, having been ordered by the hypnotist to perform an action too much at variance with his own conscience, he breaks off the hypnotic relation and wakes up from his trance. This state of things, which is *manifest* where the patient responds to an interpretation with an actual outbreak of anxiety or one of its equivalents, may be *latent* where the patient shows no response. And this latter case may be the more awkward of the two, since it is masked, and it may sometimes, I think, be the effect of a greater overdose of interpretation than where manifest anxiety arises (though obviously other factors will be of determining importance here and in particular the nature of the patient's neurosis). I have ascribed this threatened collapse of the analytic situation to an overdose of interpretation: but it might be more accurate in some ways to ascribe it to an *insufficient* dose. For what has happened is that the second phase of the interpretative process has not occurred: the phase in which the patient becomes aware that his impulse is directed towards an archaic phantasy object and not towards a real one.

The Second Phase of Interpretation

In the second phase of a complete interpretation, therefore, a crucial part is played by the patient's sense of reality: for the successful outcome of that phase depends upon his ability, at the critical moment of the emergence into consciousness of the released quantity of id-energy, to distinguish between his phantasy object and the real analyst. The problem here is closely related to one that I have already discussed, namely that of the extreme lability of the analyst's position as auxiliary super-ego. The analytic situation is all the time threatening to degenerate into a 'real' situation. But this actually means the opposite of what it appears to. It means that the patient is all the time on the brink of turning the real external object (the analyst) into the archaic one; that is to say, he is on the brink of projecting his primitive introjected imagos on to him. In so far as the patient actually does this, the analyst becomes like anyone else that he meets in real life—a phantasy object. The analyst then ceases to possess the peculiar advantages derived from the analytic situation; he will be introjected like all other phantasy objects into the patient's super-ego, and will no longer be able to function in the peculiar ways which are essential to the effecting of a mutative interpretation. In this difficulty the patient's sense of reality is an essential but a very feeble ally; indeed, an improvement in it is one of the things that we hope the analysis will bring about. It is important, therefore, not to submit it to any unnecessary strain; and that is the fundamental reason why the analyst must avoid any real behaviour that is likely to confirm the patient's view of him as a 'bad' or a 'good' phantasy object. This is perhaps more obvious as regards the 'bad' object. If, for instance, the analyst were to show that he was really shocked or frightened by one of the patient's id-impulses, the patient would immediately treat him in that respect as a dangerous object and introject him into his archaic severe super-ego. Thereafter, on the one hand, there would be a diminution in the analyst's power to function as an auxiliary super-ego and to allow the patient's ego to become conscious of his id-impulses—that is to say, in his power to bring about the *first* phase of a mutative interpretation; and, on the other hand, he would, as a real object, become sensibly less distinguishable from the patient's 'bad' phantasy object and to that extent the carrying through of the *second* phase of a mutative interpretation would also be made more difficult. Or again, there is another case. Supposing the analyst behaves in an opposite way and actively urges the patient to give free rein to his id-impulses. There is then a possibility of the patient confusing the analyst with the imago of a treacherous parent who first encourages him to seek gratification, and then suddenly turns and punishes him. In such a case, the patient's ego may look for defence by itself suddenly turning upon the analyst as though he were his own id, and treating him with all the severity of which his super-ego is capable. Here again, the analyst is running a risk of losing his privileged position. But it may be equally unwise for the analyst to act really in such a way as to encourage the patient to project his 'good' introjected object on to him. For the patient will then tend to regard him as a good object in an archaic sense and will incorporate him with his archaic 'good' imagos and will use him as a protection against his 'bad' ones. In that way, his infantile positive impulses as well as his negative ones may escape analysis, for there may no longer be a possibility for his ego to make a comparison between the phantasy external object and the real one. It will perhaps be argued that, with the best will in the world, the analyst, however careful he may be, will be unable to prevent the patient from projecting these various imagos on to him. This is of course indisputable, and indeed, the whole effectiveness of analysis depends upon its being so. The lesson of these difficulties is merely to remind us that the patient's sense of reality has the narrowest limits. It is a paradoxical fact that the best way of ensuring that his ego shall be able to distinguish between phantasy and reality is to withhold reality from him as much as possible. But it is true. His ego is so weak—so much at the mercy of his id and super-ego—that he can only cope with reality if it is administered in minimal doses. And these doses are in fact what the analyst gives him, in the form of interpretations.

Interpretation and Reassurance

It seems to me possible that an approach to the twin practical problems of interpretation and reassurance may be facilitated by this distinction between the two phases of interpretation. Both procedures may, it would appear, be useful or even essential in certain circumstances and inadvisable or even dangerous in others. In the case of interpretation,* the first of our hypothetical phases may be said to 'liberate' anxiety, and the second to 'resolve' it. Where a quantity of anxiety is already present or on the point of breaking out, an interpretation, owing to the efficacy of its second phase, may enable the patient to recognize the unreality of his terrifying phantasy object and so to reduce his own hostility and consequently his anxiety. On the other hand, to induce the ego to allow a quantity of id-energy into consciousness is obviously to court an outbreak of anxiety in a personality with a harsh super-ego. And this is precisely what the analyst does in the first phase of an interpretation. As regards 'reassurance', I can only allude briefly here to some of the problems it raises.† I believe, incidentally, that the term needs to be defined almost as urgently as 'interpretation', and that it covers a number of different mechanisms. But in the present connection reassurance may be regarded as behaviour on the part of the analyst calculated to make the patient regard him as a 'good' phantasy object rather than as a real one. I have already given some reasons for doubting the expediency of this, though it seems to be generally felt that the procedure may sometimes be of great value, especially in psychotic cases. It might, moreover, be supposed at first sight that the adoption of such an attitude by the analyst might actually directly favour the prospect of making a mutative interpretation. But I believe that it will be seen

---

* For the necessity for 'continuous and deep-going interpretations' in order to diminish or prevent anxiety-attacks, see Melanie Klein's *Psycho-Analysis of Children*, pp. 58–59. On the other hand: 'The anxiety belonging to the deep levels is far greater, both in amount and intensity, and it is therefore imperative that its liberation should be duly regulated' (*ibid.*, p. 139).

† Its uses were discussed by Melitta Schmideberg in a paper read to the British Psycho-Analytical Society on February 7, 1934.

on reflection that this is not in fact the case: for precisely in so far as the patient regards the analyst as his phantasy object, the second phase of the interpretation does not occur—since it is of the essence of that phase that in it the patient should make a distinction between his phantasy object and the real one. It is true that his anxiety may be reduced; but this result will not have been achieved by a method that involves a permanent qualitative change in his super-ego. Thus, whatever tactical importance reassurance may possess, it cannot, I think, claim to be regarded as an ultimate operative factor in psycho-analytic therapy.

It must here be noticed that certain other sorts of behaviour on the part of the analyst may be dynamically equivalent to the giving of a mutative interpretation, or to one or other of the two phases of that process. For instance, an 'active' injunction of the kind contemplated by Ferenczi [1919c, this volume, Chapter 5] may amount to an example of the first phase of an interpretation; the analyst is making use of his peculiar position in order to induce the patient to become conscious in a particularly vigorous fashion of certain of his id-impulses. One of the objections to this form of procedure may be expressed by saying that the analyst has very little control over the dosage of the id-energy that is thus released, and very little guarantee that the second phase of the interpretation will follow. He may therefore be unwittingly precipitating one of those critical situations which are always liable to arise in the case of an incomplete interpretation. Incidentally, the same dynamic pattern may arise when the analyst requires the patient to produce a 'forced' phantasy or even (especially at an early stage in an analysis) when the analyst asks the patient a question; here again, the analyst is in effect giving a blindfold interpretation, which it may prove impossible to carry beyond its first phase. On the other hand, situations are fairly constantly arising in the course of an analysis in which the patient becomes conscious of small quantities of id-energy without any direct provocation on the part of the analyst. An anxiety situation might then develop, if it were not that the analyst, by his behaviour or, one might say, absence of behaviour, enables the patient to mobilize his sense of reality and make the necessary distinction between an archaic object and a real one. What the analyst is doing here is equivalent to bringing about the second phase of an interpretation, and the whole episode may amount to the making of a mutative interpretation. It is difficult to estimate what proportion of the therapeutic changes which occur during analysis may not be due to *implicit* mutative interpretations of this kind. Incidentally, this type of situation seems sometimes to be regarded, incorrectly as I think, as an example of reassurance.

'Immediacy' of Mutative Interpretations

But it is now time to turn to two other characteristics which appear to be essential properties of every mutative interpretation. There is in the first place one already touched upon in considering the apparent or real absence of effect which sometimes follows upon the giving of an interpretation. A mutative interpretation can only be applied to an id-impulse which is actually in a state of cathexis. This seems self-evident; for the dynamic changes in the patient's mind implied by a mutative interpretation can only be brought about by the operation of a charge of energy originating in the patient himself: the function of the analyst is merely to ensure that the energy shall flow along one channel rather than along another. It follows from this that the purely informative 'dictionary' type of interpretation will be non-mutative, however useful it may be as a prelude to mutative interpretations. And this leads to a number of practical inferences. Every mutative interpretation must be emotionally 'immediate'; the patient must experience it as something actual. This requirement, that the interpretation must be 'immediate', may be expressed in another way by saying that interpretations must always be directed to the 'point of urgency'. At any given moment some particular id-impulse will be in activity; *this* is the impulse that is susceptible of mutative interpretation at that time, and no other one. It is, no doubt, neither possible nor desirable to be giving mutative interpretations all the time; but, as Melanie Klein has pointed out, it is a most precious quality in an analyst to be able at any moment to pick out the point of urgency (1932, pp. 58–59).

'Deep' Interpretation

But the fact that every mutative interpretation must deal with an 'urgent' impulse takes us back once more to the commonly felt fear of the explosive possibilities of interpretation, and particularly of what is vaguely referred to as 'deep' interpretation. The ambiguity of the term, however, need not bother us. It describes, no doubt, the interpretation of material which is either genetically early and historically distant from the patient's actual experience or which is under an especially heavy weight of repression—material, in any case, which is in the normal course of things exceedingly inaccessible to his ego and remote from it. There seems reason to believe, moreover, that the anxiety which is liable to be aroused by the approach of such material to consciousness may be of peculiar severity (Klein 1932, p. 139). The question whether it is 'safe' to interpret such material will, as usual, mainly depend upon whether the second phase of the interpretation can be carried through. In the ordinary run of case the material which is urgent during the earlier stages of the analysis is not deep. We have to deal at first only with more or less far-going displacements of the deep impulses, and the deep material itself is only reached later and by degrees, so that no sudden appearance of un-manageable quantities of anxiety is to be anticipated. In exceptional cases, however, owing to some peculiarity in the structure of the neurosis, deep impulses may be urgent at a very early stage of the analysis. We are then faced by a dilemma. If we give an interpretation of this deep material, the amount of anxiety produced in the patient may be so great that his sense of reality may not be sufficient to permit of the second phase being accom-plished, and the whole analysis may be jeopardised. But it must not be thought that, in such critical cases as we are now considering, the difficulty can necessarily be avoided simply by not giving any interpretation or by giving more superficial interpretations of non-urgent material or by attempting reassurances. It seems probable, in fact, that these alternative procedures may do little or nothing to obviate the trouble; on the contrary, they may even exacerbate the tension created by the urgency of the deep impulses which are the actual cause of the threatening anxiety. Thus the anxiety may break out in spite of these palliative efforts and, if so, it will be doing so under the most unfavourable conditions, that is to say, outside the mitigating influences afforded by the mechanism of interpretation. It is possible, therefore, that, of the two alternative procedures which are open to the analyst faced by such a difficulty, the interpretation of the urgent id-impulses, deep though they may be, will actually be the safer.

'Specificity' of Mutative Interpretations

I shall have occasion to return to this point for a moment later on, but I must now proceed to the mention of one further quality which it seems necessary for an interpretation to possess before it can be mutative, a quality which is perhaps only another aspect of the one we have been describing. A mutative interpretation must be 'specific': that is to say, detailed and con-crete. This is, in practice, a matter of degree. When the analyst embarks upon a given theme, his interpretations cannot always avoid being vague and gen-eral to begin with; but it will be necessary eventually to work out and inter-pret all the details of the patient's phantasy system. In proportion as this is done the interpretations will be mutative, and much of the necessity for apparent repetitions of interpretations already made is really to be explained by the need for filling in the details. I think it possible that some of the delays which despairing analysts attribute to the patient's id-resistance could be traced to this source. It seems as though vagueness in interpretation gives the defensive forces of the patient's ego the opportunity, for which they are always on the look-out, of baffling the analyst's attempt at coaxing an urgent id-impulse into consciousness. A similarly blunting effect can be produced by certain forms of reassurance, such as the tacking on to an interpretation of an ethnological parallel or of a theoretical explanation: a procedure which may at the last moment turn a mutative interpretation into a non-mutative one. The apparent effect may be highly gratifying to the analyst; but later experience may show that nothing of permanent use has been achieved or even that the patient has been given an opportunity for increasing the strength of his defences. Here we have evidently reached a topic discussed not long ago by Edward Glover in one of the very few papers in the whole literature which seriously attacks the problem of interpretation [1931, this volume, Chapter 23]. Glover argues that, whereas a *blatantly* inexact interpretation is likely to have no effect at all, a *slightly* inexact one may have a therapeutic effect of a non-analytic, or rather anti-analytic, kind by enabling the patient to make a deeper and more efficient repression. He uses this as a possible explanation of a fact that has always seemed mysterious, namely, that in the earlier days of analysis, when much that we now know of the characteristics of the unconscious was still undiscovered, and when interpretation must therefore often have been inexact, therapeutic results were nevertheless obtained.

Abreaction

The possibility which Glover here discusses serves to remind us more generally of the difficulty of being certain that the effects that follow any given interpretation are genuinely the effects of interpretation and not transference phenomena of one kind or another. I have already remarked that many patients derive direct libidinal gratification from interpretation as such; and I think that some of the striking signs of abreaction which occasionally follow an interpretation ought not necessarily to be accepted by the analyst as evidence of anything more than that the interpretation has gone home in a libidinal sense.

The whole problem, however, of the relation of abreaction to psychoanalysis is a disputed one. Its therapeutic results seem, up to a point, undeniable. It was from them, indeed, that analysis was born; and even to-day there are psycho-therapists who rely on it almost exclusively. During the War, in particular, its effectiveness was widely confirmed in cases of 'shell-shock'. It has also been argued often enough that it plays a leading part in bringing about the results of psycho-analysis. Rank and Ferenczi, for instance, declared that in spite of all advances in our knowledge abreaction remained the essential agent in analytic therapy (Ferenczi and Rank, 1924, p. 27). More recently, Reik has supported a somewhat similar view in maintaining that 'the element of surprise is the most important part of analytic technique' [Reik 1933, this volume, Chapter 26]. A much less extreme attitude is taken by Nunberg in the chapter upon therapeutics in his text-book of psycho-analysis.* But he, too, regards abreaction as one of the component factors in analysis, and in two ways. In the first place, he mentions the improvement brought about by abreaction in the usual sense of the word, which he plausibly attributes to a relief of endo-psychic tension due to a discharge of accumulated affect. And in the second place, he points to a similar relief of tension upon a small scale arising from the actual process of becoming conscious of something hitherto unconscious, basing himself upon a statement of Freud's that the act of becoming conscious involves a discharge of energy (1920, p. 28). On the other hand, Radó appears to regard abreaction as opposed in its function to analysis. He asserts that the therapeutic effect of catharsis is to be attributed to the fact that (together with other forms of non-analytic psycho-therapy) it offers the patient an artificial neurosis in exchange for his original one, and that the phenomena observable when abreaction occurs are akin to those of an hysterical attack (Radó 1925). A consideration of the views of these various authorities suggests that what we describe as 'abreaction' may cover two different processes: one a discharge of affect and the other a libidinal gratification. If so, the first of these might be regarded (like various other procedures) as an occasional adjunct to analysis, sometimes, no doubt, a useful one, and possibly even as an inevitable accompaniment of mutative interpretations; whereas the second process might be viewed with more suspicion, as an event likely to impede analysis—especially if its true nature were unrecognised. But with either form there would seem good reason to believe that the effects of abreaction are permanent only in cases in which the predominant ætiological factor is an external event: that is to say, that it does not itself bring about any radical qualitative alteration in the patient's mind. Whatever part it may play in analysis is thus unlikely to be of anything more than an ancillary nature.

* *Allgemeine Neurosenlehre auf psychoanalytischer Grundlage* (1932), pp. 303–304. This chapter appears in English in an abbreviated version as a contribution to Lorand's *Psycho-Analysis To-day* (1933). There is very little, I think, in Nunberg's comprehensive catalogue of the factors at work in analytic therapy that conflicts with the views expressed in the present paper, though I have given a different account of the interrelation between those factors.

Extra-Transference Interpretations

If we now turn back and consider for a little the picture I have given of a mutative interpretation with its various characteristics, we shall notice that my description appears to exclude every kind of interpretation except those of a single class—the class, namely, of *transference* interpretations. Is it to be understood that no extra-transference interpretation can set in motion the chain of events which I have suggested as being the essence of psycho-analytical therapy? That is indeed my opinion, and it is one of my main objects in writing this paper to throw into relief—what has, of course, already been observed, but never, I believe, with enough explicitness—the dynamic distinctions between transference and extra-transference interpretations. These distinctions may be grouped under two heads. In the first place, extra-transference interpretations are far less likely to be given at the point of urgency. This must necessarily be so, since in the case of an extra-transference interpretation the object of the id-impulse which is brought into consciousness is not the analyst and is not immediately present, whereas, apart from the earliest stages of an analysis and other exceptional circumstances, the point of urgency is nearly always to be found in the transference. It follows that extra-transference interpretations tend to be concerned with impulses which are distant both in time and space and are thus likely to be devoid of immediate energy. In extreme instances, indeed, they may approach very closely to what I have already described as the handing-over to the patient of a German-English dictionary. But in the second place, once more owing to the fact that the object of the id-impulse is not actually present, it is less easy for the patient, in the case of an extra-transference interpretation, to become directly aware of the distinction between the real object and the phantasy object. Thus it would appear that, with extra-transference interpretations, on the one hand what I have described as the first phase of a mutative interpretation is less likely to occur, and on the other hand, if the first phase *does* occur, the second phase is less likely to follow. In other words, an extra-transference interpretation is liable to be both less effective and more risky than a transference one.* Each of these points deserves a few words of separate examination.

It is, of course, a matter of common experience among analysts that it is possible with certain patients to continue indefinitely giving interpretations without producing any apparent effect whatever. There is an amusing criticism of this kind of 'interpretation-fanaticism' in the excellent historical chapter of Rank and Ferenczi (1924, p. 31). But it is clear from their words that

* This corresponds to the fact that the pseudo-analysts and 'wild' analysts limit themselves as a rule to extra-transference interpretations. It will be remembered that this was true of Freud's original 'wild' analyst (Freud 1910c).

what they have in mind are essentially extra-transference interpretations, for the burden of their criticism is that such a procedure implies neglect of the analytic situation. This is the simplest case, where a waste of time and energy is the main result. But there are other occasions, on which a policy of giving strings of extra-transference interpretations is apt to lead the analyst into more positive difficulties. Attention was drawn by Reich* a few years ago in the course of some technical discussions in Vienna to a tendency among inexperienced analysts to get into trouble by eliciting from the patient great quantities of material in a disordered and unrelated fashion: this may, he maintained, be carried to such lengths that the analysis is brought to an irremediable state of chaos. He pointed out very truly that the material we have to deal with is stratified and that it is highly important in digging it out not to interfere more than we can help with the arrangement of the strata. He had in mind, of course, the analogy of an incompetent archæologist, whose clumsiness may obliterate for all time the possibility of reconstructing the history of an important site. I do not myself feel so pessimistic about the results in the case of a clumsy analysis, since there is the essential difference that our material is alive and will, as it were, re-stratify itself of its own accord if it is given the opportunity: that is to say, in the analytic situation. At the same time, I agree as to the presence of the risk, and it seems to me to be particularly likely to occur where extra-transference interpretation is excessively or exclusively resorted to. The means of preventing it, and the remedy if it has occurred, lie in returning to transference interpretation at the point of urgency. For if we can discover which of the material is 'immediate' in the sense I have described, the problem of stratification is automatically solved; and it is a characteristic of most extra-transference material that it has no immediacy and that consequently its stratification is far more difficult to decipher. The measures suggested by Reich himself for preventing the occurrence of this state of chaos are not inconsistent with mine; for he stresses the importance of interpreting *resistances* as opposed to the primary id-impulses themselves—and this, indeed, was a policy that was laid down at an early stage in the history of analysis. But it is, of course, one of the characteristics of a resistance that it arises in relation to the analyst; and thus the interpretation of a resistance will almost inevitably be a transference interpretation.

But the most serious risks that arise from the making of extra-transference interpretations are due to the inherent difficulty in completing their second phase or in knowing whether their second phase has been completed or not. They are from their nature unpredictable in their effects. There seems,

---

* Bericht über das 'Seminar für psychoanalytische Therapie' in Wien," *Internationale Zeitschrift für Psychoanalyse* 13 (1927): 241–244. This has recently been republished as a chapter in Reich's volume upon *Charakteranalyse* (1933), which contains a quantity of other material with an interesting bearing on the subject of the present paper.

indeed, to be a special risk of the patient not carrying through the second phase of the interpretation but of projecting the id-impulse that has been made conscious on to the analyst. This risk, no doubt, applies to some extent also to transference interpretations. But the situation is less likely to arise when the object of the id-impulse is actually present and is moreover the same person as the maker of the interpretation.* (We may here once more recall the problem of 'deep' interpretation, and point out that its dangers, even in the most unfavourable circumstances, seem to be greatly diminished if the interpretation in question is a transference interpretation.) Moreover, there appears to be more chance of this whole process occurring silently and so being overlooked in the case of an extra-transference interpretation, particularly in the earlier stages of an analysis. For this reason, it would seem to be important after giving an extra-transference interpretation to be specially on the *qui vive* for transference complications. This last peculiarity of extra-transference interpretations is actually one of their most important from a practical point of view. For on account of it they can be made to act as 'feeders' for the transference situation, and so to pave the way for mutative interpretations. In other words, by giving an extra-transference interpretation, the analyst can often provoke a situation in the transference of which he can then give a mutative interpretation.

It must not be supposed that because I am attributing these special qualities to transference interpretations, I am therefore maintaining that no others should be made. On the contrary, it is probable that a large majority of

* It even seems likely that the whole possibility of effecting mutative interpretations may depend upon this fact that in the analytic situation the giver of the interpretation and the object of the id-impulse interpreted are one and the same person. I am not thinking here of the argument mentioned above—that it is easier under that condition for the patient to distinguish between his phantasy object and the real object—but of a deeper consideration. The patient's original super-ego is, as I have argued, a product of the introjection of his archaic objects distorted by the projection of his infantile id-impulses. I have also suggested that our only means of altering the character of this harsh original super-ego is through the mediation of an auxiliary super-ego which is the product of the patient's introjection of the analyst as an object. The process of analysis may from this point of view be regarded as an infiltration of the rigid and unadaptable original super-ego by the auxiliary super-ego with its greater contact with the ego and with reality. This infiltration is the work of the mutative interpretations; and it consists in a repeated process of introjection of imagos of the analyst—imagos, that is to say, of a real figure and not of an archaic and distorted projection—so that the quality of the original super-ego becomes gradually changed. And since the aim of the mutative interpretations is thus to cause the introjection of the analyst, it follows that the id-impulses which they interpret must have the analyst as their object. If this is so, the views expressed in the present paper will require some emendation. For in that case, the first criterion of a mutative interpretation would be that it must be a transference interpretation. Nevertheless, the quality of urgency would still remain important; for, of all the possible transference interpretations which could be made at any particular moment, only the one which dealt with an urgent id-impulse would be mutative. On the other hand, an extra-transference interpretation even of an extremely urgent id-impulse could never be mutative—though it might, of course, produce temporary relief along the lines of abreaction or reassurance.

our interpretations are outside the transference—though it should be added that it often happens that when one is ostensibly giving an extra-transference interpretation one is implicitly giving a transference one. A cake cannot be made of nothing but currants; and, though it is true that extra-transference interpretations are not for the most part mutative, and do not themselves bring about the crucial results that involve a permanent change in the patient's mind, they are none the less essential. If I may take an analogy from trench warfare, the acceptance of a transference interpretation corresponds to the capture of a key position, while the extra-transference interpretations correspond to the general advance and to the consolidation of a fresh line which are made possible by the capture of the key position. But when this general advance goes beyond a certain point, there will be another check, and the capture of a further key position will be necessary before progress can be resumed. An oscillation of this kind between transference and extra-transference interpretations will represent the normal course of events in an analysis.

Mutative Interpretations and the Analyst

Although the giving of mutative interpretations may thus only occupy a small portion of psycho-analytic treatment, it will, upon my hypothesis, be the most important part from the point of view of deeply influencing the patient's mind. It may be of interest to consider in conclusion how a moment which is of such importance to the patient affects the analyst himself. Mrs. Klein has suggested to me that there must be some quite special internal difficulty to be overcome by the analyst in giving interpretations. And this, I am sure, applies particularly to the giving of mutative interpretations. This is shown in their avoidance by psycho-therapists of non-analytic schools; but many psycho-analysts will be aware of traces of the same tendency in themselves. It may be rationalized into the difficulty of deciding whether or not the particular moment has come for making an interpretation. But behind this there is sometimes a lurking difficulty in the actual *giving* of the interpretation, for there seems to be a constant temptation for the analyst to do something else instead. He may ask questions, or he may give reassurances or advice or discourses upon theory, or he may give interpretations—but interpretations that are not mutative, extra-transference interpretations, interpretations that are non-immediate, or ambiguous, or inexact—or he may give two or more alternative interpretations simultaneously, or he may give interpretations and at the same time show his own scepticism about them. All of this strongly suggests that the giving of a mutative interpretation is a crucial act for the analyst as well as for the patient, and that he is exposing himself to some great danger in doing so. And this in turn will become intelligible when we reflect that at the moment of interpretation the analyst is in fact deliberately evoking a quantity of the patient's id-energy while it is alive and actual and unambiguous and aimed directly at himself. Such a moment must above all others put to the test his relations with his own unconscious impulses.

Summary

I will end by summarizing the four main points of the hypothesis I have put forward:

1. The final result of psycho-analytic therapy is to enable the neurotic patient's whole mental organization, which is held in check at an infantile stage of development, to continue its progress towards a normal adult state.
2. The principal effective alteration consists in a profound qualitative modification of the patient's super-ego, from which the other alterations follow in the main automatically.
3. This modification of the patient's super-ego is brought about in a series of innumerable small steps by the agency of mutative interpretations, which are effected by the analyst in virtue of his position as object of the patient's id-impulses and as auxiliary super-ego.
4. The fact that the mutative interpretation is the ultimate operative factor in the therapeutic action of psycho-analysis does not imply the exclusion of many other procedures (such as suggestion, reassurance, abreaction, etc.) as elements in the treatment of any particular patient.

CHAPTER 25

RICHARD F. STERBA

THE FATE OF THE EGO
IN ANALYTIC THERAPY

INTRODUCTORY NOTE

In the International Journal of Psycho-Analysis *of 1934, the articles of Sterba and Strachey follow each other. They are poles apart in what they consider crucial for the success of psychoanalytic treatment. Sterba stresses the rational aspects of the therapeutic relationship. He is the first to speak of what will later be called "the therapeutic alliance." The term "therapeutic ego dissociation" has been coined by him, and is the basic term of this paper, as indicated in the editors' introduction, p. 38. Psychoanalytic ego psychology begins with the formulation of the problem of anxiety (Freud 1926a), but Sterba's paper can be seen as the first contribution of this new ego psychology to technique.*

The approach Sterba advocates depends heavily on the capacity of the analysand to identify himself with the analyst. This identification is based on a narcissistic satisfaction resulting from the participation in the intellectual work of the analysis. Kaiser (1934, this volume, Chapter 27) regards such a narcissistic satisfaction as detrimental to intrapsychic change.

In an earlier paper Sterba (1929) wrote: "The possibility of the identification with the analyst—so necessary for the interpretation—is a conditio sine qua non *for analytic treatment" (p. 372). Some psychoanalysts besides Kaiser also see identification in a less positive light as a defense against the full development of a transference neurosis.*

Originally entitled "Das Schicksal des Ichs im therapeutischen Verfahren," *Internationale Zeitschrift für Psychoanalyse* 20 (1934):66–73. English translation in *International Journal of Psycho-Analysis* 15 (1934):117–126. Reprinted by permission of the Institute of Psycho-Analysis.

THAT PART of the psychic apparatus which is turned towards the outside world and whose business it is to receive stimuli and effect discharge-reactions we call the ego.* Since analysis belongs to the external world, it is again the ego which is turned towards it. Such knowledge as we possess of the deeper strata of the psychic apparatus reaches us by way of the ego and depends upon the extent to which the ego admits it, in virtue of such derivatives of the Ucs as it still tolerates. If we wish to learn something of these deeper strata or to bring about a change in a neurotic constellation of instincts, it is to the ego and the ego alone that we can turn. Our analysis of resistances, the explanations and interpretations that we give to our patients, our attempts to alter their mental attitudes through our personal action upon them—all these must necessarily start with the ego. Now amongst all the experiences undergone by the ego during an analysis there is one which seems to me so specific and so characteristic of the analytic situation that I feel justified in isolating it and presenting it to you as the 'fate' of the ego in analytic therapy.

The contents of this paper will surprise you by their familiarity. How could it be otherwise, seeing that it is simply an account of what you do and observe every day in your analyses? If, nevertheless, I plead justification, it is because I believe that, in what follows, adequate recognition is given for the first time to a factor in our therapeutic work which has so far received too little attention in our literature. The nearest approach to my theme is to be found in a paper on character-analysis by Reich,† in which he talks of 'isolating' a given character-trait, 'objectifying' it and 'imparting psychic distance' to it, referring thereby no doubt to that therapeutic process which I shall now present in a much more general form.‡

For the purposes of our incomplete description it will suffice if we regard the ego in analysis as having three functions. First, it is the executive organ of the id, which is the source of the object-cathexis of the analyst in the transference; secondly, it is the organization which aims at fulfilling the demands of the super-ego and, thirdly, it is the site of experience, i.e. the institution which either allows or prevents the discharge of the energy poured forth by the id in accordance with the subject's previous experiences.

In analysis the personality of the analysand passes first of all under the domination of the *transference*. The function of the transference is twofold. On the one hand, it serves to satisfy the object-hunger of the id. But, on the other, it meets with opposition from the repressive psychic institutions—the super-ego, which rejects it on moral grounds, and the ego, which, because of

---

* [The belief that only the ego is accessible to psychoanalytic influence is a distinguishing mark of psychoanalytic ego psychology; see A. Freud 1936, pp. 4, 6.]

† ["Über Charakteranalyse"], *Internationale Zeitschrift für Psychoanalyse* 14 (1928):180–196 [later incorporated in Reich's book *Character Analysis* (1933)].

‡ [At this date, Sterba is not yet critical of Reich (1933, this volume, Chapter 18). See also Sterba's evaluation of Reich (1953, this volume, Chapter 20).]

unhappy experiences, utters a warning against it. Thus, in the transference-resistance the very fact of the transference is utilized as a weapon against the whole analysis.

We see, then, that in the transference a dualistic principle comes into play in the ego: instinct and repression alike make themselves felt. We learn from the study of the transference-resistance that the forces of repression enter into the transference no less than the instinctual forces. Anti-cathexes* are mobilized as a defence against the libidinal impulses which proceed from the Ucs and are revived in the transference. For example, anxiety is activated as a danger-signal against the repetition of some unhappy experience that once ensued from an instinctual impulse, and is used as a defence against analysis. Here the repressive forces throw their weight on the side of the transference because the revival of the repressed tendency makes it the more imperative for the subject to defend himself against it and so put an end to the dreaded laying bare of the Ucs.

In order to bring out the twofold function of the transference let me sketch a fairly typical transference-situation such as arose at the beginning of one of my analyses.

A woman patient transferred to the analyst an important object-cathexis from the period of early childhood. It represented her love for a physician to whom she was frequently taken during her fifth year on account of enlarged tonsils. On each occasion he looked into her mouth, without touching the tonsils, afterwards giving her some sweets and always being kind and friendly. Her parents had instituted these visits in order to lull her into security for the operation to come. One day, when she trustfully let the doctor look into her mouth again, he inserted a gag and, without giving any narcotic or local anæsthetic, removed the unsuspecting child's tonsils. For her this was a bitter disillusionment and never again could she be persuaded to go to see him.

The twofold function of the transference from this physician to the analyst is obvious: in the first place it revived the object-relation to the former (a father-substitute), but, in the second place, her unhappy experience with him gave the repressive forces their opportunity to reject the analyst and, with him, the analysis. 'You had much better stay away, in case he hurts you', they warned her, 'and keep your mouth shut!' The result was that the patient was obstinately silent in the analysis and manifested a constant tendency to break it off.

This typical example shows how the ego manages in the transference to rid itself of two different influences, though in the shape of a conflict. For the establishment of the transference is based on a conflict between instinct and repression. Where the transference-situation is intense, there is always the danger that one or other of the conflicting forces may prevail: either the

* [For a discussion of the terms cathexis and anticathexis, see editors' introduction, p. 22.]

analytic enterprise may be broken up by the blunt transference demands of the patient, or else the repressive institutions in the mind of the latter may totally repudiate both analyst and analysis. Thus we may describe the transference and the resistance which goes with it as the conflict-laden final result of the struggle between two groups of forces, each of which aims at dominating the workings of the ego, while both alike obstruct the purposes of the analysis.

In opposition to this dual influence, the object of which is to inhibit the analysis, we have the corrective influence of the analyst, who in his turn, however, must address himself to the *ego*. He approaches it in its capacity of the organ of perception and of the testing by reality. By *interpreting* the transference-situation he endeavours to oppose those elements in the ego which are focussed on reality to those which have a cathexis of instinctual or defensive energy. What he thus accomplishes may be described as a *dissociation* within the ego.*

We know that dissociations within the ego are by no means uncommon. They are a means of avoiding the clash of intolerable contradictions in its organization. 'Double consciousness' may be regarded as a large-scale example of such dissociation: here the left hand is successfully prevented from knowing what is done by the right. Many parapraxes† are of the nature of 'double consciousness', and abortive forms of this phenomenon are to be found in other departments of life as well.

This capacity of the ego for dissociation gives the analyst the chance, by means of his interpretations, to effect an alliance with the ego against the powerful forces of instinct and repression and, with the help of one part of it, to try to vanquish the opposing forces. Hence, when we begin an analysis which can be carried to completion, the fate that inevitably awaits the ego is that of *dissociation*. A permanently unified ego, such as we meet with in cases of excessive narcissisms or in certain psychotic states where ego and id have become fused, is not susceptible to analysis. The therapeutic dissociation of the ego is a necessity if the analyst is to have the chance of winning over part of it to his side, conquering it, strengthening it by means of identification with himself and opposing it in the transference to those parts which have a cathexis of instinctual and defensive energy.

* It may be doubted whether 'dissociation' is an appropriate term for non-pathological processes in the ego. This point is answered by the following passage in Freud's *New Introductory Lectures on Psycho-Analysis*, a work which has appeared since this paper was read: 'We wish to make the ego the object of our study, our own ego. But how can that be done? The ego is the subject *par excellence*: how can it become the object? There is no doubt, however, that it can. The ego can take itself as object; it can treat itself like any other object, observe itself, criticize itself, do Heaven knows what besides with itself. In such a case, one part of the ego stands over against the other. The ego can, then, be split; it becomes dissociated during many of its functions, at any rate in passing. The parts can later on join up again' (p. 80) [Freud 1933].
† [Parapraxes are slips of the tongue or pen.]

The technique by which the analyst effects this therapeutic dissociation of the ego consists of the explanations which he gives to the patient of the first signs of transference and transference-resistance that can be interpreted. You will remember that in his recommendations on the subject of technique Freud says that, when the analyst can detect the effects of a transference-resistance it is a sign that the time is ripe for interpretation. Through the explanations of the transference-situation that he receives the patient realizes for the first time the peculiar character of the therapeutic method used in analysis. Its distinctive characteristic is this: that the subject's consciousness shifts from the centre of affective experience to that of intellectual contemplation. The transference-situation is *interpreted*, i.e. an explanation is given which is uncoloured by affect and which shows that the situation has its roots in the subject's childhood.* Through this interpretation there emerges in the mind of the patient, out of the chaos of behaviour impelled by instinct and behaviour designed to inhibit instinct, a *new point of view of intellectual contemplation.* In order that this new standpoint may be effectually reached there must be a certain amount of positive transference, on the basis of which a transitory strengthening of the ego takes place through identification with the analyst. This identification is induced by the analyst. From the outset the patient is called upon to 'co-operate' with the analyst against something in himself. Each separate session gives the analyst various opportunities of employing the term 'we', in referring to himself and to the part of the patient's ego which is consonant with reality. The use of the word 'we' always means that the analyst is trying to draw that part of the ego over to his side and to place it in opposition to the other part which in the transference is cathected or influenced from the side of the unconscious. We might say that this 'we' is the instrument by means of which the therapeutic dissociation of the ego is effected.

The function of interpretation, then, is this: Over against the patient's instinct-conditioned or defensive behaviour, emotions and thoughts it sets up in him a principle of intellectual cognition, a principle which is steadily supported by the analyst and fortified by the additional insight gained as the analysis proceeds. In subjecting the patient's ego to the fate of therapeutic dissociation we are doing what Freud recommends in a passage in *Beyond the Pleasure Principle* (p. 18): 'The physician ... has to see to it that some measure of ascendancy remains [in the patient], in the light of which the apparent reality [of what is repeated in the transference] is always recognized as a reflection of a forgotten past.'

The question now suggests itself: What is the prototype of this therapeutic ego-dissociation in the patient? The answer is that it is the process of *super-ego formation.* By means of an identification—of analysand with

* [This view is in keeping with Freud's emphasis on the genetic aspects of psychoanalysis (1913c); see editors' introduction, p. 4.]

analyst—judgements and valuations from the outside world are admitted into the ego and become operative within it. The difference between this process and that of super-ego-formation is that, since the therapeutic dissociation takes place in an ego which is already mature, it cannot well be described as a 'stage' in ego-development: rather it represents more or less the opposition of one element to others on the same level. The result of super-ego-formation is the powerful establishment of moral demands; in therapeutic ego-dissociation the demand which has been accepted is a demand for a revised attitude appropriate to the situation of an adult personality. Thus, whilst the super-ego demands that the subject shall adopt a particular attitude towards a particular tendency in the id, the demand made upon him when therapeutic dissociation takes place is a demand for a balancing contemplation, kept steadily free of affect, whatever changes may take place in the contents of the instinct-cathexes and the defensive reactions.

We have seen, then, that in analysis the ego undergoes a specific fate which we have described as therapeutic dissociation. When analysis begins, the ego is subject to a process of 'dissimilation' or dissociation, which must be induced by the analyst by means of his interpretation of the transference-situation and of the resistance to which this gives rise.

As the analysis proceeds, the state of 'dissimilation' in the ego is set up again whenever the unconscious material, whether in the shape of instinctual gratification or of defensive impulses, fastens on the analyst in the transference. All the instinctual and defensive reactions aroused in the ego in the transference impel the analyst to induce the therapeutic process of ego-dissociation by means of the interpretations he gives. There is constituted, as it were, a standing relation between that part of the ego which is cathected with instinctual or defensive energy and that part which is focussed on reality and identified with the analyst, and this relation is the filter through which all the transference-material in the analysis must pass. Each separate interpretation reduces the instinctual and defensive cathexis of the ego in favour of intellectual contemplation, reflection and correction by the standard of reality.

However, once the analyst's interpretations have set up this opposition of forces—the ego which is in harmony with reality versus the ego which acts out its unconscious impulses—the state of 'dissimilation' does not last and a process of *'assimilation'* automatically begins. We owe to Hermann Nunberg our closer knowledge of this process, which he calls 'the synthetic function of the ego' [Nunberg 1930, this volume, Chapter 22]. As we know, this function consists in the striving of the ego, prompted by Eros, to bind, to unify, to assimilate and to blend—in short, to leave no conflicting elements within its domain. It is this synthetic function which, next to therapeutic dissociation of the ego, makes analytic therapy possible. The former process enables the subject to recognize intellectually and to render conscious the claims and the content of his unconsciousness and the affects associated with these, whilst when that has been achieved, the synthetic function of the ego enables him to incorporate them and to secure their discharge.

Since there are in the transference and the transference-resistance two groups of forces within the ego, it follows that the ego-dissociation induced by the analyst must take place in relation to each group, the ego being placed in opposition to both. At the same time the interpretations of defensive reactions and instinctual trends become interwoven with one another, for analysis cannot overcome the defence unless the patient comes to recognize his instinctual impulses, nor put him in control of the latter unless the defence has been overthrown. The typical process is as follows: First of all, the analyst gives an interpretation of the defence, making allusion to the instinctual tendencies which he has already divined and against which the defence has been set up. With the patient's recognition that his attitude in the transference is of the nature of a defence, there comes a weakening in that defence. The result is a more powerful onslaught of the instinctual strivings upon the ego. The analyst then has to interpret the infantile meaning and aim of these impulses. Ego-dissociation and synthesis ensue, with the outcome that the impulses are corrected by reference to reality and subsequently find discharge by means of such modifications as are possible. In order that all these interpretations may have a more profound effect, it is necessary constantly to repeat them; the reason for this I have explained elsewhere ('Zur Dynamik der Bewältigung des Übertragungswiderstandes,' *Internationale Zeitschrift für Psychoanalyse.* Bd. XV, 1929 [translated into English in 1940]).

Now let us return to the case I cited before and see how it illustrates what I have just said. The patient's resistance, which began after a few analytic sessions, took the form of obstinate silence and a completely negative attitude towards the analyst. Such meagre associations as she vouchsafed to give she jerked out with averted head and in obvious ill-humour. At the close of the second session an incident occurred which showed that this silence and repellent attitude were a mode of defence against a positive transference. At the end of the hour she asked me if I had not a cloakroom where she could change her clothes as they were all crumpled after she had lain on the sofa for an hour. The next day she said to me in this connection that, after her analysis, she was going to meet a woman friend, who would certainly wonder where the patient had got her dress so crushed and whether she had been having sexual intercourse. It was clear that, as early as the second session, her ego had come under the influence of the transference and of the defence against it. Of course, she herself was completely unconscious of the connection between her fear of being found out by her friend and the attitude of repudiation which she assumed in analysis.

The next thing to do was to explain to the patient the *meaning* of her defence. As a first step, the defensive nature of her attitude was made plain to her, for of this, too, she was unconscious. With this interpretation we had begun the process which I have called therapeutic ego-dissociation. When the interpretation had been several times repeated the patient gained a first measure of 'psychic distance' in relation to her own behaviour. At the start her gain was only intermittent and she was compelled almost at once to go on acting her instinctual impulses out. As, however, the positive transference was sufficiently strong, it gradually became possible to enlarge these islands of intellectual contemplation or observation at the expense of the process of acting the unconscious impulses out. The result of this dissociation in the ego was that the patient gained an insight into the defensive nature of her attitude in analysis, that is to say, she now began to work over preconsciously the material which had hitherto been enacted unconsciously in her behaviour. This insight denoted a decrease in the cathexis of those parts of the ego which were carrying on the defence.

Some time afterwards there emerged the memory of her visits to the kind throat-specialist and of the bitter disillusionment in which they had ended. This recollection was in itself a result of the synthetic function of the ego, for the ego will not tolerate within itself a discrepancy between defence and insight. The effect of the infantile experience had, it is true, been felt by the ego, but this effect had been determined from the unconscious; it now became incorporated in the preconscious in respect of its causal origin also. It is hardly necessary for me to point out that the discovery of this infantile experience of the patient with the physician was merely a preliminary to the real task of the analyst, which was to bring into consciousness her experiences with her father and especially her masochistic phantasies relating to him.

In overcoming the transference-defence by the method of therapeutic ego-dissociation we were not merely attacking that part of the ego which was using the patient's unhappy experience with the physician in her childhood to obstruct the analysis; we were, besides, counteracting part of the super-ego's opposition. For the defensive attitude was in part also a reaction to the fear that her friend might find out that the patient had been having sexual intercourse. Now she had developed an obvious mother-transference to this particular friend, and the mother was the person who had imposed sexual prohibitions in the patient's childhood. By means of the therapeutic ego-dissociation a standpoint of intellectual contemplation, a 'measure of ascendancy', had formed itself in her mind, in opposition to her defensive behaviour: in that dissociation the 'reality' elements in the ego were separated not only from those elements which bore the stamp of that unhappy experience and signalled their warning, but also from those other elements which acted as the executive of the super-ego.

In the case we are considering, the next result of the analysis was that the positive transference began to reveal itself, taking more openly possession of the ego and manifesting itself in the claims which the patient made on the analyst's love. Once more, dissociation had to be induced in the ego, so as to separate out of the processes of dramatic enactment an island of intellectual contemplation, from which the patient could perceive that her behaviour was determined by her infantile experiences in relation to her father. This, naturally, only proved possible after prolonged therapeutic work.

I hope that this short account may have sufficed to make clear what I believe to be one of the most important processes in analytic therapy, namely, the effecting of a dissociation within the ego by interpretation of the patient's instinctually conditioned conduct and his defensive reaction to it. Perhaps I may say in conclusion that the therapeutic dissociation of the ego in analysis is merely an extension, into new fields, of that self-contemplation which from all time has been regarded as the most essential trait of man in distinction to other living beings. For example, Herder expressed the view that *speech* originated in this objectifying process which works by the dissociation of the mind in self-contemplation. This is what he says about it: 'Man shows reflection when the power of his mind works so freely that, out of the whole ocean of sensations which comes flooding in through the channel of every sense, he can separate out, if I may so put it, a single wave and hold it, directing his attention upon it and being conscious of this attention. . . . He shows reflection when he not only has a vivid and distinct perception of every sort of attribute, but can acknowledge in himself one or more of them as distinguishing attributes: the first such act of acknowledgment yields a clear conception; it is the mind's first judgment. And how did this acknowledgment take place? Through a characteristic which he had had to separate out and which, as a characteristic due to conscious reflection, presented itself clearly to his mind. Good! Let us greet him with a cry of "eureka"! This first characteristic due to conscious reflection was a word of the mind! With it human speech was invented!' (*Über den Ursprung der Sprache.*)*

In the therapeutic dissociation which is the fate of the ego in analysis, the analysand is called on 'to answer for himself'† and the unconscious, ceasing to be expressed in behaviour, becomes articulate in *words*. We may say, then, that in this ego-dissociation we have an extension of reflection beyond what has hitherto been accessible. Thus, from the standpoint also of the human faculty of speech, we may justly claim that analytic therapy makes its contribution to the humanizing of man.

---

* [*On the Origin of Language,* by Jean-Jacques Rousseau and Johann G. Herder.]
† [German: "*zur Rede gestellt*"; literally, "is put to speech"—trans. note.]

CHAPTER 26

THEODOR REIK

NEW WAYS IN PSYCHO-ANALYTIC TECHNIQUE

INTRODUCTORY NOTE

In this paper Reik develops his original contributions to psychoanalytic technique. His emphasis on surprise—interpretation that is correct and given at the right moment as distinct from the application of a standard formula—comes as a surprise, to the patient as well as to the psychoanalyst. In literature, Isaac Babel held a similar conviction. He believed that the essence of art is a surprise attack upon the reader's habitual assumptions. Reik developed this thesis, at length, in Surprise and the Psychoanalyst *(1935). He elaborated the same idea again in his most popular book,* Listening with the Third Ear *(1948). Freud had said on the same subject: ". . . the most successful cases are those in which one proceeds, as it were, without any purpose in view, allows oneself to be taken by surprise by any new turn in them, and always meets them with an open mind, free from any presuppositions" (1912b). Thus, what Freud said* en passant, *became to Reik the cornerstone to a whole structure.*

Originally published in *International Journal of Psycho-Analysis* 14 (1933):321–334. Reprinted by permission of Arthur Reik and the Institute of Psycho-Analysis. This paper was read before the Twelfth International Psycho-Analytic Congress, Wiesbaden, September 6, 1932.

ALTHOUGH I have practised psycho-analysis for twenty years, this is the first time that I have made bold to address you on the subject of psycho-analytic technique.* I have been deterred from any earlier venture into publicity by two personal idiosyncrasies. In the first place, I have a very marked incapacity for learning from the mistakes of others. No aphorism, no advice, no warning is of any use: if I am to learn anything from other people's errors I must first of all commit them myself. Having made them my own I may then perhaps rid myself of them. And besides this obstinate lack of mental pliancy I have another peculiarity: I am almost incapable of learning from my own mistakes until I have repeated them several times. Only occasionally, when I have been clumsy in certain minor points of technique, have I been able to avoid my error the second time I recognized it.

Having thus prefaced my remarks, I will pass on to the discussion of certain fundamental problems of analytic technique. Let us direct our attention especially to its heuristic significance. The therapeutic use of analytic technique is a special case of analysis in this its wider aspect. Therefore I do not propose to discuss the psychotherapeutic application of the conclusions at which we shall arrive. This will form the subject of a later, separate discussion. I must acknowledge from the outset that the fundamental features of the technique described here claim a *programmatic* character. Some of them are new, while others are here formulated, or expressed in scientific terms, for the first time; all alike are derived from analytic practice and accordingly look to verification or criticism only from the standpoint of experience with the living psychological material which we are studying.

Let us begin with a provisional statement of the nature of our technique, which will be not so much a definition as a description. Psycho-analysis is a scientific method based on what survives of our belief in magical effects. As you know, the point in question is the belief in the magical effect of words. And here already I have to pause, for these introductory remarks to our discussion threaten to turn into a discussion themselves. You are aware that some of our colleagues say that they cannot set so high a value upon words. But I am afraid they would be very much at a loss if you asked what psychological material, other than words, they had at their disposal. An occasional look, a gesture, a movement—otherwise nothing but words, words, words! what these say and what they leave unsaid.

The essence of the analytic process consists in the series of shocks experienced as the subject takes cognizance of his repressed processes, the effect of which makes itself felt for long afterwards. I have purposely used the word 'take cognizance' [*Kenntnisnahme*] and not 'apprehend' [*Erkenntnis*], because the former term implies something deeper and more comprehensive than a mere intellectual appreciation. In this connection there really is a

---

* [Strictly speaking, the claim is not accurate; see Reik 1915.]

confusion of idioms such as Ferenczi refers to.* Else how could we so readily overlook the difference, for one example, between *telling* the analyst every-thing and *confiding* everything to him? We say light-heartedly that so-and-so has 'had' an analysis, but it makes a great difference whether anyone merely passes through an analysis or whether it is a living experience to him.

If I now try to determine what is the nature of the peculiar psychic shock which is specific to psycho-analysis, if I seek to define its particular character as generally and yet as precisely as possible, I should say that it is essentially a *surprise*.

I beg that you will defer your objections for the moment and allow me to explain my meaning by a reference to a theory which I tried to set out in 1926, in a book entitled *Der Schrecken* [see Reik 1929]. I said there that the element of 'surprise' lies in the *encounter, at an unexpected moment or in unexpected circumstances, with a fact of which the expectation has become unconscious.*

This definition holds good of both external and internal perception, of material as well as psychological facts. In either case we can see that surprise is a defence-reaction, directed against the suggestion that we should disregard what we are accustomed to and re-discover in what is new something from our earliest past which is no longer recognized by us. In other words: *surprise is the expression of our struggle against any call upon us to acknowledge something long known to us which has become unconscious.* This means, when it relates to analysis: our struggle against recognizing a part of the ego, once known to us but now become unconscious. The affective manifestation of this attitude has become familiar to you in analysis as resistance. Naturally, the surprise will be greatest when there is involved the confirmation or fulfil-ment of expectations which have been repressed.

In analytic practice, those pieces of new insight will be most effective which contain this element of surprise. The effect will be progressive, in pro-portion to the depth of the strata to which that insight penetrates, and it will be most lasting if the old expectation, to which the analysis has found its way, had been repressed. This effect can easily be accounted for psychologically if we consider not only the topographical change, so often discussed, but also the economic and dynamic characteristics of the process. The *economic* dis-placement depends on that saving in expenditure of energy on suppression or inhibition which results from the analyst's interpretation or reconstruction. A whole series of ideas and affects which we normally need as a kind of psychic transition is, as it were, cleared at a bound and becomes superfluous. The analysand experiences something of the pleasure which results from such an economy and resembles that of a person who hears a witticism. The *dynamic*

---

* [The reference is to Ferenczi's 1933 paper, where he contrasted the need of the child for tender love with the blockage of this capacity in many adults who can only respond to the child with passion. Such adults can only be seductive.]

effect arises in that through the analytical insight a piece of psychic reality which has been unconsciously repressed is seen to coincide with a piece of material reality.* Let us take quite a simple example which you can observe every day in your work. Let us suppose that, in one of your patients, violent aggressive impulses against a near relative have emerged and have led to an unconscious wish to murder him. You have detected these processes from certain obsessional symptoms which represent the patient's defence against the murderous impulses and whose secret significance can perhaps be seen to be that of an expiatory ceremonial. What I am now trying to describe is the process set up in the patient's mind by your explanation of the latent meaning of his symptoms and their hidden connection with his unconscious murderous wish. By putting that secret significance into words you have brought a piece of psychic reality into contact with a piece of external reality and caused them to coincide in the most terrifying way. By uttering it aloud in words you have freed the unconscious process from its dumbness and oppressive weight. The unconscious has its own means of expression; all that our explanation has done has been to exchange them for others, more familiar to us. We may say that we have translated something from the language of symptoms into that of words. Now it is this translation or transposition which signifies the coincidence—the concurrence—of psychic with material reality. Here we have one of the most instructive instances of the surprise so frequently and variously repeated in the phenomenon of belief in the omnipotence of thought. Let me give you a commonplace instance of this belief, for comparison with those mental processes in the patient which I have already described, very different though they appear to be. Let us suppose that he tells you how often it happens that he has just been thinking hard about a certain acquaintance when he quite unexpectedly meets him. Such a coincidence between thought and external fact has given him a slight psychic shock. This situation is very similar to that which I have described as taking place during analysis: the patient's unconscious thoughts were occupied with the murder of his relative; suddenly that possibility which was merely a thought took bodily form and came to meet him as words from your lips. It is as though in the broad daylight of consciousness he had experienced a miracle which seemed to confirm his belief in the omnipotence of thought. It is not merely his trait of secret thoughts becoming vocal and turning into material reality which links the analytic 'translation' with the powerful impression made by such phenomena of omnipotence upon the obsessional neurotic. There is a second element of 'surprise' which establishes a bridge between these and the verbal grasp and expression of psychic processes in the analytic hour. You have doubtless discovered in many cases that at the bottom of this belief in omnipotence there were evil or grossly egoistic wishes and that every apparent

* [For a discussion on psychoanalytic metapsychology, see editors' introduction, p. 8.]

confirmation of the belief suggested the possibility of their fulfilment, implied, in fact, that they were already fulfilled in thought. The conflict between his wishes and his counteracting tendencies very often explains the obsessional neurotic's peculiar behaviour: it is designed to ward off the least approach of the realization of his desires. He is struggling in this strenuous way against the possibility of his secret wishes becoming a reality. Strange as it may seem, however, when the analyst puts these repudiated wishes into words, this signifies a partial fulfilment of them: verbal reality has been conferred upon the patient's thoughts, and this is the nearest approach to a material realization of what he desired, yet repudiates. That is the second way in which analytic explanation resembles those phenomena which seem to confirm the belief in the omnipotence of thought: by means of embodiment in words it gives a certain measure of gratification to forbidden impulses. But this fulfilment connotes at the same time some degree of psychic mastery over instinct. Therefore in this function, too, analysis testifies to the magical influence of words.

The type of surprise experienced by the patient in analysis when mental and material reality coincide is seen in its most natural and simple form when he says things which surprise himself. He did not know that he had such thoughts, cherished such feelings and harboured such impulses. In this situation, when the external perception of his own words forces inward perception upon him with a shock of surprise, when material reality is, so to speak, suddenly overpowered by psychic reality, we see most clearly that analytic insight represents the confirmation of a repressed expectation. We can check this fact when an erroneous interpretation or a faulty construction is offered.* In such a case a psychic effect is certainly produced, but there is no surprise in our sense of the term, *i.e.* that of a reaction to the confirmation of an unconscious expectation.

If it be true that surprise presupposes an expectation which has become unconscious, we can see that there is a far-reaching psychological affinity between the technique of the analytic method of investigation and the technique of wit. I will take a quite elementary example from analytic practice. A patient dreams of a charlatan and her associations, when she tells the dream,

---

* [There are various ways of differentiating between interpretation and construction. Freud defined the difference as follows: "Interpretation applies to something that one does to some single element of the material, such as an association or a parapraxis. But it is a 'construction' when one lays before the subject of the analysis a piece of his early history that he has forgotten!" (Freud 1937a, p. 261).

In subsequent psychoanalytic writings it has become customary to designate as construction a hypothetical event that cannot be remembered but must have taken place in order to account for a group of dreams and fantasies that otherwise have no explanation. By contrast to a construction, an interpretation brings together ideas, wishes, affects and thoughts that belong together but could not be recognized as such on account of repression or isolation. For further consideration of the subject, see Rosen 1955 and Bergmann 1968.]

point in a definite direction: her thoughts have been occupied with an acquaintance named Charles. Does not such an interpretation sound like a rather feeble pun? Please do not reply that this is an isolated case, for this case of analytic interpretative technique follows the lines of thousands of other cases. In the discovery of the hidden meaning and the mechanisms of neurotic symptoms our technique will make use of the same means as are characteristic of the technique of wit: condensation, displacement, *double entendre*, omission, etc. And we shall use the same peculiar technique when we pass from single symptoms and particular psychological traits to the hidden purpose of a neurosis, the secret trend of the psychic forces underlying some special development or the determining psychic traits in a character. We are obliged to use this technique because otherwise we have no prospect of divining the latent psychic processes, just as in a foreign country we have to learn the language in order to understand what is said. Thus, one of the most important tasks of the analyst is to learn to guess the meaning of allusions, just as it is so necessary to do in the play of wit.

At this point you will no doubt all be thinking of the way in which Freud was led to investigate the technique of wit: he was told that his interpretations of dreams were so funny. However, in trying to formulate psychologically the similarity between the technique of wit and the analytic technique whereby we divine unconscious processes in general I am not simply stressing this characteristic, I am saying a good deal more.

It is easy to point out the apparent silliness or absurdity of such an interpretation as that of the dream-element 'charlatan' and to go on to argue plausibly that analytic interpretation in general is of the nature of a pun. This sort of 'scientific' objection loftily ignores the fact that the interpretation is not left to the good pleasure of the analyst: that it simply reproduces what he has been able to divine from various intimations and allusions, and has to adapt itself, both as regards the thought it contains and its verbal expression, to the peculiarities of the unconscious and especially to its infantile character. The analyst can be held responsible for the apparent absurdity and silliness of his interpretation just as far, and with the same justice, as the Egyptologist for the contents of a piece of hieroglyphic script which he has deciphered. It is highly improbable that the Egyptologist himself believes in a goat-headed god named Chnumu, nor is it likely that he personally supposes that, if he prays to the goddess Hathor with her cow's head, he will be cured of his rheumatism.

With the exception of Freud, no one, so far as I am aware, has as yet recognized the importance for the evaluation of practical analytic technique of the investigation of the technique of wit. The same technique reveals the meaning of the unconscious in both cases and shows 'the poodle's kernel',* whether this makes us laugh or sympathize. I may make reference here to

* *Faust*. [The Devil enters Faust's cell disguised as a poodle. When the Devil reveals himself, Faust comments humorously "on the Devil as the kernel of the poodle."]

earlier studies in which I drew attention to the psychological significance of the factor of *surprise* in wit. Let me turn now to consider a serious objection which analysts in particular will very likely raise. They will point out that analytic technique does not consist simply in giving surprising interpretations and reconstructions: it includes, besides, logical explanations, dissections and other kinds of discussion. True, but I do not think that this in any way invalidates my theory about surprise: what this affirms is that in every case of analytic investigation the most important pieces of insight are of the nature of surprises. I am not describing one characteristic side by side with others; I am defining the essential nature of psycho-analytic technique. My theory is that analysis is essentially a series of confirmations of unconscious expectations. A house does not consist of bricks alone, but also of cement, iron, wood, etc. Nevertheless bricks are the chief material in the building of a house.

If my thesis is correct, we may draw the following conclusion with regard to practical analytic technique: the analyst must approach the psychic material with a conscious openness of mind. I hold that this is a sine qua non of analytic research. Students of analysis cannot be too strongly warned against setting out to investigate the unconscious psychic processes with any definite ideas of what they will find, ideas probably derived from their conscious theoretical knowledge. No great harm would be done if it were merely 'ridiculous' to interpret every association connected with an umbrella as the thought of a penis, or to 'unmask' the returning Œdipus wish behind any friendly impulse on the part of an elderly lady! If a man has not the courage to make himself 'ridiculous' when a vital point is at stake, or has not sufficient intellectual independence to maintain what he has correctly recognized, in face of the laughter of the educated rabble, he had better look round for another profession than that of analyst. But such a schematic application of knowledge is wrong. In a large number of cases it might even conceivably be correct in point of content and yet wrong in point of technique. For an interpretation in itself correct might at that particular point be devoid of all significance. Here we have a specially clear instance of the crucial difference between the kind of analytic knowledge which I should call 'pigeon-hole' knowledge and the *knowledge derived from one's own unconscious*. As in the smallest things, so in the greatest: such a misunderstanding, such a lack of understanding, of the essential nature of analytic insight will continue to make itself felt when it is no longer a question of interpreting a dream-element, an association or a symptomatic action but of grasping the most important unconscious purposes of a neurosis, and here it will lead to consequences even more unpleasant. As one looks back on one's own cases and listens to the account of other people's, how often one has the impression that, under the influence of such pigeon-hole knowledge, the gangway to reality has been hauled in much too soon and the ship has far too quickly put out into the ocean of theory. In practice, I generally find myself becoming the more confused the more I think of familiar analytic theory when I am treating a patient, and I find my bearings again only in the chaos of the living psychic processes.

I should have liked to work out in detail the comparison between analytic technique and the technical processes in wit, but this would take us far beyond the scope of the present paper. There is just one other point to which I wish to refer; it is specially apposite here. It arises here because, in our examination of the psychic processes on which our technique is based, we have passed unawares from a discussion of the situation of the analysand to that of the analyst. Strangely enough, the analyst does here seem to draw very close psychologically to somebody who produces a witty remark. However reluctant we may be to admit such a comparison, there is one vital point in which the psychic processes of these two persons coincide, very different though their aims be. In a witty person the process is as follows: an impression (or, it may be, more than one) is for a moment committed to the unconscious for elaboration and the product of this plunge into the unconscious is then seized upon by conscious apprehension. The mental process of the analyst is similar, different as are its conditions and its aims. To put it in a nutshell: I hold that the more important discoveries and insights in an analysis come as a surprise to analysand and analyst alike.

Here both the scientific and the non-scientific resistances to my 'surprise' theory will probably become most clamorous. But it is just here that I am loth to concede any points, however many objections can be brought and must be brought against my thesis. These objections are easy to understand psychologically, for is it not actually maintained in this theory that the analytic method differs in its whole principle from other scientific methods? This is in fact precisely what is maintained. It is true that even scientists in other fields are prepared to meet with surprises in the course of their investigations —or, at any rate, quickly accept this fact when they do so. But at least they know where they are going and are not throughout dependent scientifically upon unforeseen occurrences, even though on particular points they may meet with surprises, perhaps even great surprises. But is there any other method of research which, from the outset and by its very nature, counts on surprises? Is there any other kind of diagnostic or heuristic enterprise so devoid of any fixed plan, so unsystematical, so lacking in foresight or care for what is to come? Can such a method really be called scientific at all? The psychologist, when he undertakes a piece of scientific investigation, will concentrate all his mental faculties on a definite, precisely delimited problem, and direct all his attention to that single point. The analyst, on the other hand, when he wishes to discover a psychological truth, behaves as strangely as if he were a pupil of the witch:

Theodor Reik

> Who takes no thought,
> To him knowledge comes,
> Without a care he wins it.*

As I said, it is psychologically only too understandable that, in the face of such regrettably casual ways, a new tendency should appear, which expects and demands that psycho-analysis shall conduct its heuristic work systematically. This new tendency, which, curiously enough, originated from within analysis, may be said to demand that we should make an energetic attack on our patients' complexes, penetrate rapidly and vigorously to the central infantile conflict and march according to plan into the realm of the unconscious. The attitude of passive waiting is condemned; henceforth the divining of unconscious relationships is not to depend on the uncontrollable, so to speak 'intuitive' ideas which occur to the analyst. Under the new procedure chaotic situations will no longer arise.† Such a programme not only promises to the budding analyst that his work will be systematic and assured: it also follows that it will be very considerably shortened. In fact, we may view the new programme as a kind of One Year Plan in our psychic economics.

This programme sounds excellent, but, when we examine it more closely, we shall begin to feel serious doubts as to whether there really exists a fixed route for the march into the unconscious. Personally, I think it most improbable that in carrying through an analysis we can avoid all chaotic situations, and I do not find it credible that, at all times and in every phase of the investigation, we can fully understand the structure of a neurosis. Such a view is informed by an unjustified optimism as to the extent and depth of our knowledge of man's psychic life, and fails to take into account that, in spite of all our researches, this is still the most obscure field of which we have any knowledge, more unknown and more difficult to penetrate than anything on earth, in heaven or under the earth. The systematic tactics proposed for our advance into this field may in themselves be admirable, but they are so grandiose that they proceed as if they can afford to ignore the nature of the ground on which the decisive battles are fought. It is easy to be energetic, less easy to know whither our energy is to be directed. In reality there is no material so unsuitable for this vaunted incisive procedure as are the unconscious mental processes. In contrast to this systematic and militant kind of psycho-analysis—though indeed I doubt if it can still be correctly called psycho-analysis—let me commend the deliberate discarding of order and forced regulation in our technique, the lack of all system, the absence of every definite plan; let me be permitted to declare myself the opponent of any and every mechanization of analytic technique.‡ This recommendation to intro-

---

* *Und wer nicht denkt,/Dem wird sie geschenkt,/Er hat sie ohne Sorge—Faust*
† [The reference is to Wilhelm Reich, who wanted at all costs to avoid a chaotic situation.]
‡ [Reik here reiterates Freud's recommendation of 1912b: "The technique, how-

duce order into the things of the mind is very like the labour of many a housemaid who ruthlessly 'tidies up' everything on your writing-table, but in her systematic way stupidly and ignorantly mislays or destroys the fruit of years of arduous work.

The first thing we have to do in analysis is to trust ourselves to the unconscious and to recognize that the only fit governing principle for our technique is to allow ourselves to be surprised. Here he who does not seek, finds. One cannot too strongly recommend to students and to our young colleagues, who are conducting their first analyses, to discard conscious directing ideas in analytic work and to surrender themselves without resistance to the guidance of the unconscious. The one guide can find his way even in the dark: the other loses it in broad daylight. We must impress upon those who are training as analysts that their fundamental attitude in their heuristic work should not differ from that of the analysand and that they are no more at liberty than their patients to reject what occurs to them in connection with the material, even though their ideas seem futile, illogical, meaningless, irrelevant or of no importance. In fact one is sometimes inclined to feel that the analyst needs this warning even more than the patient, for it is made so much easier for the latter to give himself up to what comes into his mind without any sort of expectation, and he does not feel that he has to find an explanation for so much that is obscure in our psychic life. The temptation for the analyst in this situation to follow the laws of conscious thinking is the greater inasmuch as it is his purpose to understand and his business to recognize hidden connections and to explain apparently meaningless symptoms, thoughts, inhibitions, etc. Surely this is the very place for the exercise of intellect, and here intelligence, reason and perspicacity will certainly celebrate their triumphs? Or if not here, where everything turns on understanding, where else are these faculties in place? To this we must reply: even if they are in place everywhere else, they are most certainly out of place here, for intellect is a completely unsuitable instrument for the investigation of the unconscious mental processes. Conscious intellectual effort is not the key to the realm of the dread mothers.* Therefore our advice to the future analyst is not to cling to conscious thinking but, like the patient, to let associations occur to him. In the analytic process reflection amounts to an interruption of the activity of *seeking*, for the sake of checking and testing. 'To think is to lose one's thread', said a French poet (Paul Valéry). If the analyst remains deaf to this warning and pursues the train of his conscious thoughts, yielding prematurely to his craving for logical connections, while the associations are being given, it may well happen that

ever, is a very simple one. As we shall see it rejects the use of any special expedient (even that of taking notes). It consists simply in not directing one's notice to anything in particular and in maintaining the same 'evenly-suspended attention' (as I have called it) in the face of all that one hears" (p. 111).]

* *Faust*, Part II

he will force the patient artificially in a wrong direction; so too, if, instead of following his own unconscious, he looks for what his theoretical knowledge has taught him to expect and prefers 'pigeon-hole' information to living insight. The patient then often has difficulty in guiding the analyst, who is trying to bring him to reason, back to the only correct point of view, namely, that of free association. The situation is very like that which at times wrings a lament from the individual members of an orchestra: 'It is terribly hard', they sigh, 'to keep a conductor together'. The difficulty which the analyst has overcome in his own analysis confronts him again when he comes to analyse others. For a rational frame of mind it is much more difficult to leap in thought from 100 to 1000 than to pass from 100 to 101, thence to 102 and so on. To require such a peculiar and unwonted procedure, in which the censorship is eliminated, is something like expecting a sedate citizen to take his daily constitutional in a series of wild bounds instead of his usual measured pace.

One has constantly to emphasize the fact that here at once the way of analysis diverges from that of other methods of research. The first principle of these is that the investigator shall reflect, criticize, scrutinize, and concentrate rigorously on a clearly defined goal towards which his efforts are directed. The analyst's first principle, on the other hand, is that he must yield to free association, abandon the idea of a preconceived goal and only later, by way of a testing process, work over his material, ordering and scrutinizing it and arranging it in logical sequence. All other methods seek to exclude any intrusion of unconscious factors as detrimental to the investigation, while for analysis it is the intrusion of conscious factors which constitutes an impediment to its heuristic work.

The basis of analysis is the establishment of an understanding between the unconscious of one person and that of another; conscious thought and speech serve only to clarify this kind of wireless telegraphy, and indicate more often where the trouble lies than where are the forces which operate to produce it. This duologue between one unconscious and another goes on, as it were, as an accompaniment to the other 'official' communications—a secret conversation, inaudible but genuinely recognizable and recognized. Such being the psychic conditions of analytic investigation, it follows that there will be a different kind of evidence for our results. Every other method primarily seeks for objective certainty; psycho-analysis has to aim at a particular kind of subjective conviction. When our material leads us to a certain conjecture, we have to test this repeatedly and critically by what has gone before and what comes after, until we are so convinced of it that it is equivalent to an objective certainty. In psychological experiments a definite piece of knowledge is arrived at by means of strenuous intellectual work; in analysis it is a question of making discoveries. There is indeed an inveterate scientific prejudice to the effect that discoveries come by chance. The fact is, however, that they merely come as a surprise. In analysis also psychic work is most certainly going on—work which makes peculiarly exacting demands on the heart and brain of the investigator—but it has no affinity with the strenuous exertions of conscious intelligence: it goes on in the unconscious. No conscious effort, no conscious application of knowledge, no theoretical reflection is of any use in dealing with this unique kind of material. If an analyst be interrogated as to the nature of his work, he may answer like a certain poet, to whom a lady remarked that it must be very difficult to write a poem: 'Either you do it very easily or you don't do it at all'. Moreover, it must not be forgotten that the main part of the psychic work was done before the analysis of the patient began: it consisted in opening the way to the analyst's own unconscious.

It is, nevertheless, astonishing to see how often an analyst thinks it enough to follow with uniform attentiveness the associations of the analysand. But listening is not enough: the analyst must hear what the patient says, but, at the same time, he must hear what his own inner voice says, and he must have the courage to understand it, even if the connections do not become plain to him until much later. Of course, he need not say what comes into his mind nor surrender himself passively to it, but he must learn to pay attention to it and keep hold of it. He must have the courage to understand; but also he must have the courage not to understand what his own need for logical connections, his common sense and his conscious knowledge try to thrust upon him. He must have the courage not to understand, even when analytic theory suggests to him certain expectations, when he seems to be going upon conscious psychological knowledge. He must maintain an attitude of steadfast scepticism in the face of the lure of analytic thoughts which only too readily and willingly oppose themselves to the current of the unconscious. Many of the interpretations which throng into his mind from recollections of what he has read and heard, learnt and taught himself, are comparable to the unconscious derision so often present in the analytic interpretations offered by patients. We have reason to fear such analytic interpretations, presented, as it were, on a salver, as the seer of old feared the gifts of the Greeks. The analyst's attitude towards them, for all their specious pretensions, will be that of St. Joan in Bernard Shaw's play, when she is told that the Church prescribes this and that: 'But my voices do not tell me so'. The technique of analysis cannot be learnt in the abstract: it can only be won from living experience. The courage to understand and the courage not to understand— these are not intellectual qualities, but a matter of character, an expression of moral courage, an issue of inner sincerity—manifested in spite of and often in opposition to the ego.

Whence comes, then, that psychological insight with its element of surprise which we maintain to be the most important part of analytic technique? If it is not conscious knowledge, laboriously acquired by means of systematic work, it can only be a gift due to intuition. Can it be that our technique throws us back on this uncontrollable faculty, this will-o'-the-wisp, a mystic fount or a chimera whose nature we cannot discern? No; this insight has its source in knowledge which has become unconscious. The element of surprise derives from this very unconscious possession: we do not here need to find something new, but something we had lost; we do not discover, we remember. The source of that unconscious knowledge is in the reservoir of our own suffering, which thus bears fruit. That which comes into our mind when we listen to a patient rises from those psychic depths in which our personal suffering has learnt to understand that of others. I purposely use the word 'suffering'. Not misfortune, not mere calamity, nor yet disastrous or sad experience thus bears fruit. A burnt child dreads the fire as a neurotic dreads the power of his instincts, but in the case of fire there are many things one can do besides burning oneself. It is true that the hurts which we sustain teach us caution, but suffering, consciously experienced and mastered, teaches us wisdom.

CHAPTER 27

HELLMUTH KAISER

PROBLEMS OF TECHNIQUE

INTRODUCTORY NOTE

Kaiser, in this article, writes as a disciple of Wilhelm Reich—a disciple with a distinctly original mind. The "defiant" or "sulking" character he describes reminds us more of Kafka or Camus than of the patients delineated by Reich as having a "character armor." He carried resistance-analysis to the very extreme and therefore to the very border of psychoanalysis itself. The reasoning is often obsessive, and yet every section brings something new. Kaiser's views did not prevail. Until now the English speaking reader knew about him only from the criticism leveled at him by Fenichel (1935, this volume, Chapter 30), and Alexander (1935, this volume, Chapter 29). His ideas deserved better. One need not follow his footsteps to learn from his approach.

Noteworthy also is Wilhelm Reich's criticism of this paper. It appeared a year after the publication of Kaiser's paper in Reich's Lucerne address; incidentally, this was the last time he spoke as a member of the International Psycho-Analytic Association:

> *. . . if Kaiser says that consistent resistance analysis not only makes any interpretation superfluous but a mistake, I could agree with him only on the ground of theoretical principles. In saying so, he forgets that my formulation of the "interpretation at the end" is practically necessary as long as character-analytic technique is not perfected to such a degree that we no longer have any difficulty at all in finding our way in the maze of defenses. His contention, then, is correct only for the ideal case of character-analytic work. I must admit that I am still far from that ideal and that I still find the dissolu-*

Originally entitled "Probleme der Technik," *Internationale Zeitschrift für Psychoanalyse* 20 (1934):490–522. First English translation in this volume by Margaret Nunberg. This translation is based on an earlier unpublished translation by Hellmuth Kaiser, given to the editors by T. A. Munson, M.D. Reprinted by permission of Louis B. Fierman.

tion of the defense formation difficult work, particularly with regard to con-
tactlessness and the interlacing of defenses. What makes character-analytic
work so difficult is a consideration which Kaiser overlooks, the sex-economic
consideration; this makes it necessary to work in such a fashion that the
maximum amount of sexual excitation becomes concentrated on the genital
and thus appears as orgasm anxiety *(Reich 1933, p. 284).*

I / Introduction

IN THE COURSE of my practice of analysis, my attention was drawn especially to a group of patients singularly resistant to the technique I employed.
At the beginning of treatment the difficulties which each patient exhibited
appeared distinctly different from those of the others. The more their analyses
progressed, however, or at least as the number of hours increased, the apparent differences between the patterns of these difficulties lessened, and at last I
gained the conviction that the core of resistance which had made my analytic
effort unproductive was actually the same in all these cases. The patients
behaved quite differently, and neurosis stunted their lives in varying degrees
(at least to more casual observation). In spite of the apparent differences
among them, I observed that the resistance which they developed during their
analyses invariably produced the same kind of feeling reaction in me. The first
progress I made with one of these patients followed my attempt to interpret
the essence of his resistance as a kind of defiance or stubborness. Soon afterward I noticed that this same interpretation helped me to clear away the
resistance in the other patients of the group.

This experience led me to a series of studies into two different problems.
First I tried to clarify the concept of the "defiant character," to give an exact
description of the common denominator of this character structure, to understand its genesis theoretically as far as possible, and to differentiate it from
other character types. The second problem was to determine why these patients were so little influenced by the kind of analytic technique which helped
others to varying degrees by producing considerable improvement if not complete cures. Beginning my investigation with the most general theoretical
fundamentals of technique, I arrived at a formulation of several principles for
the analytic treatment of the "defiant character" and studied the application
of these principles in practice.

The conclusions I reached about this are contained in a paper which I
finished some months ago under the title, "The Defiant Character,"* but

---

* [The German "Defiant Character" could also be translated as "sulking character"
or "stubborn character." To the best of our knowledge the essay referred to has never
been published.]

space for its publication has not yet become available. The aim of the present paper is to discuss some thoughts about the theory of technique which resulted from my study of the "defiant character" and its typical resistances. These, I believe, are relevant to the development of analytic technique even outside the context of this special character formation.

I shall begin with some general remarks about analytic technique and will refer to the special mechanisms of the "defiant character" only in those places where the experience I gained in treating such patients proved to be particularly instructive.

II / *Basic Problems of Technique*

There is agreement among analysts that the patient's recovery, in successful cases, is achieved in the following way: instinctual impulses that since early childhood have been warded off by the patient's ego through the use of more or less complex mechanisms, and have been kept away from the means of discharge or modification offered by the system preconscious, are freed from their occlusion; consequently they can take part in forming the total personality and can obtain the means of expression that are characteristic of the healthy individual.

We are accustomed in analytic treatment to observing a concrete correlate of the liberation of such repressed impulses. For the purposes of the following discussion, I shall refer to this phenomenon as "a genuine breakthrough of instinct";* in order to make a sure identification with his own observations possible for a reader, I will immediately describe this phenomenon.

The patient speaks in an agitated, often oddly quavering tone of voice. He is highly excited; his musculature is loosened, though this relaxation is frequently interrupted by spasmodic contractions or momentary periods of stiff posture which express anxiety, revulsion, or shame. He speaks now hurriedly, now hesitatingly; the expression, tone, and modulation of his voice, his facial expression and gestures, and his diction and grammar strike the observer as being natural. He gives an impression of spontaneity, and his behavior carries the same conviction as does the manner of a normal person who is moved by strong affects. The content of what he says relates to an impulse which is arising in him, and refers at one and the same time to the analyst and to an object of his childhood. At the beginning of the instinctual

* [The emphasis on discharge as the crucial factor in cure is close to the thinking of Wilhelm Reich and Fenichel (see Fenichel 1937a, p. 133).]

breakthrough, the childhood object is still fused with the person of the analyst. As the breakthrough continues it becomes separated from the analyst and presents itself as a definite personality from the childhood days of the patient. The patient's speech oscillates between the present and past tenses, between addressing the analyst as the previous object and relating recollections.

Following the breakthrough, the patient feels relaxed, exhausted, and greatly relieved. This is especially the case when there is time during the analytic hour for the breakthrough to run its full course. Then the patient is often in a relieved and intensely happy mood.

The analyst's feeling reaction toward a genuine instinctual breakthrough is similar to that evoked by witnessing a magnificent display of natural forces such as a severe thunderstorm, a storm tide, or the eruption of a volcano. He feels shaken. The impression such a genuine breakthrough leaves is so strong that he is rarely doubtful of the genuineness of the phenomenon. It may happen fairly often in the beginning that the analyst takes a bogus breakthrough (expressing a certain kind of resistance) for the real thing, but usually he has some doubt about it and sooner or later realizes that he has been taken in.

We cannot tell with certainty whether these instinctual breakthroughs are merely indicators of the liberation of instincts that has been previously accomplished, or whether the liberation of instincts actually takes place during this phenomenon. However, the changes in the patient which indicate progress in the direction of a real cure seem only to be observed after the occurrence of such a phenomenon. We expect that analysts will generally agree on this point too.

The basic problem of technique, namely, how and by what means it is possible to liberate the repressed instincts, thus seems to coincide with the problem of finding the means by which one can achieve such an instinctual breakthrough. The most general answer to this problem has been given by Freud: Repressed impulses in themselves tend to break through into consciousness. What prevents them from doing so are the defense mechanisms established in childhood and since then tremendously developed. The analyst experiences the effect of these mechanisms in the so-called resistances, the intensity and number of which increase during analysis if analysis itself makes new defenses necessary for the patient. It follows then that the liberation of the repressed impulses has to be accomplished by overcoming the resistances.

The basic problem of technique can now be formulated thus: how to overcome or remove the resistances of the patient. The answers which Freud has given to this question are brief and far less precisely formulated than the majority of those dealing with different problems contained in his technical papers. Strangely enough, his solutions have not been reformulated, supplemented, or interpreted since they were offered about twenty years ago. The most important and pertinent passages are:

> Finally, the present-day technique evolved itself, whereby the analyst abandons concentration on any particular element or problem, contents himself with studying whatever is occupying the patient's mind at the moment, and employs the art of interpretation mainly for the purpose of recognizing the resistances which come up in regard to this material and making the patient aware of them. A rearrangement of the division of labor results from this; the physician discovers the resistances which are unknown to the patient; when these are removed the patient often relates the forgotten situations and connections without any difficulty (1914c, p. 147).
>
> The first step in overcoming the resistance is made, as we know, by the analyst's discovering the resistance, which is never recognized by the patient, and acquainting him with it. . . . One must allow the patient time to get to know this resistance of which he is ignorant, to 'work through' it, to overcome it, by continuing the work according to the analytic rule in defiance of it (1914c, p. 155).

We can find in these statements of Freud no completely satisfying answer to our basic question about technique, but this is not because they are too poor in content. On the contrary, they are immensely rich, so rich that the full development of the theories, observations, suggestions, and problems raised in them would fill volumes. They are in this sense similar to the "Three Essays on the Theory of Sexuality" in that one could say that much of the analytic literature can be viewed as an explicit elaboration of the basic concepts which are presented there in concentrated form.

It is certainly an interesting question, though one which lies outside the scope of the present paper, why technical therapeutic problems, in spite of their obviously immense practical significance, have had such a relatively poor developmental elaboration, much poorer than that of many theoretical questions.

Freud's formulations are neither so precise nor so clearly presented as to admit of only one interpretation. One evidence of this is the fact that different analysts employ very different methods, while each gives us to understand that his embodies Freud's essential principles and should be adopted by all.

There is one interpretation of Freud's remarks that I want to discuss prior to the others, because it falls in a special category; namely, that Freud's intention was only to advise that one should overcome the patient's resistances, that he did not say anything about how to do this for the simple reason that rules and fixed methods are out of the question—everything has to be left to the intuition of the individual analyst. This opinion is approximately that presented by Reik in his last lecture to the Congress at Lucerne.* Such an

* [The reference must be to Reik's paper of 1933, this volume, Chapter 26. However, the footnote to the paper says it was read at the Wiesbaden Congress and not at the Lucerne Congress.]

opinion could hardly make the further development of the theory of technique superfluous. For if this interpretation did justice to Freud's opinion—which I do not believe—and even if such an opinion were correct, the question of how it is possible that the resistances should be so structured as to permit their own dissolution or overthrow would still require answering. Also, the further questions, what should the specific role of intuition be in this process and on what grounds could we at least dissuade the inexperienced candidate from telling the patient all of his intuitive ideas? As long as these questions remain unanswered, the opinion that no theory of technique is possible can hardly claim scientific basis.

III / *The Resolution of Resistances*

We are now going to pose a basic question without committing ourselves in advance to any definite theory. How is it possible to remove, overcome, or resolve the resistances of the patient? Asking this question implies that one has a definite idea about the nature of the resistances, for we can hardly hope to find an opponent's vulnerable point before we have studied him thoroughly. The implication of this at once arouses our anxiety. The forms, kinds, and classes of resistances show such overwhelming richness. It seems hopeless to attempt to find a common denominator on the basis of which it would be possible to say something in general about their resolution.

I propose to start, therefore, with a far more modest task. Let us look at a single concrete example of the resolution of resistance. We shall choose one as favorable and instructive as possible and will try to understand precisely what happens in this single case. During the analytic hour the patient reports somewhat angrily that this morning he had intended to meet his brother-in-law, his sister's husband, who is on a trip and had a layover of one hour in the patient's home town. The patient was to have met him at the railway station but did not make it. He felt very sorry about this because he liked his brother-in-law very much. Asked how this had happened, he answers that he had been late but this had been through no fault of his own. Just at the moment he was to leave the house in order to take the streetcar, the telephone had rung. The analyst asks, "Well, was there nobody else in the house to take the call?" The patient hesitates a moment, then says, "Well, there was someone. It wasn't even so late, only the telephone conversation dragged on so long." The analyst asks, "Was it something very important?" "No, not exactly," answers the patient, somewhat hesitatingly, "but the other party wouldn't stop talking, and I couldn't be so impolite as to interrupt him." While saying this the patient's voice becomes a bit hoarse; he continues in a more irritated tone of voice. "And just at the moment that I got down to the street, off went the streetcar." "And then?" "Well, I went to my office." "But couldn't you take a taxi?" "This morning," answers the patient in a low voice, "that didn't occur to me." The analyst asks, "Why do you sound so angry?" "My God, why are you needling me? So I missed the damn guy, so what?" Immediately after this answer the patient turns around to the analyst, his face is relaxed and his glance conveys that inside he understands what was going on in him.

We contend that this scene, though certainly a very trite one, contains a key to the resolution of resistance. The repressed impulse is a hostile feeling toward the brother-in-law, repressed by the patient because it originates in an incestuous fixation on his sister, of which he, of course, is unaware. This impulse was reactivated by the brother-in-law's passing through the patient's home town and the demand made on the patient to meet his brother-in-law at the railway station. The defense against the impulse was insufficient; though it did not break through into consciousness, it showed itself in a series of parapraxes which prevented the rendezvous. It was also visible in the angry mood of the patient at the beginning of the hour. What was the mechanism which prevented the prohibited impulse from coming to consciousness at that time? It seems obvious that much of it is contained in the patient's conviction that he was not to blame for his being late. This conviction was maintained by a number of thoughts which were subjected by the patient to very little scrutiny on his part, perhaps we could even say, preconscious thoughts. One of these, for instance, was "to interrupt the other party on the phone would be impolite. To be impolite is unthinkable, simply not done." If this highly intelligent patient had focused his attention on these thoughts, he could hardly have maintained his "conviction" which prevented the repressed impulse from becoming conscious. What the analyst did was draw the patient's attention in an appropriate manner to the faulty thinking at the basis of his conviction and so make it possible for the intellect of the patient to recognize it as faulty, and thus to destroy his conviction that his being late was due only to external circumstances. As the process of showing up the faultiness of the thinking is further pursued, more of the forbidden impulse comes to light. At the very moment when the patient begins to recognize the hostile character of the impulse, new means of defense are attempted, now limited to casting obscurity around the object of the patient's hostile striving. The patient interprets the questions of the analyst as needling, and hence, is able immediately to motivate an irritation which allegedly is aimed at the analyst. In other words, defense by means of transference is tried, but, alas, too late. The swearing at the brother-in-law, which slips involuntarily from his lips, is proof that the game is lost, the defense cannot be maintained.

It must be stressed (and we shall return to this point) that the dynamic element in the defense of the patient is not contained in the faulty thoughts, but in the narcissistic libido mobilized by the associative link between the concept of his own aggression and the idea of damage to his ego. The unimpaired intelligence of the patient would have refuted the faulty thoughts, which imply such incomplete testing of reality; the impairment of the intelligence is engendered by the narcissistic libido.*

In order to clarify as far as possible the development and delineation of this brief scenario, the phenomena which take place in the psyche of the patient are listed as follows in a schematic fashion, starting from the deepest layer:

> Aggressive impulse—idea of injury to the ego (perhaps castration anxiety).
>
> Mobilization of narcissistic libido—repression of the impulse-arousing stimulus (announcement of the pending arrival of the brother-in-law).
>
> Activation of the impulse—symptoms of the breakthrough (parapraxes) —impairment of the self-observation by the narcissistic libido (rationalization of the parapraxes).

From the topographic viewpoint, the last two phenomena occur in the realm of the preconscious; all the antecedents, in the realm of the unconscious.

How is therapeutic intervention effective in resolving the resistance? It has already been pointed out that the analyst's purpose is to draw the patient's attention to the erroneous thoughts which stand in the way of his self-observation. This description adequately treats the situation topographically, but gives no dynamic understanding. In fact, it would seem as if no effort at all were necessary, as if the patient would immediately unmask and reject his erroneous thoughts once his attention were drawn to them, without further ado from the analyst. But in writing this sentence we realize at once that a dynamic element has been omitted. That an otherwise very intelligent patient should think erroneous thoughts and give them inadequate attention can only be explained on the basis of an inhibiting force. Reference has already been made to this force, namely, the narcissistic libido (more specifically a certain amount of narcissistic libido). It is possible to direct the attention of the patient to his erroneous thoughts only if the force of this attention-inhibiting libido is overcome. In other words, a counterforce is needed, the origin of which, in this particular case, is close at hand. Our patient clearly has a certain amount of contact with his analyst. He answers the analyst's questions and wants to be understood. The force which opposes and limits the effect of the narcissistic libido so that the rationalizations of the patient are brought into the focus of attention is none other than the transference attachment, more exactly his tender, reasonable attachment to the analyst.

Indeed, it is not always possible to direct the patient's attention where we wish. Another patient might easily have answered the question about his

* [This is an example of a typical jump from a clinical observation to a metapsychological assumption; see editors' introduction, p. 11.]

lateness with, "Why this sudden curiosity? You don't usually ask questions," and might have continued perhaps with further scornful remarks about analysis. It is also apparent in the example that the analyst employs the technique of leading the conversation by questions much as one might in everyday conversation, yet he does so with a certain deliberate skill. He takes one step at a time in putting his questions, taking into account the impaired rather than the unimpaired intellect of the patient. His tone of voice is neither inquisitional nor ironical.

It is now necessary to amplify whatever understanding has been reached of the sequence of happenings in the example by pursuing a further question pertaining to the energic or (more familiarly) economic situation. So far we have considered the dynamics, but not the transformation of energy.* One thing is certain—the possibility of discharge of the dammed-up hostile impulse against the brother-in-law is now greater than the previous narrow channeling into the parapraxes. This discharge is no longer limited to verbal aggression against the brother-in-law during the analytic hour, but will inevitably find greater scope once the affect has been admitted to consciousness, the range depending on the cultural level of the patient (for example, one possible action would be a polite but cool letter of apology to the brother-in-law).

What happens to the narcissistic libido which brought about the defense but which has yielded to the strength of the transference relationship? A certain field of action has been excluded from its domain, but there is no reason to suppose that the entire force was demobilized in this analytic hour. On the contrary, analysis has liberated a part of the repressed impulse, thereby increasing the danger that the core of the impulse, the ancient death wish against the father stemming from the Oedipal situation, may flood into consciousness. Increased counter-cathexis is necessary to forestall this happening. For example, the narcissistic libido could now be used to discover some innocent sounding rationalization for the hostility against the brother-in-law, and by cautious steering of attention, to protect this new formation from the intellect of the patient. The analyst will have to reckon with the sudden appearance of a new resistance, this time one step closer to the core of the neurosis. In the example given, our understanding of the effectiveness of the action of the analyst in resolving the resistance is that the latter made use of rationalizations, the irrationality of which the patient could recognize once his attention had been drawn to these thoughts (which we call resistance-thoughts) by the help of the existing transference relationship.

---

* [Here, once more, Kaiser expects to solve a problem on a purely metapsychological level.]

IV / Can Resistances Be Approached in Ways Other Than Those Already Illustrated?

It may be agreed that the example demonstrates the resolution of a resistance (though certainly only a *single* resistance) and provides data for an understanding of this resolution. However, we must now proceed one step further and pose the question of whether the conduct of the analyst in this example was the only mode of intervention open to him.

Clearly it was not. Even if the function of the analyst were formulated narrowly in terms of the special task of making the patient conscious of his resistance-thoughts, diverse methods may be used. Moreover, it must be expected that the individuality of the analyst will find expression in his method; leeway must be allowed for this factor. One could imagine an analyst, for instance, whose characteristic way of expressing his aggression is by means of persistent questions aimed at cornering his opponent. In this case it would not be possible for the analyst to use the tactic of the given example, since he would intuitively know that, for him, questions tend to be put sharply and inquisitively, to the detriment of the analytic work. However, this analyst might arrive at the same goal in an equally effective manner by giving the patient a description of all that happened and drawing his attention to certain inconsistencies in the way he (the patient) gave vent to his disappointment about the missed rendezvous. If the patient has understood the analyst's mere question, "Were you entirely innocent of being late?" possibly this would be sufficient to destroy the patient's resistance-thoughts.

Leaving the example aside for a moment, it might be added that the destruction of the resistance-thoughts is frequently brought about not by words but by the actions or attitudes of the analyst. If, for instance, the resistance-thought of the patient is that the analyst will despise him because of his sexual fantasies (which are therefore kept out of the analysis), then the mere calmness and frankness of the analyst in regard to sexual topics may suffice to make this resistance-thought untenable.

However, there is a more important question. It is the general one of whether the resolution of a resistance may be effected in any other way than by leading the patient to refute the resistance-thoughts. It would seem that one such method is to inform the patient directly or tacitly of his repressed impulse. In the case of the example, the method would consist of telling the patient, at some appropriate moment, that he obviously has some hostile impulse against his brother-in-law.

The data of the example are insufficient to indicate what effect giving such information would have on the patient. Every analyst will be familiar with instances during analyses where he achieved with such interpretations much the same success as did the analyst in the example. Supposing that such a direct method had succeeded in the given example too, the question of how it worked would arise.

An immediate answer comes to mind. If someone secretly places in my room an object that is strange to me, it may happen that for days on end I do not notice it. If, however, I am then told: "Think over whether you have not seen such and such a thing in your room," I will often say: "Oh yes, I vaguely remember having seen such an object there." Similarly, it might be explained that the patient becomes conscious of his repressed impulse only if he is given the anticipatory idea of finding within himself an impulse of this or that type, and if he is urged to look for it.* However, this explanation does not stand up to close scrutiny. A repressed impulse exists neither in the system Cs. nor in the system Pcs. Even were the patient given a very precise anticipatory idea about the object of his search, he has no more chance of finding the impulse than a person who searches the wrong room has of retrieving a lost object; the patient cannot look into his unconscious.

If, in the hypothetical example, as a consequence of telling the patient, "You have hostile feelings toward your brother-in-law," the impulse in question comes vividly into consciousness, some different explanation will evidently be needed. Telling the patient this may, under favorable conditions, suffice to cause the patient to examine critically his resistance-thoughts. Such an examination will be induced if the information about the hostility to the brother-in-law puts into his possession a more convincing explanation of the errors than his previous, inadequate rationalizations. If this explanation is correct, then the apparently alternative methods of imparting information about the repressed impulse destroy the resistances also by bringing about an invalidation of the resistance-thoughts. The method merely uses a special technical tactic in order to draw the patient's attention to his resistance-thoughts.

But this question concerning the effect of giving the patient information about repressed impulses is merely introduced, rather than settled, by the discussion so far. Before it can be pursued further we must dispense with the limit imposed by the particular example and state our position more generally. To do so puts the views advocated here in a very exposed position, but provides a much wider basis for their explanation and illumination.

* [See editors' introduction on anticipatory ideas, p. 24.]

V / *The Structure of Resistances*

It is now proposed to extend the set of problems under discussion by contending that the little example of the resolution of resistance may claim a general validity. In other words, it is contended not only that all resistances have in essence a common structure with that of the example, but also that in principle their resolution is possible using the same method (under basically the same conditions) as in the illustrative example. More explicitly, it is maintained that every resistance contains a faulty thought or act of thinking (at variance with reality), that the dynamic aspect of the resistance is furnished by a given quantity of narcissistic libido which keeps these thoughts out of range of the critical attention of the adult ego of the patient, and that the resolution of resistances is possible in no other way than by the analyst's drawing the attention of the patient to the resistance-thoughts. The motivating force on the side of the patient stems from his tender attachment to the analyst. The patient's interest in recovery as such does not play a role.

Two answers tend to be given, with about equal frequency, to this contention—first that it is trivial, second that it is fantastically bold. The accusation of triviality seems to be based on the fact that the contention is actually only a somewhat more explicit presentation of certain basic thoughts of Freud, expressed, for instance, in the two quotations on p. 387. (There is further consideration of this at the end of this paper.) The second accusation probably relates to the fact that if my contention is logically thought through to its end, it leads to certain compelling conclusions and directions concerning technique, conclusions to which analysts subscribe with widely varying degrees of conviction.

It is, however, a usual and permissible procedure to determine by examination whether or not a concept applicable to a special case has more general applicability; but before this examination is pursued, an apparently formidable argument against the general contention must first be faced.

The argument or, perhaps better, the reproach, might be expressed as follows: "The very idea! Why, Freud had already proved very plausibly over twenty years ago that the mere transmission of knowledge has no therapeutic effect in the analytic sense of the word, for which reason he repeatedly stressed the dynamic nature of the resistances. Now, someone calling himself a follower of Freud comes out with the statement that the resistances are based on errors, and that the correcting of the errors effects the cure. Surely a return to prepsychoanalytic notions! Virtually approaching the tenets of Christian Science!"

But the reproach misses by far what actually is maintained. It was neither stated nor believed that the neurotic becomes cured by increasing his knowledge, no matter whether this is knowledge of his resistances or of forgotten childhood experience. Rather it is believed that his becoming cured is dependent upon change in his libidinal economy.* This can occur only by way of resolution of resistances. In turn, the resolution of resistances is accomplished only by the imparting of insight (which, as already explained, is not necessarily identical with the communication of opinions). Here again the dynamic situation calls for consideration, and there will be further explicit discussion of the thoroughness with which the task must be pursued in order to bring about the economic change which is intended. If the procedure which has been outlined is called intellectualistic, there can be no objection; but on this account it should not be confused with the procedure wherein the attempt is made to give the patient knowledge about the etiology of his neurosis and the contents of his unconscious in the expectation of a successful outcome.

The task of justifying the central contention of the paper is, however, of considerable magnitude; there is little hope that it can be accomplished as completely as might be wished. The forms of resistance are too great to allow each type of resistance to be dealt with in a relatively limited space. Above all, the phenomenon of resistance (which could equally well be called defense) is a rather complex one; to understand it fully we would need a greater knowledge of ego-psychology than is yet at our disposal. What is aimed at is no more than convincing the reader that it would be worth his while to examine our contention's correctness and to test its theoretical and practical usefulness.

VI / *Transference-Resistances*

An essential step in the development of the present thesis would be the demonstration that the so-called transference-resistances can be conceptualized as analogous to those of the example; furthermore, that the same technical principle should be applicable in their resolution.

We say that a transference-resistance is present when the patient directs a repressed infantile impulse onto the analyst in such a manner that the analyst becomes the object of this impulse. However, this formulation is neither very exact nor logical. The infantile impulse has precise determination both as to its instinctual aim and its object; this is an inseparable correlate with the fact of its being repressed. For example, the impulse may be to bite

---

* [Cure is described in metapsychological terms.]

off the father's penis, but the father in question is the father as he appeared at a precise phase of the patient's development—e.g., in the third year of his life. This instinctual wish remains intact (with all its characteristics) until liberated by removal of its repression by analysis. The repressed wish was actually experienced in all its intensity at one point during development, and at that same point was repressed. The wish retains all its original characteristics even when it enters the transference or becomes transferred to the analyst. When we say the repressed wish is manifest in the transference, what we mean is that the patient has acted toward the analyst (understanding action in the broadest sense to include specific acts as well as speech and attitudes) in such a way that a genetic connection between the action and the repressed impulse becomes apparent. To return to the example, let us suppose that the patient suddenly began to call the analyst "Dr. X" instead of "Herr Dr. X."* It would be misleading to call this phenomenon a resistance, and yet it appears in the analysis only after a certain resistance has been overcome. Of course, the fact that the repressed impulse itself does not enter consciousness is the effect of the resistance.

The phenomenon of transference-action is regularly accompanied by certain characteristic features; but before discussing these features it must be added that this proposition is based on clinical experience. It is in no way contradictory to what we know of ego psychology, but the latter is insufficiently developed to allow the proposition to be derived from it in a purely deductive way.

The first of these characteristic features is the particular form which the transference-action assumes in the example, namely, that the patient either does not notice that his mode of addressing the analyst changes or else accounts for the change with some rationalization (he may say, for instance, that the address *Herr* is incompatible with his general non-bourgeois outlook on life).

In either case (as well as in corresponding phenomena in other examples) something happens which shields (or protects) the patient's conscious ego from being confronted with an ego-alien action. Of course, it would be a mistake to assume that the transference-action occurs before the ego becomes protected. Rather, the transference-action is only possible while the ego-protecting measure is in force. We wish to designate this ego-protecting measure as rationalization. In the second case of the hypothetical example, the designation is a bit forced in the first case when applied to the scotomization of the transference-action, but perhaps it can be tolerated in the interest of brevity of expression, especially since we understand that this scotomization of the transference-action keeps the ego unaware of an irrational action.

It is now possible to complete the contention that the transference-action

* [In polite German usage *Herr* (Mister) precedes almost any title—trans.]

as such is not a resistance by stating that the resistance comes into appearance only in the accompanying rationalizations.* The dynamic factor again lies in that quantity of narcissistic libido which protects the ego by turning the attention of the patient away from the rationalizations.

The resistances are again resolved by the analyst's drawing the patient's attention to his rationalizations. If the transference-action is scotomized, the patient either immediately stops this particular action, only to start a little later some other mode of action stemming from the same repressed impulse, or alternatively he produces rationalizations in the narrower meaning of the word, i.e., the *resistance-thoughts* previously discussed. In the first instance nothing new is presented. In the second case the analyst should, in some suitable manner, bring it about that the patient is induced to think through these resistance-thoughts (he may give evidence of doing so either in words or in his attitudes or actions) so that the resistance disappears. In very rare cases a true instinctual breakthrough may follow in the wake of the dissolving transference-action. More usually a new transference-action makes its appearance, now closer to the repressed affect, more filled with emotion. The rate of dissolution of the resistance-thoughts (and whether it occurs at all) depends partly on the technical skill of the analyst (about which more will be said later), and partly on the strength and suitability of the patient's attachment to the analyst.

However, the above account was deliberately oversimplified and a certain elaboration is needed. There are other means of protecting the ego which accompany the transference-action than scotomization or the development of resistance-thoughts. Not infrequently the patient does observe his transference-action, but he disowns it, attributing it to an inexplicable irresistible compulsion, foreign to his nature. Close scrutiny shows, however, that this phenomenon falls in the second of the classes of rationalization mentioned previously. The patient's allegation that he has no reasons for his action but is driven by a superior force is just as erroneous as the corresponding disclaimers of obsessive-compulsive neurotics. A very careful examination always shows that behind the feeling of being compelled hides a well-developed rationalization covered up by the apparently fatalistic submission to an irresistible compulsion, and thus protected from the criticism of the intellect (which consideration leads to certain speculations concerning the treatment of obsessive-compulsive neurotics; speculations which are, however, somewhat away from the present theme and can best be dealt with in another paper).

---

* [This is a most valuable distinction. Freud originally equated transference with resistance, because the patient failed to follow the basic rule when overpowered by transference feelings.]

VII / *Character-Resistances*

Before dealing with the *character-resistances* (using much the same line of argument as for the transference-resistances) something must be said about the meaning attached to this term. What we have in mind are the same phenomena to which Reich, in his very important book on character analysis, gives this designation. (The content of this book will be discussed toward the end of the paper, to the extent necessary for the present purpose. My approach diverges somewhat from Reich's.)

The character, and here I am in complete agreement with Reich, appears in the analysis of every patient as resistance. I also agree with Reich's contention that the character is a system of permanent modes of reaction of the ego which form a whole and afford protection against the breakthrough of instinctual drives on the one hand and against hostile responses of the outer world to affective breakthrough on the other. I believe, however, that the behavioral phenomena, which we group together under the concept of *character* or *character-resistance* so as to form a psychologically completely satisfying gestalt, are best conceptualized within the framework and organization around the primacy of feeling for the self or self-esteem. The more the self-perception of a person is distorted by neurotic libidinal development and by compensatory regression to an archaic (infantile) style, the more his character appears as peculiar, odd, or warped, and the more his behavior in the analytic situation takes on the appearance of character-resistance.

Character-resistance is distinguishable phenomenologically from transference-resistance in that it appears impersonal and lacking in affective vitality. If a patient whose character-resistance is unresolved tries to attack his analyst, the analyst feels untouched in a peculiar way; the impression which is made is not that some inhibited or veiled impulse reaches expression, but that the aggression aims at some invisible target. If an aggressive attack, which is perhaps quite similar in its content, appears as an element of the transference-resistance, the analyst feels far more strongly that he is the real object of the aggression, even if he knows (and can prove it by the wording of the aggressive act) that the patient really means his father. The kind of *character-armor* in which many single but interrelated features are shaped into a system not only has a history in which the genetic development of each element may be pursued, but it also forms something like a thought system (comparable with a paranoid delusional system), the core of which expresses the particular style by which the self-esteem is regulated.

The "defiant character" is distinguished by a type of preconscious ideology which might be expressed as follows: everything depends on the obliteration of, or compensation for, a humiliating blemish which blackens and damns the patient forever. The blemish or taint can be neutralized only by incessant observance of the law. The observance is dominated by pedantry and by being undaunted in the face of any sacrifice. By following the law to the letter, justice or even temporary redemption may be attained. Each individual observance of the law affords a partial redemption in itself.* By following the law, the pleasures of life are spoiled and suffering produced, but a moral superiority in regard to the law and the fantasied law-giver is gained. Some sort of claim is staked out against the unknown (the unknown, for the most part remains nebulous, unnamed, impersonal). The claim alone has value, not its fulfillment, which would immediately upset the moral balance. Such is the basic scheme (perhaps slightly caricatured) to which various individual features are added in the case of a particular patient. The patient knows practically nothing of this thought system when he comes into treatment. Within the realm of consciousness it is represented by short remarks, shorn of elaboration, which at first appear entirely trivial and colorless, and which emerge only when the patient speaks of some actual event. He may say, "I have also been invited along"; the "also . . . along" implies "but I am really one of the damned." Or, "Play tennis? No, that's out of the question"; "that's out of the question" means "that is against the law; to accept would be presumptious, claiming something I do not deserve. In the moment of doing it I would be humiliated in the most terrible fashion, and the reward for unending abnegations would be lost."

Patients come into analysis with such character-armor. They also come into analysis during a symptom-breakthrough and quickly reestablish this character-armor. In both cases the resistance which emanates from this armor appears in the fact that the patient is almost completely absorbed in his striving for justification and has no interest available for contact and communication with the analyst. The analysis, like everything else in life, is simply another opportunity to show his obedience to the law and to prove that such obedience brings neither success nor reward. Here it must be stated that the character-resistance ultimately is based on a transference. The patient transfers the experience and attitudes of the nursery to his entire surroundings including the analytic situation. If, nevertheless, we use the concepts of transference-resistance and character-resistance as logically independent concepts, we do so because the concept of transference-resistance is very closely associated with transference of affect, while the outstanding feature of character-resistance is precisely the unusual paucity of affect. The manifestations of character-resistance are not attempts of an id impulse to force discharge insofar as censorship permits; rather they express the narcissistic tendency to support an endangered feeling of self.†

* [In 1931, Kaiser published an essay entitled "Franz Kafka's Inferno." The description of the pedantry exhibited by the defiant character reminds one of Kafka's heroes in *The Castle* or *The Trial*.]

† [In 1936, Anna Freud called this mechanism "transference of defense"; see

It is now possible to draw the parallel mentioned above between the transference-resistance and the character-resistance. The transference-resistance is made manifest by the transference-action; the corresponding phenomena for the character-resistance are both the artificial, affectless behavior which yields to the demands of an archaic form of regulation of the feeling of self and self-esteem, and the kind of actual remarks of the patient quoted on p. 399. The transference-action in its turn is supported by rationalizations, while the behavior typical of the character-resistance is justified by the preconscious ideology (described above for the defiant character) and by what we ventured to interpret as the real meanings of the patient's remarks. Just as a transference-action disappears when its rationalizations have become transparent to the patient and devalued, so the character-resistance very gradually disappears to the degree to which the resistance-thoughts surrounding its preconscious ideology have become accessible to the critical attention of the patient. The process is set in motion and brought about by the analyst by essentially the same means used in the dissolution of the transference-resistance, but in the case of the character-resistance the attachment of the patient to the analyst is of a far less favorable kind.

VIII / *The Resolution of Resistances (cont'd.)*

Further discussion is now needed concerning effectiveness in resolving resistances of the method by naming the repressed impulse to the patient. To anyone unfamiliar with analytic practice it may seem odd to return repeatedly to this problem. Everything in the paper so far suggests that we already possess a useful (and theoretically plausible) method of resolving resistances, the means being to make conscious the preconscious resistance-thoughts. Moreover, since the publication of Freud's technical papers (quoted above) almost every work on analytic technique has repeated the warning against giving depth interpretations without first analyzing the resistance. But every professional analyst will at once appreciate the need to return to this question. The fact is (no matter whether it stems from the historical development of analytic therapy or from the very nature of analysis) that in most clinical cases the repressed impulse (and the childhood situation to which it belongs) is discernible earlier and more easily than the resistance-thoughts by which

editors' introduction, p. 39. However, Kaiser's understanding of this mechanism goes deeper. Compare also with Abraham 1919b (this volume, Chapter 4) and Riviere 1936 (this volume, Chapter 28) when endangered feelings of self are discussed within a different context.]

the patient defends against it. For example, the analyst may become aware that the patient's attitude toward him is determined by an aggressive impulse much sooner than he understands how the patient succeeds in remaining unaware of this fact. For instance, an analyst observes over many weeks that the patient's superficial attitude toward him is extremely peaceful and submissive, but he notes as well the unconscious hostility which shines through every crack. Let us suppose the analyst has done everything he can to understand and to bring the inconsistencies in expression to the attention of the patient, but without success, and that nevertheless he (the analyst) remains thoroughly convinced that even the blindest person would recognize the hostility which lurks behind the facade of innocent compliance. If the analyst now keeps in mind the principle of making interpretations only a single step ahead of the patient, he will probably be very tempted to say to the patient, "Look here! You are saying this and that. You are behaving in such and such a manner. Can this mean anything else than the effect of an aggressive impulse thinly veiled behind your friendly behavior?" My own experience would suggest that the analyst would indeed be so tempted. And in fact I have succumbed to this temptation repeatedly, convinced that I was making the proper interpretation. Such experience should be kept in mind in order to evaluate properly the significance of the following thoughts.

The above-mentioned action of the analyst is justified theoretically only when the interpretation has the effect of illuminating for the patient his own resistance-thoughts. Experience suggests that such occasions are so rare as to be negligible. In all other cases the giving of the interpretation not only is incompatible with the theory (which might perhaps mean only an argument against the theory) but does not have the intended effect (i.e., the instinctual breakthrough does not take place).

Even if this is admitted, it might be contended that analysis is a difficult art, that the analyst should not be expected to be able to cope with every single problem arising in the course of treatment, and that the analyst may indeed do no more on this occasion than extract himself gracefully from a situation he cannot completely master in order to achieve therapeutic success at some later point. Such thinking seems very reasonable and tolerant and to be inspired by good common sense, but it must be opposed rather emphatically since it is based on a wrong assumption. The procedure of content interpretation (that is the method of telling the patient which unconscious wishes are hidden behind his defense) is harmful in all but two cases. One has already been mentioned, namely, the rare occasion when the procedure coincidentally has the same effect as the correct procedure. The second case is when the patient rejects the interpretation without, however, any show of affect (in which case the analytic situation does not change and the analyst continues to be confronted with the same problem). In all other cases the interpretation does damage—i.e., in all those instances when the patient finds the interpretation plausible (no matter whether he remains affectless or shows emotion), and even in those instances where he refutes the interpretation with a certain amount of affect. In all such cases the action of the analyst evokes a special kind of resistance within the patient, a resistance which may not be insurmountable but which is harder to resolve (and takes longer to resolve) than the initial resistance.

This phenomenon is especially impressive in the case of the defiant character, as the following example clearly illustrates. I once had a patient of this type whose character-resistance I had loosened enough for him to be able to transfer onto me an aggressive impulse. After prolonged attempts at destroying his rationalizations, I said in effect, "Look at the way you behave toward me," and described to him his provocative actions which had given me an overwhelming impression of his aggressivity. I asked if it was possible to understand all this except in terms of his being mad at me, though attempting to conceal this affect in certain specific ways. The patient, who characteristically tried to hide all his affects, was amazed. His eyes dilated, and he visibly paled. In a hoarse and tremulous voice he said, "Yes, that seems right—how dreadful!" I was very satisfied with this success. By the following day the patient had calmed down. At some point during the hour he referred in a collected manner to the rage he had felt against me, and I was still satisfied. It seemed to me that this transference-resistance had been fully resolved; but it hardly needs saying that sometime later I realized that I had completely deceived myself. The agitation which the patient had shown was not rage (as would have been the case if a real breakthrough had occurred), but rather anxiety and defense. He had subsequently calmed down not because part of the transference had been resolved with liberation of affect, but because a new and more effective defense had become established. This defense might have been rationalized as follows: "Although I felt rage, this should not be taken too seriously. After all, such feelings do come up in analysis. They belong to the analysis and are of no great significance. In real life (that is outside the analysis) I am still a decent person devoid of vicious impulses." It was a difficult task to demolish this defense; step by step this whole sequence of thoughts (approximately as quoted) was presented to the attention of the patient. Then, and only then, the unmistakable outburst of rage against me took place with the desirable therapeutic result (i.e., bringing with it the transition from the transference situation to the infantile prototype).

It might be thought that content interpretation could be helpful (or at least innocuous) in the case of less severe neurotic illnesses. My experience has not allowed me to be definite on this point, but it is my impression that even when a good therapeutic result had been achieved in the case of a less severe neurosis (and content interpretations had been given), the instinctual eruptions were never as intense as in the case of severely defiant characters with narcissistic armor in whose treatment content interpretation had been abandoned. I am convinced that the affective reactions which I provoked by content interpretations were not genuine in the sense that the emergence of the impulse was due not to the dissolution of the resistance but to suggestion. The impulse lost its sting because I brought it to the attention of the patient. The patient did not experience the impulse itself, but only a laboratory replica created by the analyst. My belief is that therapeutic successes brought about in this way are transference cures, resulting from a shift in the symptomatology which sometimes, however, does not work out unfavorably for the patient.

It follows from these considerations that the aim of analytic treatment is not achievable by means of the technique of content-interpretations. Instead, the attempt should be made solely to resolve the rationalizations, and only this procedure is compatible with the aim. It might seem that this precept is only a more pedantic and rigorous version of the advice given the trainee in his courses, technical seminars, and supervisory analyses against making too early or too deep interpretations. Nobody believes that Reich goes too far when, in his seminars and papers, he insists that every resistance should be attacked from the side of the ego, that the patient should be first shown his means of defense and then that his defenses must be resolved. Is it merely a theoretical radicalism to contend that there is any crucial difference between the advice to give content-interpretations as the final step after the resolution of the appropriate resistances, and the belief that content-interpretations should be completely eliminated? Our belief is that there is a significant difference between the two, and that all the relevant theoretical points which speak in favor of the more radical position have been presented, albeit in too condensed a manner. However, if the position against content-interpretation is to be completely understood, its consequences must be examined in the light of a detailed and concrete example; that is the intention of the next section.

IX / *The Practice of Consistent Resistance-Analysis*

The theoretical difference between consistent resistance-analysis and analysis which employs content-interpretation as the final step seems comparatively slight, but experientially (from the point of view of the analyst) the difference is much more striking than would be expected on the basis of the conceptual difference. This experiential difference does not, of course, become fully apparent at the moment one decides to drop content-interpretations.

In the first place, once this decision has been made, the analyst only slowly becomes aware of the very impressive difficulties which are encountered in bringing to light the rationalizations which go into making the resistance situation seem reasonable. He has to develop the habit of expanding his "evenly suspended attention" so that a much greater part of the present ego of the patient falls more sharply into the field of examination. There is as much to learn about observing the nuances of the present ego as there is about unconscious instinctual structures. Indeed a year's training in both is about equal.

I recollect listening once to a discussion between two analysts. Analyst A had had some experience in consistent defense analysis while analyst B had not. B was describing the great difficulties he was encountering in the treatment of one of his women patients. He gave a short account of the patient's symptoms, including her affectless and inaccessible behavior in the analysis. He demonstrated her movements, how she lay on the couch, how she talked; and he described his various efforts to master the situation. Then he began to think aloud about the patient. His hypotheses about the instinctual development of the patient as it related to the analytic situation were to the point and, for the most part, convincing. "But I can't tell her that now," he added each time with resignation. We could but agree. He went on with novel and quite plausible explanations of the behavior of the patient in terms of childhood experience and instinctual patterns, but none of the speculations pointed to what should be done now. Analyst A then asked further questions about the behavior of the patient in the analytic situation, especially her expressed attitude toward the treatment as such and her reasons for continuing treatment although she appeared to be convinced of its fruitlessness. Analyst B was very well able to answer these questions and to describe vividly a number of scenes from the analysis. However, he had disregarded the conscious and intellectual attitudes of the patient, believing them not to be trustworthy and only the expression of the patient's fear of her aggressivity. In the latter thought he was absolutely right, but he had overlooked the fact that the contradictions in the patient constituted the first bulwark of defense. He had, as it were, been studying the inner works of the defensive fortress of resistance through his well-trained depth-psychological telescope, without, however, being able to approach them directly. This fortress analogy leads to a problem that plays a conspicuous part in consistent resistance-analysis, a discussion of which may illuminate further the application of the technique. The problem in question we wish to designate as the problem of immediacy.

X / *The Problem of Immediacy*

The comparison between the neurosis and a fortress which must be taken by storm would seem to suggest that there is a definite order in which the resistances should be tackled, just as in a siege, the various lines of defense, such as trenches, bulwarks, ramparts, and wire entanglements must be overrun one by one (though this analogy may stem from my ignorance of matters of siege warfare). But this is not so, as the following hypothetical example will show. A patient starts the hour by being silent. The analyst has the impression that this silence expresses the patient's unwillingness to take the responsibility for speaking first and a wish to shift this responsibility onto the analyst. The analyst tells the patient this. The patient responds by seeming to ignore the analyst's remark; in a slightly aggressive way he says, "That's what you said yesterday, isn't it? You always say the same thing. Well, perhaps that's a rule of analysis." Many indications (among them the patient's tone of voice) point to the fact that the patient wishes to avoid the delicate question of who takes responsibility, and covers the anxiety which this problem arouses by aggression. In turn the edge is taken off this aggression since the patient weakens his initial personal attack ("You are always saying the same thing") by changing it into a criticism of analysis in general ("Well, perhaps that's a rule of analysis"). The original object of attack, however, is still recognizable in the touch of irony.

At this point there may be some doubt as to whether the analyst, after saying something about the patient "avoiding," should come back to the point of departure (the patient's silence) or should focus only on the last remark the patient made. The latter possibility may seem to carry the disadvantage that the patient may again slip by a statement of the analyst, move off on one tangent after another in such a way that the analyst will be forced to follow him through innumerable topics without the patient coming to grips with any one of them. Experience testifies that this is a real possibility. It may indeed happen that the patient changes the topic in response to each remark of the analyst, no matter how appropriate the remark may be. It may even happen that when the analyst calls the attention of the patient to this jumping from one topic to another the patient merely takes another jump. In this case, the analyst is unable to hit any one of the successive resistances of the patient in such a way that the patient has no option but to recognize the resistance-quality of his utterances. However, it is not always wrong to do something which does not immediately promise success. Let us consider the other possibility of the analyst continuing to talk about the first resistance phenomenon (the silence) before the second (or the most recent) is really worked through. This alternative offers even less prospect of success. It may happen that the patient attends to what the analyst has to say about the initial manifestation of resistance, but only because this first problem has in the meantime lost its significance for the patient—it is no longer immediate (in the sense of being the point at issue in an affective meaningful way).

We might account for this theoretically in the following manner. Resistances are broken down when the patient can be made to examine his resistance-thoughts. For this purpose it is, however, insufficient to repeat to the patient the part of his resistance-thoughts that he has already put into words and to show him what further thoughts must lie behind the resistance. He may possibly listen to all this without feeling himself intellectually responsible for it (if this expression be permitted), and without experiencing the truth or falsity of the thoughts. A healthy adult taking part in a discussion in which he is not much involved emotionally can generally be made to think through any thought which he or the other discussant has introduced and to detect implied discrepancies. All that is needed is to make the steps in thought sufficiently small so as to bring them within the grasp of understanding. But the conditions are far less favorable when it comes to making a patient think through his resistance-thoughts. In that case a certain force (the narcissistic counter-cathexis),* which tends to deflect the attention of the patient away from where we wish it to focus, has to be dealt with. As already demonstrated, this force can be counteracted by the available tender attachment of the patient to the analyst. Suppose in the example that the analyst wished to make the patient think through his (the analyst's) remark about the meaning of the patient's silence. If the tender attachment of the patient for the analyst were sufficiently strong, it might be sufficient for the analyst to reply to the patient's tangential and ironical comment by saying, "But you are sidestepping the issue!" However, this is not necessarily simply the case. In fact, the data of the example suggest that the patient is still not capable of such an attachment to the analyst. This is certainly so where severe character-resistances have not yet been pierced—e.g., in the defiant character. Then the analyst is unable to activate in the patient a train of thought which he (the analyst) introduced and help him think it through, judging its truth. Rather, the analyst is forced to deal only with that resistance-thought of the patient which has immediate life owing to the defensive need of the moment. In that situation, every remark of the analyst, which misses the immediate or most topical resistance-formation of the patient is ineffective. It would probably be incorrect to speak of such a patient having no tender attachment toward the analyst, since in that case his analysis would be impossible altogether. Rather, one should probably speak of a weak or attenuated attachment. The question arises whether some element in the narcissistic transference-relation, which is the inevitable concomitant of an unfolding of character-resistance, may not be

* [The term "narcissistic counter-cathexis" is not usually used in psychoanalytic metapsychology. Only the ego is conceptualized as capable of employing counter-cathexis.]

utilizable in the analysis, but this question will not be pursued further here. In general, it may be said that the stronger the resistance the weaker the tender attachment; and that in this situation the analyst must attend even more closely to what is immediate.

In rather severe cases the analyst cannot do justice to this requirement unless his state of attention has been influenced by it. It might be objected that this would militate against his attention being evenly suspended. However, the latter concept does not mean that the analyst should abandon his feeling for the deeper psychological meaning of what he hears from the patient (a feeling based on his endowment, his experience, and his thinking), but rather that he should give free range to this feeling, unhampered by consciously enforced direction of his attention. Each technical insight deduced from theoretical considerations (such as this one about the demand for immediacy) disturbs his evenly suspended attention, but in this case also the disturbing effect will vanish as quickly as the effect of other considerations which influence the technique.

Within the limits of this paper it is quite impossible to give a complete picture of the practice of consistent resistance-analysis. The promise to give a more detailed and vivid presentation can be made good only in an imperfect manner. But, as far as space will allow, the intention is now to discuss a frequently encountered analytic situation the handling of which often adds a characteristic touch to the process. The discussion will not introduce any new theory.

XI / Speaking in Allusions

The situation in mind is the following: Some resistances have already been resolved, but no real instinctual breakthrough has occurred. The patient is somewhat more relaxed. Fairly vivaciously he now begins to express a series of associations which, although only flimsily held together from the point of view of logic, could be understood collectively as allusions to a certain childhood affect and to the experience linked with the emotion.

Three possible procedures are open to the analyst, each of which can be found in current analytic practice.

The first method—that of content-interpretation—proceeds as follows: The analyst reviews the various thoughts of the patient in some suitable order and shows the patient how admirably they fit together into the picture of a certain scene with definite emotional content. He infers that the devil must have had a hand in the situation if such apparent interrelation were due to chance, and suggests that the patient once actually did experience such a situation, or fantasied it, at which the patient will probably respond with pleasure and interest. Grateful for such ingenuity exerted on his behalf, he will eagerly add some thoughts or recollections which confirm the analyst's interpretation; but the therapeutic effect will be about zero.

The second method, which we might call the method of intimation, is preferred by the analyst who feels that the first is too crude. It is essentially the same procedure as the first method, but it differs from it in two respects. The analyst carefully puts the material together in the most advantageous order, giving it a well-chosen emphasis which hints at the interpretation. The analyst draws no conclusions himself but simply waits for the patient's reaction. One of two things may then happen. Either the patient cannot make head or tail of what is put before him and everything remains as it was before the analyst's intervention, or he understands what is meant and shows his ability to solve the riddle. In the latter case, the patient is even happier than if the first method had been used; he has reason to be proud of his acumen. But again the therapeutic effect is virtually nil. The patient has not experienced his repressed impulse; he merely infers that he must have had such an impulse just as he would draw the same conclusion concerning some stranger who expressed himself similarly. It becomes apparent that the resistance is not manifest simply in the fact of having forgotten. There are also intellectual or reflected acts. This is clear in the case of so-called screen memories, where the repression shows itself not only in the forgetting of the real emotional experience, but also in the remembering of some substitute with an unwarranted vividness.

The third method—and it is the contention of this paper that it is the only effective method—is that of resistance-analysis. It may consist, for instance, in drawing the patient's attention to the fact that his associations lack cohesion. Or, for example, it may be possible to show him that he wants to be figured out by the analyst, or to present the analyst with a riddle. In this connection an interesting and surprising sequence in the analysis is worth mentioning. The analyst may succeed in confronting the patient with the element of resistance in his behavior—e.g., the wish to present the analyst with a problem—but the emotional material or impulses which then come to the patient's mind are only rarely those to which his associations seemed to point at the time of the intervention. The latter appears much later; meanwhile some quite different material comes to expression. One learns from such experience that the defense against a certain impulse or emotional recollection often makes use defensively of an even more deeply repressed impulse.

It follows that the fascinating work of reconstructing scenes from the patient's past on the basis of a few widely scattered traces or allusions has no immediate therapeutic application though it may be of importance in helping the analyst privately to outline an approximate picture of the development of the neurosis.

One is perhaps left wondering if a consistent resistance-analysis allows the analyst any opportunity to talk directly about the infantile instinctual wishes of the patient. The remark is frequently heard that the patient must be given some explanation of what he expressed after a genuine instinctual breakthrough has taken place; but this seems to originate in a misconception of a "genuine breakthrough."

When this phenomenon takes place, the patient understands so completely what he says that nothing is left for the analyst to explain. However, there is something which needs to be done if all the potential therapeutic gain of a genuine breakthrough is to be realized. The analyst should give the patient a review of the resistances, the dissolution of which led to the instinctual or affective breakthrough; and he should, if possible, make it clear to the patient just how a particular resistance hindered the emergence of a given impulse or reminiscence. The farther back in the analysis the analyst extends his review, the better—of which fact we have some understanding. It is not necessary to resolve every single resistance-thought with the same thoroughness in order to bring about a genuine breakthrough. The pressure of the impulse suffices to effect the final eruption once the chief strands of the web of resistance are torn. The consequence is that residues of resistance remain unresolved so that resistances can be reconstituted, unless with comparatively little effort these residues are wiped out in the manner described.

This is the end of my presentation proper. The brief historical critique which follows was left to this point so as not to confuse the presentation itself. I am aware that my discussion of technique has involved a wide range of questions and theoretical problems, and is most fragmentary. My hope is to have aroused interest rather than agreement in the problems discussed. Since I have taken a firm stand in favor of a very radical, even very strict, technical procedure, I feel I owe the reader a further remark. Any strictness of technical prescription is legitimate only if it is necessary for the therapeutic aim. There is always the danger that a method which is reasonable in itself can become a ritual demanding allegiance, not only as a means but as an end in itself. Abstract technical principles are not to be applied as such, but only to be understood in order to refine and enrich the tact and instinct of the analyst. In the actual analytic situation it is the analyst's feeling, tact, and instinct which must have the last word.

XII / *Reich's Theory of Analytic Technique*

The most decisive contribution to the theory of analytic technique following the papers of Freud was made by Wilhelm Reich. I do not wish to omit discussion of the bearing of Reich's thought on the problem of this paper for the special reason that the development of my own thinking concerning the theory of technique is based directly on his. I am deeply grateful and indebted to him for the stimulating ideas and suggestions derived from his seminars.

It is not possible to recapture here the fertility of ideas and richness of formulation which characterize Reich's technical teaching. It seems to me that two main theses stood out in his seminars with even greater clarity and impressiveness than in his writings.

The first major line of thought runs as follows: The neurosis is an organic growth with a certain structure. The analytic work should, therefore, follow a definite order corresponding with the neurotic structure which it is attacking layer by layer. On the basis of this premise Reich justifiably concludes that the analyst's choice of topic from among all the material the patient brings into the hour is not a matter of indifference. He believes that the very nature of the neurosis determines that one particular layer (or, more exactly, the representative of one layer) must be singled out as the starting point of the therapeutic procedure.

The question arises, "By what criterion can the analyst decide the correct point of attack on which to focus?" Reich gave two very different answers to this question. His first answer is that the analyst cumulatively builds up a picture of the structure of the neurosis as the analysis proceeds, and with the help of this picture decides which is the uppermost layer. One should begin with the representatives of this layer. Here I cannot agree with Reich's point of view. The recommendation is theoretically correct, but it is inapplicable. Even though the neurosis of every patient is structured, and although it is possible with sufficient experience and intuitive gift to understand rather early the broad outline, the finer structure of the neurosis is far too complex to comprehend before one has begun therapeutic intervention. These finer points are, to my mind, the more important. Thus, I maintain that Reich too quickly applies a fruitful theoretical idea to clinical practice.

Reich's second answer is more empirical. The analyst must subject the picture presented by the patient in a session to the most intensive and sensitive examination and use his own feelings to detect where the patient exposes the uppermost, most recent, most alive, and therefore most accessible layer. Here I am in full agreement with Reich and I believe that he is here close to what I consider to be the second principle point of his teaching.

This is the important tenet that one must attack every resistance from the side of the ego. Reich has rendered us an extraordinary service by repeatedly and forcefully stressing this principle. Beyond this he has provided an abundance of illustration of the application of the technique. The great emphasis Reich places on close observation of the patient's posture, behavior, manner of speaking, and, above all, tone of voice, and the emphasis he places on making the patient conscious of these, represent the important, practical consequences of his teaching.

It seems to me that the essence of his teaching leads to the theoretical conclusions presented in sections III and V of the present paper centering around the concept of the resistance-thoughts. Since Reich did not take this further theoretical step, a peculiar consequence ensues in the theory of his technique. In his book *Character Analysis* (1933) is the prescription, "First point out to the patient that he has resistances, then his means of resistance, and finally what the resistances are directed against." This can only be a direction finally to interpret content. This would be a logical procedure only if something remains to be done in resolving the resistance after the means by which the patient defends himself are fully analyzed, a contradiction which is given no further explanation by Reich.

I wish to mention here a third and most important teaching of Reich, the character analysis proper, only in order to state that nearly everything he says is in agreement or is compatible with my points of view, or could be construed as a consequence of the theories of this paper.

However, concerning a single point which is theoretically inessential but not unimportant practically, I must contradict him decisively. Reich thinks that the inexperienced analyst should be dissuaded from the practice of character analysis. This seems to me as mistaken as it would be to teach a medical student first to practice surgery on the living patient with a blunt knife, and only later with a sharp one. Concern for the patient is justified, but it would be better advice to limit the beginner to the less severe and dangerous cases, and to increase his supervision.

XIII / *Freud's Formulations*

Careful examination of the principle of modern technique (quoted on p. 387) shows that it contains the insight which is the basis of the thoughts presented in this paper—namely, that the main task of analysis is not making guesses or drawing conclusions about the forgotten or repressed experiences of the patient, the communication of these guesses or conclusions to the patient depending merely on the ability of the analyst in expressing himself.* Rather the essential point is to produce a change in the patient on the strength of which the patient himself finds access to these insights. Freud defined this change in the patient as the resolution of resistances.

His formulation even gives some information on the way in which this task is to be accomplished. The analyst should recognize the resistances of the patient, and make the patient conscious of them.

If we put aside for the moment questions that might be called tactical and turn to the strategic problem of analysis, we note that here too Freud gave direction. The recommendation continually to observe what is occupying the surface of the patient's mind implies an ordered procedure on the part of the analyst, forcing him to follow the lawful development of the patient. In other words, the analyst must take as his guide the continuous changes at the surface of the psyche of the patient, not his own progressive understanding of the deeper layers.

The translation of this recommendation into the tactical procedure of analysis is the rejection of content-interpretation. This consequence is unacceptable only to those who would include under the concept of "surface" everything the patient tells during the analytic hour, but such an interpretation of the meaning of surface seems unpsychological. For example, a patient who is dominated by a tendency to make the analyst impatient tells of an experience from his fourth year of life in an affectless, boring manner; only the affectless, boring manner belongs to the surface, not what is told. If the patient hides his aggressive impulses against the analyst behind a kind of superpolite behavior, only the super-politeness belongs to the surface and not the aggressive impulses. Freud's formulation does not say that one should start from the surface and then proceed to the depths, but that one should study the surface as it is presented on each occasion and should not touch the layer beneath until it has come to the surface.

However, Freud's positive direction regarding the tactical procedure does not seem sufficient to determine completely the behavior of the analyst. The formulation "to make the resistance conscious to the patient" implies that one should show the patient that he is defending himself, but it clearly expresses neither that one should make him conscious of his means of defense, nor that one should avoid mention of the drive against which he is defending. I tried to show above that the latter negative prescription follows as a consequence the recommendation to study the immediate psychological surface of the patient, in which case we should interpret "to make conscious the resistance" as to make the fact of defense and its means known to the patient. Nevertheless, we must admit that Freud's formulation does not lend itself to completely certain interpretation. What is missing from this formula-

---

* [Statements by Freud can be used to support both sides of this argument. For the one in opposition to Kaiser, see Freud 1937a.]

tion of Freud is the sharp distinction which needs to be drawn between the procedure of naming a repressed impulse and the technique by which a pre-conscious mechanism of resistance (resistance-thoughts according to our terminology), which is eluding the attention of the patient, is brought to his awareness. The words "to make conscious" are, from a topographic point of view, applicable to two very different events, namely the transmission of an impulse from the system Ucs. into the system Cs., and the transmission of a thought or act of reflection by which we mean archaic beginnings of adult thinking from the system Pcs. to the system Cs.

The last sentence of Freud's formulation (1914c, p. 147) does not seem unequivocal either. It would seem that the sharp division he makes between uncovering of resistances on the part of the physician and the relating of "the forgotten situations and connections" on the part of the patient can best be understood in the sense of our point of view, that is, the analyst *must not* present the forgotten and repressed material which he has been able to infer to the patient. However, when Freud describes the patient's remembering as being "without any difficulty" it would seem that this account does not include the genuine instinctual breakthrough.

I do not believe that a further minute examination of Freud's views looking for arguments for and against my view is warranted. Freud did not write an elaborate code of analytic technique, but rather gave an elastic expression to the living development of analysis. What I did wish to demon-strate with these last arguments was that my thoughts on technique do not run counter to the direction in which Freud has developed his recommendations on technique.

CHAPTER 28

JOAN RIVIERE

A CONTRIBUTION
TO THE ANALYSIS
OF THE NEGATIVE
THERAPEUTIC REACTION

INTRODUCTORY NOTE

In this essay, Riviere explores Freud's concept of the negative therapeutic reaction within the theoretical frame of reference of Melanie Klein. She attempts a synthesis between the ideas Abraham delivered in his 1919 paper, (this volume, Chapter 4) and the views of Melanie Klein. She throws new light on the narcissistic resistances of the patients who resist cure most strenuously. However, her synthesis will seem persuasive to the reader depending on his or her attitude toward the basic formulation of Klein. Of particular interest are her observations on those patients who cannot accept help because they feel that the analyst should spend his time on sicker patients or more worthy ones. She also describes how a patient, in order to defend himself against the depressive position, may psychologically exploit the analyst rather than being cured by him.

Originally published in *International Journal of Psycho-Analysis* 17 (1936):304–320. Reprinted by permission of Diana Riviere and the Institute of Psycho-Analysis. This paper was read before the British Psycho-Analytical Society, October 1, 1935.

IN THIS CONTRIBUTION my aim is to draw attention to the important bearing recent theoretical conclusions have on the practical side of the problem of the negative therapeutic reaction. I mean the latest work of Melanie Klein and in particular her Lucerne Congress paper on the depressive position (1934).

To start with, it is necessary to define what is meant by the negative therapeutic reaction. Freud gave this title to something that he regarded as a specific manifestation among the variety of our case-material, though he says that in a lesser measure this factor has to be reckoned with in very many cases. When I referred to Freud's remarks on this point, I was interested to find that actually they are not exactly what they are generally remembered and represented as being. The negative therapeutic reaction, I should say, is generally understood as a condition which ultimately precludes analysis and makes it impossible; the phrase is constantly used as meaning unanalysable. Freud's remarks on the point are almost all in *The Ego and the Id*, the last eighteen pages of which deal with the problem of the unconscious sense of guilt.* He says, 'Certain people cannot endure any praise or appreciation of progress in the treatment. Every partial solution that ought to result, and with others does result, in an improvement or temporary suspension of symptoms produces in them for the time being an exacerbation; they get worse instead of better'. This last sentence might imply that they are unanalysable; but he does not actually say so, and has just said the exacerbation is for *the time being*. He says the obstacle is 'extremely difficult to overcome'; 'often there is no counteracting force of similar intensity'; and that 'it must be honestly confessed that here is another limitation to the efficacy of analysis'—but he does not say a preventive. Clearly the point is merely one of degree, and he might concur in the general attitude taken up. He is not, however, actually as pessimistic about it as people incline to suppose; and this interested me, because it is not intelligible why one reaction should be thought more unanalysable than another. The eighteen pages in *The Ego and the Id* are in fact part of his contribution towards analysing it; and our understanding of it has now been very greatly advanced by Melanie Klein.

Freud's title for this reaction, however, is not actually very specific; a negative therapeutic reaction would just as well describe the case of any patient who does not benefit by a treatment; and it would describe those psychotic or 'narcissistic' patients whom Freud still regards as inaccessible to psycho-analysis. It seems to me that this specific reaction against a cure described by him may not differ so very greatly in character from those more general cases of therapeutic failures I mentioned, and that the difficulty may be due to some extent to the analyst's failure to understand the material and to interpret it fully enough to the patient. The common assumption is that

* [The negative therapeutic reaction is discussed in Chapter 5, Freud (1923b).]

even when the analyst has fully understood and interpreted the material, the super-ego of certain patients is strong enough to defeat the effects of analysis. I shall try to show that other factors are at work in this severity of the super-ego that until recently have not been fully understood and therefore cannot have been sufficiently or fully interpreted to our patients.

It will be clear now that what I propose to talk about is in fact the analysis of specially refractory cases. I do not think I can go much further in defining the type of case to which my remarks refer, partly because any one analyst's experience is necessarily limited, even of refractory cases; moreover, my expectation is that similar unconscious material may probably exist in other difficult cases of a kind I have not personally met with. I would say this, however, that the cases in which I have made the most use and had greatest advantage from the new understanding have been what we call difficult character-cases. The super-ego of the transference neurotic, it must be remembered, has always been placated by his sufferings from his sense of guilt, and by his symptoms, which are a real cause of inferiority and humiliation to him whatever epinosic gain he has from them; the character-case has never placated his super-ego in these ways; he has always maintained the projection that 'circumstances' have been against him. After some analysis he may guess that he has punished others all his life and feel that what he now deserves is not 'cure', but illness or punishment himself; and he unconsciously fears that that is what analysis may bring him if he submits to it. Of course we find these motives for or against co-operation in all cases; I merely suggest that in character-cases they may have peculiar strength.

With reference to this matter of character-resistances I shall recall to your minds a paper of Abraham in which he described and commented on a certain type of difficulty in analysis, that he virtually names the *narcissistic type* of character-resistance [1919b, this volume, Chapter 4]. He tells us that such analyses are very lengthy and that in no such case did he obtain complete cure of the neurosis, and we can see that the degree of negative therapeutic reaction in this type is what led him to distinguish it. The narcissistic features of this type are, shortly: that they show a chronic, not merely occasional, inability to associate freely, in that they keep up a steady flow of carefully selected and arranged material, calculated to deceive the analyst as to its 'free' quality; they volunteer nothing but good of themselves; are highly sensitive and easily mortified; accept nothing new, nothing that they have not already said themselves; turn analysis into a pleasurable situation, develop no true positive transference, and oust the analyst from his position and claim to do his work better themselves. Under a mask of polite friendliness and rationalization they are very mean, self-satisfied and defiant. Abraham shows the relation of all these features to anal omnipotence, and he especially emphasizes the *mask of compliance*, which distinguishes this type of resistance from an open negative transference and renders it more difficult to handle than the latter. And 'These patients', he says, *'shut their eyes to the fact that the object of the treatment is to cure their neurosis'*. Incidentally, I do not suppose that Abraham was guilty of it, but I feel that analysts themselves are not always incapable of shutting their eyes to a fact too, namely, that when a patient does not do what he ought, the onus still remains with the analyst: to discover the cause of his reaction. In my opinion the patient was entirely in the right who said, 'Yes, doctor, when you have removed my inhibitions against telling you what is in my mind, I will then tell you what is in my mind', and the situation is similar in regard to getting well.

This paper of Abraham's suggests what I take to be a generally valid proposition, that in specially long and difficult analyses the core of the problem lies in the patient's narcissistic resistances. One surmises, further, that this narcissism may not be unconnected with the inaccessibility to treatment of the 'narcissistic neuroses', as Freud has called certain psychoses. There is nothing very new, or immediately helpful, in the idea that narcissism is the root of the problem—for what is narcissism? I will mention only two general points in this connection. One—the old one—is that any marked degree of narcissism presupposes a withdrawal of libido from external objects into the ego, and secondly, the newer point, that ego-libido can now be recognized, especially in the light of Melanie Klein's more recent work, to be an extremely complex thing. Freud speaks of the secondary narcissism derived from the ego's 'identifications', which most of us here now regard as including the ego's *internal objects*. And Melitta Schmideberg (1931) suggests that love for the introjected objects is a part of narcissism.* And now the significance of the ego's relations to its internalized objects shows clearly that this great field of object-relations within the ego, within the realm of narcissism itself, needs much further understanding; and it is my belief that more light in this direction will do much to explain such hitherto inexplicable analytic resistances as the narcissistic ones of Abraham and the super-ego one of Freud.

The concept of *objects* within the ego, as distinct from identifications, is hardly discussed in Freud's work; but it will be remembered that one important contribution of his to the psychology of insanity is built up almost entirely on this conception—I mean of course his essay on 'Mourning and Melancholia', dealing with the problems of *depressive* states. His discussion in *The Ego and the Id* of the unconscious sense of guilt, too, is closely interwoven with aspects of the melancholic condition. This brings me to my second point. Observations have led me to conclude that where narcissistic resistances are very pronounced, resulting in the characteristic lack of insight and absence of therapeutic results under discussion, these resistances are in fact part of a highly organized system of defence against a more or less

---

* [Hartmann (1950) defined narcissism as cathexis of the self. He, as well as Jacobson (1964), do not regard the cathexis of the internal object or object representations as belonging to narcissism, but Melanie Klein and her followers do.]

unconscious depressive condition in the patient and are operating as a mask and disguise to conceal the latter.*

My contribution to the understanding of especially refractory cases of a narcissistic type will therefore consist in the two proposals (a) that we should pay more attention to the analysis of the patient's inner world of object-relations, which is an integral part of his narcissism, and (b) that we should not be deceived by the positive aspects of his narcissism but should look deeper, for the depression that will be found to underlie it. That these two recommendations are not unconnected might be guessed from Freud's paper, which links the two, and from Melanie Klein's view that the internal object-situation in this position is of supreme importance. The depressive position might be described as a miscarriage of introjection, she says; and *this* is the unconscious anxiety-situation that our narcissistic patients are defending themselves against and that should be the true objective of analysis in such cases.†

Now this particular anxiety-situation, the depressive, has its own special defence-mechanism, the manic reaction, of which Melanie Klein also gives a general outline.‡ The essential feature of the manic attitude is omnipotence and the *omnipotent denial of psychical reality*, which of course leads to a distorted and defective sense of external reality. Helene Deutsch (1934) has pointed out the inappropriate, impracticable and fantastic character of the manic relation to external reality. The *denial* relates especially to the ego's object-relations and its *dependence on its objects*, as a result of which *contempt* and depreciation of the value of its objects is a marked feature, together with attempts at inordinate and tyrannical *control and mastery of its objects*. Much could be written about the manic defence, and I hope will be, for in my opinion the future of psycho-analytic research, and therefore of all psychology, now depends on our belated appreciation of the immense importance of this factor in mental life. It is true that we have known of many of its manifestations and even had a name which would have represented it, if we had known how to apply it—the word omnipotence—but our knowledge and understanding of the factor of omnipotence has never yet been organized, formulated and correlated into a really useful theoretical unit. Omnipotence has been a vague concept, loosely and confusedly bandied about, hazily interchanged with narcissism or with phantasy-life, its meaning and especially its functions not clearly established and placed. We ought now to study this

---

* [This is an original view of the author.]

† [See editors' introduction, pp. 35–36.]

‡ [To this point, Klein said, "When the depressive position arises, the ego is forced (in addition to earlier defences), to develop methods of defence which are essentially directed against the 'pining' for the loved object. These are fundamental to the whole ego-organization. I formerly termed some of these methods *manic defences*, or the *manic position* because of their relationship to the manic-depressive illness" (1940, p. 316).]

omnipotence and particularly its special development and application in the manic defence against depressive anxieties.

It will not be difficult to see how characteristic the most conspicuous feature of the manic attitude, omnipotent denial and control by the ego over all objects in all situations, is of our refractory patients with their narcissistic resistances. Their inaccessibility is one form of their *denial*; implicitly they deny the value of everything we say. They literally do not allow us to do anything with them, and in the sense of co-operation they do nothing with us. *They* control the analysis, whether or not they do it openly. If we are not quick enough to be aware of it, too, such patients often manage to exert quite a large measure of real control over the analyst—and can even do this when we are quite aware of it. So far, it seems to me, we have not known, or not known enough, exactly where to place this tendency or how to relate it to the rest of the analytic context and so—we have not been able to analyse it. We have tended to see it as a negative transference and as an expression of aggressive attitudes towards the analyst. We have understood these as defences against anxiety, but we have not realized that a *special* fear lay beneath this special way of attaining security. I think Abraham's whole description, with every detail of the 'narcissistic' resistances he describes, in fact presents an unmistakable picture of various expressions of the manic defence—the omnipotent control of the analyst and analytic situation by the patient—which yet, as he points out, is often enough extremely cleverly masked. The conscious or unconscious refusal of such patients to produce true 'free associations', their selection and arrangement of what they say, their implicit or explicit denials of anything discreditable to themselves, their refusal to accept any alternative point of view or any interpretation (except with lip-service), their defiance and obstinacy, and their claim to supersede the analyst and improve on his work all show their determination to keep the upper hand and their anxiety of getting into the power of the analyst. Free association would expose them to the analyst's 'tender mercies'; love for the analyst, a positive transference, would do the same; and so would any admission of failings in themselves. Along with their self-satisfaction and megalomanic claims, their egotism is shown in pronounced meanness, and often in an absence of the most everyday acknowledgements or generosity. Certain patients of this type especially withhold from us all 'evidence' of an indisputable character in support of our interpretations. They leave us with dreams, symbols, voice, manner, gesture; no statements, no admissions from themselves. So we can say what we like, nothing is proved—yet of course they accept the help they get, but refuse us all help and all acknowledgements. Abraham interprets this trait as anal omnipotence. Beside this connection, it signifies especially their need to reserve and preserve everything of any value, all good things, to themselves, for various reasons, and especially for fear that others (the analyst) will gain in power by means of them. Above all, however, the trait of *deceptiveness*, the mask, which conceals this subtle reservation of all control under intellectual rationalizations, or under feigned compliance and superficial politeness, is characteristic of the manic defence. This mask owes its origin undoubtedly to the specialized dissimulation of the paranoiac; but it is exploited in the manic position not as a defence in itself but as a cover for the defence of securing exclusive control. To this description of this type of patient I would here add an important further detail: they show a quite special sensitiveness on the point of consciously feeling any anxiety; it is quite apparent that they have to keep control so as not to be taken unawares by, and not to be exposed to, a moment's anxiety. Abraham comments on their lack of affect, and this in my view is to be taken first as a dread of *anxiety*-affect. But their complete incapacity for any feeling of guilt is equally astonishing and is of course one of their most psychotic traits in its lack of the sense of reality; they deal with guilt-situations entirely by projection, denial and rationalization.

Now it might be objected here that no analyst worth his salt has failed to interpret these manifestations in precisely this way time and again in his practice and this of course is true; but in my view there is all the difference in the world between what may be called single isolated interpretations, however correct and however frequent they may be, and the understanding and interpretation of such detailed instances as part of a general *organized system of defence* and resistance, with all its links and ramifications spreading far and wide in the symptom-picture, in the formation of character and in the behaviour-patterns of the patient. Analysis has to concern itself with daily details because only the immediate detail of the moment has affect and significance for the patient, but the analyst has to be careful not to become too affectively interested in working out detailed interpretations: he has to be careful not to lose sight of the wood for the trees. He must aim, not merely at understanding each detail in itself but at knowing where to place it in the general scheme of the patient's mental make-up and in the continuous context of the analytic work. Of course, what have been called 'spot-analyses' or snapshot interpretations have long been condemned, and Ella Sharpe, for instance, once led a crusade against meaningless *ad hoc* symbol-interpretations which do not form part of a whole picture.* What I am urging now is only a further application of this principle. I suggest that the common tendency we often see in patients to control the analysis and the analyst is even more widespread than we suppose, because it is largely masked and disguised

---

* [The reference is to Ella Sharpe (1930, p. 35): "We should not overemphasize the significance of dreams. They have their place. We get an immense help from them, but work that degenerates into symbol hunting into dream material, is not analysis."

Anna Freud's warning (1936, p. 27) is more precise: "A technique which confines itself too exclusively to translating symbols would be in danger of bringing to light material which consisted too exclusively of id contents."]

by superficial compliance, and that it forms part of an extremely important general defensive attitude—the manic defence—which has to be understood as such.

Now what is the specific relation between this special line of defence and the negative therapeutic reaction; why does the need to control everything express itself so particularly in not getting well? There are certain obvious answers to this, all of which would show that not getting well is an unavoidable indirect result of these resistances. For instance, I have just suggested that hitherto these tendencies in patients to usurp all control have been regarded as expressions of a negative transference and hostility to the analyst. This interpretation, so far as it goes, is certainly correct; the patient is extremely hostile; but that is not all. Things are not so simple. The very great importance of analysing aggressive tendencies has perhaps carried some analysts off their feet, and in some quarters is defeating its own ends and becoming in itself a resistance to further analytic understanding. Nothing will lead more surely to a negative therapeutic reaction in the patient than failure to recognize anything but the aggression in his material.

The question why the defence by omnipotent control leads so characteristically to the negative therapeutic reaction cannot be answered fully until we consider the anxiety-situation underlying this defence; but I think there is one direct connection between the two which may be stated here. There actually is a kind of wish in the patient not to get well and this wish is itself partly in the nature of a defence. It comes from the desire to preserve a *status quo*, a condition of things which is proving bearable. It is built upon many compromises; the patient does not finish the analysis, but neither does he break it off. He has found a certain equilibrium and does not intend it to be disturbed. To my mind, this is an important general explanation of the phenomenon Freud comments on. He says 'A few words of praise or hope or even an interpretation bring about an unmistakable aggravation of their condition' (1933). If the patient is changing, or is being changed, he is losing control; the equilibrium he has established in his present relation with the analyst will be upset; so he has to reinstate his former condition, and regain his control of things. Actually, this anxiety-reaction to the idea of making progress often disappears on being itself interpreted; and of course not interpreted only in this general way, but the detailed connection of the immediate resistance to the immediate anxiety made clear. Incidentally, there are many ways in which this aspect of the defence by control (namely, that of prolonging and maintaining the *status quo*) verges on and merges into the obsessional technique of prolonging in time and preserving in space certain distances, always maintaining a relative, never an absolute or a final relation. But the connection between the manic and the obsessional forms of defence is not part of my subject here.

If the patient desires to preserve things as they are and even sacrifices his cure for that reason, it is not really because he does not wish to get well. The reason why he does not get well and tries to prevent any change is because, however he might wish for it, he has no faith in getting well. What he really expects unconsciously is not a change for the better but a change for the worse, and what is more, one that will not affect himself only, but the analyst as well. It is partly to save the analyst from the consequences of this that he refuses to move in any direction. Melitta Schmideberg said something of the same kind in the paper quoted already: 'Inaccessibility in patients is due to a fear of something "even worse" happening'. Now what is the still worse situation which the patient is averting by maintaining the *status quo*, by keeping control, by his omnipotent defences? It is the danger of the *depressive position* that he is guarding himself and us against;* what he dreads is that that situation and those anxieties may prove to be a reality, that that psychical reality in his mind may become real to him through the analysis. The psychic truth behind his omnipotent denials is that the worst disasters have actually taken place; it is this truth that he will not allow the analysis to make real, will not allow to be 'realized' by him or us. He does not intend to get any 'better', to change, or to end the analysis, because he does not believe it possible that any change or any lessening of control on his part can bring about anything but the realization of disaster for all concerned. I may say at once that what this type of patient ultimately fears most of all—the kernel, so to speak, of all his other fears—is his own suicide or madness, the inevitable outcome, as he feels it unconsciously, if his depressive anxieties come to life. He is keeping them still, if not dead, as it were, by his immobility. Patients I have analysed have felt this dread of losing the manic defence quite consciously during the analysis of it, have both threatened and implored me to leave it alone and not 'take it away', and have foreseen that its removal would mean chaos, ruin to himself and me, impulses of murder and suicide: in other words, the depression that to some extent supervenes as the defence weakens. But I need hardly say the analyst has not this despair, for as the capacity to tolerate the depression and its anxieties gradually increases, very notable compensations accompany it and the capacity for love begins to be released as the manic stranglehold on the emotions relaxes.

The content of the depressive position (as Melanie Klein has shown) is the situation in which all one's loved ones *within* are dead and destroyed, all goodness is dispersed, lost, in fragments, wasted and scattered to the winds; nothing is left *within* but utter desolation. Love brings sorrow, and sorrow brings guilt; the intolerable tension mounts, there is no escape, one is utterly alone, there is no one to share or help. Love must die because love is dead. Besides, there would be no one to feed one, and no one whom one could feed,

* [That the negative therapeutic reaction is in essence a defense against the depressive position is Riviere's basic position in this paper. For a contrasting view, see Chapter 14.]

and no food in the world. And more, there would still be magic power in the undying persecutors who can never be exterminated—the ghosts. Death would instantaneously ensue—and one would choose to die by one's own hand before such a position could be realized.

As analysis proceeds and the persecutory projection defences, which are always interwoven with the omnipotent control position, weaken along with the latter, the analyst begins to see the phantasies approximating to this nightmare of desolation assuming shape. But the shape they assume is that of the patient, so to speak; the scene of the desolation is himself. External reality goes on its ordinary round: it is *within himself* that these horrors dwell. Nothing gives one such a clear picture of that inner world, in which every past or present relation either in thought or deed with any loved or hated person still exists and is still being carried on, as the state of a person in depression. His mind is completely and utterly preoccupied and turned inward; except in so far as he can project something of this horror and desolation, he has no concern with anything outside him. To save his own life and avert the death of despair that confronts him, such energy as he has is all bent on averting the last fatalities within, and on restoring and reviving where and what he can, of any life and life-giving objects that remain. It is these efforts, the frantic or feeble struggles to revive the others within him and *so* to survive, that are manifested; the despair and hopelessness is never, of course, quite complete. The objects are never actually felt to be dead, for that would mean death to the ego; the anxiety is so great because life hangs by a hair and at any moment the situation of full horror may be realized.*

But struggle as he may and does under his unconscious guilt and anxiety to repair and restore, the patient has only a slenderest belief unconsciously in achieving anything of the kind; the slightest failure in reality, the faintest breath of criticism and his belief sinks to zero again—death or madness, his own and others', is ever before the eyes of his unconscious mind. He cannot possibly regenerate and recreate all the losses and destruction he has caused and if he cannot pay this price his own death is the only alternative.

I think the patient's fear of being forced to death himself by the analysis is one of the major underlying factors in this type of case and that is why I put it first. Unless it is appreciated many interpretations will miss their mark. All his efforts to put things right never succeed *enough*; he can only pacify his internal persecutors for a time, fob them off, feed them with sops, 'keep them going'; and so he 'keeps things going', the *status quo*, keeps some belief that 'one day' he will have done it all, and *postpones* the crash, the day of reckoning and judgement. One patient had woven this into a lifelong defensive pattern: his death would be exacted, yes, but he would see to it that this was postponed until his normal span had elapsed. He had reached a position of

* [This paragraph and the preceding one illustrate how the theoretical position of the analyst determines his line of interpretation.]

success and recognition in his own department of the world's work, so in old age his obituary notices would eventually serve him still as last and final denials and defences against his terrible anxieties and his own fundamental disbelief in any real capacity for good within himself.

I said before that understanding of these refractory cases lay on the one hand in our recognizing that the narcissistic and omnipotent resistances were masking a depressive position in these patients. This has been my own experience, but I might substantiate this theoretically in a simple way. The patient does not get well. The analysis has no effect on him (or not enough), because he resists it and its effect. Why? Now analysis means unmasking and bringing to light what is in the depths of his mind; and this is true in the sense both of external conscious reality and of internal psychic reality. What he is resisting, then, is precisely this: becoming aware consciously of what is in the depths of his mind. But this is a truism; all of us and all patients do this, you will say. Of course that is true; only these patients do it *more* than the others, for the simple reason that in them the underlying unconscious reality is more unbearable and more horrible than in other cases. Not that their phantasies are more sadistic; Glover has often reminded us that the same phantasies are found in everybody. The difference is that the *depressive position* is relatively stronger in them; the sense of failure, of inability to remedy matters is so great, the belief in better things is so weak: despair is so near. And analysis means unmasking, that is, to the patient, displaying in all its reality, making real, 'realizing', this despair, disbelief and *sense of failure*, which then in its turn simply means death to the patient. It becomes quite comprehensible why he will have none of it. Yet, with what grains of hope he has, he knows that no one but an analyst ventures to approach even to the fringes of these problems of his; and so he clings to analysis, as a forlorn hope, in which at the same time he really has no faith.

The patient's inaccessible attitude is the expression, then, of his denials of all that the analyst shows him of the unconscious contents of his mind. His megalomania, lack of adaptation to real life and to the analysis are only superficially denials of external reality. What he is in truth concerned to deny is his own *internal reality*. Here we come to my second point: the internal object-relations which are an integral part of his narcissism.*

When we come to close quarters with the importance of the internalized objects in this connection, one general aspect of the situation will at once become clear in view of what has already been said about the depressive position. The patient's conscious aim in coming for analysis is to get well himself: unconsciously this point is relatively secondary, for other needs come first. Unconsciously his aim is: (1) on the paranoid basis, underlying his depressive position, his task is something far more urgent than getting well; it

---

* [See editors' footnote on narcissism, p. 417.]

is simply to avert the impending death and disintegration which is constantly menacing him. But more even than this (for the paranoid aspect of things is not the most unbearable), unconsciously his chief aim must be (2) to cure and make well and happy all his loved and hated objects (all those he has ever loved and hated) before he thinks of himself. And these objects now to him are within himself. All the injuries he ever did them in thought or deed arose from his 'selfishness', from being too greedy, and too envious of them, not generous and willing enough to allow them what they had, whether of oral, anal or genital pleasure—from not loving *them* enough, in fact. In his mind every one of these acts and thoughts of selfishness and injury to others has to be reversed, to be made good, by sacrifices on his own part, before he can even be sure that his own life is secure—much less begin to think about being well and happy himself. Our offer of analysis to make him well and happy is unconsciously a direct seduction, as it were, a betrayal; it means to him an offer to help him to abandon his task of curing the others first, to conspire with him to put himself first again, to treat his loved objects as enemies, and neglect them, or even defeat and destroy them instead of helping them. On his paranoid level, this is all very well, and he wants nothing better; but there is always something more than the paranoid position; there is the only good thing he has, his buried core of love and his need to think of others before himself at last, to make things better for them and so to make himself better. And the analyst's offer to help him seems to him unconsciously a betrayal of them—of all those others who deserve help so much more than he. In addition, he does not for a moment believe that any good person really would be willing to help him before all the others who need it so greatly; so his suspicions of the analyst, and of his powers and intentions, are roused. One might suppose one could perhaps allay these suspicions by emphasizing how others will benefit by his cure; but on this point of technique I must here make an important digression. It will have struck you how incongruous and contradictory this picture of the patient's unconscious aim—one of them—(to make all his objects well and happy) is compared with his manifest egoistic behaviour. But its incongruity is of course no accident; the terrific contrast of extreme conscious egotism as against extreme unconscious altruism is one of the major features of the defence by denial. In order to disprove one underlying piece of reality, he parades its opposite extreme. So I have to remind you that his unconscious aims are really *unconscious* and that we cannot use them directly as a lever to help on the analysis. We cannot say 'What you really want is to cure and help other people, those you love, and not yourself', because that thought is precisely the most terrible thought in all the world to him; it brings up at once all his despair and sense of failure—all his greatest anxieties. Any such imputation, if at all plainly and directly expressed, has the immediate effect of producing a paranoid resistance as a defence; because, when we see through his denials, the manic defence has failed him. We have

to be as guarded about directly imputing any altruistic motives to such patients as about imputing sadism or aggression to a hysteric. Nevertheless,
when we know the unconscious situation, we know how to watch our steps;
and even if we cannot use this lever ourselves for a very long time at least, we
know it is there and can bring into play any indications of it there are, in
subtle, indirect and gradual ways which do not rouse instant and unmanageable resistances.

This difficulty—that the patient unconsciously feels himself utterly unworthy of analytic help and, moreover, feels he is betraying the only good side
of himself in accepting analysis, the side which would devote his life to
making his loved ones happy—can only be got over in one way, namely,
through the possibility that analysis, by making him better, will in the end
make him at last capable of achieving his task for others—his loved ones. His
true aim is the other way round—to make *them* better first and so to become
well and good himself; but that is indeed impossible, both externally and
internally, for his sadism is still unmanageable. The nearest hope is this
reversal, again on the lines of a contradiction, or this compromise—to be cured
himself in order then to cure others. It is only on this understanding, so to
speak, unconsciously, and by placing all responsibility on the analyst, that
such patients accept analysis at all; and I think this hope, and this only, is the
ultimate source of the endless time, suffering and expense that such patients
will bring to continue analysis. We have to recognize that they do this much,
even if they do not get well. Why they do it has not hitherto been fully
understood. This single unconscious motive then, that he is to be cured in
order at last to be capable of fulfilling his task to others, and not for his own
ends, is the one slender positive thread on which the analysis hangs. But we
can see at once how impotent this motive can remain, how it is weakened,
obstructed and undermined by innumerable counteracting forces. For one
thing the patient does not for a moment believe in it; his fear of his own id
and its uncontrollable desires and aggression is such that he feels no sort of
security that he would eventually use any benefits obtained through analysis
for the good of his objects; he knows very well, one might say, he will merely
repeat his crimes and now use up the analyst for his own gratification and add
him to the list of those he has despoiled and ruined. One of his greatest
unconscious anxieties is that the analyst will be deceived on this very point
and will allow himself to be so misused. He warns us in a disguised way
continually of his own dangerousness.

Further, over and above this anxiety of accepting analysis on false pre-
tences and deceiving and betraying his good objects again by it, there is an
even greater fear, one which concerns the ego's fear for itself again, and links
up with the fear of death unconsciously so strong in his mind. This is the
dread that if he were cured by analysis, faithfully and truly, and made at last
able to compass the reparation needed by all those he loved and injured, that

the magnitude of the task would then absorb his whole self with every atom of all its resources, his whole physical and mental powers as long as he lives, every breath, every heartbeat, drop of blood, every thought, every moment of time, every possession, all money, every vestige of any capacity he has—an extremity of slavery and self-immolation which passes conscious imagination. This is what cure means to him from his unconscious depressive standpoint, and his uncured *status quo* in an unending analysis is clearly preferable to such a conception of cure—however grandiose and magnificent in one sense its appeal may be.

I hope that while I have spoken of the patient's unconscious aim of making others well and happy before himself, you will have borne in mind that the others I refer to always are the loved ones *in his inner world*; and these loved ones are also at the same moment the objects of all his hatred, vindictiveness and murderous impulses! His egoistic self-seeking attitude corresponds accurately enough to one side of things in his unconscious mind—to the hatred, cruelty and callousness there; and it represents his fears for his own ego if the love for his objects became too strong. We all fear the dependence of love to some extent.

I have spoken, too, of the contrast and incongruity of his love and need to save with his egoism, his tyranny, his lack of feeling for others. This egoism *is* his lack of a sense of reality. For his object-relations are not to real people, his object-relations are all within himself; his inner world is *all* the world to him. Whatever he does for his objects he does for himself as well; if only he could do it! he thinks; and in *mania* he thinks he *can*. So it is the overwhelming importance of the inner world of his emotional relations that makes him in real life so egocentric, asocial, self-seeking—so fantastic!

The unconscious attitude of love and anxiety for others in the patient is not identical with Freud's unconscious sense of guilt, though the feeling that the patient deserves no help till his loved ones have received full measure corresponds to it. This unworthiness finds atonement, as Freud says, in the illness, but only some atonement; the illness or the long analysis are compromises. To my mind it is *the love for his internal objects*, which lies behind and produces the unbearable guilt and pain, the need to sacrifice his life for theirs, and so the prospect of death, that makes this resistance so stubborn. And we can counter this resistance only by unearthing this love and so the guilt with it. To these patients if not to all, the analyst represents an internal object. So it is the positive transference in the patient that we must bring to realization; and this is what they resist beyond all, although they know well how to parade a substitute 'friendliness', which they declare to be normal and appropriate and claim ought to satisfy us as 'not neurotic'. They claim that their transference is resolved before it has been broached. We shall be deluded if we accept that. What is underneath is a *love* (a craving for absolute bliss in complete union with a perfect object for ever and ever), and this love is bound up with an uncontrollable and insupportable fury of disappointment, together with anxiety for other love-relations as well.

In Freud's remarks on the difficulties of the negative therapeutic reaction he has a footnote which in this connection is extremely interesting. He says that this unconscious sense of guilt is sometimes a 'borrowed' one, adopted from some other person who had been a love-object and is now one of the ego's identifications. And, 'if one can unmask this former object-relation behind the unconscious sense of guilt, success is often brilliant'.* This is the view I have just stated; the love for the internal object must be found behind the guilt (only Freud regards the love as past and over). He adds a link, too, with the positive transference. 'Success may depend, too', he says, 'on whether the personality of the analyst admits of his being put in the place of the ego-ideal'. But Freud's suggestion that the guilt is 'adopted' from a now internal object shows us that the brilliant success rests on a *projection* (or localization) *of the guilt on to an object, though an internal one*; and this is an extremely common feature of the manic defence (which may of course have been built up on some facts in experience). And his suggestion that the personality of the analyst determines whether or not he plays the part of ego-ideal indicates that consciousness and external circumstances are being allowed to blur the issue—exactly as the manic patient employs them to do if he can. The analyst *is* unconsciously the ego-ideal, or prototype of it, already to these patients; if they can rationalize their overmastering love and idealize it, then they can to some extent realize it without analysis; and this is in part a reparation, of course. The true aggressive character of their love, and their unconscious guilt of that, is still denied. Freud admits that this is a 'trick method' which the analyst cannot use. But the patient tries his utmost to trick us in this way. A great deal of our therapeutic success in former years in my opinion actually rested, and still may do, on this mechanism, without our having understood it.† The patient exploits us in his own way instead of being fully analysed; and his improvement is based on a manic defensive system. Nowadays I regard this possibility as a danger, even if it was not so formerly; for the analysis of primitive aggression now rouses severe anxieties, while recognition and encouragement by the analyst of the patient's attempts at reparation (in real

---

* [The reference is to Freud's statement (1923b, p. 50): "One has a special opportunity for influencing it when his unconscious sense of guilt is a 'borrowed' one—when it is the product of an identification with some other person who was once the object of an erotic cathexis. A sense of guilt that has been adopted in this way is often the sole remaining trace of the abandoned love-relation and not at all easy to recognize as such. (The likeness between this process and what happens in melancholia is unmistakable.) If one can unmask this former object-cathexis behind the unconscious sense of guilt, the therapeutic success is often brilliant, but otherwise the outcome of one's efforts is by no means certain." By contrast to Riviere, Freud stresses in the passage quoted above that the guilt resulted from the abandonment of the former love object.]

† [Riviere touches here on the problems of the inexact interpretation raised by Glover 1931, this volume, Chapter 23.]

life) allay them merely by the omnipotent method of glossing over and deny-
ing the internal depressive reality—his feeling of failure. The result is that the
patient may develop a manic defensive system—a denial of his illness and
anxieties—instead of a cure, because the depressive situation of failure has
never been opened up. In my experience the true analysis of the love and guilt
of the depressive anxiety-situation, because they are so deeply buried, is far
the hardest task we meet with; and the instances of success Freud quotes seem
to be last-minute evasions of it by the patients' chosen methods of projection
and denial.

The most important feature to be emphasized in these cases is the degree
of unconscious falseness and deceit in them. It is what Abraham comments
on; he, however, did not connect it with an unconscious sense of guilt. To us
analysts both the full true positive and true negative transference are difficult
to tolerate, but the *false* transference,* when the patient's feelings for us are
all insincere and are no feelings at all, when ego and id are allied in deceit
against us, seems to be something the analyst can see through only with
difficulty. A false and treacherous transference in our patients is such a blow
to our narcissism, and so poisons and paralyses our instrument for good (our
understanding of the patient's unconscious mind), that it tends to rouse strong
depressive anxieties in ourselves. So the patient's falseness often enough meets
with denial by us and remains unseen and unanalysed by us too.

---

* [The concept of false transference is Riviere's bridge to Wilhelm Reich. Anna
Freud's concept of transference of defense is more precise and not as pejorative as false
transference. If a patient defends himself, the analyst must accept that this is all he can
do, until he receives further help from the analyst.]

CHAPTER 29

FRANZ ALEXANDER

THE PROBLEM OF

PSYCHOANALYTIC

TECHNIQUE

INTRODUCTORY NOTE

According to Alexander, psychoanalytic cure is brought about by the interaction of three different psychological agents: emotional abreaction, intellectual insight, and making conscious repressed early memories. He explained the controversies in the field of technique by the relative emphasis various analysts attach to each one of the three components. This tripartite subdivision of the process of cure enables him to see the field from a perspective of his own.

In this article the reader will encounter a very different point of view from the one expressed ten years earlier (this volume, Chapter 6). Now, Alexander is strongly under the influence of Nunberg's synthetic function of the ego (1930, this volume, Chapter 22). He is critical of Rank, Ferenczi (1930, this volume, Chapter 21), Reich (1933, this volume, Chapter 18), and Kaiser (1934, this volume, Chapter 27). There is further no indication as yet that in ten years he will become the advocate of brief psychotherapy (see Alexander and French 1946).

First published in *The Psychoanalytic Quarterly* 4 (1935):588–611. Reprinted in *The Scope of Psychoanalysis: 1921–1961, Selected Papers of Franz Alexander,* © 1961 by Basic Books, Inc., Publishers. Reprinted by permission of Basic Books, Inc., Publishers.

THE GENERAL PRINCIPLES of psychoanalytic technique, as formulated by Freud in his five articles between 1912 and 1914, have often been subjected to careful reconsideration by various authors. Yet, and it is remarkable, these authors have failed to make any important innovation or modification. Many of the authors in developing their ideas of technique do so with the honest conviction that they are suggesting radical improvements over the standard technique. Others, more modest, maintain that their discussion calls attention to certain principles developed by Freud but for some reason or other neglected by the majority of analysts in their practical daily work.

There is an obvious reason for this constant urge to improve upon the analytic technique. Psychoanalytic therapy is extremely cumbersome, consumes the time and energy of patient and analyst, and its outcome is hard to predict on the basis of simple prognostic criteria. The desire to reduce these difficulties and increase the reliability of psychoanalytic treatment is only too intelligible. The difficulties, the time and energy consuming nature of psychoanalytic therapy, are by no means disproportionate to its ambitious aim: to effect a permanent change in an adult personality which always was regarded as something inflexible. Nevertheless, a therapist is naturally dissatisfied, and desires to improve upon his technique and to have precise definite rules of technique in place of indefinite medical art. The unremitting search to reform the technique therefore needs no special explanation; what needs explanation is the frequency with which pseudo-reforms are presented by their authors, under the illusion that they are discovering something new. This illusion originates in the complex nature of the psychoanalytic method. Psychoanalytic technique cannot be learned from books. The psychoanalyst must, so to speak, rediscover in his own experience the sense and the details of the whole procedure. The complex behavior of the patient as it is presented to the therapist simply cannot be described in all details, and the understanding of what is going on emotionally in the patient's mind is based on an extremely refined faculty usually referred to as intuition. In a former article I tried to deprive this faculty of the mystical halo which surrounds it by defining it as a combination of external observation with the introspective knowledge of one's own emotional reactions (1931).

Freud's articles on technique were published between 1912 and 1914,* at least fifteen years after he had started to treat patients with the method of free association and they may therefore be considered a resumé of at least fifteen years of clinical experience. These technical discoveries, for which a genius needed fifteen years, every student of psychoanalysis must recapitulate on the basis of his own experience. Though his study is now facilitated by general and simple formulations and by the precise description of those psychological processes which take place during the treatment, nevertheless

* [Alexander is not quite accurate; the papers on technique (*Standard Edition,* Vol. 12) extend from 1911 to 1915.]

the material which presents itself in every case is so complex and so highly individual that it takes many years for the student to achieve real mastery of the technique. Transference, resistance, acting out, removal of the infantile amnesia—these things he learns to appreciate only gradually. In consequence, he will be especially prone to emphasize those particular points of technique whose validity and importance are beginning to impress him. This alone can explain so many tedious repetitions and reformulations of the principles of technique that are, moreover, usually one-sided and much less judicious and clear than Freud's original formulations.

The general principles of the standard technique are consistent adaptation to the psychological processes which are observed during treatment: the phenomena of transference, resistance, the patient's increasing ability to verbalize material previously unconscious and the gradual removal of infantile amnesia. In the procedures that deviate from the standard, either one or another of these phenomena is overrated from the standpoint of therapeutic significance and is dealt with isolated from the others. The controversy is always centered around the therapeutic evaluation of (1) *emotional abreaction*, (2) *intellectual insight*, (3) *appearance of repressed infantile memories*. Those who consider emotional abreaction as the most important therapeutic factor will emphasize all those devices that may produce emotional eruptions resembling the abreactions in cathartic hypnosis: certain manipulations of the resistance, or the creation of emotional tensions in the patient, for example by avoiding interpretation of content. Those who believe that the best permanent therapeutic result comes from the patient's complete insight into the nature of his emotional conflicts will stress technical devices which have this aim; they will concentrate upon the analysis of content. Finally those who consider the most effective therapeutic factor to be the removal of infantile amnesias will be inclined to stress the reconstruction of the infantile history. Now in reality all these therapeutic factors are closely interrelated and dependent upon one another. For example, the occurrence of infantile memories is often, though not always, connected with emotional abreaction; intellectual insight on the other hand may prepare the way for emotional abreaction and recollections; and emotional experience, if not overwhelmingly intense, is the only source of real insight. Without recollection and emotional abreaction, intellectual insight remains theoretical and ineffective. The close interrelation of these three factors is clearly recognized in Freud's papers on technique, and his technical recommendations are based upon knowledge of these interrelationships.

All innovations up to today consist in an undue emphasis upon one or another of these factors—an overemphasis which is based on an insufficient insight into the dynamics of therapy.

One can roughly differentiate between three trends in technique: (1) neocathartic experiments, (2) reconstruction and insight therapy, and (3) resistance analysis. It should be stated, however, that none of these innovations or technical procedures have ever found general acceptance, and I suspect that the actual technique used by most of the innovators themselves in their daily work remained closer to the original than one would assume from their publications. Most psychoanalysts expect progress in technique to come not from one-sided overemphasis of one technical device but from an increasing precision in our knowledge, especially our quantitative knowledge of mental processes. Such greater knowledge should make possible a more economic procedure which will spare us much wasted time—the greatest weakness of our therapy. I shall try to evaluate critically some of these technical procedures in the perspective of the development of the technical concepts of psychoanalysis.

The therapeutic efficiency of abreaction of emotions in connection with recollection during hypnosis was the starting point of psychoanalysis both as a therapy and as a psychological theory. This led Freud to assume that the symptom disappeared because the dynamic force which sustained it had found another outlet in the hypnotic abreaction. The next step in the development of therapy was derived from the observation that emotional abreaction has no permanent efficacy, because the phenomenon of abreaction does not alter the constant tendency of the ego to eliminate certain psychic forces from motor expression. The state of hypnosis only temporarily created a situation in which such an outburst of emotionally loaded tendencies could take place, but this abreaction was dependent upon the state of hypnosis and the disappearance of the symptom depended upon the emotional relationship of the patient to the hypnotist. From this Freud came to recognize the phenomenon of resistance and discovered the technical device of free association. To eliminate one of the most important manifestations of the resistance he devised the basic rule, namely, the involuntary directing of the train of thought away from the repressed material. The last step in the development of the technique consisted in the recognition of the role of the patient's emotional attitude toward the analyst. What appeared on the surface as the patient's confidence in the analyst revealed itself as the repetition of the dependent attitude of the child on its parents, which by correct handling allows expression of deeply repressed material.

The insight gained from experience with cathartic hypnosis and then later with the method of free association may briefly be summarized as follows: The mere expression of the unconscious tendencies which sustained the symptom is not sufficient to secure a lasting cure. The rehearsal of individual traumatic situations of the past during treatment is not as important as the building up of the ego's capacity to deal with those types of tendencies which it could not face and deal with in the pathogenetic childhood situations. The original repressions create certain repression patterns, according to which, in later life, tendencies related to the original repressed ones become victims of repression. The cure consists in a change in the ego itself, an increase in its power—one might say its courage—to deal with certain emotional problems which it could not deal with early in life. The expression, an increase in the courage of the ego, is appropriate; for as we know now, fear is the motor of repression and courage is the faculty of overcoming fear.

But another expression requires explanation. What do we mean by increasing the ego's capacity to deal with repressed tendencies? A symptom obviously is not cured by the fact that the tendency which produced it enters consciousness. The mere fact of its conscious appearance cannot be of curative value unless we assume that when the preconscious and ultimately unconscious content becomes conscious, the process of becoming conscious consumes the same amount of energy as was represented by the symptom itself. That this is not the case is clearly seen by the fact that to become conscious of the formerly unconscious content does not always or necessarily relieve the symptom. Gradually it became clear that the appearance of a repressed tendency in consciousness is only one necessary condition of the cure; it opens a new outlet for the symptom-bound energy, namely, the outlet through voluntary innervations. Whether the process itself by which an unconscious tendency becomes conscious consumes at least a portion of the repressed energy quantum is still an open question. The dynamic equation of the process of cure is that the energy bound in a symptom before analytic treatment equals the energy spent in certain voluntary motor innervations afterward. It is possible that a smaller amount of symptom-bound energy is consumed in the process of its becoming conscious, that is to say, in the psychological processes which constitute conscious thinking.

The dynamic formulation that the energy which was bound in the symptom, after treatment takes up a new dynamic allocation needs further qualification. The new appropriation of energy, in voluntary innervations, must be in harmony with the forces already residing within the ego. If this condition is not fulfilled, a conflict is created within the ego which inhibits the free disposal of the formerly symptom-bound energy. This harmonizing or integrating function of the ego, however, is generally considered a faculty on which the analyst has to rely but to which he cannot contribute much by his therapeutic activity. This limits the indication of psychoanalysis to patients who possess an ego of sufficient integrating power, because the process of integration and its end result, a conflictless disposal of formerly symptom-bound energy, must be left to the patient himself.

Nunberg subjected this integrating or synthetic function of the ego and its role in therapy to a careful investigation and showed that the process of a repressed content's becoming conscious itself represents an integrating process in the ego [1930, this volume, Chapter 22]. I shall return to this problem later. It is certain that with or without the analyst's cooperation the formerly repressed energy, which during the process of the treatment becomes a part of the dynamic inventory of the ego, must become reconciled and harmonized with the already existing forces in the ego.

The fundamental validity of this formulation of the process of therapy has been corroborated in particular by recent developments which have shifted the emphasis from the analysis of symptoms to the analysis of character or of the total personality. We have learned that apart from neurotic symptoms, in many patients an even more important expression of repressed tendencies takes place in so-called neurotic behavior. This is a more or less stereotyped automatically fixed and unconsciously determined way of behavior, which in contrast to voluntarily guided behavior is beyond the control of the conscious ego. There is even a group of neurotic personalities whose sickness consists mainly or exclusively in such impulsive or stereotyped behavior without any pronounced symptoms. Gradually it became an aim of our therapeutic endeavors not only to cure neurotic symptoms but to extend the ego's administrative power over this automatic and rigidly fixed expression of instinctive energies.

The aim of the therapy can thus be defined as the extension of conscious control over instinctual forces which were isolated from the conscious ego's administrative power, either as symptom or as neurotic behavior. We may now investigate by what means those who deviate from the standard technique hope to achieve this aim. In order to evaluate these deviations, we must consider the part played by the three therapeutic factors in the analytic process, *abreaction, insight,* and *recollection.* We saw that abreaction without insight is insufficient. We understand now why. The process of integration, by which the repressed tendency becomes an organized part of the ego, does not take place without insight; insight is the condition—perhaps the very essence —of this integrating process. Equally obvious on the other hand, insight without emotional experience, that is to say, without abreaction, is of little value. Something which is not in the ego cannot be integrated into it, and emotional experience is the sign that the tendency is becoming conscious. Therefore theoretical knowledge of something which is not experienced emotionally by the patients is perforce therapeutically ineffective, though it must be admitted that in certain situations a merely intellectual insight may prepare the way for abreaction. It is not advisable to think of these processes too schematically. Abreaction without insight and insight without abreaction are two extremes, between which in practice there are all degrees of combination and analyses do in fact consist of such differently graded mixtures of insight and emotional experience. Abreactions, small in quantity, take place throughout any analysis conducted by the standard technique and each successive abreaction is attended by more and more insight.

Whereas there is considerable agreement concerning the relation of insight to emotional experience, there is much controversy about the effectiveness of infantile recollections. The concept that the energy contained in a symptom can simply be transformed and absorbed by the process of recollection, is obviously erroneous. Nevertheless, recollection seems to be an indispensable precondition if a repressed tendency is to be thoroughly integrated into the ego system, in that it is recollection which connects the present with the past. Although the direct therapeutic value of the process of recollection may be questioned, the removal of the infantile amnesia must be considered as a unique indicator of the successful resolving of repressions. Therefore the removal of infantile amnesia might be required as a sign of a fully successful analysis, even though a cure and the removal of infantile amnesia may not necessarily be directly causally related.

We see now that all three factors, abreaction, insight, and recollection, are required in order to obtain the goal of psychoanalytic procedure, which is, the removal of certain repressions and the subsequent integration of the formerly repressed tendencies which makes their ego-syntonic disposal possible. Whereas insight and abreaction are in direct relationship to the process of relieving repressions and of integration of the repressed forces, the importance of recollection may be a more indirect one. It serves as an indicator of the removal of repressions.

A brief survey will illustrate our point that the divergence from the standard procedure usually is a one-sided overemphasis of one of these three factors. So far as one can reconstruct the evolution of analytic technique, Freud, after he gave up hypnosis, began to lay more and more stress on insight and the reconstruction of the infantile history. This was quite natural. He tried to reproduce in the waking state the phenomenon he and Breuer observed during hypnosis, namely, the patient's recollection of forgotten traumatic situations. The main goal became to make the patient remember during the process of free association, and, so far as this was not fully possible, to complete the gaps in memory through intellectual reconstructions. Around 1913, however, when Freud first formulated systematically the principles of the technique as we use it today, we see that he was already fully in the possession of the above described dynamic concepts and considered analysis by no means a merely intellectual procedure. Yet once he had recognized the importance of the patient's intellectual insight as precondition of the integrating activity of the ego, in contrast to many of his followers, he never lost sight of its significance.

It seems that at some time between the introduction of the method of free association and the publication of the technical recommendations of Freud in 1912, 1913 and 1914, there must have been a period in which analysts overrated the importance of an intellectual reconstruction of the infantile history.* This can be seen from the fact that even after Freud's

---

* [The emphasis on reconstruction of infantile history to the exclusion of other considerations can be seen most strikingly in the case history of The Wolf Man (Freud

publications on technical recommendations many analytic pioneers apparently persistently overintellectualized the analytic process, and stressed the interpretation of content and reconstruction of infantile history, overlooking the more dynamic handling of resistance and transference. This explains the joint publication by Ferenczi and Rank of *Entwicklungsziele der Psychoanalyse*, which may be regarded as a reaction against this overintellectualized analysis (1924). Ferenczi and Rank, as I tried to show when their pamphlet appeared in print, went to the other extreme.* According to them the whole analysis consists in provoking transference reactions and interpreting them in connection with the actual life situation. The old abreaction theory began to emerge from the past. Ferenczi and Rank thought that after the patient had reexperienced his infantile conflicts in the transference neurosis, there was no need to wait for infantile memories; they believed that insight was possible without recollection merely through the understanding of the different transference situations which are modeled upon the forgotten conflictful childhood experiences. Much of the originally repressed material they held had never been verbalized in the child's mind, and therefore one could not always expect real recollection of those situations upon which the transference reactions are modeled. Assuming that Ferenczi and Rank were right, and that one does not need to wait for the infantile amnesia to be dispelled, the obvious practical value of their concept would be a considerable abbreviation of the treatment. In this concept obviously the ego's integrating function is neglected, together with the corresponding technical device, the working through. The tedious task of helping the patient to bring his transference manifestation into connection both with the actual situation and with his former experiences plays a less important role in this technique. After the transference manifestation becomes clearly expressed and understood by the patient, even though the connection with the original patterns of the transference is not established, the analysis could be terminated on a date set by the analyst.

The further developments are well known. Rank more and more centered his attention on the actual life situation, and considered insight into the infantile history as merely a research issue with no therapeutic significance whatsoever. Ferenczi, however, soon discovered that the artificial termination of the analysis did not work out therapeutically, dismissed it from his technique and tried to enhance the effectiveness of the therapy by increasing emphasis upon the abreaction factor. Though he did not return to the method of cathartic hypnosis, he frankly admitted that he considered abreaction, as it takes place during cathartic hypnosis, to be the really effective therapeutic factor, and he tried to reproduce it in the method of free association by

---

1918); see also *The Wolf Man by the Wolf Man: The Double Story of Freud's Most Famous Case* (Gardiner 1971).]

    * F. Alexander, review of Ferenczi and Rank's *Development of Psychoanalysis, Internationale Zeitschrift für Psychoanalyse* 11 (1925).

creating artificial emotional tensions, at first through his active technique ([1919c, this volume, Chapter 7]; 1920, 1930 [this volume, Chapter 21]; 1931; 1934), later through his relaxation method.* With the help of the ingenious technical device of relaxation, in certain cases he succeeded in creating semihypnotic states, in which the patient in a twilight state repeated his infantile emotional conflicts in a dramatic fashion.

Both the joint attempts of Rank and Ferenczi and Ferenczi's later technical experiments can be classified as abreaction therapies, in which the element of insight, that is to say, the process of integration, is neglected. These technical reforms imply a regression back toward cathartic hypnosis with a reintroduction of all the therapeutic deficiencies of this period. They represent an emphasis of intensive transference analysis and neglect of the intellectual integrating side of therapy, the working through.

Another technical trend is represented by Reich's resistance and later analysis (1933). According to Reich the aim of therapy is the transformation mainly into orgastic genitality of energy bound in neurotic symptoms and character trends. The discussion of this narrow theoretical concept does not lie within the scope of this study. Our present interest is his technical motto, the stress on certain hidden manifestations of resistance, which according to him are not recognized by most psychoanalysts, and his strict distinction between interpretation of resistance and interpretation of content. According to Reich certain hidden manifestations of resistance must first be analyzed and only afterwards can the analysis deal with the content which the patient's ego is resisting.† The important things are not the familiar open manifestations of resistance, but those secret manifestations which the patient expresses only in a very indirect way in characteristic behavior, for example, in pseudo-cooperativeness, in overconventional and overcorrect behavior, in affectless behavior, or in certain symptoms of depersonalization. The emphasis on hidden forms of resistance is unquestionably of great practical value. Glover (1927) mentions in his treatise on technique the importance of these hidden forms of resistance which one easily overlooks, and Abraham in one of his classical contributions described the pseudo-cooperative attitude of certain patients as a specific form of hidden resistance [1919, this volume, Chapter 4]. Reich's emphasis on understanding the patient's behavior apart from the content of his communications is largely a typical example of the rediscovery of one of the many therapeutic revelations that every analyst encounters during his development, as he gradually becomes more and more sensitive to

* [In his 1930 paper, Ferenczi suggested that "the analysis of resistance can be preceded by a kind of comforting preparatory treatment" (p. 122). This he called the "principle of indulgence," which when the analysand suffered traumatic experiences as a child takes precedence over the principle of frustration (this volume, Chapter 21) For further details of Ferenczi's development, see the biographical sketch on him, this volume, pp. 48–49.]

† [Reich 1933, this volume, Chapter 18.]

the less obvious, more indirect manifestations of the unconscious. However, Reich's distinction between resistance which is expressed in the patient's communications and that expressed by his gestures and general manner of behavior is quite artificial. All of these expressions complement each other and constitute an indivisible unity.

Reich's other principle of the primacy of resistance interpretation over content interpretation is based upon a similarly artificial and schematic distinction. As Fenichel has correctly pointed out, the repressing tendencies and repressed content are closely connected.* They constitute one psychic entity and can only be separated from each other artificially. The patient's resistance, for the careful observer, always displays at least roughly the content against which the resistance is directed. There is no free-floating resistance. At least the general content of the repressed can be recognized at the same time as the fact of the resistance itself. The more the analyst is able to help the patient to understand his resistance in connection with what it is directed against, the sooner the resistance itself can be resolved. Mostly the verbalization of what the patient is resisting diminishes the resistance itself. Strachey has convincingly described this reassuring effect of correct and timely interpretations, which can best be witnessed in child analysis [1934, this volume, Chapter 24]. It is true, as Fenichel states in his critical discussion of Reich's technique, that in the interpretation of the content the analyst can go only slightly beyond what the patient himself is able to see alone at any given moment. Yet every resistance should preferably be interpreted in connection with what it is directed against, provided of course that the content interpretation corresponds to the status of the analysis.

Reich's concept of layer analysis is similarly a product of his over-schematizing tendency. That unconscious material appears in layers is a familiar observation. Freud operates with this concept as far back as the "History of an Infantile Neurosis" [1918], and in *Totem and Taboo* he shows that the primary aggressive heterosexual phase is as a rule concealed by an overdomestication of these tendencies, by a masochistic passive homosexual phase. Following Freud's lead, I tried, in an early paper, "Castration Complex and Character" [1921], to reconstruct the history of a patient's neurosis as a sequence of polar opposite phases of instinctual development.

The existence of certain typical emotional sequences, such as: early oral receptivity leading under the influence of deprivations to sadistic revenge, guilt, self-punishment, and finally to regression to a helpless dependence are

* Otto Fenichel, "Zur Theorie der psychoanalytischen Technik," *Internationale Zeitschrift für Psychoanalyse* 24 (1935). As a matter of fact Fenichel mentions this argument as expressing not his own views, but those of the advocates of content interpretations, including Freud. He writes: "They [these advocates] think that because of the persistent interweaving of defensive forces and rejected tendencies, it is impossible to verbalize the ones without at the same time verbalizing the others." (Author's paraphrase) [this volume, Chapter 30].

generally known. The validity of such typical emotional sequences, which make the material appear in "layers," is sufficiently proven, and every analyst uses this insight as a useful orientation in the chaos of unconscious reactions. This, however, does not change the supreme rule that the analyst cannot approach the material with a preconceived idea of a certain stratification in the patient, for this stratification has individual features in different patients. Though certain general phases in the individual's development succeed others with universal regularity, the different emotional attitudes do not necessarily appear during the treatment in the same chronological order as they developed in the patient's past life history. Moreover, the pathogenetic fixations occur at different phases in different cases, and the fixation points determine what is the deepest pathogenetic layer in any given case. Often we find an early period of sadism leading to anxiety and covered consecutively by a layer of passivity, inferiority feelings, and secondary outbreak of aggression. In other cases we see that the deepest pathogenetic layer is a strong fixation to an oral dependent attitude, compensated then by reaction formations of over-activity and aggressiveness, which in turn are covered by a surface attitude of helpless receptivity. It is not uncommon that a patient in the course of the first two or three interviews reveals in his behavior and associations a sequence of emotional reactions belonging to different phases of his development. As Abraham many years ago emphasized during a discussion in the Berlin Psychoanalytic Society, it is not advisable to regard the different emotional reactions as they appear during the treatment in a too literal, too static sense, as though they were spread out one layer over the other, for in the unconscious they exist side by side. During development, it is true, they follow each other in temporary sequences, one emotional phase being the reaction to the preceding one. During treatment, however, probably due to as yet unknown quantitative relationships, they do not repeat exactly their historical chronological order. I have often observed in more advanced stages of an analysis—sometimes even in the early stages—that patients during one analytic session display almost the whole history of their emotional development. They may start with spite and fear, then take on a passive dependent attitude, and end up the session again with envy and aggression. The analyst can do no better than follow the material as it presents itself, thus giving the lead to the patient, as Horney (1935) has again recently emphasized. Reich's warning against premature deep interpretations is correct, to be sure; Freud emphasized this point in his technical recommendations, and it is implicit in the general principle that interpretation should always start from the surface and go only as deep as the patient has capacity for comprehending emotionally. But in Reich's overschematic procedure, the danger resides in that the analyst instead of following the individual stratification of emotional reactions in the patient, approaches the material with an overgeneralized diagram of layers, before he is in a position to decide which emotional attitude is primary and which should be considered as reaction. The chronological order of the appearance is by no means a reliable criterion. An observation of Roy Grinker and Margaret Gerard in the Department of Psychiatry of the University of Chicago clearly demonstrates that the order in which the transference attitude of a patient appears is determined also by factors other than the chronological order in which it developed in the patient's previous history. As an interesting experiment they had a female schizophrenic patient associate freely for a few days alternately in the presence of a male and a female psychoanalyst; they observed that the patient's attitude was influenced by this difference of the analyst's sex. When the male analyst conducted the session, the patient was constantly demanding and aggressive; to the female analyst she complained and was more confiding, seeking for reassurance. This experiment clearly shows that the chronological sequence of transference attitudes does not follow rigidly a historically predetermined stratification of infantile attitudes, and is determined also by other factors.*

The slogan of the primacy of resistance interpretation over content interpretation found its most consistent expression in an extreme distortion of the analytic technique, in Kaiser's resistance analysis, from which every interpretation of content is pedantically eliminated. The analysis is reduced to an extremely sterile procedure of pointing out to the patient his resistance manifestations [Kaiser 1934, this volume, Chapter 27].

After Fenichel's excellent critical analysis of this technique, there is little call for comment. Its most paradoxical feature consists in the fact that Kaiser, who limits the therapeutic agent of analysis to dramatic abreactions entirely, reminding us of the latest experimentation of Ferenczi, attempts to achieve such abreactions by a merely intellectual procedure—namely, by convincing the patient of the irrationality of his resistance behavior and resistance ideas. This intellectual insight, Kaiser thinks, can break down the resistance itself and allow the repressed material to appear in a dramatic fashion. In order to create strong emotional tensions, he carefully avoids every interpretation of content and goes so far as to condemn every indirect allusion of the analyst to preconscious material, even if this is so near to consciousness that it needs only verbalization in order to appear on the surface. It is not the intellectual insight into the resistance, but the avoidance of all content interpretation, that creates in the patient such tensions as to provoke dramatic abreactions. The reassuring effect of verbalizing preconscious material, which encourages further expression of repressed material, has been mentioned above. To call the child by its name divests much of the patient's fear of the uncanny tension that comes from the pressure of preconscious material when it is merely felt as some unknown danger. The analyst's objective discussion of such material eliminates the infantile fear of the condemning parents and of their inner

* I wish to thank Drs. Grinker and Gerard for permission to refer to this interesting observation.

representative, the harsh super-ego. Verbalization of repressed content has for the patient the meaning of a permission; careful avoidance of it means condemnation.

Obviously here the fascination of the analyst by the fireworks of emotional rockets is what leads to such a distortion of the analytic technique, which is quite without logical justification and contradicts our dynamic concepts of the analytic procedure. The ideal of the standard technique is just the opposite—a permanent, steady, uninterrupted flow of repressed material, undisturbed by sudden dramatic advances that necessarily lead to new regressions, which often neutralizes the effect of many weeks' or many months' work. This steady flow can, however, only be obtained by a judicious economic use of resistance and content interpretations in such connections as they appear, by helping the patient connect the emerging material with the rest of his conscious mind and with his past and present experience.

Without attempting to advance any radical reforms or lay down new technical rules, I shall try in the following to investigate the question as to how far and in what way the analytic method aids the integrating or synthetic process in the ego, which, as Nunberg has correctly claimed, is an integral part of the analysis.

The process of the cure we described as the combination of two fundamental psychological processes, (1) the inviting of unconscious material into consciousness and (2) the assimilation of this material by the conscious ego. To the first phase our literature refers by different expressions: *emotional experience, abreaction, transformation of unconscious into conscious material;* the second phase is called *insight, digestion or assimilation of unconscious material by the ego* or *synthesis and integration.* Seen in this perspective it is obvious now that the technical reforms and innovations which we have been discussing in detail are all primarily concerned with the first phase and are reactions to an early overintellectualized period of psychoanalytic treatment, in which intellectual insight was overstressed, and in which reconstructions and interpretations were made by the analyst upon material which had not yet appeared in the consciousness of the patient. Ferenczi and Rank stress the emotional experience in the transference, and Reich's and Kaiser's main interest is focused upon methods of mobilizing unconscious material by manipulating and interpreting the resistance. In all these experiments with technique the first problem, the mobilization of unconscious material, is considered the crucial one; the assimilation of the unconscious material is left to the integrating powers of the psychoneurotic's fairly intact ego. The question is now in which way this ultimate aim of the therapy, the integration in the ego of the material previously unconscious, can be supported by the correct handling of our technique.

Nunberg's analysis of the process by which unconscious content becomes conscious clearly shows that this process itself is an integrating act of the ego

[1930, this volume, Chapter 22]. The quality of consciousness in itself involves an integrating act: a psychological content in becoming conscious becomes included in a higher, richer, more complex system of connections. The preconscious material's becoming conscious has long been considered by Freud as the establishment of a new connection: that between object images and word images. Obviously what we call abstractions, or abstract thinking, represent again a higher grade of synthesis between word images. Although we do not yet know much about its details, what we call conscious thinking consists mainly in the establishment of new connections between conscious contents. It must be remembered, however, that these new connections of higher grade cannot be established arbitrarily by the ego. The connections must be correct, that is to say, they must be in conformity with the results of the reality testing of the ego. Therefore generalizations, the establishing of connections between different conscious elements is permanently counteracted by the critical or distinguishing faculty of the ego, which it uses, however, only under the pressure of reality. Without the pressure of the reality testing functions, the synthetic function would run amuck as it does in many philosophical systems. Nunberg convincingly demonstrates all this and considers the delusional system in paranoia to be the result of such a faulty overstressed synthetic effort of the ego, by which it desperately tries to bring order into a personality chaotically disorganized by the psychotic process.

Nunberg also called attention to the fact that every neurotic system and most psychotic symptoms are synthetic products. In fact all unconscious material, as it presents itself to us during the treatment in its *status nascendi* of becoming conscious appears in certain synthetic units; fear together with guilt and hate, receptive wishes and dependence overreacted to by aggression appear to us as two Janus faces of the same unit. We discover the synthetic nature of the unconscious material also in such generalizations as connect or identify the objects of sexual impulses in the unconscious. The extension of the incest barrier over all individuals of the other sex is the simplest and best known example of this generalizing tendency of the mental apparatus. The process by which an unconscious content becomes conscious consequently consists in the disruption of primitive synthetic products and the reassembling of the elements in the higher synthetic system of consciousness, which is more complex, more differentiated and consequently more flexible. Thomas M. French's recent studies of consecutive dreams clearly demonstrate that during the course of the treatment a progressive breaking up of primitive emotional patterns takes place, together with a building up of new more complex relationships between the elements. This new synthesis allows behavior more flexible than the rigid automatic behavior which is determined by unconscious synthetic patterns. It is the ego's function to secure gratifications of instinctive needs harmoniously and within the possibilities of the existing external conditions. Every new experience requires a modification in the previously established patterns of instinct gratification. The unconscious consists of psychological units, expressing more primitive, usually infantile connections between instinctual needs and external observations. These primitive units as we know are not harmonized with each other, nor do they correspond to the external conditions of the adult. Therefore they must undergo a new integrating process into higher systems: a new adjustment between instinctual needs and external reality must be accomplished, in which process the ego plays the part of a mediator. The establishment of these new connections, however, necessitates the breaking up of the old units—in other words, of symptoms or fixed behavior patterns which correspond to earlier phases of the ego development. What must be given emphasis, however, is the fact that all unconscious material appears in synthetic units, which constitute certain primitive patterns that connect instinctual demands with the results of reality testing.

According to this concept the process by which an unconscious content becomes conscious corresponds to a recapitulation of ego development, which consists in a gradual building up of more and more complex and flexible systems of connections between different instinctual needs and sense perceptions.

We are now prepared to discuss the technical question: in what way does our technique contribute to the breaking up of the primitive psychological units as they exist in the unconscious and help their elements to enter into the new, more diversified connections in the conscious ego? The main function of psychoanalytic interpretations obviously consists precisely in the establishment of new correct connections and in the breaking up of old overgeneralized and more primitive connections. The effect of interpretation can most simply be compared with the process of the child's learning to connect and differentiate objects. At first, when the child learns the word "stick," it begins to call every longitudinal object a stick, and then gradually learns to differentiate between stick, pencil, poker, umbrella, etc. When a neurotic patient learns to differentiate between incestuous and nonincestuous objects, that is to say, to react differently toward them, he essentially repeats the same process.

In his current systematic study of patients' consecutive dreams during the process of cure, French subjects this learning process to a thorough investigation, from which we expect to learn much of the nature and details of this learning process.* At present we know only its general principle, namely, that it consists in a gradual establishment of new and more differentiated connections between the psychic representatives of instinctual needs and the data of sense perception.

---

* [Alexander's emphasis on the similarity between psychoanalytic insights and learning can be attributed to French's influence; see "A Clinical Study of Learning in the Course of a Psychoanalytic Treatment," 1936. This view underrates the significance of repression and fails to explain why patients should defend so strongly their mistaken primitive overgeneralizations.]

What does this insight teach us with regard to our analytic technique? It is obvious that our interpretations must fulfill both purposes: they must break up the primitive connections and help to establish new, more differentiated ones that are in harmony with the reality with which the adult is confronted. The standard technique, as it was described in its basic principles by Freud about twenty years ago, still serves this double purpose better than any of those reform procedures, which neglect to give aid to the synthetic functions of the ego and take into account only the mobilization of unconscious material. What we call "working through" has the function of aiding the integrating process. Its therapeutic value is sufficiently proven by experience. My contention, however, is that every correct interpretation serves both purposes: mobilization of unconscious material and its integration into the system of consciousness. The *synchronization* of the two functions of interpretation into one act, inducing abreaction and insight at the same time, may be considered a fundamental technical principle, which I should like to call the *integrating principle of interpretation.** I disagree with every attempt which tries artificially to isolate these two processes, most extremely represented in Kaiser's technique, because the best means still of overcoming a resistance is the correct interpretation of its not yet verbalized background. The basis of the ego's resistance is its inability to master or to assimilate unconscious material. Everything which the patient can understand, that is to say, everything which he can connect with other familiar psychological content of which he already is master, relieves fear. In other words, every new synthesis within the ego, by increasing the ego's ability to face new unconscious material, facilitates the appearance of new unconscious material. The longer the patient is exposed to material which puzzles him, which seems strange, and appears to him as a foreign body, the longer the analysis will be retarded and the appearance of new unconscious material blocked. The ideal we strive for in our technique is that whatever unconscious material appears in consciousness should be connected at the same time with what is already understood by the patient. This makes of the analysis a continuous process. Therefore, whenever it is possible interpretations should refer to previous insight. To be sure, as has already been emphasized, interpretation does not consist merely in the creation of new connections but also in the breaking up of primitive infantile connections. This can be done only if the material as it appears in its totality is exposed to the patient's critical judgment. Umbrella, walking cane, poker, lead pencil, must be demonstrated together in order to break up their faulty identification and generalization as a stick. The interpretations must point out these connections, formed by the mind in infancy as they appear in the presenting material. We know that these connections, as they occur in symbols, for example, often seem extremely strange to the mind of the adult, who has forgotten and

* [This is an important point clarifying the theory of interpretation. For a similar emphasis, see Hartmann 1939, this volume, Chapter 31.]

overcome this primitive language of the unconscious. It is too much to expect that the patient will be able to recognize without help the infantile generalizations as something self-evident. I do not doubt, however, that after the old primitive connections are broken up, the patient, because of the integrating power of his ego, would in time establish the new syntheses alone. Here, however, is the place where the analyst can help and accelerate the integrating process. Interpretations which connect the *actual life situation* with *past experiences* and with the *transference situation*—since the latter is always the axis around which such connections can best be made—I should like to call *total interpretations*. The more interpretations approximate this principle of totality, the more they fulfill their double purpose: they accelerate the assimilation of new material by the ego and mobilize further unconscious material.

This principle of totality should not be misunderstood and used in a different sense than it is meant. Totality does not mean, for example, that all deep overdeterminations in a dream should be interpreted. Totality does not mean an unlimited connection of material which though in fact related is still far from the surface. It means totality not as to depth but as to extension—the connecting with each other and with previous material of elements which belong together. It cannot be emphasized too much, however, that these connections should center around the emotionally charged material, usually the transference manifestations. Fenichel's formulation regarding the penetration of the depth is valid, to wit, that the interpretation can only contain just a little more than the patient is able to see for himself at the moment.

The supreme requirement for the correct handling of the technique, however, more important than any principles and rules, is the precise understanding in detail of what is going on at every moment in the patient. It is needless to say that all the formulations here given should be considered not as rules but as general principles to be applied always in accordance with the individual features of the patient and the situation.

The isolation of resistance from content interpretation is not a desirable aim though at times it is necessary, in particular when the tendency against which the patient has resistance is not yet understood by the analyst. Probably the only effective way of permanently overcoming resistance consists in helping the ego to integrate, that is to say, to understand new material. Therefore in the long run all those technical experiments which aim at sudden abreactions of great quantities of unconscious tendencies fail. These techniques expose the ego not to a continuous flow but to sudden eruptions of new material and necessarily must cause new repressions, repression being a phenomenon which Freud has explained as resulting from the infantile weak ego's inability to deal with certain instinctual needs. The reproduction of such an inner traumatic situation in which the ego is exposed to overpowerful stimuli cannot be a sound principle of our technique. Many, not all, roads lead to Rome. In analytic therapy our main allies are the *striving of uncon-*

scious forces for expression and the integrating tendency of the conscious ego. Even if we do nothing else, if we do not interfere with these two dynamic forces, we will be able to help many patients, and if we succeed without therapeutic activity in aiding and synchronizing both of these two fundamental agents, we will increase the efficiency of our technique.

Nunberg's thesis that the psychoanalytic treatment is not only an analytic but simultaneously a synthetic process as well is fully valid. It has often been maintained that psychoanalysis consists mainly in the mobilization of unconscious material and that the integration of this material must be left to the patient's ego. The standard technique, as it is used since Freud's technical recommendations, consisting in interpretations centering around the transference situation, really involves an active participation of the analyst in the integrating process. Through our interpretations, without fully realizing it, we actually do help the synthesis in the ego. Doing it consciously and understanding this integrative function of our interpretation may contribute to developing the art of analysis into a fully goal conscious, systematically directed procedure. Always keeping in mind the function which our interpretations fulfill in the treatment eventually will help to bring us nearer to the ultimate goal, the abbreviation of the psychoanalytic treatment.*

---

* [This last line foreshadows Alexander's future development as an advocate of the drastic shortening of psychoanalytic treatment.]

CHAPTER 30

OTTO FENICHEL

CONCERNING THE THEORY OF PSYCHOANALYTIC TECHNIQUE

INTRODUCTORY NOTE

The controversy between Fenichel and Kaiser is of interest because it raises many fundamental questions about psychoanalytic technique, particularly the question of content interpretation versus resistance interpretation. With the publication in this volume (Chapter 27) of Kaiser's paper, the English-speaking reader can himself decide where he stands in this controversy.

One can see from this paper how embattled the position of the early psychoanalytic ego psychologists was in the 1930s. On the one hand, Fenichel argues against Theodor Reik and Melanie Klein, who were all too ready to make interpretations about the id without regard to the readiness of the ego to absorb these interpretations. On the other hand, ego psychologists argue against Reich, and even more against Kaiser, who wished to limit all interpretations to resistance interpretations. The current paper by Fenichel was written one year after Sterba's paper (1934, this volume, Chapter 25), and one year before the publication of Anna Freud's book. It is a forceful exposition of the principles of psychoanalytic ego psychology in the field of technique.

Originally entitled "Zur Theorie der psychoanalytischen Technik," *Internationale Zeitschrift für Psychoanalyse* 21 (1935):78–95. English translation reprinted from *The Collected Papers of Otto Fenichel*, First Series. Collected and edited by Dr. Hanna Fenichel and Dr. David Rapaport. By permission of W. W. Norton & Company, Inc. Copyright 1953 by W. W. Norton & Company, Inc. This paper is a discussion of H. Kaiser's "Probleme der Technik" (this volume, Chapter 27).

I

IN DISCUSSING THE "theory of technique" it is unfortunately still necessary to discuss the justification of this concept. There exist some views about the "irrational" nature of psychoanalytic technique which oppose any attempt at constructing a theory of its technical principles. One of these views, for instance, was expressed recently by Reik [1933, this volume, Chapter 26]. Since the instrument of psychoanalytic technique is the unconscious of the analyst (the "relay" conception), and since intuition is indispensable for apprehending what goes on in the patient, he wants to leave everything in that technique to the unconscious and to intuition. But in view of the fact that the subject matter of psychoanalysis is the irrational, such conceptions must in the final analysis lead to that method itself being regarded as irrational, losing every characteristic of science and becoming pure art.

Per contra, we argue as follows. We have a dynamic-economic conception of psychic life. Our technique, too, which aims at a dynamic-economic change in the patient, must follow dynamic-economic principles. It must always remain true to the mode of thinking underlying psychoanalysis, and must order our behavior issuing from intuition (which is of course indispensable) according to rational directives.

Freud was the originator of the concepts "dynamics" and "economics" in psychic life. His whole method of studying neurotic phenomena, as well as his papers on technique (1911a, 1912a, 1912b, 1913a, 1914c, 1915a), leaves no doubt that he considered analytic interpretation, as well as the procedure of the analyst in general, as an intervention in the dynamics and economics of the patient's mind, and thus he demanded of interpretations more than that they should be correct as to content. It was he who asserted that only a procedure which used *resistances* and *transference* could be called psychoanalysis (1914a); that is, only a procedure which intervened in the dynamics and did not merely give "translations" of the patient's allusions, as soon as the analyst understood to what they alluded. The formula that the analyst should make the unconscious conscious might lead to such a misunderstanding. Indeed, it is possible that the statement of a symptom's meaning will at times make it disappear; but it need not. Mere topological* conceptions do not suffice to explain what determines whether an interpretation does or does not have this effect. Whether it does or not is determined by whether or not a repression (more correctly, an instinctual defense implying continued expenditure of energy) is actually eliminated. But what eliminates a repression? The dynamic conception views the psyche as a continuous struggle between mental trends which seek discharge and the defensive and selective forces of

* [By "topological" Fenichel refers to surface versus depth interpretation.]

the ego, between the instinctual cathexes and the anti-cathexes of the ego. That the latter too arose from the former does not interest us here. In reviewing an already existing neurotic conflict, we see that it takes place between an unconscious instinctual demand and the defensive forces of the ego, which are supplied with "anti-cathexes," and which manifest themselves in the treatment as resistances. (Kaiser's designation of the energy used by the forces defending against drives as "narcissistic libido," is *per se* correct, but apt to be misleading. To avoid discussions about the genesis of this libido, it is probably better to use for these energies the term "anti-cathexis" introduced by Freud.) What we have to do is to intervene in this interplay between drive and resistance. In so intervening, we need not, and even cannot, reinforce the drive. The repressed drive is our ally in our work; it strives of itself toward consciousness and motility. Our task is only to see that no resistances bar its way. Were it possible to brush aside the resistances, the repressed would appear on its own. This *dynamic* conception of interpretation—that our task is to seek out resistances and to uncover them so that the repressed manifests itself—must be supplemented by the economic conception that our task is to tackle the economically most important and strongest resistance in order to achieve an actual and economically decisive liberation of libido, so that what was tied up so far in the struggle of repression shall be available to real gratification. The infantile sexual impulses which have been repressed then find contact with the ego and change, for the greater part, into genitality capable of orgasm, the rest becoming capable of sublimation.

The "theory of technique" is but a commentary on these propositions. These propositions should be taken seriously; and it cannot be denied that there are many factors in the analytic situation which tempt the analyst not to take them seriously, but, sooner or later, to "drift" and to use the only-an-art conception of analytic technique, making the inevitability of a certain lack of system in the analysis an excuse for letting himself float along in a planless way—that is, letting himself interpret purely intuitively what occurs to him, or at best, to his patient.

II

It is Reich's merit to have especially warned us against this procedure. His proposals for the reform of the technique derive mostly from a serious view of the *economic* conception, namely, from an insight into the fact that our task is to liberate the energy tied up in the repressive struggle and to change repressed infantile sexuality into adult sexuality capable of orgasm, by eliminating repression (1933).

There are a number of "technical formulas" transmitted by tradition from Freud, inspection of which shows that what Reich's proposals on technique are saying is, "Consider whether you are really always applying the true Freudian technique." One of these formulas is, "Work always where the patient's affect lies at the moment." To the thoughtful analyst this clearly does not mean: "Work where the patient *believes* his affects lie." The analyst must always *seek out* the points where at the moment the conflict is decisively centered. Another example: "Interpretation begins always at what is on the surface at the moment." Taken correctly, this can only mean that it makes no sense to give "deep interpretations" (however correct they might be as to content) as long as superficial matters are in the way. For this reason one cannot, as Melanie Klein wants, "get into direct contact with the unconscious of the patient," because to analyze means precisely to come to terms with the patient's *ego*, to compel the patient's ego to face its own conflicts. If we, for instance, know that a compulsion neurosis has regressed, out of castration anxiety, from the genital oedipus conflict to the anal-sadistic stage, we cannot use this knowledge for discussing "immediately" the genital oedipus conflict; the only way to it is by working through the anal-sadistic conflicts. This is obvious. But it is also necessary always to keep in view the hundreds of analogies of everyday life. The defensive attitudes of the ego are *always* more superficial than the instinctual attitudes of the id. Therefore, before throwing the patient's instincts at his head we have first to interpret to him that he is afraid of them and is defending himself against them, and why he does so.

Here is another formula: "Interpretation of resistance goes before interpretation of content." Every resistance hinders the patient from digesting a contentual interpretation, that is, an utterance of his unconscious trends so that it effects a dynamic change. Thus there is no point in trying to do this before the obstacle is out of the way. Since not all resistances are manifest, however, the analyst must continuously seek out and work on the momentarily acute resistances, first, by separating the patient's judging ego from his resistance-determined behavior; secondly, by getting the patient to experience the latter as arising from his resistance; thirdly, by finding the occasions of the resistance; fourthly, by explaining why the resistance takes precisely this form; and fifthly and lastly, by telling him against what it is directed. Freud has, moreover, repeatedly discussed and demonstrated by examples that not only the content of what the patient says but his modes of behavior, his "accidental" actions, and his manner and bearing are all also the subject matter of the analysis.

There are, it is true, some other traditional formulas as well, which at first sight seem to contradict Reich's views. There is, for instance, Freud's warning against making a kind of "stock-taking" of the situation from time to time during the course of the analysis, in order to clarify the structure of the case for oneself, because in this way one only gets a biased view of it; one should rather respond to the patient's unconscious with one's own unconscious and wait until a structural picture arises of itself (1912b). Again, there is his comparison of analysis with a jig-saw puzzle in which one piece after another is observed at random as each presents itself "by accident," until one finds how they fit together (1896). Then, too, there is the formula: "The patient determines the theme of the session."

The apparent contradictions of such formulas are resolved if we keep in mind that the psychoanalytic technique is a *living art*, in which rules never have more than a relative validity. Surely Freud's views, if represented correctly, mean that the analytic technique must guard against two extremes, both equally incorrect: on the one hand, one must not analyze too much according to a rational plan, by intellect alone (the concept of "relay," the analyst's own unconscious as his instrument—"he who wants to analyze must be analyzed"); and on the other hand, one must not be too irrational, because to analyze means to subject the irrational in man to reason. (Otherwise psychoanalytic technique would be unteachable. The frequently used comparison between analytic and surgical technique is here indeed in place; for analytic technique, too, one needs endowment and intuition, but these, without training, do not suffice in surgery either.)

Reich's view is that interpretation has never been as yet consistently thought through and followed out as a dynamic-economic process. Instead of using their insight into the dynamics and economics of psychic processes to build up their technique in a planned and systematic way, analysts succumb to indolence and lack any system. In their work, despite their better knowledge they take the task of interpreting resistances to mean that they are to interpret whatever the patient is talking about. In Reich's opinion, the reason why the analyst usually fails to work "where the affect really is" is that it simply does not occur to him to look for the affect where it should be sought—namely, in characterological behavior. Characterological behavior acts as a kind of armor-plating which covers the real conflicts, and this aspect of it is not taken seriously enough. Indeed, were one to take the rule "to work where the affect actually lies" seriously, it would mean that as long as the leading characterological resistance was unbroken, one would work on no other subject matter, and discuss no other theme with the patient but this. The more "the affect" is "frozen" into an "attitude," the less does the patient know about it, and the more important is it to work on this *first*, so that the contentual interpretations which the analyst will make later on shall not be wasted beforehand. Though the comparison with the jig-saw puzzle is correct, that game too can be played systematically and according to a plan, by not examining the pieces as they accidentally present themselves, but looking every time for the pieces which would fit. The psychic material of the patient has a certain stratification. His resistance attempts to conceal this stratification. The analyst,

nevertheless, must discover this stratification and follow it exactly in his interpretations, and he must recognize when material whose content belongs to a
deeper stratum emerges only in order to ward off more superficial material.
Otherwise, he will be confronted with the dreaded "chaotic situation," in
which material from every stratum of the mind is produced in disordered
comminglement. Thus Reich thinks the principle "begin always at the surface," too, should be taken more seriously and carried out more consistently
than hitherto. Such a consistent procedure demands primarily that material—
including dreams—which does not serve the momentary purpose should be
left untouched, in order not to "fire away" uselessly the work of interpretation. "The patient determines the theme of the session," not by what he says,
but by showing the analyst where his economically crucial resistance lies. This
theme the analyst must then compel the patient to work through, even if the
latter would rather talk about something else.

I should like to insert a few critical comments at this point. With those
principles of the so-called "Reichian technique" which I have attempted to
represent here, I am in complete agreement—qua *principles*. I consider them
as correctly deduced from Freud's theoretical and technical views.* I also
agree with Reich that in our everyday work all of us often infringe on these
principles, and that in this respect no amount of self-control is too much. The
contradictory judgments which have been made in analytic circles about the
so-called "Reichian technique"—some saying, "It is nothing new, but only
exactly what Freud does," and others saying "It is so different from Freud's
analysis, that it ought not to be called psychoanalysis"—can be explained in
this way: In so far as these principles are merely elaborations of Freud's
views, they are "nothing new"; in so far as they are *consistent elaborations* of
it, they *are* something new.

The agreement in principle with Reich which I have just expressed is
only limited, on the one hand, by two minor theoretical objections, and on the
other, by objections, not against the essence of his views and principles, but
simply against the manner of their application on particular points.

The two minor theoretical objections are:

(1) The psychic material in the patient does not have an orderly stratification. Reich's assertion to the contrary is schematic and disregards compli-
cating details. The regularity of the stratification is just as regularly broken
through—in different people to different degrees—even when there have been
no incorrect analytic interpretations. The phenomenon which geology calls
"dislocation" is a general one and consists in materials originally layered
over, or side by side with each other, being mixed into each other by various
natural events; consequently, the sequence in which the material presents itself

---

* They have been excellently formulated already by W. Reich in his study "On the
Technique of Interpretation and of Resistance Analysis," Chapter 3 of *Character Analysis*
[1933].

to the geologist who drills into the earth is not identical with the age of the layers in question. The material is only "relatively" ordered. In the same way, to my mind, the relatively correct views of Reich on the "consistency" of interpretation must not be taken for absolute either. For there are *spontaneous chaotic situations* too; indeed, there are people whose character neurosis presents a picture which cannot be diagnosed by any other term but "chaotic situation." Moreover, "dislocation" continues to take place during psychoanalytic treatment. And the fluctuations of everyday life, which cannot be disregarded, also lessen this "consistency" to some extent.

(2) If we are to pay particular attention to "frozen resistances,"* to habitual actions and attitudes, we must not only know that they express resistance, but get to understand their meaning. Naturally, even only to call the patient's attention to his resistive attitude is better than to overlook it completely. But there is no doubt that we shall succeed the more easily, the more completely we understand the concrete meaning of such a resistive attitude. The discovery of this meaning in turn will be facilitated by every piece of knowledge we gain about the individual patient's past history. We are thus faced with a vicious circle: His past history becomes accessible only through resolving these attitudes—and resolving these attitudes requires knowledge of that history. In my opinion, this vicious circle is best resolved by the analyst's setting out from the beginning to learn for his own information (without major "interpretations") as much as possible about the patient's past. I think that it is always a good thing to use the first period of analysis for *collecting material*. The more *information* one has, the better armed one goes into the actual struggle with the resistances. We do not always

---

* [The controversy as to whether the so-called "frozen defenses" should be analyzed before the unfrozen ones became an important point of controversy between Anna Freud and Wilhelm Reich. The relevant passage reads:

"*Permanent defence-phenomena.* Another field in which the ego's defensive operations may be studied is that of the phenomena to which Wilhelm Reich refers in his remarks on the 'consistent analysis of resistance.' Bodily attitudes such as stiffness and rigidity, personal peculiarities such as a fixed smile, contemptuous, ironical and arrogant behavior—all these are residues of very vigorous defensive processes in the past which have become dissociated from their original situations (conflicts with instincts or affects) and have developed into permanent character traits, the 'armor-plating of character' (as Reich calls it). When in analysis we succeed in tracing these residues to their historical source, they recover their mobility and cease to block by their fixation our access to the defensive operations upon which the ego is at the moment actively engaged. Since these modes of defence have become permanent, we cannot now bring their emergence and disappearance into relation with the emergence and disappearance of instinctual demands and affects from within or with the occurrence and cessation of situations of temptation and affective stimuli from without. Hence, their analysis is a peculiarly laborious process. I am sure that we are justified in placing them in the foreground only when we can detect no trace at all of a present conflict between ego, instinct and affect. And I am equally sure that there is no justification for restricting the term 'analysis of resistance' to the analysis of these particular phenomena, for it should apply to that of all resistances" (Anna Freud 1936, pp. 35–36).]

succeed in such an initial collection of material; nor is such failure any reason for giving up an analysis. I believe, however, that we should not deliberately by-pass occasions for collecting such material. It seems to me that Reich, in the intention of doing nothing else but work consistently upon the point around which for the moment everything turns, often leaves aside material which, if he had regarded it, would precisely have helped him to understand the point in question. I have especially often had this impression in connection with "by-passed dream analyses." In free associations, we often have the experience to which Freud has called attention (1912b), viz., that what the patient says becomes comprehensible only from what follows after it. Therefore, until what is to come after *has* come after, we cannot, I think, know what material we ought to leave aside.

Naturally, there are many situations in which absolutely *every* interpretation of a dream is contraindicated, namely, when "dream interpretation" *per se* has some other unconscious meaning for the patient which the analysis has not yet apprehended. But where this is not the case, I believe that it is precisely through correct dream interpretation that an attitude of the patient can often be understood. After all, the dream is a commentary on the patient's ego-attitudes of the previous day.* Among the latent dream thoughts there are always some which are close to the conscious attitude but yet contain an additional element or show the attitude in a relationship which the patient has not thought of on account of his repressions. To interpret a dream does not mean telling the patient, "You want to sleep with your mother"; it may also be "to infer latent dream thoughts and by means of them to show the patient the actual nature of his present behavior and its intentions." Latent dream thoughts, however, cannot be inferred without getting the patient's associations to the elements of the manifest dream. If the patient does not associate, we give up the attempt at interpreting and try to apprehend this resistance. If, however, he does associate, his attention is thereby not necessarily fatefully diverted from consideration of his characterological behavior of the moment; it should rather be possible to use his associations precisely to lead him to that point.

I have called these two objections "minor objections," because they do not undermine Reich's principles, but only make them less absolute. The question now is how these principles are applied. This will vary with each case, and particularly with the personality of the analyst. In spite of Reich's assertions that there is no such danger, I believe that the "shattering of the armor-plating" could be done in a very aggressive way, but that both the

---

* [Fenichel refers here to the day residue that appears in dreams. We are, however, accustomed to interpret dreams as telling what remained repressed and was struggling to emerge into consciousness the day before, rather than to scrutinize dreams for the ego attitudes of the previous day. One could wish that Fenichel had explained his approach to dream interpretation in greater detail.]

aggression and the consequent disintegration of the armor can be *dosed*, and indeed, that it is the task of the physician to make this procedure as little unpleasurable as possible for the patient. The first thing we must be clear about is that the consistent tackling of the patient's character traits wounds his narcissism much more than any other analytic technique. Not only does the degree to which patients can tolerate such wounds vary, but also the degree to which analysts can or should inflict them. As analysts we should in principle certainly not be afraid of "crises" (the surgeon isn't afraid of blood either when he cuts); but that is no reason for inviting such "crises" in every case. On the contrary, I believe that our aim ought to be the *gradual* reduction of the existing insufficient neurotic equilibrium. We are familiar with the resistance of some patients, who long for a "trauma" and expect cure not from a difficult analysis, but from the magic effect of a sudden explosion. There is an analogous longing for a trauma on the part of the analyst also. Let us beware of it!

The conviction that a consistent working through of character resistance is the one and only correct method may make one overlook the fact that experiencing this kind of analysis may itself become for the patient a *transference resistance*. This would naturally be an even more superficial one than the "character resistance" and would have to be dealt with first. In one case, a patient, who in his fear of experiencing sexual excitation, always fell away, at a certain height of excitation, from his active masculinity into a receptive orality ("at this point I dare go no further, you do it for me"), experienced and enjoyed the "activity" with which the analyst pursued his current "attitudes," etc., as the fulfillment of his receptive longing, and this he did without the analyst noticing it. In another case the unconscious content of the neurosis of a woman patient was her rebellion against her father, who throughout her childhood reproached her for her traits and mimicked them. To have begun her treatment with an attempt at a "consistent attitude analysis"—a procedure which became highly necessary later on—would have led to an immediate breaking off by her of the analysis.

III

Kaiser's study, "Probleme der Technik" [1934, this volume, Chapter 27], agrees in many points with Reich's views, so that the agreements and criticisms so far expressed here apply, for the greater part, to it also. I also agree with some objections which Kaiser makes against Reich; for instance, as to the impossibility of advising beginners not to use character analysis if it is the right technique.

Kaiser, however, goes further and states that analysis should be carried out entirely without "content interpretations." This proposition seems to me to be inimical. In one place, Kaiser says that if one has first given a warning against "too early" or "too deep" interpretations, there is not so very much difference between whether one does in the end allow a "content interpretation" or not. But the difference is very great, if this "in the end" is correctly understood. In order to do this we must first discuss the theory of "content interpretations."

According to Kaiser, "content interpretation" contradicts our understanding that the aim of psychoanalytic work should be to eliminate resistances, not to reinforce unconscious drives. He maintains that if our task is to work on resistances, then we must not interpret anything but resistances. If the repressed does not appear after an interpretation of resistances has been made, then our job is not to "name it," but to consider this a proof that the interpretation has failed, that it was not differentiated enough. Though he does not deny that under certain circumstances "content interpretations" may also eliminate repressions, he believes that theoretically this cannot be explained except in a round-about way, namely, that without the analyst's knowledge and intention, such an interpretation too can direct the patient's attention to "resistance ideas," and can correct them by this change in the direction of his attention. He denies that an "anticipatory idea"* might be at work here, or more correctly, that an interpretation could have the same effect as the indication given by the teacher of histology to the student as to what he will see in the microscope, without which the student's eye, which is not yet set for microscopic vision, would not see anything—which is actually Freud's view of the essence of interpretation. He considers this impossible (though his "resistance analysis," too, consists in nothing but calling the patient's attention to something whose presence has so far eluded his attention), and argues as follows: "A repressed impulse is neither in the Pcpt.-Cs. system, nor in the Pcs. system. Even the most exact and most apposite anticipatory idea we might give to the searcher could not facilitate his search as long as he is searching in a space which does not contain the object sought; and the patient cannot look into his unconscious."

This argument collapses if we realize that "content-interpretation" does not designate unconscious instinctual impulses, but preconscious derivatives of them.

But what are the contents of consciousness and of the preconscious?

---

* [The importance of giving the analysand anticipatory ideas was discussed by Freud in his papers of 1910a and 1937a. In 1910 he said: "The treatment is made up of two parts—what the physician infers and tells the patient, and the patient's working-over of what he has heard. . . . We give the patient the conscious anticipatory idea (the idea of what he may expect to find) and he then finds the repressed unconscious idea in himself on the basis of its similarity to the anticipatory one" (1910a, pp. 141–142.]

First of all, perceptions (and the phenomena of feeling accompanying them); then memory traces (ideas, etc.) mobilized by new perceptions—these memory traces differing according to the instinctual excitation of the moment. That is, they consist of instinctual impulses. The defense against impulses consists in this: that they—or rather, their undistorted ideational representatives—are thwarted in their striving toward consciousness by deep strata of the ego by means of anti-cathexes of every kind. The anti-cathexis creates a barrier in front of the preconscious. The defensive struggle demands a *permanent expenditure* of psychic energy. The impulse that has been warded off continuously produces *substitutive formations*; that is, it uses other ideas (impulses) that are associatively connected with it, which break through to consciousness, in order to discharge its energies. It *reinforces* preconscious formations that are innocent in themselves, thus forming them into those "mixed formations" between preconscious and unconscious, which have been discussed by Freud. Against these—as derivatives of the repressed—the defensive trends of the ego can still be directed just as much as against the unconscious impulse proper (Freud 1915b). The fate of these derivatives— that is, whether they become conscious or are also repressed—depends on a variety of factors; namely, on all those which influence the dynamics and economics of the interplay between drive and defense. In general, one may say that the greater the degree of its distortion the more easily does the derivative become conscious. Analytic therapy may be described as a general education of the ego toward tolerating increasingly less distorted derivatives.

We cannot ever "interpret the unconscious." Stekel and his followers, who "fired interpretations" at their patients, attempted this.* A "too early" or "too deep" interpretation, one which names something unconscious, but no preconscious "derivative"—something, therefore, that the patient cannot find however much he searches—is no interpretation. The adherents of "content interpretation"—and among them Freud—do not interpret repressed drives, but their preconscious derivatives. Moreover, since defense and what is defended against are forever interwoven, they do not believe that they can always name what is being defended against without also naming with it the defense. The details of defense are just as unconscious to the patient as the details of what is being defended against. In "interpreting a resistance," too, one cannot reveal more to the patient than what he is able to discover in himself by self-observation. What is remarkable, however, about psychoanalytic interpretations from the economic angle is that, whether one interprets a resistance or a derivative of the repressed, one does not interpret only that portion of what is being defended against which has already penetrated into the preconscious, but just a *little bit more*—a bit that the analyst already

---

* [Wilhelm Stekel, among the early followers and later dissenters from Freud, was known for his interpretation of dreams directly from the manifest content dispensing with free associations.]

senses but the patient not yet. How does it happen that by so being interpreted, this little bit more actually breaks through? Naturally, this procedure, too, has not increased the "cathexis of the repressed" but weakened the "anticathexis of resistance." It remains true that one cannot do anything but break the resistance. But this can be done in various ways, and "naming," even if it is only by the "method of indication," seems to us the *via regia*. Of course, this naming has to take place at the right point and in the right way. Where is this point, what is this way, and how does the naming become effective?

In order to understand this, let us consider first the theory of the basic rule of psychoanalysis. What do we want to achieve by it? In every person there is an uninterrupted struggle between instinctual impulses which strive toward consciousness and motility, and the forces of the ego, which refuse untimely instinctual demands and, directed by purposive ideas, admit only what is pertinent to present action or speech. By means of the basic rule, we want to eliminate first of all the thousands of "resistances" of everyday, which on all ordinary occasions are what make life and mutual communication possible at all. But if it were really possible to eliminate all "purposive ideas" and concentrate attention only on noting what emerges of itself, we should not even then get to see the repressed. True, most "resistances" would be eliminated, but precisely not the resistances of repression, which are the strongest in intensity; for these are by definition not amenable to the conscious will of the ego. What we see are the results of the struggle between unconscious impulses and unconscious defenses of the deep layers of the ego. What do we do then? When we "interpret resistances" to the patient—no matter whether we do so by shifting certain resistive ideas into the focus of his attention, or whether we are only able to tell him at first that something in him rebels against abiding by the basic rule, or against the interests of his analysis in general—we are always demanding of him that he should *discover* something in himself; that is, we do exactly as the teacher of histology does. If the indication "you have a resistance" (an indication which is naturally not given in these words) leads the patient, who did not know this till then, to notice that something in him is indeed rebelling against the analysis, then we have given an effective interpretation. For this reason, such a statement to a patient does not seem to me by any means as ridiculous as some recent discussers of technique consider it to be. What is important is to give this indication in such a way as not to be a reproach against the patient, but a direction of his attention to something preconscious which he had not before noticed (and to a "something more"). The resisting ego, however, rebels against accepting such indications of the real nature of preconscious derivatives. This is what Freud has in mind when he says that our daily practice continually proves that besides the main censorship, between unconscious and preconscious, there is yet a second censorship between preconscious and conscious. *To begin with, we only work against this second censorship.* We have various means of doing this. These means are the same as those at the disposal of people for inducing other people to do something disagreeable: first, by convincing the patient that what is disagreeable is *useful* (the patient wants to be cured and the therapist explains to him that it is necessary for the cure); second, by making use of his libidinal tie to the maker of the demand (his "transference of affection"). Freud says with good reason in his *General Introduction to Psychoanalysis* that we use all means of suggestion (which is nothing but "using the libidinal tie") to induce the patient to produce and recognize "derivatives" (1916–1917). The defensive ego, which, limited in its reality function, cannot itself notice what is going on—which does not "direct" its "attention" to it—is as a rule bribed to behave in this way by a "secondary gain," by a sort of premium (Radó 1928). In that case we must try to uncover these premiums. If we cannot do that, we must at least let the patient experience the insufficiency of his ego's reality function. (This is the "directing attention to resistive ideas.") This happens in that the ego's observing part is made to stand away from its experiencing part and is thus able to condemn it as irrational, and such a condemnation results in a change of the dynamics of defense. These events and the reason for them seem to me to have been best described in a study by Sterba (1940). I readily agree with Kaiser that this standing away and this calling to attention can be better achieved if one has learned to study ever more sharply and exactly all the details of conscious defensive attitudes. Yet it must happen that as the patient's experiences of his defending anxiety become accessible to him, his attention is also directed to the *contents* of his anxiety, which gradually appear in consciousness in the form of derivatives. The latter, however, originate in early days, and can no longer be separated from what is being warded off. If the derivatives of the warded-off material are named as well, and *at the right place*—that is, where the naming leads to the patient's being really able to discover the derivatives in himself—then this, by producing an external perception which is in consonance with those derivatives in their nascent state, brings about a removal of the resistance and an entrance of the derivatives into consciousness. The "something" which we add here is thus also swept by this consonance into consciousness.

The assertion that content interpretations should only be given "at the end" means, to my mind, merely that they should not be given so long as the patient cannot discover their representatives in his preconscious, because obstructing resistances prevent their entrance there. I cannot convince myself that the interpretation quoted by Kaiser: "Note please how you behave toward me . . . Can you understand it any other way but that you are actually very angry with me . . . ?" is as wrong in principle as he believes. Kaiser argues that what is brought to consciousness by such an interpretation is not the warded-off affect which we seek, but an "innocent version of it," something about which one can think and talk without disquiet; and that such an interpretation, instead of mediating the experience of the affect, mediates the experience of a reassurance, which takes away the real seriousness of the affect and is thus a resistance. No doubt this can happen. If it does, we are then dealing with an "obstructing resistance," which must be recognized and interpreted as such. But I cannot see why this should always or even often be the case. A "reassurance" need not, by any means, always have this resistive meaning. A certain "taking distance" from the affect—which has the air of "Here I can permit myself to let all the affects come, because I do it only as an experiment and I am not really aiming at the analyst about whom I feel the emerging affect"—need not, by any means, imply a lack of seriousness in the whole discovered affect; on the contrary, it can increase the tolerance of the ego and thereby considerably facilitate the discovery of the presence of very serious affects. I believe that on this point Kaiser is guilty of *an underestimation of reason*. He is right when he warns against analytic pseudo-interpretations, which offer the patient more or less the same as the reading of Freud's writings, namely, dynamically ineffective knowledge. But he exaggerates this concern, when, because of it, he refuses to *use reason* in self-observation. One can make quite effective discoveries in oneself, discoveries which leave no doubt as to the reality and affective vividness of the discovered material, without wanting to *act it out* in reality at the moment. A few words should be said here about the nature of "acting out." It is often welcome as a means to the patient's self-knowledge, and at times it is even indispensable. But if it is not followed by a sufficient analytic "working through" it is also a great danger and a resistance. A "reliving" of affects without the self-judging ego taking a sufficiently detached attitude is at least as dangerous as that isolation which limits the experience to the judging ego's "cognizing" the affect. The impulses which were repressed must be *experienced* as actually existing (and as, at the moment, inappropriate—as in the "transference"); but they need not for this reason be acted out. We often succeed in subjecting the experience of them to intellectual judgment while they are still in *statu nascendi*. Indeed, I believe that to succeed in this is the real goal of psychoanalytic interpretation—"to remember with affect, and to recognize what is remembered as truly operative in the present."

The difference must be somehow related to what Kaiser calls the "genuine break-through of instinct." There is no doubt that the patient must make affect-filled new discoveries, or rather, rediscoveries in himself. Transmitting of "knowledge concerning the unconscious" is no analysis. But must the elimination of the judging ego go as far as Kaiser demands, when he gives the following as a characteristic example of what he is after: "Then he addresses the analyst by the name of a person out of his past . . ."? Can we not read here between the lines a "longing for trauma" on the part of the analyst, to which reference has already been made? I think so, and I think it although in another place Kaiser warns against "sham break-throughs" and knows very well that the task of analysis is not to entice out unconscious material, but to get the ego to work over the unconscious which it has recognized. Therefore it makes no sense to interpret unconscious drives when there is no ego capable of digesting it; in such cases we have first to establish such an ego. For this reason all sham break-throughs, just like all "acting out" in general, must be unmasked to the patient as manifestations of resistance as soon as it is once more possible to discuss things reasonably with him. Kaiser thinks that the phrase used by Freud to demonstrate the success of a psychoanalytic intervention, "Thus the patient often relates without any difficulty . . . ," does not show that the phenomenon of genuine instinctual break-through is being indicated here. But I believe that that is precisely what is being indicated; or perhaps, more correctly, what is indicated is not a phenomenon of an instinctual break-through but of a real removal of repressions, in that the patient now recognizes impulses and affects which were so far barred for him, and can judge and control them. It seems to me that Kaiser underestimates, in favor of a "traumatic" conception, the gradual increase of extent to which "derivatives" become admissible, owing to persistent interpretation of these "from the ego side" and the concomitant gradual increase of the power of the ego, when he writes: "At any rate, it is a fact that changes in the patient which may be considered as an advance in the direction of real cure, are observed only after the occurrence of phenomena of the described sort," namely, after more or less traumatic instinctual break-throughs. Therefore, the fact that when he "gave content-interpretations with apparently good effect" he "never achieved comparably intensive break-throughs of drives" as when he "refrained from using any content-interpretations," does not seem to me a proof that that good effect was only "apparent." Kaiser believes that nothing is gained therapeutically if one brings a patient to convince himself that he has criminal impulses, but that he must experience this impulse, and experience it in such a way that he has to suppress it quite consciously if he is not to become criminal. We, on the contrary, believe that it suffices if he experiences it to such an extent that he has no doubt about its original reality. The actual situation, however, may be so evident to him at the moment of experiencing the impulse that it needs no particular suppression to avoid criminal behavior.

I cannot agree with Kaiser's criticism of the "procedure of interpretation and indication" either. To my way of feeling, there is nothing artificial in the analyst's behavior if, when he wants to demonstrate something to his patient, he does not simply assert it, but puts the evidence for it so clearly before the patient that he must draw the correct inference himself. If the patient answers this with an "Aha, you're thinking again that . . . ," then that is a new resistance which must be analyzed; that this can happen is, however, not a sufficient reason for rejecting the procedures of interpretation and "indication" in general. Kaiser's assertion that "the therapeutic effect will be just about nil" is altogether contrary to my observations. Undoubtedly there is such a thing as a discovery of affect which is the opposite of experiencing, but there is also a form of discovery which leads to experiencing, and indeed may be identical with a certain kind of experiencing. If the "indicatory procedure" makes it possible for the patient to fend off the affect anew, as Kaiser describes it, then we recognize this and must not make the interpretation before that resistance is removed. But there is a "taking distance" from the affect which seems to me, in contrast to Kaiser, to be desirable. The judging ego of the patient should stand at a distance from its affect and should recognize it as untimely, while remembering its origins affectively. A break-through of affect without such "taking distance" is—as Freud once well put it—"an outright mishap." Kaiser's comment, that "after a 'genuine instinctual break-through' there is nothing left for the analyst to explain or clarify or add to the contents expressed by the patient" makes us suspect that he actually does not recognize this "mishap" for what it is, that he neglects the process of "working through" and does not understand its essential role in the true elimination of repressions, so that he is aiming at a sort of "neo-catharsis" instead of analysis. "Working through" is an essential constituent of psychoanalytic work and consists in rediscovering what one has found in one place in many others. For instance, in resistance analysis we have to undermine every single one of the many resistance positions; we do this by stressing the characteristics which are common to them all, unmasking them as variants of a single core. The "attitude" in question is represented in various single complexes of ideas, just as is the unconscious instinctual demand discovered in analysis. The undermining of these complexes by working through is no different from what the work of mourning does, since the idea of the mourned-for lost object is also represented in many connecting ideas (Freud 1917b). For Kaiser the demand to "start always from what is at the moment on the surface" also entails a contraindication of any "content interpretation." He says "Only he will escape drawing this conclusion who counts as belonging to the surface the 'contents' brought forth by the patient in any given analytic hour. Such a conception of the surface, however, appears extremely unpsychological. Supposing a patient, impelled by an urge to make the analyst impatient, relates an experience from his fourth year of life in an affectless and boring manner, then it is the affectless and boring manner of presentation, not the content of the narrative, which belongs to the surface." Very good; but if the patient at a later stage of his analysis relates an experience from the fourth year of life with appropriate affect, then the content, as well, of this narrative can belong to the surface. And if he misjudges a detail of this experience, the correct judging of which might lead, for instance, to the understanding of a present attitude, so that the naming of it takes the patient by surprise, then this content belongs to the surface.

Kaiser is thus unwilling to use reason on the two points where it is legitimate to use it: namely, in order to make certain experiences accessible to the patient on the one hand, and, on the other, to get him to judge his affects and stand at a proper distance from them. This *underestimation* of reason is in contrast to Kaiser's *overestimation* of it at other points. "To interpret resistance" is for him identical with indicating the existence of resistance ideas. But this undoubtedly necessary indication of a resistance does not amount to an interpretation of it if the patient cannot also understand why his attention was diverted from the mistaken ideas—namely, because he had anxiety—and why he had anxiety and when and how he acquired it. To indicate the mistaken ideas is not yet to correct them, as long as the cause which made the patient think mistakenly is not eliminated. The reduction of character resistances to "mistakes in thought" is the best example there is of the overestimation of reason in the work of interpretation. This becomes also particularly clear where Kaiser equates "compulsion" and "rationalization." The incorrect thinking of the compulsion neurotic is not a consequence of his ego's having been "distracted" by some forces from the act of thinking, so that this might be altered by his attention being called to the incorrectness of his thought. It is because he has wanted to escape from instincts which are unpleasurable to him, into thought which is remote from instincts; but that which was defended against broke into the defense and the function of thought itself became "sexualized" and thereby pathological and distorted. The nature of "transference resistance," too, seems to me to have been misrepresented by Kaiser. Transference becomes resistance not only by being "rationalized" but, in general, because its transference character is not conscious. The tendency to satisfy a repressed drive on a substitute object is, in and of itself, a compromise between the drive, which seeks gratification, and a resistance, which prevents it from getting to the object proper. In analytic therapy, therefore, transference (except for its positive, affectionate form, which to begin with facilitates the overcoming of other resistances) becomes fundamentally a resistance, and must be recognized and worked through as such.

What Kaiser does is nevertheless in keeping with the principles of Freudian analysis, and he corrects many mistakes made by other analysts through an insufficient regard for the dynamic and economic factors. But his study exaggerates in its distribution of emphasis, and it contains a latent danger, which if this emphasis is continued might become manifest: the neglect of the factor which is crucial for Freud's psychoanalysis, namely, the unconscious and its specific characteristics.

The history of psychoanalysis brought it about that we became acquainted with the unconscious before the conscious and with the repressed before the ego. Nowadays the psychology of the ego stands in the center of our investigation. All those fine differences in the consciousness of man, which have been studied by the non-analytic schools while they have so far, of necessity, been neglected by psychoanalysis, now come within the sights of analysis, too. No doubt the psychoanalytic technique stands to gain much from this through the refinement of "resistance interpretations." Let us remember, for instance, how convincing Kaiser is when he constantly emphasizes, side by side with the instinctual needs of the id, the need of the ego to maintain its level of self-regard. But let us not forget, either, that the preceding exploration of the unconscious makes it possible for the psychoanalyst to approach the phenomena of the ego, the differences of consciousness and the phenomena of self-regard, in a fundamentally different manner from other schools: psychoanalysis must explain these phenomena too as arising from an interplay between unconscious—and in the final analysis, biological—instinctual tendencies, and influences of the external world.

Kaiser contrasts "resistance analysis" with "interpretative procedure" (and "indicatory procedure"). What he means by "interpretative procedure" is clear: he means that "content interpretation" which he rejects and which we have discussed at sufficient length here. Nevertheless, "interpretative procedure" is applied all around. But anyone who would really not admit any interpretative procedure of any sort could not, I believe, be called an analyst. Because if the means of analysis consist in a surmounting of resistance and a use of transference, then the principles by which it sets these means into motion are to be found in the fundamental rule of analysis and in interpretation.

CHAPTER 31

HEINZ HARTMANN

THE EGO AND THE PROBLEM OF ADAPTATION

An Excerpt

INTRODUCTORY NOTE

Hartmann's great influence on psychoanalytic thinking belongs to the post–World War II period and therefore lies outside the scope of this volume. The seminal work from which this excerpt has been taken was, however, written in 1939. But at that time it evoked little interest in psychoanalytic circles. Note, for example, how little space Fenichel (1945) assigns to Hartmann.

The passage extracted here is of great significance. Following Freud's emphasis on archaeology as the model for psychoanalysis, psychoanalysts tended to see their work essentially as a reconstruction of what has once existed and was buried by the repression. By contrast, Hartmann sees the work of interpretation not only, or not even primarily, as that of reconstruction but rather as the establishment of a new connection, and therefore as a new creation.

DEFENSES (typically) not only keep thoughts, images, and instinctual drives out of consciousness, but also prevent their assimilation by means of thinking. When defensive processes break down, the mental elements defended against and certain connections of these elements become amenable to

Originally entitled "Ich-Psychologie und Anpassungs problem," *Internationale Zeitschrift für Psychoanalyse* 24 (1939):62–135. Excerpted here from *Ego Psychology and the Problem of Adaptation* (New York: International Universities Press, 1939), pp. 63–64.

recollection and reconstruction. Interpretations not only help to regain the buried material, but must also establish correct causal relations, that is, the causes, range of influence, and effectiveness of these experiences in relation to other elements. I stress this here because the theoretical study of interpretation is often limited to those instances which are concerned with emerging memories or corresponding reconstructions. But even more important for the theory of interpretation are those instances in which the causal connections of elements, and the criteria for these connections, are established. We cannot assume that the ways in which children connect their experiences, and which later become conscious in the course of psychoanalysis, could satisfy the requirements of the mature ego, not to speak of the requirements of a judgment which has been sharpened by psychoanalytic means of thinking. This holds quite generally, and not just for the defense by isolation. The mere reproduction of memories in psychoanalysis can, therefore, only partly correct the lack of connection or the incorrect connection of elements. An additional process comes into play here which may justly be described as a scientific process. It discovers (and does not rediscover), according to the general rules of scientific thinking, the correct relationships of the elements to each other. Here, the theory of interpretation touches on the theory of mental connections and particularly on the distinction between meaning connections and causal connections [cf. Hartmann 1927]. Clearly, I do not concur with the often-voiced idea that the unconscious basically "knows it all" and that the task is merely to make this knowledge conscious by lifting the defense.

REFERENCES

Abraham, Hilda (1974). "Karl Abraham: An Unfinished Biography." *Int. Rev. Psycho-Anal.* 1:72.

Abraham, Hilda C. and Freud, Ernst L., eds. (1965). *A Psycho-Analytic Dialogue: The Letters of Sigmund Freud and Karl Abraham, 1907–1926,* trans. Bernard Marsh and Hilda C. Abraham. New York: Basic Books.

Abraham, Karl (1909). "The Significance of Intermarriage Between Close Relatives." In *Clinical Papers and Essays on Psychoanalysis: The Selected Papers of Karl Abraham,* ed. Hilda C. Abraham, trans. Hilda C. Abraham and D. R. Ellison. New York: Basic Books, 1955, pp. 21–28.

——— (1913a). "On Neurotic Exogamy." In *Clinical Papers and Essays on Psychoanalysis: The Selected Papers of Karl Abraham,* ed. Hilda C. Abraham, trans. Hilda C. Abraham and D. R. Ellison. New York: Basic Books, 1955, pp. 48–50.

——— (1913b). "Should Patients Write Down Their Dreams?" In *Clinical Papers and Essays on Psychoanalysis: The Selected Papers of Karl Abraham,* ed. Hilda C. Abraham, trans. Hilda C. Abraham and D. R. Ellison. New York: Basic Books, 1955, pp. 33–35.

——— (1916). "The First Pregenital Stage of the Libido." In *Selected Papers on Psychoanalysis,* trans. Douglas Bryan and Alix Strachey. New York: Basic Books, 1960, pp. 248–279.

——— (1917). "The Spending of Money in Anxiety States." In *Selected Papers on Psychoanalysis,* trans. Douglas Bryan and Alix Strachey. New York: Basic Books, 1960, pp. 299–302.

——— (1919a). "The Applicability of Psycho-Analytic Treatment of Patients at an Advanced Age." In *Selected Papers on Psychoanalysis,* trans. Douglas Bryan and Alix Strachey. New York: Basic Books, 1960, pp. 312–317.

——— (1919b). "A Particular Form of Neurotic Resistance Against the Psycho-Analytic Method." In *Selected Papers on Psychoanalysis,* trans. Douglas Bryan and Alix Strachey. New York: Basic Books, 1960, pp. 303–11.

——— (1924a). "A Short Study of the Development of the Libido Viewed in the Light of Mental Disorders." In *Selected Papers on Psychoanalysis,* trans. Douglas Bryan and Alix Strachey. New York: Basic Books, 1960, pp. 418–501.

——— (1924b). "Oral Eroticism and Character." In *Selected Papers on Psychoanalysis,* trans. Douglas Bryan and Alix Strachey. New York: Basic Books, 1960, pp. 393–406.

——— (1926). "Psycho-Analytic Notes on Coué's Method of Self-Mastery." *Int. J. Psycho-Anal.* 7:190–213.

——— (1955). *Clinical Papers and Essays on Psychoanalysis: Selected Papers of Karl Abraham,* ed. Hilda C. Abraham, trans. Hilda C. Abraham and D. R. Ellison. New York: Basic Books.

———— (1960). *Selected Papers on Psychoanalysis*, trans. Douglas Bryan and Alix Strachey. New York: Basic Books.

Alexander, Franz (1921). "The Castration Complex in the Formation of Character." In *The Scope of Psychoanalysis, 1921–1961: Selected Papers of Franz Alexander*. New York: Basic Books, 1961, pp. 3–30.

———— (1925). "A Metapsychological Description of the Process of Cure." In *The Scope of Psychoanalysis, 1921–1961: Selected Papers of Franz Alexander*. New York: Basic Books, 1961, pp. 205–224.

———— (1927). *The Psychoanalysis of the Total Personality: The Applicability of Freud's Theory of the Ego to the Neuroses*, trans. B. Glueck and B. D. Lewin. New York: Nervous and Mental Diseases Publishing Co., 1930.

———— (1931). "Psychoanalysis and Medicine." In *The Harvey Lectures, 1930–1931*. Baltimore: Williams and Wilkins, pp. 88–111.

———— (1935). "The Problem of Psychoanalytic Technique." In *The Scope of Psychoanalysis, 1921–1961: Selected Papers of Franz Alexander*. New York: Basic Books, 1961, pp. 225–243.

———— (1948). *Fundamentals of Psychoanalysis*. New York: W. W. Norton.

———— (1956). *Psychoanalysis and Psychotherapy*. New York: W. W. Norton.

———— (1960). *The Western Mind in Transition*. New York: Random House.

———— (1961). *The Scope of Psychoanalysis, 1921–1961: Selected Papers of Franz Alexander*. New York: Basic Books.

———— (1965). *Psychosomatic Medicine: Its Principles and Application*. New York: W. W. Norton.

Alexander, Franz, Eisenstein, Samuel, and Grotjahn, Martin, eds. (1966). *Psychoanalytic Pioneers*. New York: Basic Books.

Alexander, Franz and French, Thomas Morton (1946). *Psychoanalytic Therapy Principles and Applications*. New York: The Ronald Press.

Alexander, Franz and Staub, Hugo (1931). *The Criminal, the Judge, and the Public*. New York: Collier, 1962.

Arlow, Jacob A. and Brenner, Charles (1964). *Psychoanalytic Concepts and the Structural Theory*. New York: International Universities Press.

Balint, Michael (1967). "Sandor Ferenczi's Technical Experiments." In *Psychoanalytic Techniques*, ed. B. B. Wolman. New York: Basic Books, pp. 147–167.

———— (1970). Introduction to *Schriften zur Psychoanalyse*, by Sandor Ferenczi. Frankfurt: Fischer.

Bergmann, Martin S. (1963). "The Place of Federn's Ego Psychology in Psychoanalytic Metapsychology." *J. Amer. Psychoanal. Assn.*, 11:97–116.

———— (1968). "Free Associations and the Interpretation of Dreams: Historical and Methodological Considerations. In *Use of Interpretations in Treatment*, ed. E. Hammer. New York: Grune & Stratton, pp. 270–279.

———— (1971). "Psychoanalytic Observations on the Capacity to Love." In *Separation-Individuation: Essays in Honor of Margaret S. Mahler*, ed. J. B. McDevitt and C. F. Settlage. New York: International Universities Press, pp. 15–40.

Block, Max (1962). "On Metaphors." In *Models and Metaphors*. Ithaca: Cornell University Press.

Blos, Peter (1972). "The Epigenesis of Adult Neurosis." In *The Psychoanalytic Study of the Child*. New York: Quadrangle, vol. 27, pp. 106–35.

Blum, Harold P. (1971). "On the Conception and Development of the Transference Neurosis." *J. Amer. Psychoanal. Assn.* 19:41–53.

Bornstein, Berta (1949). "The Analysis of a Phobic Child: Some Problems of Theory and Technique of Child Analysis." In *The Psychoanalytic Study of the Child*. New York: International Universities Press, vol. 3/4, pp. 181–226.

Brenner, Charles (1957). "The Nature and Development of the Concept of Repression in Freud's Writings." In *The Psychoanalytic Study of the Child*. New York: International Universities Press, vol. 12, pp. 19–46.

Breuer, Joseph and Freud, Sigmund (1893–1895). "Studies on Hysteria." In *The Standard Edition of the Complete Psychological Works of Sigmund Freud*, ed. James Strachey. London: Hogarth Press, 1951–, vol. 2.

Brunswick, Ruth M. (1928). "A Supplement to Freud's History of an Infantile Neurosis." *Int. J. of Psycho-Anal.* 9:439–476.

Coles, Robert (1970). *Erik Erikson: The Growth of His Work*. Boston: Atlantic Monthly Press.

Coltrera, Joseph T. (1962). "Psychoanalysis and Existentialism." *J. Amer. Psychoanal. Assn.* 10:166–215.

Coltrera, Joseph T. and Ross, Nathaniel (1967). "Freud's Psychoanalytic Technique from the Beginnings to 1923." In *Psychoanalytic Techniques*, ed. B. B. Wolman. New York: Basic Books, pp. 13–50.

Daly, Claude D. (1928). "Der Menstruationkomplex" [The Menstruation Complex]. *Imago* 14:11–75.

———— (1935). "The Menstruation Complex in Literature." *Psychoanal. Q.* 4:307–340.

———— (1943). "The Role of Menstruation in Human Phylogenesis and Ontogenesis." *Int. J. Psycho-Anal.* 24:151–170.

Deutsch, Felix (1957). "A Footnote to Freud's 'Fragment of an Analysis of a Case of Hysteria.' " *Psychoanal. Q.* 26:159–167.

Deutsch, Helene (1934). "Don Quichote und Donquijotismus." *Imago* 20:444–449.

———— (1965). *Neuroses and Character Types*. New York: International Universities Press.

Dodds, E. R. (1957). *The Greeks and the Irrational*. Boston: Beacon Press.

Eissler, K. R. (1953). "The Effect of the Structure of the Ego on Psychoanalytic Technique." *J. Amer. Psychoanal. Assn.* 1:104–143.

———— (1958). "Notes on Problems of Technique in the Psychoanalytic Treatment of Adolescents: With Some Remarks on Perversions." In *The Psychoanalytic Study of the Child*. New York: International Universities Press, vol. 13, pp. 223–254.

———— (1963). *Goethe: A Psychoanalytic Study 1775–1786*. Detroit: Wayne State University Press.

Enelow, Allen J. and Adler, Leta (1965). Biographical foreword to *Effective Psychotherapy: The Contribution of Hellmuth Kaiser*, ed. Louis B. Fierman. New York: Free Press.

Erikson, Erik H. (1950). *Childhood and Society*. New York: W. W. Norton.

———— (1954). "The Dream Specimen of Psychoanalysis." *J. Amer. Psychoanal. Assn.* 2:5–56.

———— (1956a). "The First Psychoanalyst." *Yale Review* 46:40–62.

———— (1956b). "The Problem of Ego Identity." *J. Amer. Psychoanal. Assn.* 4:56–121.

———— (1958). *Young Man Luther: A Study in Psychoanalysis and History*. New York: W. W. Norton.

———— (1959). "Identity and the Lifecycle: Selected Papers." *Psychological Issues*, vol. 1, pp. 1–171. With an introduction by David Rapaport, "A Historical Survey of Psychoanalytic Ego Psychology."

———— (1962). "Reality and Actuality: An Address." *J. Amer. Psychoanal. Assn.* 10:454–461.

———— (1964). *Insight and Responsibility*. New York: W. W. Norton.

———— (1968). *Identity: Youth and Crisis*. New York: W. W. Norton.

———— (1969). *Gandhi's Truth*. New York: W. W. Norton.

Fairbairn, W. Ronald D. (1954). *An Object-Relations Theory of Personality*. New York: Basic Books.

Federn, Paul (1952). *Ego Psychology and the Psychoses*. With an introduction by Eduardo Weiss. New York: Basic Books.

Fenichel, Otto (1938–1939). "Problems of Psychoanalytic Technique," trans. David Brunswick. In *Problems of Psychoanalytic Technique*. New York: Psychoanalytic Quarterly Press, 1941.

———— (1934). *Outline of Clinical Psychoanalysis.* New York: Psychoanalytic Quarterly Press and W. W. Norton.

———— (1935a)."Concerning the Theory of Psychoanalytic Technique." In *The Collected Papers of Otto Fenichel*, First series, ed. Hanna Fenichel and David Rapaport. New York: W. W. Norton, 1953.

———— (1935b). "A Critique of the Death Instinct." In *The Collected Papers of Otto Fenichel*, First series, ed. Hanna Fenichel and David Rapaport. New York: W. W. Norton, 1953, pp. 363–74.

———— (1937a). "On the Theory of the Therapeutic Results of Psychoanalysis." *Int. J. Psycho-Anal.* 18:133–138.

———— (1937b). "A Review of Freud's 'Analysis Terminable and Interminable.'" *Int. Rev. Psycho-anal.* 1:109–116.

———— (1941). *Problems of Psychoanalytic Technique.* New York: Psychoanalytic Quarterly Press.

———— (1945). *The Psychoanalytic Theory of Neurosis.* New York: W. W. Norton.

———— (1953). *The Collected Papers of Otto Fenichel*, ed. Hanna Fenichel and David Rapaport. New York: W. W. Norton. First series, 1953, Second series, 1954.

Ferenczi, Sandor (1909). "Introjection and Transference." In *Sex in Psychoanalysis*, trans. Ernest Jones. New York: Basic Books, 1950, pp. 35–93.

———— (1912). "Transitory Symptom-Constructions During the Analysis." In *Sex in Psychoanalysis*, trans. Ernest Jones. New York: Basic Books, 1950, pp. 193–212.

———— (1919a). "On the Technique of Psycho-Analysis." In *Further Contributions to the Theory and Technique of Psychoanalysis*, comp. John Rickman, trans. Jane Isabel Suttie et al. New York: Basic Books, 1952, pp. 177–189.

———— (1919b). "The Phenomena of Hysterical Materialization." In *Further Contributions to the Theory and Technique of Psychoanalysis*, comp. John Rickman, trans. Jane Isabel Suttie et al. New York: Basic Books, 1952, pp. 89–104.

———— (1919c). "Technical Difficulties in the Analysis of a Case of Hysteria." In *Further Contributions to the Theory and Technique of Psychoanalysis,* comp. John Rickman, trans. Jane Isabel Suttie et al. New York: Basic Books, 1952, pp. 189–197.

———— (1920). "The Further Development of an Active Therapy in Psycho-Analysis." In *Further Contributions to the Theory and Technique of Psychoanalysis*, comp. John Rickman, trans. Jane Isabel Suttie et al. New York: Basic Books, 1952, pp. 198–217.

———— (1921a). *Psychoanalysis and the War Neuroses.* London: International Psycho-analytic Press.

———— (1921b). "Psycho-Analytical Observations on Tic." In *Further Contributions to the Theory and Technique of Psychoanalysis*, comp. John Rickman, trans. Jane Isabel Suttie et al. New York: Basic Books, 1952, pp. 142–174.

———— (1924a). "On Forced Fantasies: Activity in the Association-Technique." In *Further Contributions to the Theory and Technique of Psychoanalysis*, comp. John Rickman, trans. Jane Isabel Suttie et al. New York: Basic Books, 1952, pp. 68–78.

———— (1924b). *Thalassa: A Theory of Genitality.* New York: W. W. Norton, 1968.

———— (1925). "Contra-Indications to the 'Active' Psychoanalytic Technique." In *Further Contributions to the Theory and Technique of Psychoanalysis*, comp. John Rickman, trans. Jane Isabel Suttie et al. New York: Basic Books, 1952, pp. 217–232.

———— (1927). "The Problem of the Termination of the Analysis." In *Final Contributions to the Problems and Methods of Psychoanalysis*, ed. Michael Balint, trans. Eric Mosbacher et al. New York: Basic Books, 1955, pp. 77–86.

———— (1928a). "The Elasticity of Psycho-Analytic Technique." In *Final Contributions to the Problems and Methods of Psychoanalysis*, ed. Michael Balint, trans. Eric Mosbacher et al. New York: Basic Books, 1955, pp. 87–101.

———— (1928b). "On Obscene Words." In *Sex in Psychoanalysis*, trans. Ernest Jones. New York: Basic Books, 1950, pp. 132–153.

———— (1929). "The Unwelcome Child and His Death-Instinct." *Int. J. Psycho-Anal.* 10:125–129.

———— (1930). "The Principle of Relaxation and Neo-catharsis." In *Final Contributions to the Problems and Methods of Psychoanalysis*, ed. Michael Balint, trans. Eric Mosbacher et al. New York: Basic Books, 1955, pp. 108–125.

———— (1931). "Child Analysis in the Analysis of Adults." *Int. J. Psycho-Anal.* 12:468–482.

———— (1933). "Confusion of Tongues Between Adult and Child: The Language of Tenderness and Passion." In *Final Contributions to the Problems and Methods of Psychoanalysis*, ed. Michael Balint, trans. Eric Mosbacher et al. New York: Basic Books, 1955, pp. 156–167.

———— (1934). "Gedanken über das Trauma" [Reflections on Trauma]. *Internationale Zeitschrift für Psychoanalyse.* 20:5–12.

———— (1950). *Sex in Psychoanalysis*, trans. Ernest Jones. New York: Basic Books.

———— (1952). *Further Contributions to the Theory and Technique of Psychoanalysis*, comp. John Rickman, trans. Jane Isabel Suttie et al. New York: Basic Books.

———— (1955). *Final Contributions to the Problems and Methods of Psychoanalysis*, ed. Michael Balint, trans. Eric Mosbacher et al. New York: Basic Books.

Ferenczi, Sandor and Rank, Otto (1924). "The Development of Psychoanalysis." Washington, D.C.: Nervous and Medical Diseases Publishing Company, 1925.

Fierman, Louis B., ed. (1965). *Effective Psychotherapy: The Contributions of Hellmuth Kaiser.* New York: The Free Press.

Freidemann, A. (1966). "Heinrich Meng: Psychoanalysis and Mental Hygiene." In *Psychoanalytic Pioneers*, ed. Franz Alexander, Samuel Eisenstein, and Martin Grotjahn. New York: Basic Books, 1966, pp. 333–341.

French, Thomas M. (1936). "A Clinical Study of Learning in the Course of a Psychoanalytic Treatment." *Psychoanal. Q.* 5:148–194.

Freud, Anna (1936). *The Ego and the Mechanisms of Defense.* London: Hogarth Press, 1937.

———— (1952). "The Mutual Influences in the Development of Ego and Id." In *The Psychoanalytic Study of the Child.* New York: International Universities Press, vol. 7, pp. 42–50.

———— (1954a). "Problem of Technique in Adult Analysis." *Bull. Philadelphia Association for Psychoanalysis* 4: 44–70.

———— (1954b). "The Widening Scope of Indicators for Psychoanalysis Discussion." *J. Amer. Psychoanal. Assn.* 2:607–620.

———— (1965). *Normality and Pathology in Childhood.* New York: International Universities Press.

———— (1972). "The Infantile Neurosis: Genetic and Dynamic Considerations." In *The Psychoanalytic Study of the Child.* New York: Quadrangle, vol. 27, pp. 79–96.

Freud, Sigmund (1892–1899). "Extracts from the Fliess Papers." In *The Standard Edition of the Complete Psychological Works of Sigmund Freud*, ed. James Strachey. London: Hogarth Press, 1966, vol. 1, pp. 175–279.

———— (1895a). "On the Grounds of Detaching a Particular Symptom from Neurosthenia Under Description of Anxiety Neurosis." In *The Standard Edition of the Complete Psychological Works of Sigmund Freud*, ed. James Strachey. London: Hogarth Press, 1962, vol. 3, pp. 85–115.

———— (1895b). "Project for a Scientific Psychology." In *The Standard Edition of the Complete Psychological Works of Sigmund Freud*, ed. James Strachey. London: Hogarth Press, 1966, vol. 1, pp. 295–397.

———— (1896). "The Aetiology of Hysteria." In *The Standard Edition of the Complete Psychological Works of Sigmund Freud*, ed. James Strachey. London: Hogarth Press, 1962, vol. 3, pp. 189–221.

———— (1900). "The Interpretation of Dreams." In *The Standard Edition of the Com-*

plete Psychological Works of Sigmund Freud, ed. James Strachey. London: Hogarth Press, 1953, vols. 4 and 5.

—— (1901). "The Psychopathology of Everyday Life." In *The Standard Edition of the Complete Psychological Works of Sigmund Freud*, ed. James Strachey. London: Hogarth Press, 1960, vol. 6.

—— (1905a). "Fragment of an Analysis of a Case of Hysteria." In *The Standard Edition of the Complete Psychological Works of Sigmund Freud*, ed. James Strachey. London: Hogarth Press, 1953, vol. 7, pp. 3–122.

—— (1905b). "Three Essays on the Theory of Sexuality." In *The Standard Edition of the Complete Psychological Works of Sigmund Freud*, ed. James Strachey. London: Hogarth Press, 1953, vol. 7, pp. 125–243.

—— (1908). "Character and Anal Erotism." In *The Standard Edition of the Complete Psychological Works of Sigmund Freud*, ed. James Strachey. London: Hogarth Press, 1959, vol. 9, pp. 167–175.

—— (1909a). "Analysis of a Phobia in a Five-Year-Old Boy." In *The Standard Edition of the Complete Psychological Works of Sigmund Freud*, ed. James Strachey. London: Hogarth Press, 1955, vol. 10, pp. 3–149.

—— (1909b). "Notes upon a Case of Obsessional Neurosis." In *The Standard Edition of the Complete Psychological Works of Sigmund Freud*, ed. James Strachey. London: Hogarth Press, 1955, vol. 10, pp. 153–318.

—— (1910a). "The Future Prospects of Psychoanalytic Therapy." In *The Standard Edition of the Complete Psychological Works of Sigmund Freud*, ed. James Strachey. London: Hogarth Press, 1957, vol. 11, pp. 141–151.

—— (1910b). "Leonardo da Vinci and a Memory of His Childhood." In *The Standard Edition of the Complete Psychological Works of Sigmund Freud*, ed. James Strachey. London: Hogarth Press, 1957, vol. 11, pp. 59–137.

—— (1910c). "Wild Psychoanalysis." In *The Standard Edition of the Complete Psychological Works of Sigmund Freud*, ed. James Strachey. London: Hogarth Press, 1957, vol. 11, pp. 219–227.

—— (1911a). "The Handling of Dream Interpretation in Psychoanalysis." In *The Standard Edition of the Complete Psychological Works of Sigmund Freud*, ed. James Strachey. London: Hogarth Press, 1958, vol. 12, pp. 89–96.

—— (1911b). "Psychoanalytical Notes on an Autobiographical Account of a Case of Paranoia." In *The Standard Edition of the Complete Psychological Works of Sigmund Freud*, ed. James Strachey. London: Hogarth Press, 1958, vol. 12, pp. 33–82.

—— (1911c). "Formulations on the Two Principles of Mental Functioning." In *The Standard Edition of the Complete Psychological Works of Sigmund Freud*, ed. James Strachey. London: Hogarth Press, 1958, vol. 12, pp. 213–226.

—— (1912a). "The Dynamics of Transference." In *The Standard Edition of the Complete Psychological Works of Sigmund Freud*, ed. James Strachey. London: Hogarth Press, 1958, vol. 12, pp. 97–108.

—— (1921b). "Recommendations to Physicians Practicing Psychoanalysis." In *The Standard Edition of the Complete Psychological Works of Sigmund Freud*, ed. James Strachey. London: Hogarth Press, 1958, vol. 12, pp. 109–120.

—— (1912c). "Types of Onset of Neuroses." In *The Standard Edition of the Complete Psychological Works of Sigmund Freud*, ed. James Strachey. London: Hogarth Press, 1958, vol. 12, pp. 227–238.

—— (1913a). "On Beginning the Treatment." In *The Standard Edition of the Complete Psychological Works of Sigmund Freud*, ed. James Strachey. London: Hogarth Press, 1958, vol. 12, pp. 121–144.

—— (1913b). "The Disposition to Obsessional Neurosis." In *The Standard Edition of the Complete Psychological Works of Sigmund Freud*, ed. James Strachey. London: Hogarth Press, 1958, vol. 12, pp. 313–326.

—— (1913c). "The Claims of Psychoanalysis to Scientific Interest." In *The Standard*

Edition of the Complete Psychological Works of Sigmund Freud, ed. James Strachey. London: Hogarth Press, 1955, vol. 13, pp. 165–190.

——— (1914a). "On the History of the Psychoanalytic Movement." In *The Standard Edition of the Complete Psychological Works of Sigmund Freud*, ed. James Strachey. London: Hogarth Press, 1957, vol. 14, pp. 7–66.

——— (1914b). "On Narcissism: An Introduction." In *The Standard Edition of the Complete Psychological Works of Sigmund Freud*, ed. James Strachey. London: Hogarth Press, 1957, vol. 14, pp. 67–102.

——— (1914c). "Remembering, Repeating and Working Through." In *The Standard Edition of the Complete Psychological Works of Sigmund Freud*, ed. James Strachey. London: Hogarth Press, 1958, vol. 12, pp. 145–156.

——— (1915a). "Observations on Transference Love." In *The Standard Edition of the Complete Psychological Works of Sigmund Freud*, ed. James Strachey. London: Hogarth Press, 1958, vol. 12, pp. 157–174.

——— (1915b). "The Unconscious." In *The Standard Edition of the Complete Psychological Works of Sigmund Freud*, ed. James Strachey. London: Hogarth Press, 1957, vol. 14, pp. 159–215.

——— (1915c). "Repression." In *The Standard Edition of the Complete Psychological Works of Sigmund Freud*, ed. James Strachey. London: Hogarth Press, 1957, vol. 14, pp. 141–158.

——— (1915d). "Instincts and Their Vicissitudes." In *The Standard Edition of the Complete Psychological Works of Sigmund Freud*, ed. James Strachey. London: Hogarth Press, 1957, vol. 14, pp. 111–140.

——— (1916). "Some Character Types Met With in Psychoanalytic Work." In *The Standard Edition of the Complete Psychological Works of Sigmund Freud*, ed. James Strachey. London: Hogarth Press, 1957, vol. 14, pp. 309–333.

——— (1916–1917). "Introductory Lectures to Psychoanalysis, Part III." In *The Standard Edition of the Complete Psychological Works of Sigmund Freud*, ed. James Strachey. London: Hogarth Press, 1963, vol. 16.

——— (1917a). "Mourning and Melancholia." In *The Standard Edition of the Complete Psychological Works of Sigmund Freud*, ed. James Strachey. London: Hogarth Press, 1957, vol. 14, pp. 237–258.

——— (1917b). "A Metapsychological Supplement to the Theory of Dreams." In *The Standard Edition of the Complete Psychological Works of Sigmund Freud*, ed. James Strachey. London: Hogarth Press, 1957, vol. 14, pp. 222–235.

——— (1918). "From the History of an Infantile Neurosis." In *The Standard Edition of the Complete Psychological Works of Sigmund Freud*, ed. James Strachey. London: Hogarth Press, 1955, vol. 17, pp. 3–122.

——— (1919). "Lines of Advance in Psychoanalytic Therapy." In *The Standard Edition of the Complete Psychological Works of Sigmund Freud*, ed. James Strachey. London: Hogarth Press, 1955, vol. 17, pp. 157–168.

——— (1920). "Beyond the Pleasure Principle." In *The Standard Edition of the Complete Psychological Works of Sigmund Freud*, ed. James Strachey. London: Hogarth Press, 1955, vol. 18, pp. 3–64.

——— (1921). "Group Psychology and the Analysis of the Ego." In *The Standard Edition of the Complete Psychological Works of Sigmund Freud*, ed. James Strachey. London: Hogarth Press, 1955, vol. 18, pp. 67–143.

——— (1923a). "Two Encyclopedia Articles." In *The Standard Edition of the Complete Psychological Works of Sigmund Freud*, ed. James Strachey. London: Hogarth Press, 1955, vol. 18, pp. 235–254.

——— (1923b). "The Ego and the Id." In *The Standard Edition of the Complete Psychological Works of Sigmund Freud*, ed. James Strachey. London: Hogarth Press, 1961, vol. 19, pp. 3–66.

——— (1924a). "The Loss of Reality in Neurosis and Psychosis." In *The Standard Edi-*

tion of the Complete Psychological Works of Sigmund Freud, ed. James Strachey. London: Hogarth Press, 1961, vol. 19, pp. 183–187.

——— (1924b). "The Economic Problem of Masochism." In The Standard Edition of the Complete Psychological Works of Sigmund Freud, ed. James Strachey. London: Hogarth Press, 1961, vol. 19, pp. 157–170.

——— (1925a). "Some Psychical Consequences of the Anatomical Distinction Between the Sexes." In The Standard Edition of the Complete Psychological Works of Sigmund Freud, ed. James Strachey. London: Hogarth Press, 1961, vol. 19, pp. 243–247.

——— (1925b). "Negation." In The Standard Edition of the Complete Psychological Works of Sigmund Freud, ed. James Strachey. London: Hogarth Press, 1961, vol. 19, pp. 235–239.

——— (1926a). "Inhibitions, Symptoms and Anxiety." In The Standard Edition of the Complete Psychological Works of Sigmund Freud, ed. James Strachey. London: Hogarth Press, 1959, vol. 20, pp. 77–174.

——— (1926b). "The Question of Lay Analysis." In The Standard Edition of the Complete Psychological Works of Sigmund Freud, ed. James Strachey. London: Hogarth Press, 1959, vol. 20, pp. 179–258.

——— (1928). "Dostoevsky and Parricide." In The Standard Edition of the Complete Psychological Works of Sigmund Freud, ed. James Strachey. London: Hogarth Press, 1961, vol. 21, pp. 175–194.

——— (1930). "Civilization and Its Discontents." In The Standard Edition of the Complete Psychological Works of Sigmund Freud, ed. James Strachey. London: Hogarth Press, 1961, vol. 21, pp. 59–151.

——— (1933). "New Introductory Lectures on Psychoanalysis." In The Standard Edition of the Complete Psychological Works of Sigmund Freud, ed. James Strachey. London: Hogarth Press, 1964, vol. 22, pp. 3–184.

——— (1937a). "Construction in Analysis." In The Standard Edition of the Complete Psychological Works of Sigmund Freud, ed. James Strachey. London: Hogarth Press, 1964, vol. 23, pp. 255–269.

——— (1937b). "Analysis Terminable and Interminable." In The Standard Edition of the Complete Psychological Works of Sigmund Freud, ed. James Strachey. London: Hogarth Press, 1964, vol. 23, pp. 211–253.

Gardiner, Muriel, ed. (1971). The Wolf-Man By The Wolf-Man: The Double Story of Freud's Most Famous Case. New York: Basic Books.

Gero, George (1962). "Sadism, Masochism and Aggression: Their Role in Symptom Formation." Psychoanal. Q. 31:31–42.

Gill, Merton M. (1954). "Psychoanalysis and Exploratory Psychotherapy." J. Amer. Psychoanal. Assn. 2:771–797.

Gitelson, Maxwell (1962). "The First Phase of Psycho-Analysis." Int. J. Psycho-Anal. 43:194–205.

Glover, Edward (1924). " 'Active Therapy' and Psycho-Analysis." Int. J. Psycho-Anal. 5 (July): 269–311.

——— (1925). "The Neurotic Character." In On the Early Development of the Mind. New York: International Universities Press, 1956, pp. 47–66.

——— (1927). "Lectures on Technique in Psychoanalysis." Int. J. Psycho-Anal. 8:311–358, 485–520.

——— (1929). "The Psychology of the Psychotherapist." British Journal of Medical Psychology 9:1–16.

——— (1930). "Introduction to the Study of Psycho-Analytical Theory." Int. J. Psycho-Anal. 9:471–484.

——— (1931). "The Therapeutic Effect of Inexact Interpretation: A Contribution to the Theory of Suggestion." In The Technique of Psycho-Analysis. New York: International Universities Press, 1955.

—— (1955). *The Technique of Psychoanalysis*. New York: International Universities Press.

—— (1956). *The Early Development of the Mind*. New York: International Universities Press.

—— (1966). "Metapsychology or Metaphysics: A Psychoanalytic Essay." *Psychoanal. Q.* 35:173–190.

—— (n.d.). "The Reminiscences of Edward Glover." Psychoanalytic Project, Columbia University Oral History Project. Unpublished.

Greenson, Ralph R. (1958). "Variations in Classical Psychoanalytic Technique: An Introduction." *Int. J. Psycho-Anal.* 39:200–201.

—— (1967). *The Technique and Practice of Psychoanalysis*. New York: International Universities Press.

Grossman, Carl M. and Grossman, Sylva (1965). *The Wild Analyst: The Life and Work of Georg Groddeck*. New York: George Braziller.

Hartmann, Heinz (1927). "Understanding and Explanation." In *Essays on Ego Psychology*. New York: International Universities Press, 1964, pp. 369–403.

—— (1939). "The Ego and the Problem of Adaptation." In *Ego Psychology and the Problem of Adaptation*. New York: International Universities Press, pp. 63–64.

—— (1950). "Comments on the Psychoanalytic Theory of the Ego." In *Essays on Ego Psychology*. New York: International Universities Press, 1964, pp. 113–141.

—— (1951). "Technical Implications of Ego Psychology." *Psychoanal. Q.* 20:31–43.

—— (1954). "Problems of Infantile Neurosis: A Discussion." In *The Psychoanalytic Study of the Child*. New York: International Universities Press, vol. 9, pp. 31–36.

—— (1964). *Essays on Ego Psychology*. New York: International Universities Press.

Hattinberg, Hans Von (1924). "Towards the Analysis of the Analytic Situation." *Zeitschrift für Psychoanalyse* 10:34–56.

Heimann, Paula (1963). "Obituary: Joan Riviere 1883–1962." *International Journal of Psycho-Analysis* 44:230–233.

Herold, Carl M. (1939). "A Controversy About Technique." *Psychoanal. Q.* 8:219–243.

Hitschmann, E. (1932). "A Ten Years' Report of the Vienna Psychoanalytic Clinic." *Int. J. Psycho-Anal.* 13:245–255.

Holroyd, Michael (1967). *Lytton Strachey: A Critical Biography*, vol. 1: *The Unknown Years: 1880–1910*. New York: Holt, Rinehart & Winston.

—— (1968). *Lytton Strachey: A Critical Biography*, vol. 2: *The Years of Achievement: 1910–1932*. New York: Holt, Rinehart & Winston.

—— (1972). *Lytton Strachey: A Biography*. New York: Penguin.

Hook, Sidney, ed. (1959). *Psychoanalysis, Scientific Method or Philosophy*. New York: New York University Press.

Horney, Karen (1935). "Conceptions and Misconceptions of the Analytical Method." *J. Nerv. Men. Dis.* 81:399–410.

Hug-Helmuth, Hermine von (1921). "On the Technique of Child Analysis." *Int. J. Psycho-Anal.* 2:287–305.

Jacobson, Edith (1957). "Denial and Repression." *J. Amer. Psychoanal. Assn.* 5:66–92.

—— (1959). "The 'Exceptions': An Elaboration of Freud's Character Study." In *The Psychoanalytic Study of the Child*. New York: International Universities Press, vol. 14, pp. 135–154.

—— (1964). *The Self and the Object World*. New York: International Universities Press.

Jones, Ernest (1923). "The Nature of Auto-Suggestion." *Int. J. Psycho-Anal.* 4:293–312.

—— (1953). *The Life and Work of Sigmund Freud*, vol. 1: *The Formative Years and the Great Discoveries: 1856–1900*. New York: Basic Books.

—— (1955). *The Life and Work of Sigmund Freud*, vol. 2: *Years of Maturity: 1901–1919*. New York: Basic Books.

—— (1957). *The Life and Work of Sigmund Freud*, vol. 3: *The Last Phase: 1919–1939*. New York: Basic Books.

———— (1959). *Free Associations: Memoirs of a Psychoanalyst.* New York: Basic Books.
Kaiser, Hellmuth (1930). "Kleist's Prinz von Homburg." *Imago* 16:119–137.
———— (1931). "Franz Kafkas Inferno. Eine psychologische Deutung seiner Stratphantasie" [Franz Kafka's Inferno. A Psychological Interpretation of His Fantasy of Punishment]. *Imago* 17:41–104.
———— (1934). "Probleme der Technik." *Internationale Zeitschrift für Psychoanalyse* 20:490–522.
———— (1955). "The Problem of Responsibility in Psychotherapy." In *Effective Psychotherapy: The Contributions of Hellmuth Kaiser,* ed. Louis B. Fierman. New York: Free Press, 1965.
———— (1962). "Emergency: Seven Dialogues Reflecting the Essence of Psychotherapy in an Extreme Adventure." In *Effective Psychotherapy: The Contributions of Hellmuth Kaiser,* ed. Louis B. Fierman. New York: Free Press, 1965.
———— (1965). *Effective Psychotherapy: The Contributions of Hellmuth Kaiser,* ed. Louis B. Fierman. New York: Free Press.
Kanzer, Mark (1952). "The Transference Neurosis of the Rat Man." *Psychoanal. Q.* 21: 181–189.
———— (1955). "The Communicative Function of the Dream." *Int. J. Psycho-Anal.* 36:220–266.
———— (1966). "The Motor Sphere of the Transference Neurosis." *Psychoanal. Q.* 35:522–539.
———— (forthcoming). Downstate Medical Center 25th Anniversary Volume. New York:
Kemper, Werner (1973). *In Psychotherapie In Selbstdarstellungen.* Bern: Huber.
Kernberg, Otto (1975). *Borderline Conditions and Pathological Narcissism.* New York: Jason Aronson.
Klein, George S. (1973). "Two Theories or One?" *Bull. Menninger Clin.* 37:102–132.
Klein, Melanie (1932). *The Psychoanalysis of Children.* New York: W. W. Norton.
———— (1934). "A Contribution to the Psychogenesis of Manic-Depressive States." In *Contributions to Psychoanalysis: 1921–1945.* London: Hogarth Press, 1948, pp. 282–310.
———— (1940). "Mourning and Its Relation to Manic-Depressive States." In *Contributions to Psychoanalysis 1921–1945.* London: Hogarth Press, 1948, pp. 311–338.
Klein, Melanie et al. (1955). *New Directions in Psychoanalysis.* New York: Basic Books.
Knight, Robert P. (1953). "Borderline States." *Bull. Menninger Clin.* 17:1–12.
Kohut, Heinz (1971). *The Analysis of the Self.* New York: International Universities Press.
Kris, Ernst (1951). "Ego Psychology and Interpretation in Psychoanalytic Therapy." *Psychoanal. Q.* 20:15–30.
———— (1956a). "On Some Vicissitudes of Insight in Psychoanalysis." *Int. J. Psycho-Anal.* 37:445–455.
———— (1956b). "The Recovery of Childhood Memories in Psychoanalysis." In *The Psychoanalytic Study of the Child.* New York: International Universities Press, vol. 11, pp. 54–88.
———— (1956c). "The Personal Myth: A Problem in Psychoanalytic Technique." *J. Amer. Psychoanal. Assn.* 4:653–681.
Kubie, Lawrence S. (1973). "Edward Glover: A Biographical Sketch." *Int. J. Psycho-Anal.* 54:85–95.
Kuhn, Thomas S. (1962). *The Structure of Scientific Revolutions,* 2nd edition. Chicago: University of Chicago Press, 1970.
Landauer, Karl (1914). "Spontanheilung einer Katatonie" [The Spontaneous Recovery of a Catatonic]. *Internationale Zeitschrift für ärztliche Psychoanalyse* 2:441–459.
———— (1924). " 'Passive Technik': Zur Analyse narzisstischer Erkrankungen." *Internationale Zeitschrift für Psychoanalyse* 10:415–422.
———— (1936). "On Affects and Their Development." *Imago* 22:275–291.
Lewin, Bertram (1950). *The Psychoanalysis of Elation.* New York: W. W. Norton.

—————— (1954). "Sleep, Narcissistic Neurosis and the Analytic Situation." *Psychoanal. Q.* 23:487–510.

—————— (1955). "Dream Psychology and the Analytic Situation." *Psychoanal. Q.* 24:169–199.

Lewin, Karl K. (1973–1974). "Dora Revisited." *Psychoanal. Rev.* 10:519–535.

Lipton, Samuel D. (1967). "Later Developments in Freud's Technique: 1920–1939." In *Psychoanalytic Techniques,* ed. B. B. Wolman. New York: Basic Books, pp. 51–92.

Loewald, Hans W. (1960). "On the Therapeutic Action of Psychoanalysis." *Int. J. Psycho-Anal.* 41:16–33.

—————— (1971). "The Transference Neurosis: Comments on the Concept and the Phenomenon." *J. Amer. Psychoanalytic Assn.* 19:54–66.

Loewenstein, Rudolph M. (1951). "The Problem of Interpretation." *Psychoanal. Q.* 20:1–14.

Lorand, Sandor, ed. (1933). *Psychoanalysis Today: Its Scope and Function,* 2nd ed. New York: International Universities Press, 1944.

—————— (1966). "Sandor Ferenczi: Pioneer of Pioneers." In *Psychoanalytic Pioneers,* ed. Franz Alexander, Samuel Eisenstein, and Martin Grotjahn. New York: Basic Books, pp. 14–35.

Macalpine, Ida (1950). "The Development of the Transference." *Psychoanal. Q.* 19:501–539.

McGuire, William, ed. (1974). *The Freud-Jung Letters,* trans. Ralph Manheim and R. F. C. Hull. Bollingen Series. Princeton: Princeton University Press.

Mahler, Margaret S. (1944). "Tics and Impulsions in Children: A Study in Motility." *Psychoanal. Q.* 13:430–444.

Mahler, Margaret S., in collaboration with Furer, Manuel (1968). *On Human Symbiosis and the Vicissitudes of Individuation,* vol. 1: *Infantile Psychosis.* New York: International Universities Press.

Marcus, Steven (1974). "Freud and Dora." *Psyche* 27:32–79.

Marienbad Symposium (1936). "On the Theory of the Therapeutic Results of Psychoanalysis." Papers by Edward Glover; Otto Fenichel; James Strachey; Edmund Bergler; Herman Nunberg; Edward Bibring. *Int. J. Psycho-Anal.* 18 (1937):125–189.

Meng, Heinrich (1971). *Leben als Begnung.* Stuttgart: Hippokratesverlag.

Menninger, Karl (1958). *Theory of Psychoanalytic Technique.* New York: Basic Books.

Nagel, Ernst (1959). "Methodological Issues in Psychoanalytic Theory." In *Psychoanalysis, Scientific Method or Philosophy,* ed. Sidney Hook. New York: New York University Press.

—————— (1961). *The Structure of Science.* New York: Harcourt, Brace and World.

Nunberg, Herman (1920). "On the Catatonic Attack." In *Practice and Theory of Psychoanalysis.* New York: International Universities Press, 1961, vol. 1, pp. 3–23.

—————— (1921). "The Course of the Libidinal Conflict in a Case of Schizophrenia." In *Practice and Theory of Psychoanalysis.* New York: International Universities Press, 1961, vol. 1, pp. 24–59.

—————— (1924). "State of Depersonalization in the Light of Libido Theory." In *Practice and Theory of Psychoanalysis.* New York: International Universities Press, 1961, vol. 1, pp. 60–74.

—————— (1926). "The Will to Recovery." In *Practice and Theory of Psychoanalysis.* New York: International Universities Press, 1961, vol. 1, pp. 75–88.

—————— (1928). "Problems of Therapy." In *Practice and Theory of Psychoanalysis.* New York: International Universities Press, 1961, vol. 1, pp. 105–118.

—————— (1930). "The Synthetic Function of the Ego." In *Practice and Theory of Psychoanalysis.* New York: International Universities Press, 1961, pp. 120–136.

—————— (1932). *Principles of Psychoanalysis.* New York: International Universities Press, 1955.

—————— (1948). *Practice and Theory of Psychoanalysis.* Nervous and Mental Diseases

Monographs, no. 74. New York: Coolidge Foundation. Reissued: New York: International Universities Press, 1961.

———— (1951). "Transference and Reality." *Int. J. Psycho-Anal.* 32:1–9.

———— (1965). *Practice and Theory of Psychoanalysis*, vol. 2. New York: International Universities Press.

———— (1970). *Memoirs, Reflections, Ideas.* New York: International Universities Press.

Nunberg, Herman and Federn, Ernst (1962–1975). *Minutes of the Vienna Psychoanalytic Society* (vol. 1, 1906–1908; vol. 2, 1908–1910; vol. 3, 1910–1911; vol. 4, 1912–1918). New York: International Universities Press.

Olinick, Stanley L. (1954). "Some Considerations of the Use of Questioning as a Psychoanalytic Technique." *J. Amer. Psychoanalytic Assn.* 2:57–66.

———— (1957). "Questioning and Pain, Truth and Negation." *J. Amer. Psychoanalytic Assn.* 5:302–324.

Ophuijsen, J. H. W. Van (1921). "Psychoanalytische Therapie." In *Bericht über die Fortschritte der Psychoanalyse 1914–1919.* Vienna: Internationaler psychoanalytischer Verlag, p. 131.

Radó, Sandor (1925). "The Economic Principle in Psychoanalytic Technique." *Int. J. Psycho-Anal.* 6:35–44.

———— (1928). "An Anxious Mother." *Int. J. Psycho-Anal.* 9:219–226.

Raknes, O. (1970). *Wilhelm Reich and Orgonomy.* New York: Penguin Books.

Rangell, Leo (1971). "Obituary, Elizabeth R. Zetzel." *Int. J. Psycho-Anal.* 52:229.

Rank, Otto (1923). "Zum Verständnis der Libidoentwicklung in Heilungsvorgang" [Contribution to the Understanding of the Libido in Development in the Process of Recovery]. *Internationale Zeitschrift für Psychoanalyse* 9:435–471.

———— (1924). *The Trauma of Birth.* New York: Harcourt Brace and Company, 1929.

Rapaport, David and Gill, Merton (1959). "The Points of View and Assumptions of Metapsychology." In *The Collected Papers of David Rapaport,* ed. M. M. Gill. New York: Basic Books, pp. 795–811.

Reich, Annie (1950). "On the Termination of Psychoanalysis." *Int. J. Psycho-Anal.* 31:179–183.

———— (1951). "The Discussion of 1912 on Masturbation and Our Present Day Views." In *The Psychoanalytic Study of the Child.* New York: International Universities Press, vol. 6, pp. 80–94.

Reich, Ilse Ollendorf (1969). *Wilhelm Reich: A Personal Biography.* New York: St. Martin's Press.

Reich, Peter (1973). *A Book of Dreams.* New York: Harper & Row.

Reich, Wilhelm (1927a). *The Function of the Orgasm.* New York: Farrar, Straus & Giroux, 1973.

———— (1927b). "Zur Technik der Deutung und der Widerstandanalyse: Über die gesetzmässign Entwicklung der Übertragungsneurose" [The Technique of Interpretation and of the Resistance-Analysis. The Regulated Development of the Transference Neurosis]. *Internationale Zeitschrift für Psychoanalyse* 13:142–159.

———— (1933). *Character Analysis,* trans. by Theodore Wolfe. Rangeley, Maine: Orgone Institute Press, 1945.

Reik, Theodor (1915a). "Puberty Rites among Savages." In *Ritual: Psychoanalytic Studies.* New York: W. W. Norton, 1931, pp. 91–166.

———— (1915b). "Some Remarks on the Study of Resistances." *Int. J. Psycho-Anal.* 5 (1924):141–155.

———— (1925). "The Compulsion to Confess and the Need for Punishment." In *The Compulsion to Confess: On the Psychoanalysis of Crime and Punishment.* New York: Farrar, Straus & Cudahy, 1959, pp. 175–356.

———— (1929). *Der Schrecken und andere psychoanalytische Studies* [Fright and Other Psychoanalytic Studies]. Vienna: International Psychoanalytic Publishers.

———— (1933). "New Ways in Psychoanalytic Technique." *Int. J. Psycho-Anal.* 14:321–334.

———— (1935). *Surprise and the Psychoanalyst: On the Conjecture and Comprehension of Unconscious Processes*. New York: Dutton, 1937.

———— (1941). *Masochism in Modern Man*. New York: Farrar & Rinehart; New York: Grove Press, 1959.

———— (1948). *Listening with the Third Ear: The Inner Experience of a Psychoanalyst*. New York: Farrar, Straus.

———— (1949). *Fragment of a Great Confession*. New York: Farrar, Straus.

Riviere, Joan (1936). "A Contribution to the Analysis of the Negative Therapeutic Reaction." *Int. J. Psycho-Anal.* 17:304–320.

Riviere, Joan, ed. (1952). *Developments in Psycho-Analysis*. London: Hogarth Press.

———— (1955). *New Directions in Psycho-Analysis*.

Riviere, Joan and Klein, Melanie (1937). *Love, Hate and Reparation*. London: Wolf and Hogarth Press.

Roheim, Geza (1923). "Heiliges Geld in Melanesian" [Sacred Money in Melanesia]. *Internationale Zeitschrift für Psychoanalyse* 9:348–401.

Rosen, Victor H. (1955). "The Construction of a Traumatic Childhood Event in a Case of Depersonalization." *J. Amer. Psychoanal. Assn.* 3:211–221.

Rosenfeld, Herbert (1958). "Contribution to the Discussion on Variations in Classical Technique." *Int. J. Psycho-Anal.* 39:238–239.

Sachs, Hanns (1944). *Freud: Master and Friend*. Cambridge: Harvard University Press.

———— (1948). *Masks of Love and Life*. Cambridge, Mass.: Sci-Art Publishers.

Salzburg Symposium (1925). "On Psychoanalytic Theory and Practice." *Int. J. Psycho-Anal.* 6:1–44. Introduction by Ernest Jones. The Symposium was held at the Congress on April 21, 1924. Participants: Hans Sachs; Franz Alexander; Sandor Rado.

Schimek, J. G. (1975). "The Interpretation of the Past: Childhood Trauma, Psychical Reality and Historical Truth." *J. Amer. Psychoanal. Assn.* 23:845–865.

Schmideberg, Melitta (1931). "Persecutory Ideas and Delusions." *Int. J. Psycho-Anal.* 12:331–367.

———— (1932). "The Psycho-Analysis of Asocial Children and Adolescents." *Int. J. Psycho-Anal.* 16 (1935):22–48.

———— (1938). "Intellectual Inhibition and Disturbances in Eating." *International Journal of Psycho-Analysis* 19:17–22.

Sharpe, Ella Freeman (1930). "The Technique of Psychoanalysis: Seven Lectures." In *Collected Papers on Psychoanalysis*. London: Hogarth Press, 1956, pp. 9–108.

Sherwood, Michael (1969). *The Logic of Explanation in Psychoanalysis*. New York: Academic Press.

Spitz, Rene A. (1945). "Hospitalism: An Inquiry into the Genesis of Psychiatric Conditions in Early Childhood." In *The Psychoanalytic Study of the Child*. New York: International Universities Press, vol. 1, pp. 53–74.

Sterba, Richard (1928). "An Examination Dream." *Int. J. Psycho-Anal.* 9:353–354.

———— (1929). "Dynamics of Dissolution of Transference Resistance." *Psychoanal. Q.* 9 (1940):363–379.

———— (1934). "The Fate of the Ego in Analytic Therapy." *Int. J. Psycho-Anal.* 15:117–126.

———— (1942). *Introduction to Psychoanalytic Theory of the Libido*. Nervous and Mental Disease Monograph Series, no. 68. New York: Robert Brunner, 1968.

———— (1953). "Clinical and Therapeutic Aspects of Character Resistance." *Psychoanal. Q.* 22:1–20.

———— (1970). "The Multiple Determinants of a Minor Accident." *Israeli Annals of Psychiatry* 8:111–122.

Sterba, Richard and Sterba, Editha (1954). *Beethoven and His Nephew*. New York: Pantheon Books.

Stone, Leo (1954). "The Widening Scope of Indications for Psychoanalysis." *J. Amer. Psychoanal. Assn.* 2:567–594.

———— (1961). *The Psychoanalytic Situation*. New York: International Universities Press.

Strachey, James (1934). "The Nature of the Therapeutic Action of Psycho-Analysis." In *Psychoanalytic Clinical Interpretation*, ed. Louis Paul. New York: Free Press, 1963.

———— (1963). "Obituary: Joan Riviere 1883–1962." *International Journal of Psycho-Analysis* 44:228–230.

Tausk, Victor (1919). "On the Origin of the 'Influencing Machine' in Schizophrenia." In *The Psychoanalytic Reader*, ed. R. Fliess. New York: International Universities Press, 1948, pp. 52–85.

Tolpin, Marian (1970). "The Infantile Neurosis: A Metapsychological Concept and a Paradigmatic Case History." In *The Psychoanalytic Study of the Child*. New York: International Universities Press, vol. 25, pp. 273–305.

Waelder, Robert (1936). "The Principle of Multiple Function: Observations and Over-Determination." *Psychoanal. Q.* 5:45–62.

———— (1960). *Basic Theory of Psychoanalysis*. New York: International Universities Press.

———— (1962). "Psychoanalysis, Scientific Method and Philosophy." *J. Amer. Psychoanal. Assn.* 10:617–637.

———— (1967). "Inhibition, Symptom and Anxiety: Forty Years Later." *Psychoanal. Q.* 3:1–36.

Wexler, Milton (1951). "The Structural Problem in Schizophrenia: Therapeutic Implications." *Int. J. Psycho-Anal.* 32:157–166.

Wiedeman, George (1962). "Survey of Psychoanalytic Literature on Overt Male Homosexuality." *J. Amer. Psychoanal. Assn.* 10:386–409.

Wilson, Emmett (1973). "The Structural Hypothesis and Psychoanalytic Metatheory." In *Psychoanalysis and Contemporary Science*, ed. Benjamin B. Rubinstein. New York: Macmillan, vol. 2, pp. 304–328.

Winnicott, D. W. (1969). "James Strachey: Obituary." *Int. J. Psycho-Anal.* 50:130.

Zetzel, Elizabeth R. (1956). "Current Concepts of Transference." In *The Capacity for Emotional Growth*. New York: International Universities Press, 1970, pp. 168–181.

———— (1964). "Depression and the Incapacity to Bear It." In *The Capacity for Emotional Growth*. New York: International Universities Press, 1970, 82–114.

———— (1965). "The Theory of Therapy in Relation to a Developmental Model of the Psychic Apparatus." *Int. J. Psycho-Anal.* 46:39–52.

———— (1966). "Additional Notes upon a Case of Obsessional Neurosis: Freud 1909." *Int. J. Psycho-Anal.* 47:123–129. Reprinted, under the title "An Obsessional Neurotic: Freud's Rat Man," in *The Capacity for Emotional Growth*. New York: International Universities Press, 1970.

———— (1969). "96 Gloucester Place: Some Personal Recollections." *Int. J. Psycho-Anal.* 50:717–719.

———— (1970). *The Capacity for Emotional Growth*. New York: International Universities Press.

INDEX